Geometry
Common Core Regents
Course Workbook

2017-18 Edition

Donny Brusca

www·CourseWorkbooks·com

Geometry
Common Core Regents Course Workbook

2017-18 Edition

Donny Brusca

ISBN 978-1545370391

Cover Photo: *The Dark Side of the Pyramid*, © Marco Carmassi
www.facebook.com/marcocarmassiphotographer

www·CourseWorkbooks·com

Table of Contents

Introduction

ABOUT THE AUTHOR

I have taught for about 25 years, mostly on the high school and college levels. I am currently employed as an administrator by the Williamsburg Charter High School in Brooklyn, NY. I have a B.S. in mathematics and M.A. in computer and information science from Brooklyn College (CUNY) and a post-graduate P.D. in educational administration and supervision from St. John's University. For more about the author, visit Brusca.info.

ABOUT THIS BOOK

Every topic section begins with an explanation of the Key Terms and Concepts. I have intentionally limited the content here to the most essential ideas. The notes should supplement a fuller presentation of the concepts by the teacher through a more developmental approach.

Topic sections include one or more Model Problems, each with a solution and an explanation of steps needed to solve the problem. Steps lettered (A), (B), etc., in the explanations refer to the corresponding lettered steps shown in the solutions. General wording is used in the explanations so that students may apply the steps directly to new but similar problems. However, for clarity, the text often refers to the specific model problem by using *[italicized text in brackets]*. To make the most sense of this writing style, insert the words "in this case" before reading any *[italicized text in brackets]*.

After the Model Problem are a number of Practice Problems in boxed work spaces. These numbered problems are arranged in order of increasing difficulty.

After the Practice Problems are Regents Questions on the topic, all of which have appeared on past New York State Regents exams, including *every* Geometry Common Core exam question (as of this book's publication date), plus a number of questions from previous Regents exam formats. Questions are grouped by *Multiple Choice* questions followed by *Constructed Response* questions, generally in chronological order of appearance, but with questions from Common Core exams given last and labelled by (CC).

At the end of the book are a list of the Geometry Common Core standards for New York State, a pacing calendar, and an index. An Answer Key is available at CourseWorkbooks.com and is free to organizations that purchase class sets.

My goal is to have an error-free book and answer key and I am willing to offer a monetary reward to the first person who contacts me about any error in mathematical content. Simply email me at donny@courseworkbooks.com and be as specific as possible about the error. Corrections are posted at CourseWorkbooks.com.

NEW DRAFT LEARNING STANDARDS

In September, 2016, the NY State Education Department (NYSED) released a draft of proposed changes to the Common Core mathematics learning standards, starting with the 2019 exams. After gathering feedback from the public about the proposed revisions, NYSED had planned to present a final draft of the recommended revisions to the Board of Regents for ratification in "early 2017," as stated on this web page:

www.nysed.gov/teachers/draft-standards-mathematics

However, as of the publication of this book in April, 2017, these draft standards have not yet been presented to the Board for approval. Considering the efforts made to draft these revisions, and the relatively minor impact they would have on the exam content, I have written this edition under the presumption that these changes will eventually be approved.

Therefore, this workbook accommodates both the current and revised standards. To guide the reader, I have used icons to show where these standards are expected to change.

A **circle-minus** symbol will appear at the beginning of any topics that the NYSED draft recommends should be **removed from** the standards. In the Table of Contents and in the section heading, **(–)** will be appended.

A **circle-plus** symbol will appear at the beginning of any topics that the NYSED draft recommends should be **added to** the standards. In the Table of Contents and in the section heading, **(+)** will be appended.

Where only a minor revision is recommended within a topic, a warning symbol will appear with an explanation of the proposed change.

Courses that are aligned to the current (pre-2019) standards should continue to teach the topics marked by a circle-minus and may omit the topics marked by a circle-plus.

Assuming the proposed revisions are ratified, courses aligned to the revised standards may omit topics marked by a circle-minus and should add topics marked by a circle-plus.

Appendix I contains a table comparing all of the current standards with the recommended revisions.

Chapter 1. Prerequisite Topics Review

1.1 *Simplifying Radicals*

To simplify a radical into simplest radical form:

1. Write the prime factorization of the radicand.
2. Group all pairs of factors, representing squares.
3. Remove the squares (pairs of factors) from the radicand by replacing them with their square roots (single factors outside the radical sign).
4. Multiply all factors outside the radicand, and all factors remaining inside the radicand.

Examples: $\sqrt{75} = \sqrt{3 \cdot \boxed{5 \cdot 5}} = 5\sqrt{3}$

$\sqrt{288} = \sqrt{\boxed{2 \cdot 2} \cdot \boxed{2 \cdot 2} \cdot 2 \cdot \boxed{3 \cdot 3}} = 2 \cdot 2 \cdot 3 \cdot \sqrt{2} = 12\sqrt{2}$

Model Problem:
Simplify $8\sqrt{90}$

Solution:

(A) (B) (C) (D)

$8\sqrt{90} = 8\sqrt{2 \cdot \boxed{3 \cdot 3} \cdot 5} = 8 \cdot 3\sqrt{2 \cdot 5} = 24\sqrt{10}$

Explanation of steps:

(A) Write the prime factorization of the radicand.
(B) Group all pairs of factors, representing squares.
(C) Remove the squares (pairs of factors) from the radicand by replacing them with their square roots (single factors) outside the radical sign.
(D) Multiply all factors outside the radicand, and all factors remaining inside the radicand.

1.2 Geometric Terms and Symbols

A **point** represents a location. It has no size (no dimensions). It is usually shown as a dot in a diagram. Often, a capital letter is used to name a point, as in point *P*.

• *P*

A **line** is a set of points extending infinitely in length (one dimension). Points that are on the same line are called **collinear**. Lines are usually named by two points and written using a double-arrow overbar, as in $\overleftrightarrow{AB}$, or by using a single lower case letter, as in line ℓ. Two lines that meet at a point are called *intersecting* lines.

A **plane** is a flat surface, with no thickness, extending infinitely in length and width (two dimensions). Points that are on the same plane are called **coplanar**. We can name a plane by three noncollinear points, as in plane *ABC*., or by using a capital letter, often written in cursive, as in plane $\mathcal{P}$. If two planes intersect, their intersection is always a line.

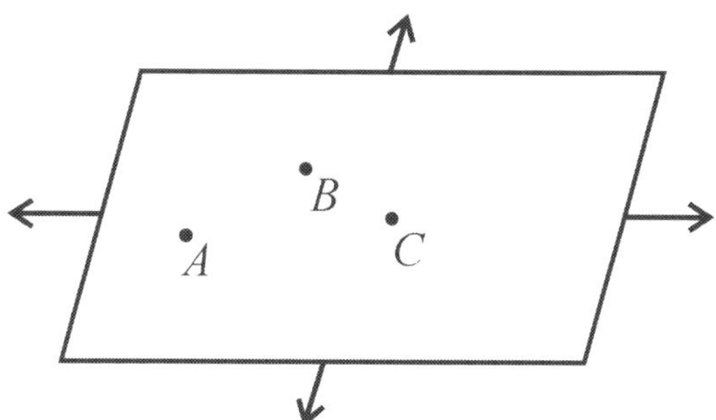

A **coordinate plane** contains two perpendicular number lines, called the *x*-axis and *y*-axis, which intersect at a point called the *origin*. Each point in a coordinate plane can be specified by an ordered pair of numbers. For example, point *P* below has coordinates (5,7).

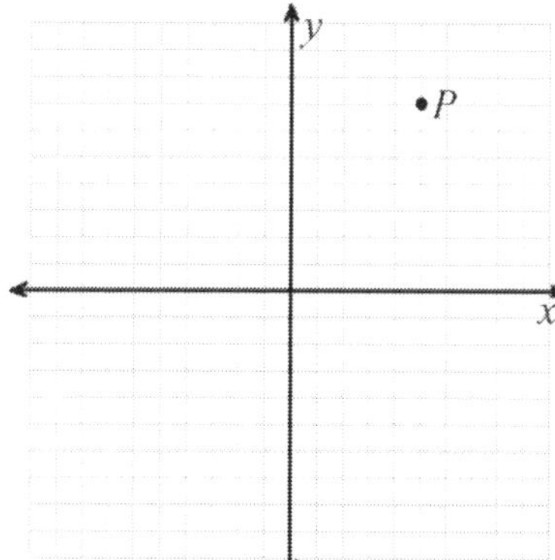

Two lines on a plane that are everywhere equidistant and never intersect are called **parallel lines.** Two lines that form right (90°) angles at their intersection are called **perpendicular lines.**

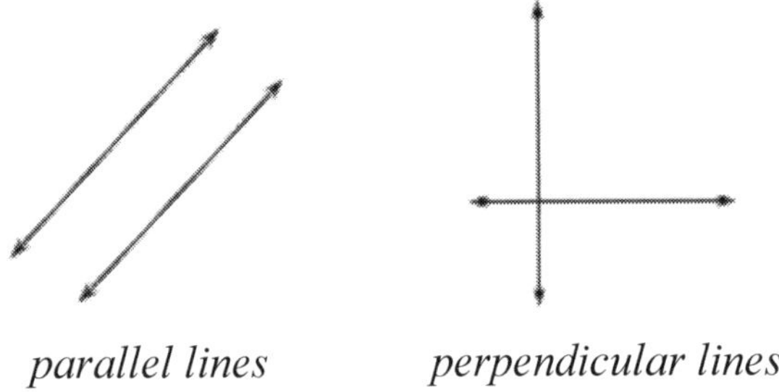

parallel lines *perpendicular lines*

The symbol for parallel is || and the symbol for perpendicular is $\perp$.

Examples: $\overleftrightarrow{AB} \parallel \overleftrightarrow{CD}$ means line AB is parallel to line CD, and $\overleftrightarrow{EF} \perp \overleftrightarrow{GH}$ means line EF is perpendicular to line GH.

A **line segment** is part of a line bounded by two endpoints. It is named by its endpoints and written with an overbar, as in $\overline{AB}$. A line segment has a finite length, which is the distance along the line between its two endpoints. We can specify the length of a line segment by omitting the overbar; for example AB is the length of $\overline{AB}$.

Lines and line segments are sometimes named by using more than two collinear points. For example, if B is a point between points A and C on the same line, the line can be called $\overleftrightarrow{ABC}$.

Similarly, if B is a point between endpoints A and C of a line segment, the line segment can be called $\overline{ABC}$.

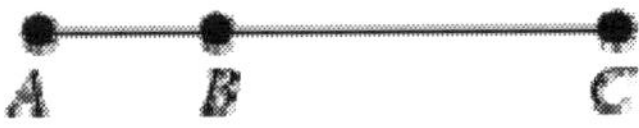

A **ray** is part of a line bounded by one endpoint and extending infinitely in one direction.

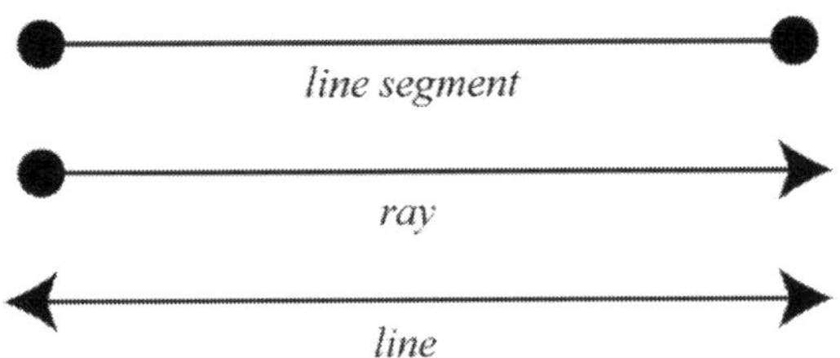

An **angle** is formed when two rays share a common endpoint, called the **vertex** of the angle. An angle is usually named by its vertex, as in $\angle A$, or by a lowercase letter or number, as in $\angle 1$. It is sometimes necessary to name an angle using three points: a point on each ray and the vertex. For example, $\angle B$ below can also be called $\angle ABC$ or $\angle CBA$. The middle letter must always be the vertex.

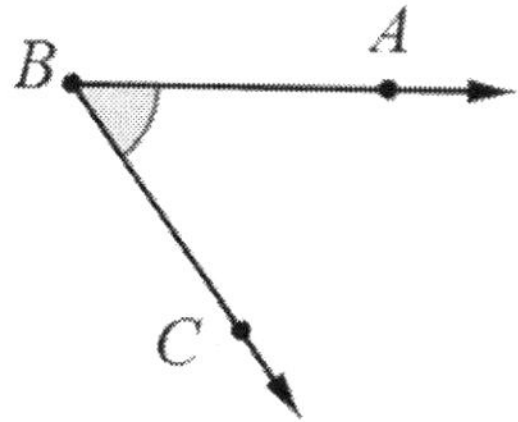

The size of an angle is usually measured in degrees. A lowercase m may be used before an angle name to specify the measure of the angle, as in $m\angle A$.

Two figures or objects are **congruent** if they have the same shape and size. The symbol for congruence is ≅.

Two line segments are congruent if they have the same length, and two angles are congruent if they have the same measure.

Examples: $\overline{AB} \cong \overline{CD}$ if $AB = CD$, and $\angle A \cong \angle B$ if $m\angle A = m\angle B$.

Note that we say that objects (such as line segments or angles) are *congruent*, but quantities (such as their lengths or measures) are *equal*.

1.3 Angles at a Point

Two angles are **congruent** if they are equal in measure.

A **right angle** is an angle that measures $90°$. *All right angles are congruent.*
A **straight angle** is an angle that measures $180°$. *All straight angles are congruent.*

Complementary angles are two angles whose measures add to $90°$.
Complements of the same angle, or of congruent angles, are congruent.

Supplementary angles are two angles whose measures add to $180°$.
Supplements of the same angle, or of congruent angles, are congruent.

Adjacent angles are two angles that have a common vertex and common side and don't overlap.

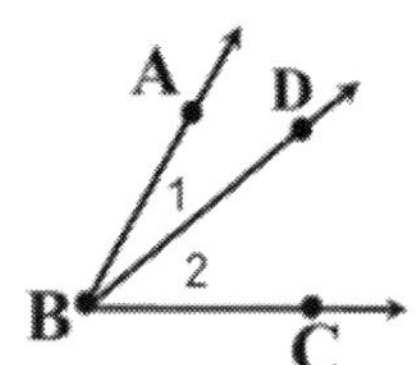

Angle Addition Postulate: the sum of the measures of two adjacent angles is equal to the measure of the whole angle that contains them both.
Example: $m\angle 1 + m\angle 2 = m\angle ABC$

A **linear pair** is a pair of adjacent angles whose non-common sides form straight line. *Linear pairs are supplementary. If two congruent angles form a linear pair, they are right angles.*

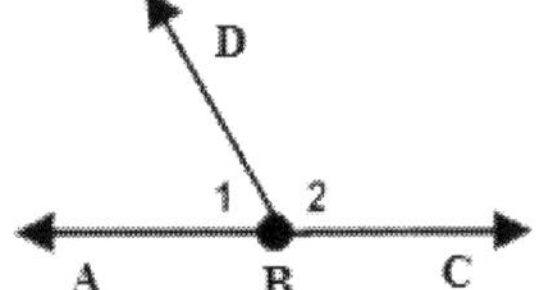

Vertical angles are two non-adjacent angles formed by the intersection of two lines. *Vertical angles are congruent.*
Example: $\angle 1$ and $\angle 3$ are a pair of vertical angles, so $\angle 1 \cong \angle 3$
$\angle 2$ and $\angle 4$ are a pair of vertical angles, so $\angle 2 \cong \angle 4$

Consecutive adjacent **angles on a straight line** add to $180°$.
Consecutive adjacent **angles around a point** add to $360°$.
Examples: $a+b+c=180°$ $x+y+z=360°$

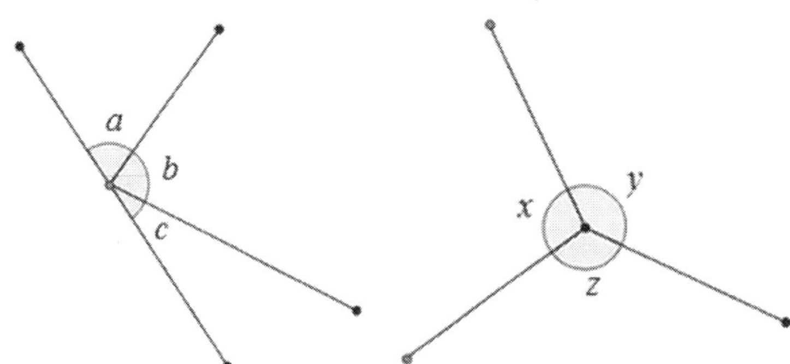

Chapter 2. Perimeter and Area

2.1 Perimeter and Circumference

Key Terms and Concepts

Perimeter is the distance around the outside of a figure. For a polygon, this is the sum of the lengths of its sides.

Example: For the perimeter of a rectangle with length l and width w, $P = 2l + 2w$.

Circumference is like perimeter, but represents the distance around the outside of a circle. The formulas for the circumference of a circle are:

$C = 2\pi r$ where r is the radius, or $C = \pi d$ where d is the diameter

For a **semicircle** or **quarter circle**, be sure to divide the full circle's circumference by 2 or 4.

Note: The formulas for the circumference of a circle are included on the Reference Sheet at the back of the Regents exam.

A **composite figure** is made up of more than one type of shape joined together. To find the perimeter of a composite figure, only add the parts of the figure that make up its outside.

An **isosceles** polygon has two equal sides. A triangle or a trapezoid may be isosceles.
An **equilateral** polygon has all sides of the same length.

Model Problem:

The "key" in a regulation size basketball court is made up of a rectangle and semicircle. The length of the rectangle measures 19 feet from the baseline to the free throw line, and the width measures 12 feet, as shown below. Find the total distance around the key to the *nearest foot*.

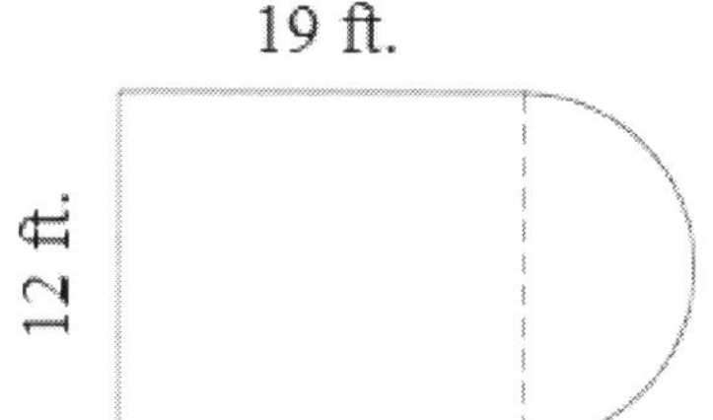

Solution:

(A) $P = 2l + w \quad + \quad \frac{1}{2}\pi d$

(B) $= 2(19) + (12) + \frac{1}{2}\pi(12)$

(C) $= 50 + 6\pi \quad \approx \quad 69$ *feet*

Explanation of steps:

(A) Write an equation that adds the parts that make up the distance around the outside of the composite figure. *[The outside of this figure uses only 3 sides of a rectangle – two lengths and a width – and half the circumference of a circle.]*

(B) Substitute the given values for the variables *[19, 12, and 12, for l, w and d]*.

(C) Simplify *[and use a calculator to find the sum rounded to the nearest foot]*.

Practice Problems

1. The second side of a triangle is two more than the first side, and the third side is three less than the first side. Write an expression, in simplest form in terms of x, for the perimeter of the triangle.

2. The plot of land illustrated in the accompanying diagram has a perimeter of 34 yards. Find the length, in yards, of *each* side of the figure.

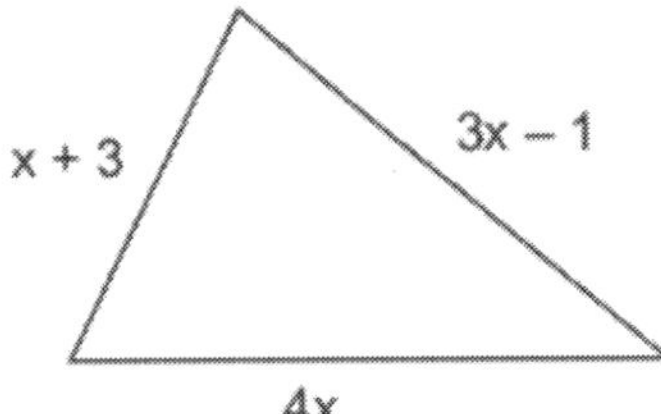

3. The figure below consists of a semicircle with radius 4 inches and a rectangle with a width of 7 inches. Find the perimeter to the *nearest tenth of an inch.*

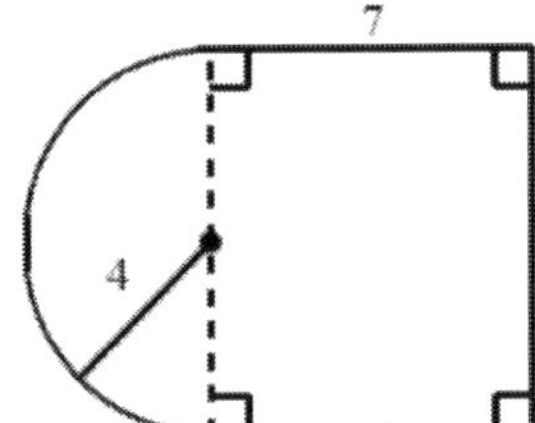

4. The figure below consists of a semicircle with radius 2 cm and a rectangle with a length of 4 cm. Find the perimeter to the *nearest tenth of a cm.*

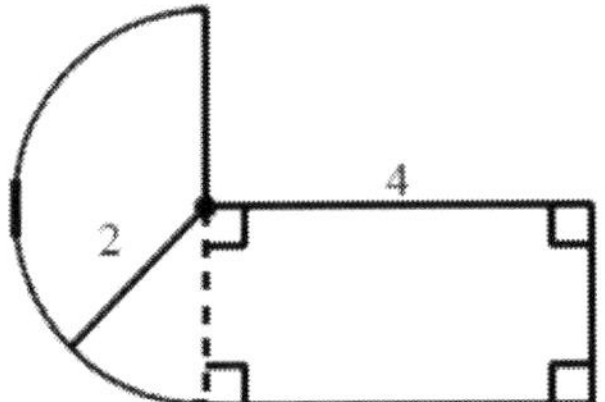

5. In the figure below, arc SBT is one quarter of a circle with center R and radius 6. If the length plus the width of rectangle $ABCR$ is 8, find the perimeter of the shaded region.

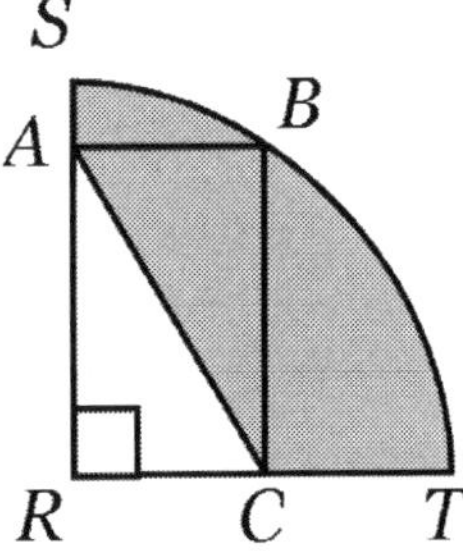

REGENTS QUESTIONS

Multiple Choice

1. A garden is in the shape of an isosceles trapezoid and a semicircle, as shown in the diagram below. A fence will be put around the perimeter of the entire garden.

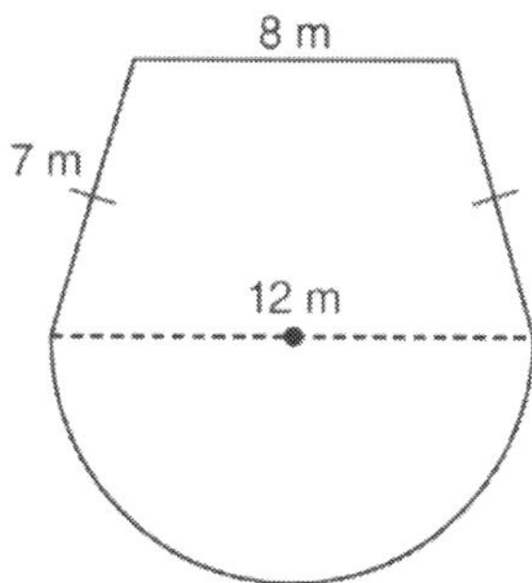

Which expression represents the length of fencing, in meters, that will be needed?

(1) $22+6\pi$ (3) $15+6\pi$
(2) $22+12\pi$ (4) $15+12\pi$

2. What is the perimeter of the figure shown below, which consists of an isosceles trapezoid and a semicircle?

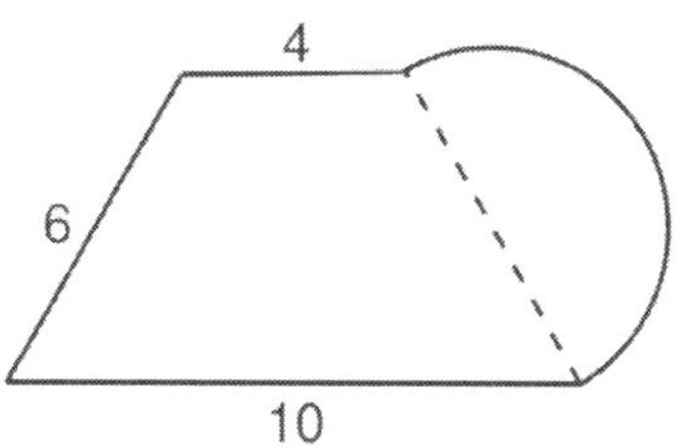

(1) $20+3\pi$ (3) $26+3\pi$
(2) $20+6\pi$ (4) $26+6\pi$

3. A designer created a garden, as shown in the diagram below. The garden consists of four quarter-circles of equal size inside a square. The designer put a fence around both the inside and the outside of the garden.

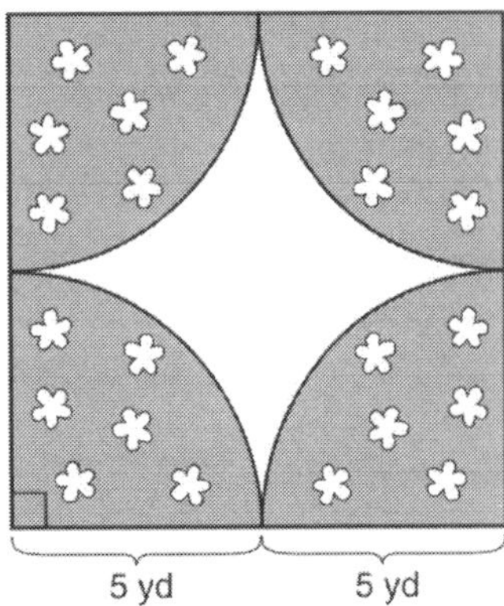

Which expression represents the amount of fencing, in yards, that the designer used for the fence?

(1) $40+10\pi$ (3) $100+10\pi$
(2) $40+25\pi$ (4) $100+25\pi$

Constructed Response

4. Ross is installing edging around his pool, which consists of a rectangle and a semicircle, as shown in the diagram below.

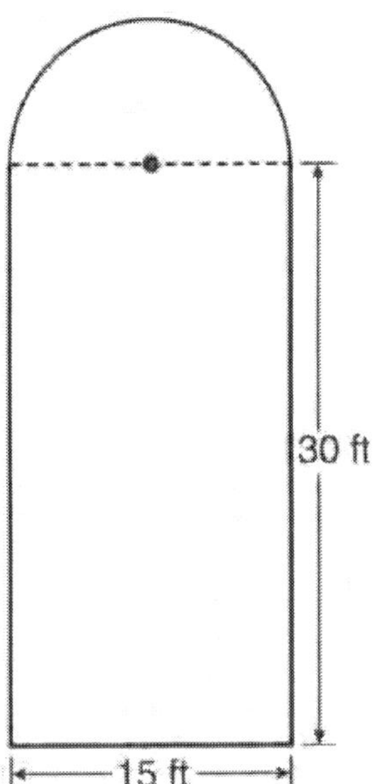

Determine the length of edging, to the *nearest tenth of a foot*, that Ross will need to go completely around the pool.

5. The diagram below consists of a square with a side of 4 cm, a semicircle on the top, and an equilateral triangle on the bottom. Find the perimeter of the figure to the *nearest tenth of a centimeter*.

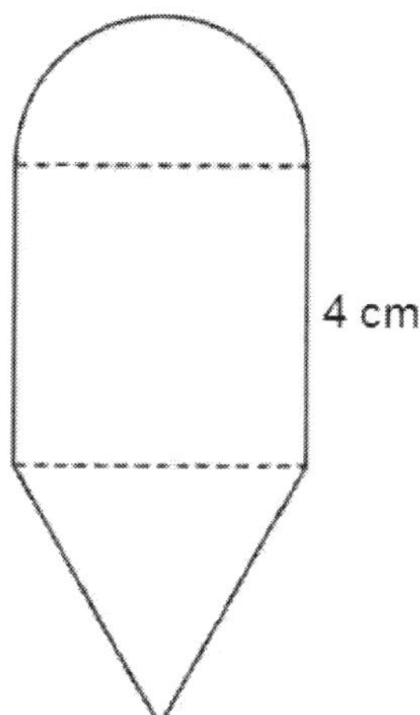

6. As shown below, polygon *ABCGFED* consists of two squares, *ABCD* and *CGFE*, and an equilateral triangle *CED*. The length of *BC* is $\sqrt{3}$ cm. Determine the perimeter of polygon *ABCGFED* in radical form.

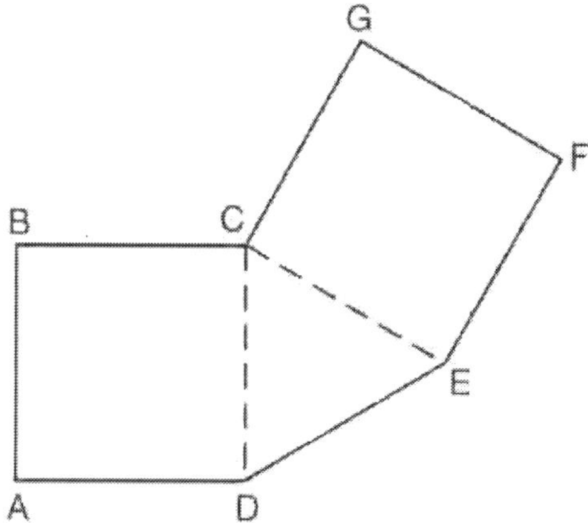

2.2 Area

Key Terms and Concepts

The **area** of a figure is the number of square units that the figure encloses.

The formulas for the areas of some common shapes include:

Square: $A = s^2$ Rectangle: $A = lw$ Parallelogram: $A = bh$

Trapezoid: $A = \frac{1}{2}h(b_1 + b_2)$ Triangle: $A = \frac{1}{2}bh$ Circle: $A = \pi r^2$

Note: The formulas for the area of a triangle, parallelogram, and circle are included on the Reference Sheet at the back of the Regents exam.

A **composite figure** is made up of more than one type of shape joined together. To find its area, add the areas of all the included shapes.

A **shaded region** is sometimes used to show part of a figure's area. To find the area of a shaded region, find the area of the larger shape that fully encloses it, then subtract any unshaded regions that are also enclosed.

If asked to express the **exact area in terms of π**, do not calculate an approximation for π. Instead, leave π in the expression that represents the area.

Model Problem 1: *area of a composite figure*

The "key" in a regulation size basketball court is made up of a rectangle and semicircle. The length of the rectangle measures 19 feet from the baseline to the free throw line, and the width measures 12 feet, as shown below. Find the exact area of the key, in terms of π.

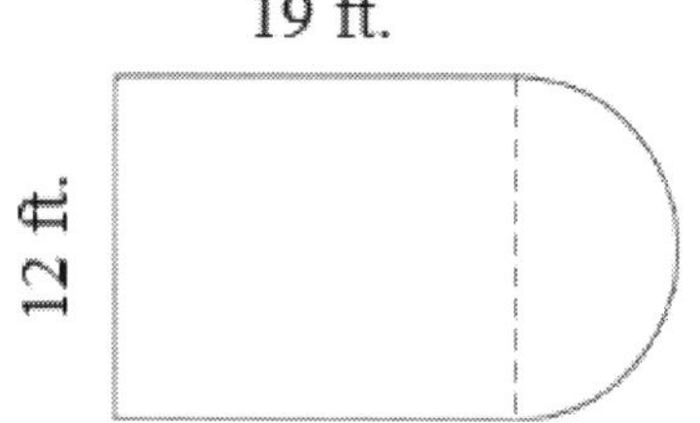

Solution:

(A) $A = lw + \frac{1}{2}\pi r^2$

(B) $= (19)(12) + \frac{1}{2}\pi(6)^2$

(C) $= 228 + 18\pi$ *square feet*

Explanation of steps:

(A) Write an equation that adds the areas of the parts that make up the composite figure *[rectangle and half of a circle].*

(B) Substitute the given values for the variables *[19, 12, and 6, for l, w and r].*

(C) Simplify. *[Write the answer in terms of π as the directions specify.]*

Model Problem 2: *area of a shaded region*

Find the area of the shaded region.

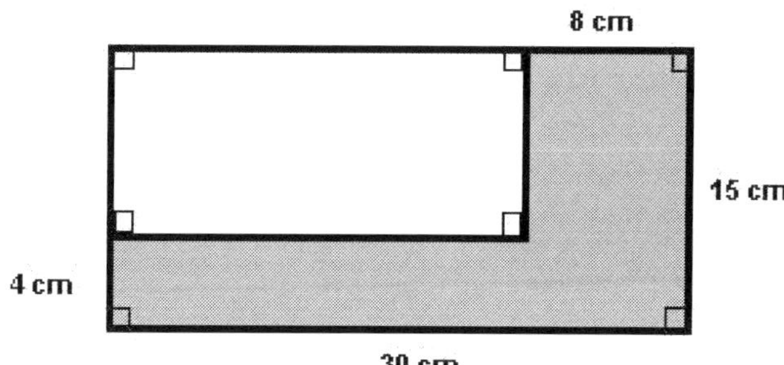

Solution:

(A) $A = A_{large} - A_{small}$

(B) $= (30)(15) - (22)(11)$

(C) $= 208$ sq.cm

Explanation of steps:

(A) Write an equation that subtracts the unshaded area *[the smaller rectangle]* from the total area *[the larger rectangle]* to find the shaded area.

(B) Calculate each area *[the sides of the larger rectangle, 30 and 15, are given, but the lengths of the sides of the smaller rectangle need to be calculated as 30 – 8 and 15 – 4].*

(C) Simplify.

Practice Problems

1. The figure below consists of two semicircles connected to a 10 foot by 20 foot rectangle. Find the area of the composite figure in terms of π.

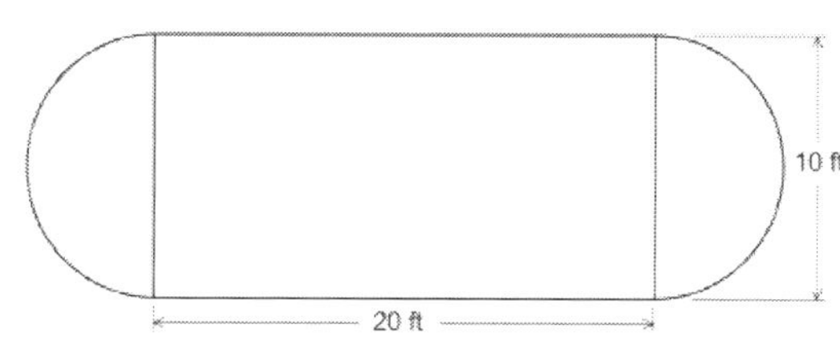

2. A circle with radius 4 is inscribed inside a square as shown. Find the shaded area in terms of π.

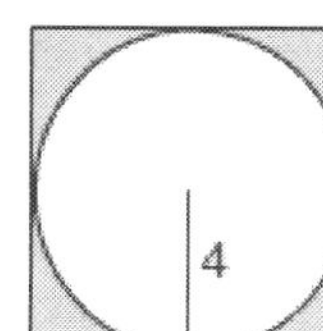

3. A circle and two semicircles are inscribed inside a 20 ft. by 10 ft. rectangle as shown. Find the shaded area in terms of π.

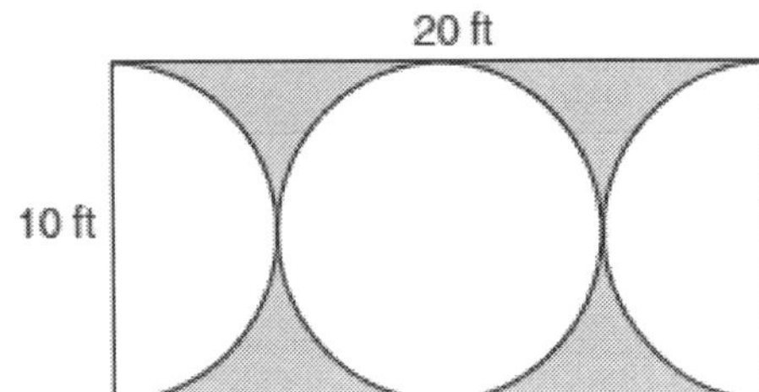

4. The diagram shows an isosceles right triangle inside a quarter-circle with a radius of 4 cm. Find the shaded area in terms of π.

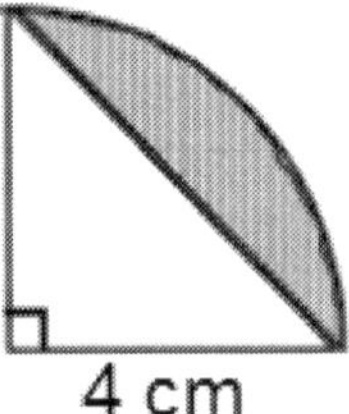

5. The diagram shows a triangle inside a rectangle with dimensions as given. Find the shaded area.

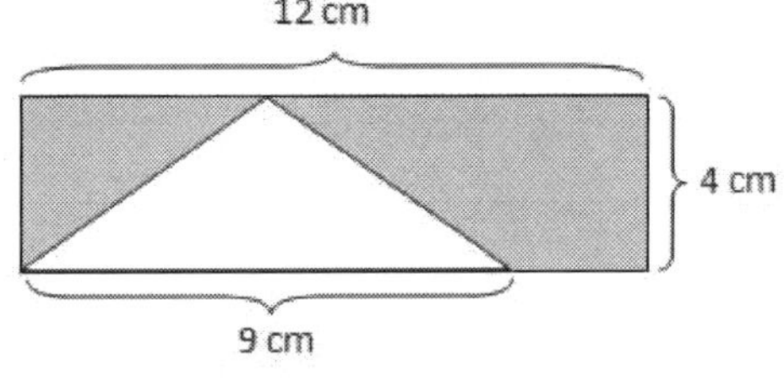

6. A designer created the logo shown below. The logo consists of a square and four quarter-circles of equal size. Express, in terms of π, the exact area, in square inches, of the shaded region.

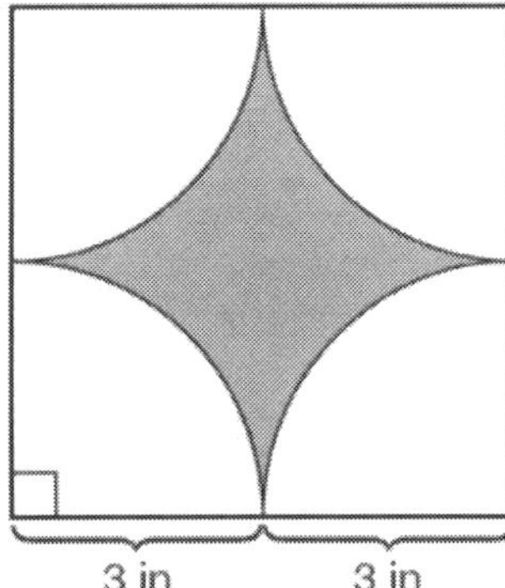

REGENTS QUESTIONS

Multiple Choice

1. The figure shown below is composed of two rectangles and a quarter circle.

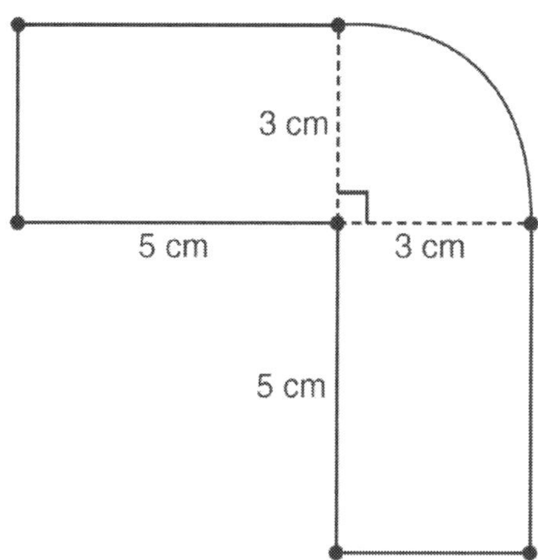

What is the area of this figure, to the *nearest square centimeter*?

(1) 33 (3) 44
(2) 37 (4) 58

2. In the diagram below, circle O is inscribed in square $ABCD$. The square has an area of 36.

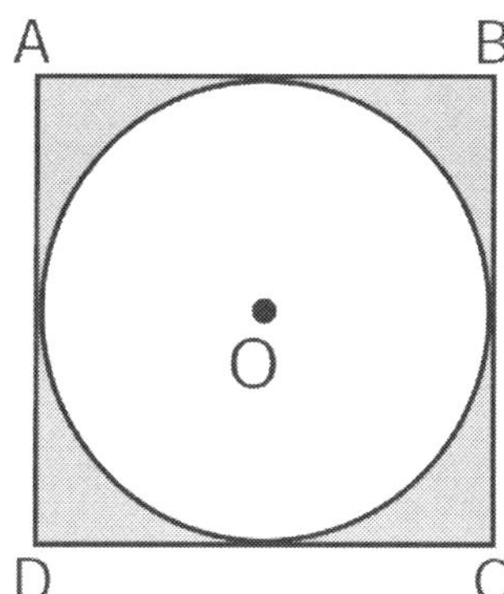

What is the area of the circle?

(1) 9π (3) 3π
(2) 6π (4) 36π

3. In the figure below, $ABCD$ is a square and semicircle O has a radius of 6.

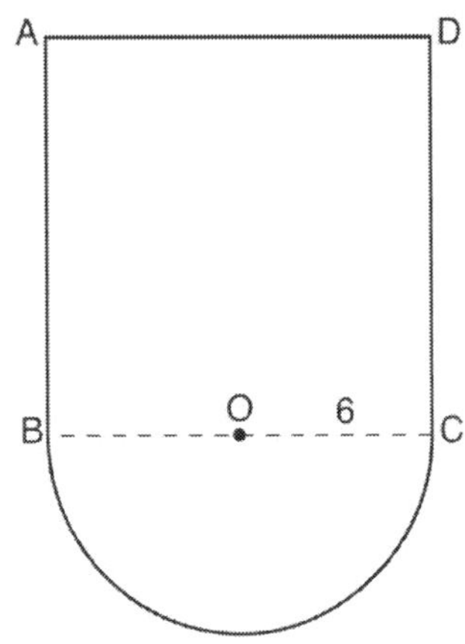

What is the area of the figure?

(1) $36+6\pi$
(2) $36+18\pi$
(3) $144+18\pi$
(4) $144+36\pi$

4. A figure consists of a square and a semicircle, as shown in the diagram below.

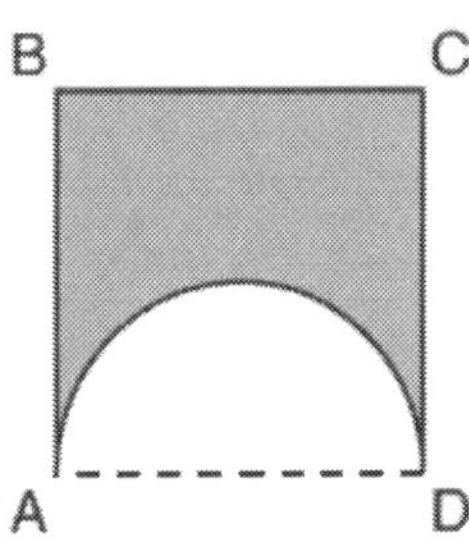

If the length of a side of the square is 6, what is the area of the shaded region?

(1) $36-3\pi$
(2) $36-4.5\pi$
(3) $36-6\pi$
(4) $36-9\pi$

5. (CC) A farmer has 64 feet of fence to enclose a rectangular vegetable garden. Which dimensions would result in the biggest area for this garden?

(1) the length and the width are equal
(2) the length is 2 more than the width
(3) the length is 4 more than the width
(4) the length is 6 more than the width

Constructed Response

6. In the diagram below of rectangle $AFEB$ and a semicircle with diameter $\overline{CD}$, $AB = 5$ inches, $AB = BC = DE = FE$, and $CD = 6$ inches. Find the area of the shaded region, to the *nearest hundredth of a square inch.*

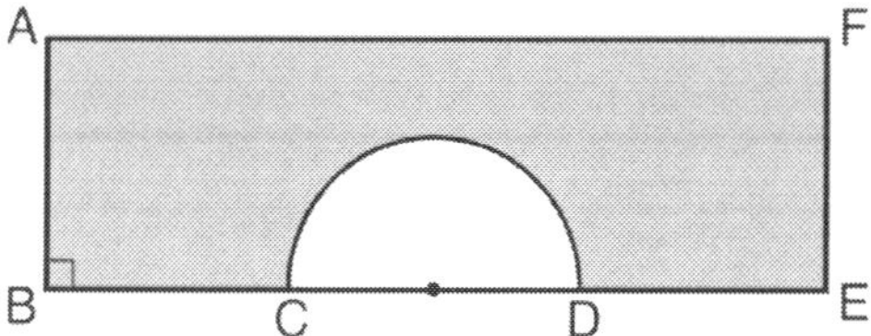

7. A patio consisting of two semicircles and a square is shown in the diagram below. The length of each side of the square region is represented by $2x$. Write an expression for the area of the entire patio, in terms of x and π.

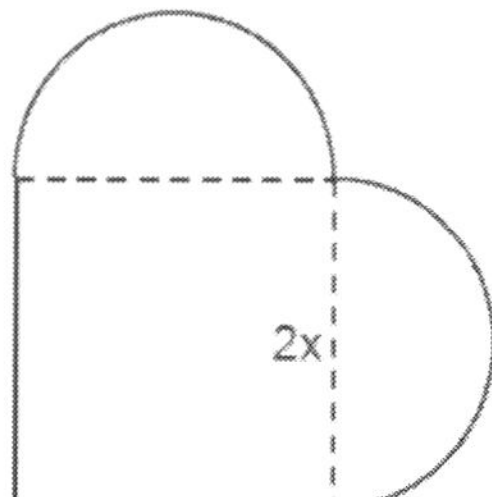

8. The Rock Solid Concrete Company has been asked to pave a rectangular area surrounding a circular fountain with a diameter of 8 feet, as shown in the diagram.

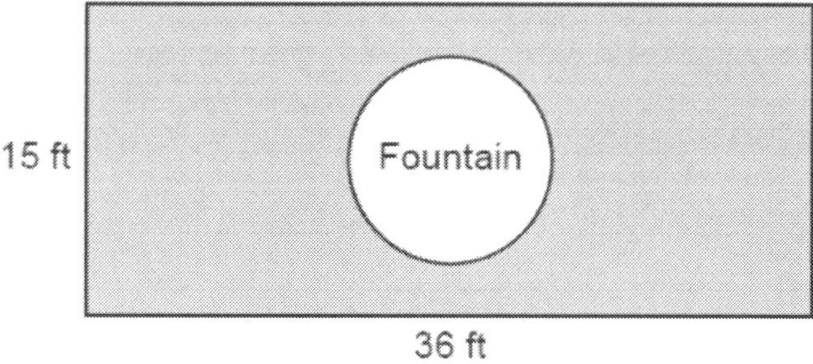

Find the area, to the *nearest square foot*, that must be paved.

Find the cost, *in dollars*, of paving the area if the Rock Solid Concrete Company charges \$8.95 per square foot.

Chapter 3. Lines, Angles and Proofs

3.1 Parallel Lines and Transversals

Key Terms and Concepts

In the diagram below, **parallel lines** r and s are both intersected by a third line, t. A line (in this case, t) that intersects two other lines at different points is called a **transversal**. In the diagram, eight angles are formed and are labelled 1 through 8.

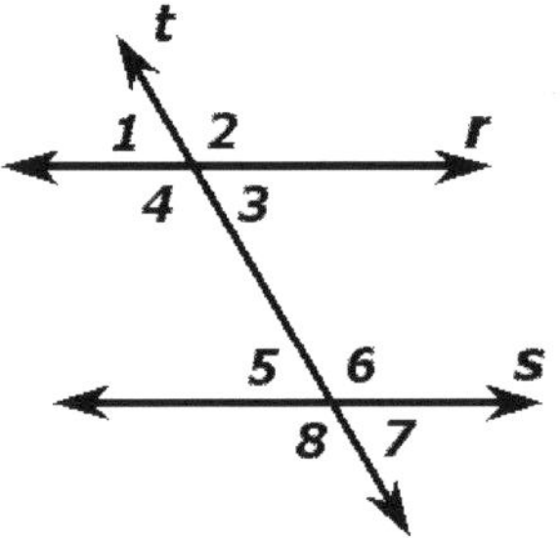

We can already see eight pairs of angles that are **linear pairs**:
1 and 2; 2 and 3; 3 and 4; 4 and 1; 5 and 6; 6 and 7; 7 and 8; 8 and 5
Linear pairs are supplementary (add to $180°$).

We can also find four pairs of **vertical angles**:
1 and 3; 2 and 4; 5 and 7; 6 and 8
Vertical angles are congruent (equal in measure).

When two lines are cut by a transversal, we also can define pairs of corresponding angles. **Corresponding angles** are a pair of angles in the same direction from each of the two points of intersection. Pairs of corresponding angles are shown in the diagrams below.

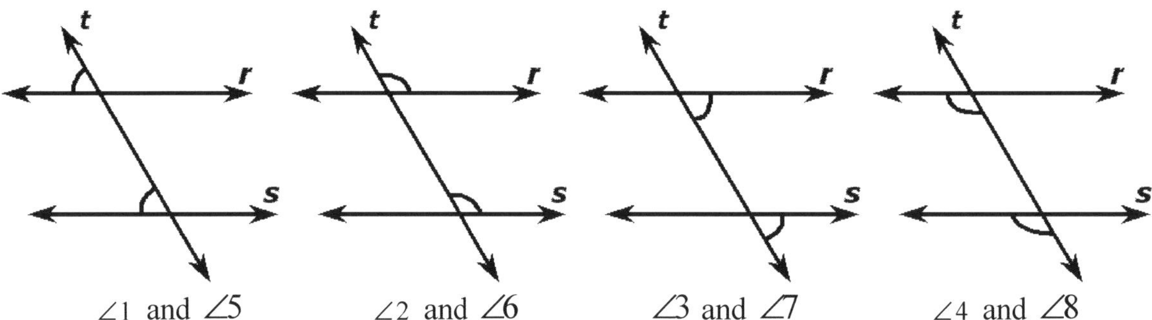

$\angle 1$ and $\angle 5$ $\angle 2$ and $\angle 6$ $\angle 3$ and $\angle 7$ $\angle 4$ and $\angle 8$

If two lines cut by a transversal are parallel, then **pairs of corresponding angles are congruent**. The converse is also true: if two corresponding angles are congruent, then the lines are parallel.

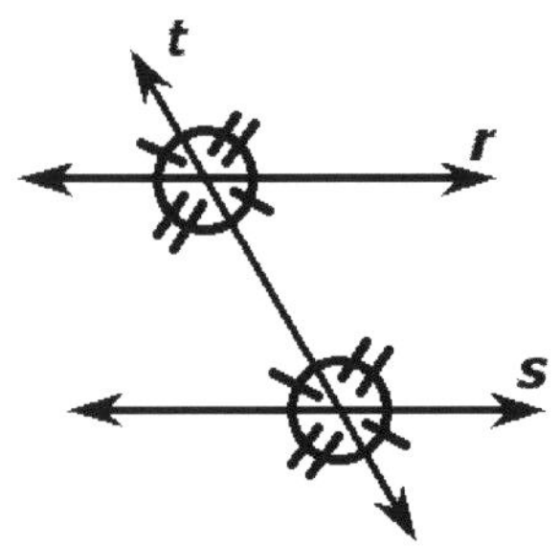

Since vertical angles are congruent and corresponding angles are congruent when $r \parallel s$, our diagram to the left really has two sets of angles that are all congruent to each other. The four angles marked by single-slashed arcs are all congruent to each other, and the four angles marked by double-slashed arcs are all congruent to each other.

Also in our diagram are two pairs of alternate interior and two pairs of alternate exterior angles. Angles are **alternate** if they are on opposite sides of the transversal. Angles are **interior** if they are between the two lines that are cut by the transversal (r and s), or **exterior** if they are not.

The following are pairs of **alternate interior angles**:

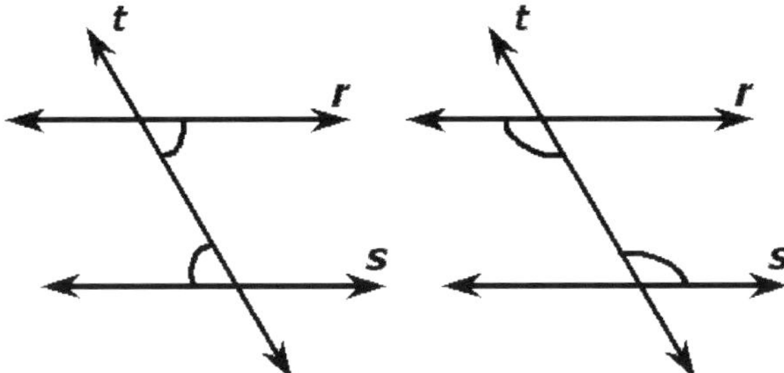

The following are pairs of **alternate exterior angles**:

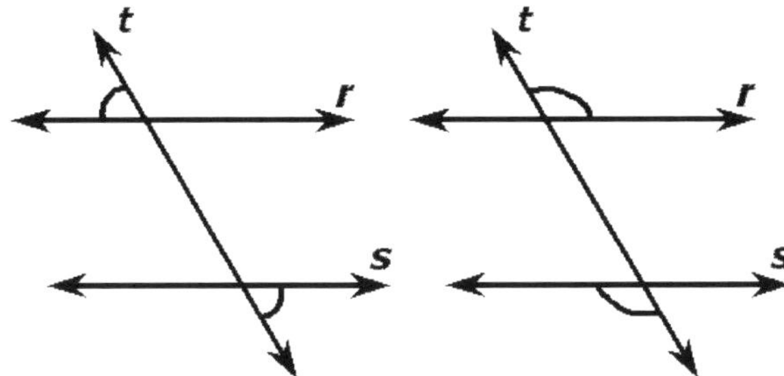

If the lines are parallel, then each pair of alternate interior angles and each pair of alternate exterior angles are congruent. If they are not congruent, then the lines are not parallel.

Also note that, when the lines are parallel, the following pairs of angles are *supplementary*:

- two interior angles on the same side of the transversal, and
- two exterior angles on the same side of the transversal

Model Problem:

Lines *s* and *t* are parallel. Find the measures of the angles labelled 1, 2, 3, and 4.

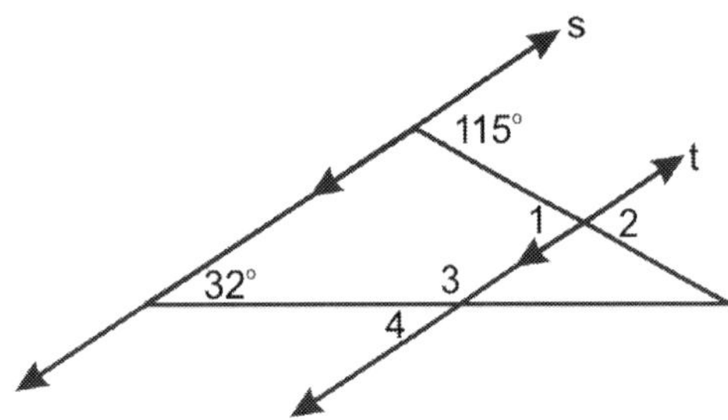

Solution:

(A) $m\angle 1 = 115°$

(B) $m\angle 2 = 115°$

(C) $m\angle 4 = 32°$

(D) $m\angle 3 = 148°$

Explanation of steps:

(A) $\angle 1$ and the angle marked as 115° are alternate interior angles, which are congruent.

(B) $\angle 1$ and $\angle 2$ are vertical angles, which are congruent.

(C) $\angle 4$ and the angle marked as 32° are alternate interior angles, which are congruent.

(D) $\angle 3$ and $\angle 4$ are a linear pair, which are supplementary (add to 180°).

Practice Problems

1. Given: $r \parallel s$

State the relationship between each pair of angles as shown in the diagram, and state whether the pair of angles are congruent or supplementary.

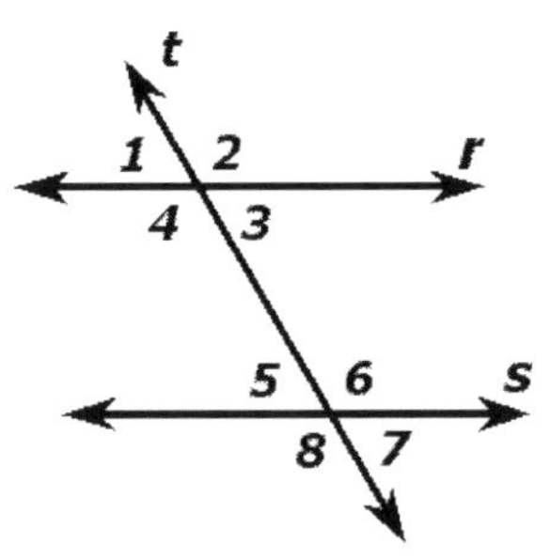

(a) $\angle 1$ and $\angle 2$

(b) $\angle 2$ and $\angle 4$

(c) $\angle 2$ and $\angle 6$

(d) $\angle 4$ and $\angle 6$

(e) $\angle 2$ and $\angle 8$

2. Lines *m* and *n* are parallel. Find the measures of the angles labelled *a*, *b*, *c*, and *d*.

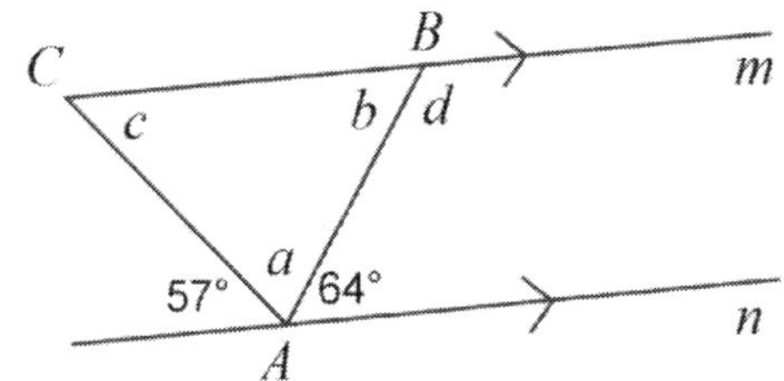

3. Given $p \parallel q$, solve for *x* and *y*.

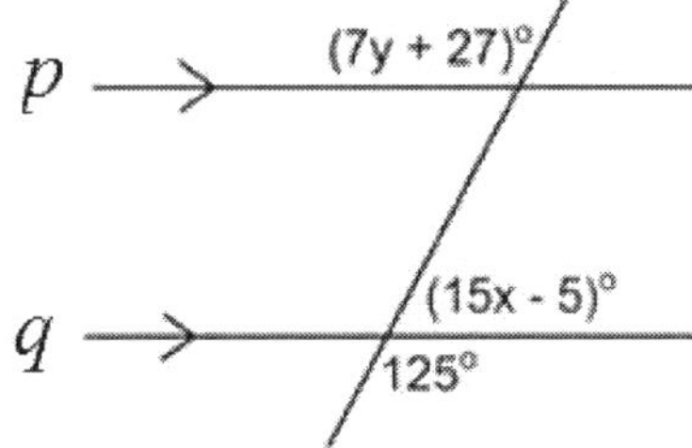

REGENTS QUESTIONS

Multiple Choice

1. In the accompanying figure, what is one pair of alternate interior angles?

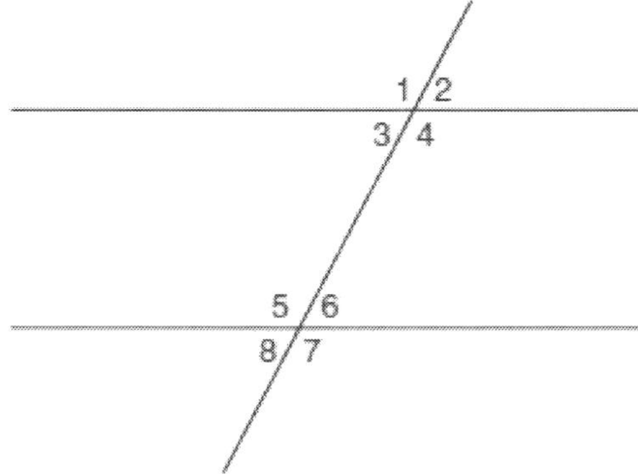

(1) $\angle 1$ and $\angle 2$
(2) $\angle 4$ and $\angle 5$
(3) $\angle 4$ and $\angle 6$
(4) $\angle 6$ and $\angle 8$

2. In the accompanying diagram, line ℓ is parallel to line m, and line t is a transversal.

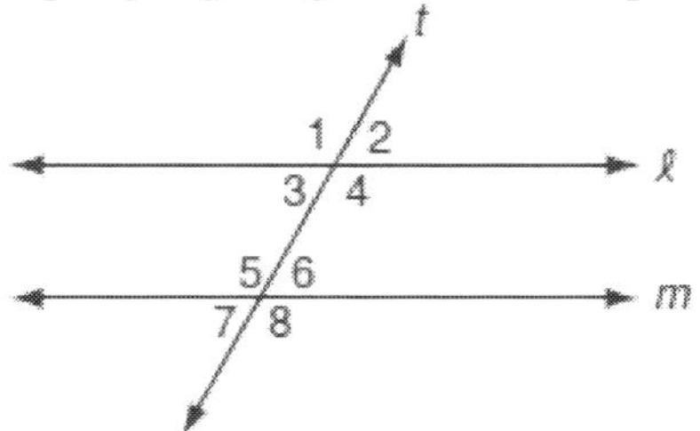

Which must be a true statement?

(1) $m\angle 1 + m\angle 4 = 180$
(2) $m\angle 1 + m\angle 8 = 180$
(3) $m\angle 3 + m\angle 6 = 180$
(4) $m\angle 2 + m\angle 5 = 180$

3. Based on the diagram below, which statement is true?

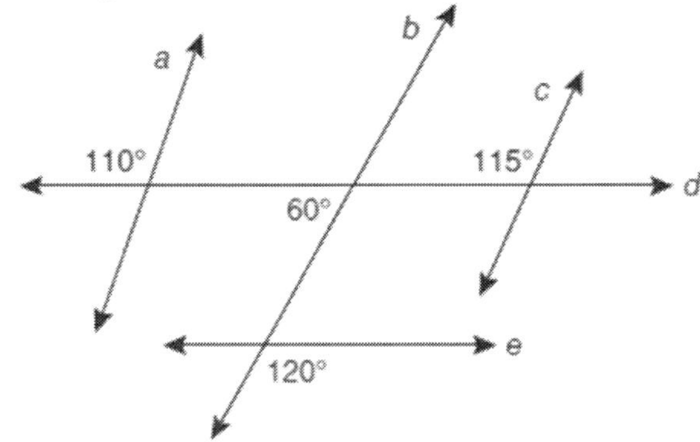

(1) $a \parallel b$
(2) $a \parallel c$
(3) $b \parallel c$
(4) $d \parallel e$

4. A transversal intersects two lines. Which condition would always make the two lines parallel?
   (1) Vertical angles are congruent.
   (2) Alternate interior angles are congruent.
   (3) Corresponding angles are supplementary.
   (4) Same-side interior angles are complementary.

5. In the diagram below, lines n and m are cut by transversals p and q.

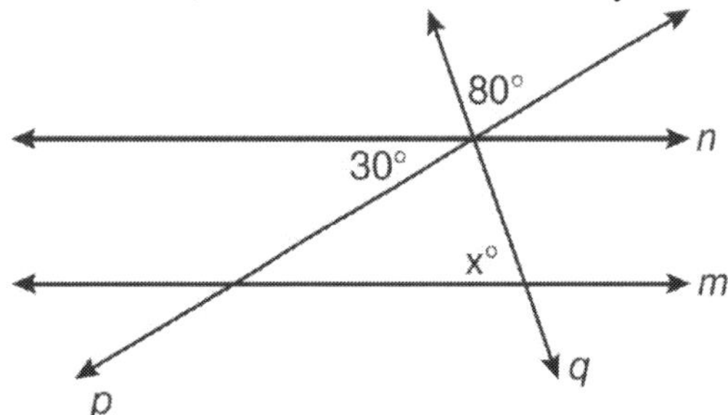

What value of x would make lines n and m parallel?
   (1) 110 (3) 70
   (2) 80 (4) 50

6. Transversal $\overleftrightarrow{EF}$ intersects $\overleftrightarrow{AB}$ and $\overleftrightarrow{CD}$, as shown in the diagram below.

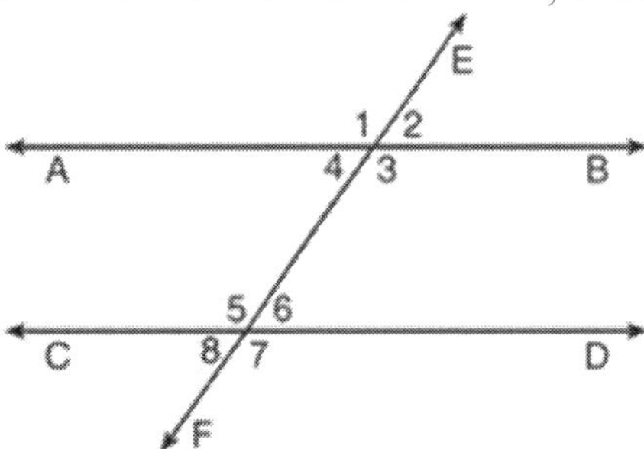

Which statement could always be used to prove $\overleftrightarrow{AB} \parallel \overleftrightarrow{CD}$?
   (1) $\angle 2 \cong \angle 4$ (3) $\angle 3$ and $\angle 6$ are supplementary
   (2) $\angle 7 \cong \angle 8$ (4) $\angle 1$ and $\angle 5$ are supplementary

7. CC Steve drew line segments $ABCD$, EFG, BF, and CF as shown in the diagram below. Scalene $\triangle BFC$ is formed.

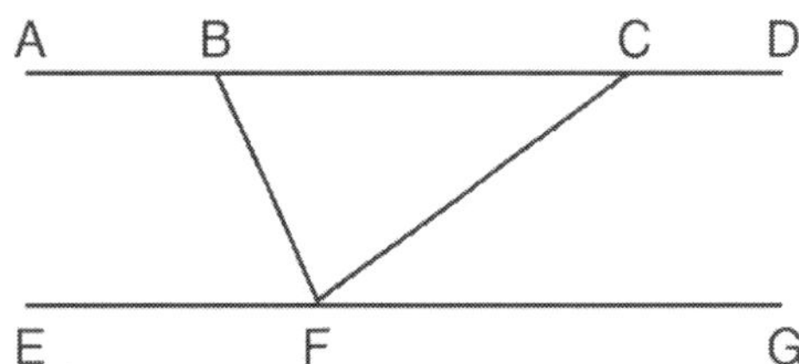

Which statement will allow Steve to prove $\overline{ABCD} \parallel \overline{EFG}$?
   (1) $\angle CFG \cong \angle FCB$ (3) $\angle EFB \cong \angle CFB$
   (2) $\angle ABF \cong \angle BFC$ (4) $\angle CBF \cong \angle GFC$

8. (CC) In the diagram below, lines ℓ, m, n, and p intersect line r.

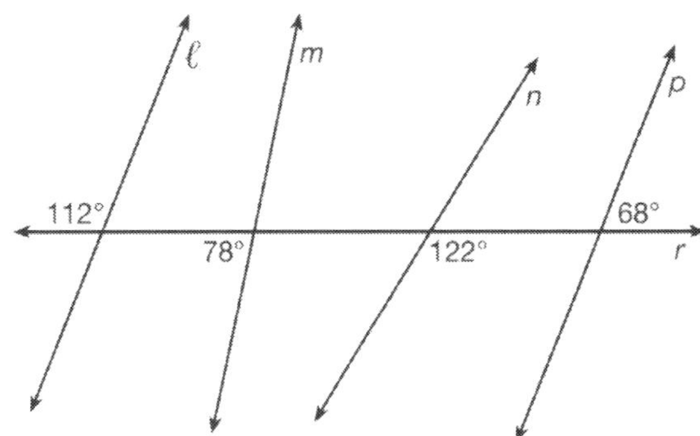

Which statement is true?

(1) $\ell \parallel n$
(2) $\ell \parallel p$
(3) $m \parallel p$
(4) $m \parallel n$

Constructed Response

9. In the accompanying diagram, parallel lines $\overleftrightarrow{AB}$ and $\overleftrightarrow{CD}$ are intersected by transversal $\overleftrightarrow{EF}$ at points X and Y, and $m\angle FYD = 123$. Find $m\angle AXY$.

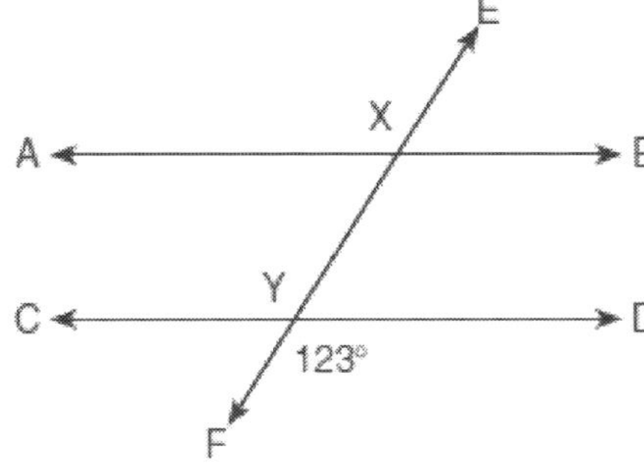

3.2 Deductive Reasoning

Key Terms and Concepts

A major aspect of geometry is the proof of statements using deductive reasoning.

The most frequently used vehicle in a deductive proof is known as the **rule of detachment** (or more commonly known by its Latin name, ***modus ponens***). It is an accepted rule of inference that states that if statement P is true, and statement P implies statement Q (often written as $P \rightarrow Q$), then statement Q is also true. The implication $P \rightarrow Q$ can be written in a variety of ways in English, but most commonly as "If P, then Q."

Example: Suppose P is the statement, "$\angle A$ is a right angle," and Q is the statement, "$m\angle A = 90°$" We can write an argument as follows.

P	$\angle A$ is a right angle.
$P \rightarrow Q$	If an angle is a right angle, then it measures $90°$.
Q	Therefore, $m\angle A = 90°$

Rather than using letters and logical symbols, we can write a proof in a table of **statements and reasons**. In the Statements column are facts such as "$\angle A$ is a right angle" or "$m\angle A = 90°$" and in the Reasons column are the reasons that each statement is accepted as true, either because it is "Given" or because it the result of an implication from an earlier statement.

Example:

Statements	*Reasons*
$\angle A$ is a right angle	Given
$m\angle A = 90°$	If an angle is a right angle, then it measures $90°$

Note how this corresponds to:

Statements	*Reasons*
P	Given
Q	$P \rightarrow Q$

We need not always write $P \rightarrow Q$ in an *if-then* format. For example, we could have written the equivalent rule, "All right angles measure $90°$," or even very simply, "Definition of right angle," since a right angle is defined as one that measure 90 degrees.

Although the Statements/Reasons table format can help to keep a proof organized, it is optional. You may prefer to write your proof in a flowchart or paragraph format, as long as every necessary statement and its corresponding reason are included in a logical sequence of steps.

When writing a proof, you may use any axioms, postulates, definitions, or theorems at your disposal, in a logical sequence of statements.

Axioms and postulates are accepted assumptions or statements of common sense that need not be proven. Examples include:

Reflexive Property: A quantity is equal (or congruent) to itself
Example: $\overline{AB} \cong \overline{AB}$

Transitive Property: If two quantities are equal to the same quantity, they are equal to each other
Example: If $m\angle 1 = m\angle 2$ and $m\angle 2 = m\angle 3$, then $m\angle 1 = m\angle 3$

Partition Postulate: A whole is equal to the sum of its parts.
Example: If given segment $\overline{ABC}$, then $AB + BC = AC$
If given $\angle ABC$ and point D, then $m\angle ABD + m\angle DBC = m\angle ABC$

Addition Property: If equals are added to equals, their sums are equal
Example: If $AB = BC$ and $CD = DE$, then $AB + CD = BC + DE$

Subtraction Property: If equals are subtracted from equals, their differences are equal
Example: If $m\angle A = m\angle B$ and $m\angle C = m\angle D$, then $m\angle A - m\angle C = m\angle B - m\angle D$

Multiplication Property: If equals are multiplied by equals, their products are equal
Example: If $m\angle ABC = m\angle DEF$, then $2(m\angle ABC) = 2(m\angle DEF)$

Division Property: If equals are divided by equals, their quotients are equal

Substitution Property: A quantity may be substituted for its equal.
Example: If $x + y = z$ and $y = w$, then $x + w = z$

Parallel Postulate: Through an external point, there is at most one line parallel to a given line. This also means that through an external point, a line may be drawn parallel to a given line.

Definitions of terms may also be used in proofs. Examples include:

- Perpendicular lines intersect to form right angles.
- A perpendicular bisector is perpendicular to a line segment at its midpoint.
- An angle bisector divides an angle into two congruent angles.

Theorems, or statements that are already accepted as true, may also be used. Examples include:

- All right angles are congruent.
- Linear pairs are supplementary.
- Supplements (or complements) of congruent angles are congruent.
- Vertical angles are congruent.
- Parallel lines cut by a transversal form congruent alternate interior angles.
- The sum of the measures of the angles of a triangle is $180°$.

To write a proof, we begin with the Given statements and follow a logical sequence of deductions until the statement we need to prove is the final statement.

Important: If we're asked to prove a theorem that we already know to be true, we cannot simply use that theorem and say we're done! If asked to prove a theorem, you must, for the sake of argument, assume that the theorem doesn't already exist.

For example, let's prove that vertical angles are congruent. Given intersecting lines m and n as shown in the diagram to the right, we'll prove that angles 1 and 3 are congruent. We can't just say they're congruent because "vertical angles are congruent." Even though this may be true, it is what we are being asked to prove. We'll need to use other reasons.

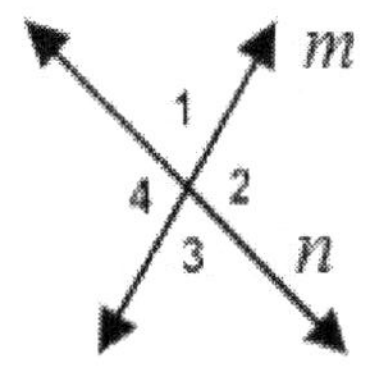

Statements	*Reasons*
m and n are intersecting lines	Given
$\angle 1$ and $\angle 2$ are a linear pair	Definition of linear pair
$\angle 1$ and $\angle 2$ are supplementary	Linear pairs are supplementary
$\angle 3$ and $\angle 2$ are a linear pair	Definition of linear pair
$\angle 3$ and $\angle 2$ are supplementary	Linear pairs are supplementary
$\angle 1 \cong \angle 3$	Supplements of the same angle are congruent.

Model Problem:

Given $m\angle PTQ = m\angle STR$, prove $m\angle PTR = m\angle STQ$.

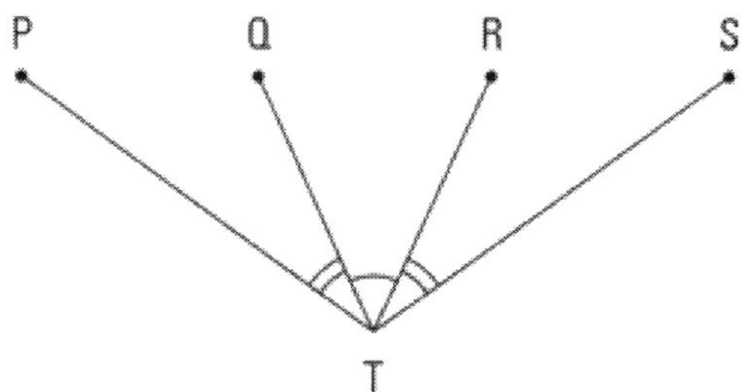

Solution:

Statements	*Reasons*
$m\angle PTQ = m\angle STR$	Given
$m\angle QTR = m\angle QTR$	Reflexive Property
$m\angle PTQ + m\angle QTR = m\angle STR + m\angle QTR$	Addition Property
$m\angle PTQ + m\angle QTR = m\angle PTR$ and $m\angle STR + m\angle QTR = m\angle STQ$	Partition Postulate
$m\angle PTR = m\angle STQ$	Substitution Property

Explanation of steps:

Start with the Given statement and follow a logical sequence of deductions until the statement we need to prove is the final statement.

[When looking at the two angles that we need to prove equal in measure, $\angle PTR$ and $\angle STQ$, we notice that they share a common angle, $\angle QTR$. Namely, we get $m\angle PTR$ by adding $m\angle PTQ + m\angle QTR$, and we get $m\angle STQ$ by adding $m\angle STR + m\angle QTR$, by the Partition Postulate. Working backwards, if we can show these angle sums are equal, then $m\angle PTR = m\angle STQ$ by Substitution. We do know the two sums are equal because we are adding equals to equals (the Addition Property): $m\angle PTQ = m\angle STR$ by the Given statement, and $m\angle QTR$ is equal to itself by the Reflexive Property.]

Practice Problems

1. Given: ∠1 and ∠2 are complementary
   ∠2 and ∠3 are complementary
   Prove: $m\angle 1 = m\angle 3$

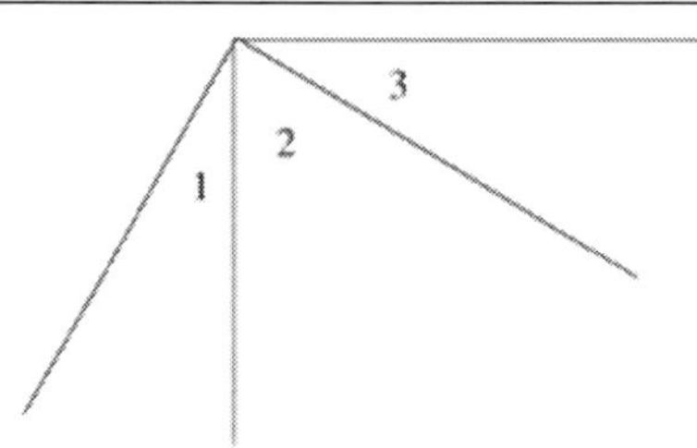

Statements	Reasons

2. Given: ∠1 and ∠3 are supplementary
   Prove: $m \parallel n$

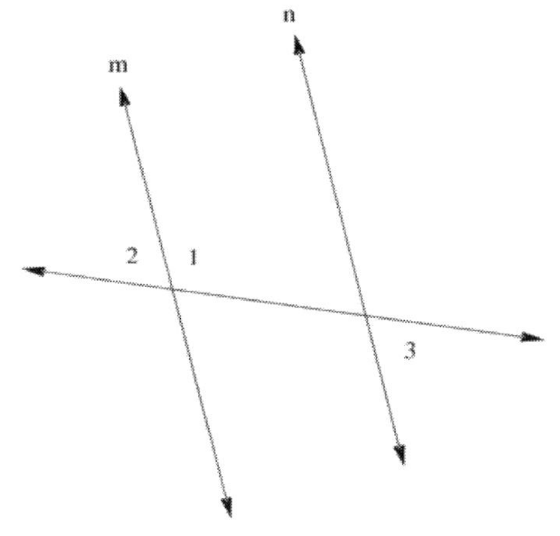

Statements	Reasons

3. Given: $\overline{PCEG}$

$\overline{PC} \cong \overline{GE}$

Prove: $\overline{PE} \cong \overline{GC}$

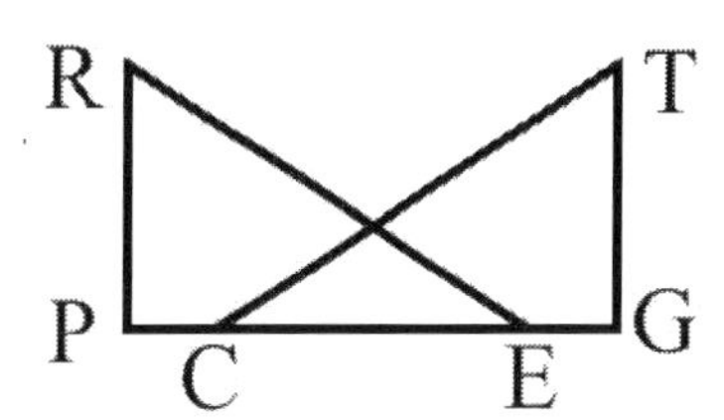

Statements	*Reasons*

REGENTS QUESTIONS

Multiple Choice

1. When writing a geometric proof, which angle relationship could be used alone to justify that two angles are congruent?
   (1) supplementary angles
   (2) linear pair of angles
   (3) adjacent angles
   (4) vertical angles

2. In the diagram below of $\overline{ABCD}$, $\overline{AC} \cong \overline{BD}$.

   Using this information, it could be proven that
   (1) $BC = AB$
   (2) $AB = CD$
   (3) $AD - BC = CD$
   (4) $AB + CD = AD$

3. In $\triangle AED$ with $\overline{ABCD}$ shown in the diagram below, $\overline{EB}$ and $\overline{EC}$ are drawn.

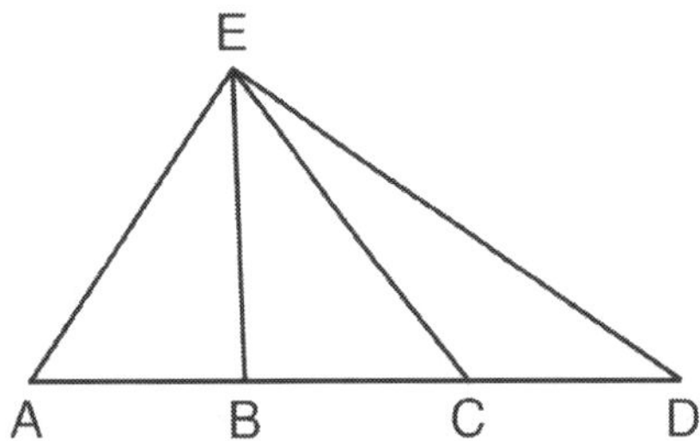

   If $\overline{AB} \cong \overline{CD}$, which statement could always be proven?
   (1) $\overline{AC} \cong \overline{DB}$
   (2) $\overline{AE} \cong \overline{ED}$
   (3) $\overline{AB} \cong \overline{BC}$
   (4) $\overline{EC} \cong \overline{EA}$

4. In the diagram of $\overline{WXYZ}$ below, $\overline{WY} \cong \overline{XZ}$.

   Which reasons can be used to prove $\overline{WX} \cong \overline{YZ}$?
   (1) reflexive property and addition postulate
   (2) reflexive property and subtraction postulate
   (3) transitive property and addition postulate
   (4) transitive property and subtraction postulate

5. (CC) In the diagram below, $\overleftrightarrow{FE}$ bisects $\overline{AC}$ at B, and $\overleftrightarrow{GE}$ bisects $\overline{BD}$ at C.

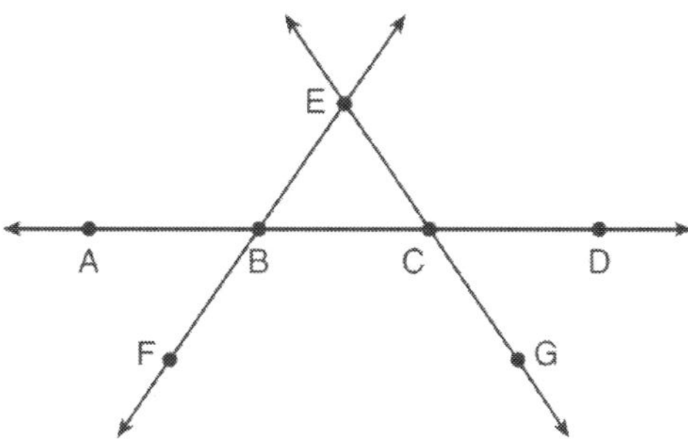

Which statement is always true?

(1) $\overline{AB} \cong \overline{DC}$

(2) $\overline{FB} \cong \overline{EB}$

(3) $\overleftrightarrow{BD}$ bisects $\overline{GE}$ at C.

(4) $\overleftrightarrow{AC}$ bisects $\overline{FE}$ at B.

Constructed Response

6. (CC) Given the theorem, "The sum of the measures of the interior angles of a triangle is 180°," complete the proof for this theorem.

Given: $\triangle ABC$
Prove: $m\angle 1 + m\angle 2 + m\angle 3 = 180°$
Fill in the missing reasons below.

Statements	Reasons
(1) $\triangle ABC$	(1) Given
(2) Through point C, draw $\overleftrightarrow{DCE}$ parallel to $\overline{AB}$.	(2) ______
(3) $m\angle 1 = m\angle ACD$, $m\angle 3 = m\angle BCE$	(3) ______
(4) $m\angle ACD + m\angle 2 + m\angle BCE = 180°$	(4) ______
(5) $m\angle 1 + m\angle 2 + m\angle 3 = 180°$	(5) ______

Chapter 4. Triangles

4.1 Angles of Triangles

Key Terms and Concepts

Triangles are sometimes classified by their angles. An **acute triangle** has all three angles that are acute (less than $90°$). A **right triangle** has one angle that measures $90°$. An **obtuse triangle** has one angle that is obtuse (more than $90°$).

Here are some important facts about the angles of triangles, some of which you may have already learned in earlier grades.

- **The sum of the angle measures in a triangle is $180°$.**

(The proof to this theorem is given as a Regents question at the end of the previous section. It involves drawing a line, through a vertex, that is parallel to the opposite side.)

- **Each angle in an equilateral triangle measures $60°$.**

An equilateral triangle is also equiangular, meaning that the angles are all equal in measure. Since the sum of the three angle measures is $180°$, each angle measures $(180 \div 3)°$.

- **In a right triangle, the two acute angles are complementary.**

The sum of the three angles is $180°$ and the right angle measures $90°$. So, if we subtract $180° - 90°$, the difference, which represents the sum of the other two angle measures, is $90°$. Therefore, the two acute angles are complementary.

- **The exterior angle of a triangle equals the sum of the two opposite interior angles.**

In the diagram below, one side is extended to create exterior angle *d*. Since *c* and *d* are a linear pair, they are supplementary, so $d = 180° - c$. As the three angles of a triangle, $a + b + c = 180°$, so $a + b = 180° - c$. Therefore, by substitution, $a + b = d$.

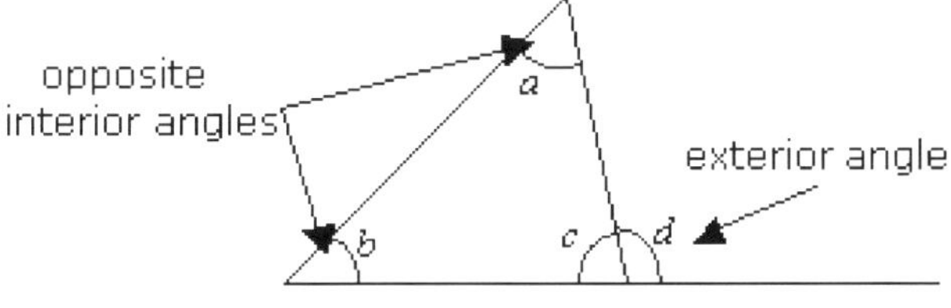

- **The sum of the measures of a set of exterior angles of a triangle is $360°$.**

A triangle has two sets of exterior angles: one set is formed by going around the triangle counterclockwise and the other by going around the triangle clockwise, as shown below.

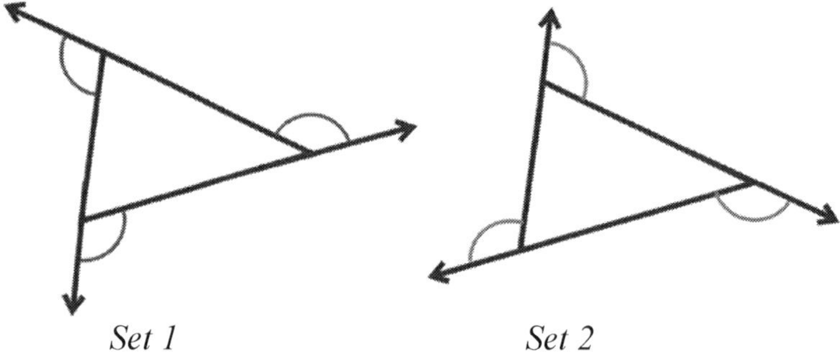

In a set of exterior angles, each exterior angle forms a linear pair with an interior angle, so in the diagram below, $d = 180 - a$, $e = 180 - b$, and $f = 180 - c$. Therefore, the sum of the exterior angles is $180 - a + 180 - b + 180 - c$, which can be rewritten as $540 - (a + b + c)$. Since the sum of the interior angle measures of a triangle equals 180, we can substitute for $a + b + c$ in the expression to get $540 - 180 = 360$. We have shown algebraically that the sum of the measures of a set of exterior angles of a triangle is $360°$.

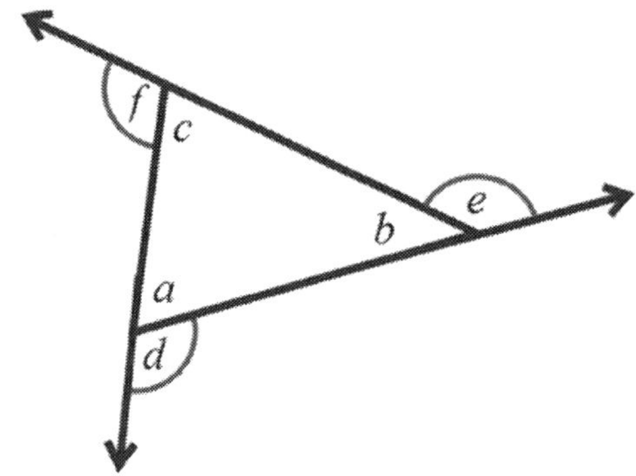

Model Problem:

In the diagram below, $\overleftrightarrow{ABCD}$ is a straight line, and angle E in triangle BEC is a right angle.

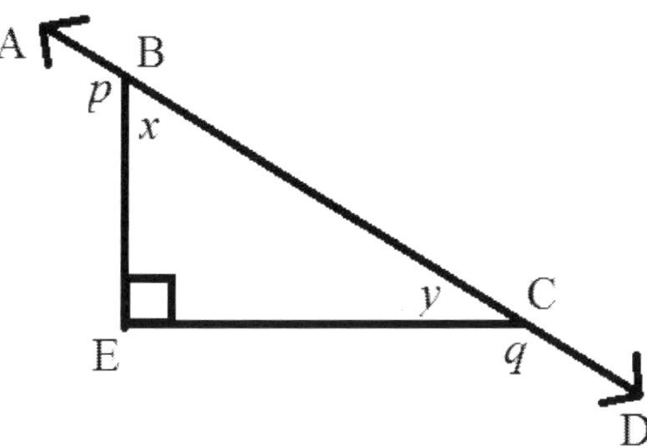

How many degrees are in $p + q$?

Solution:

(A) $p = 90 + y$

$q = 90 + x$

(B) $x + y = 90$

(C) Therefore,

$$\begin{aligned} p + q &= 90 + y + 90 + x \\ &= 180 + x + y \\ &= 180 + 90 \\ &= 270 \end{aligned}$$

So, $p + q = 270°$

Explanation of steps:

(A) The exterior angle of a triangle equals the sum of the two opposite interior angles. *[p and q are exterior angles]*

(B) In a right triangle, the two acute angles are complementary.

(C) Find the sum *[by substituting for p, substituting for q, and then substituting for x + y]*.

Practice Problems

1. What is the measure of angle C?

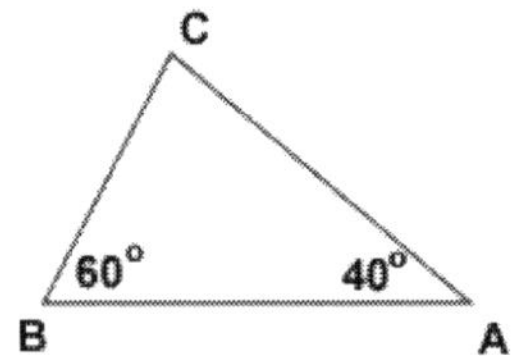

2. Solve for x.

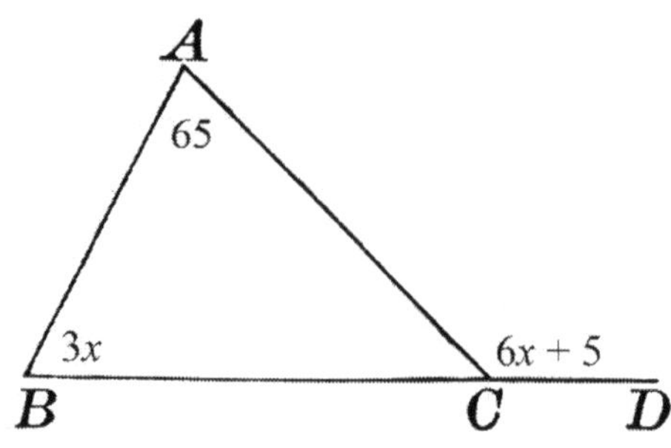

3. What is the measure of an exterior angle of an equilateral triangle?

4. The measure of the largest angle of a triangle is 5 times the smallest angle. The third angle is 12 degrees larger than the smallest angle. Is the triangle acute, right, or obtuse?

5. In the diagram below, $m\angle B = 90°$, $m\angle A = 65°$, $m\angle D = 50°$, and $m\angle DCE = 80°$. Find $m\angle x$.

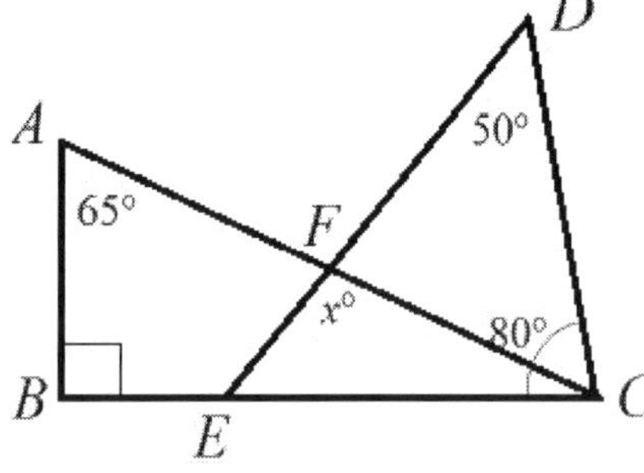

6. In the diagram below, $a \parallel b$. Find $m\angle x$.

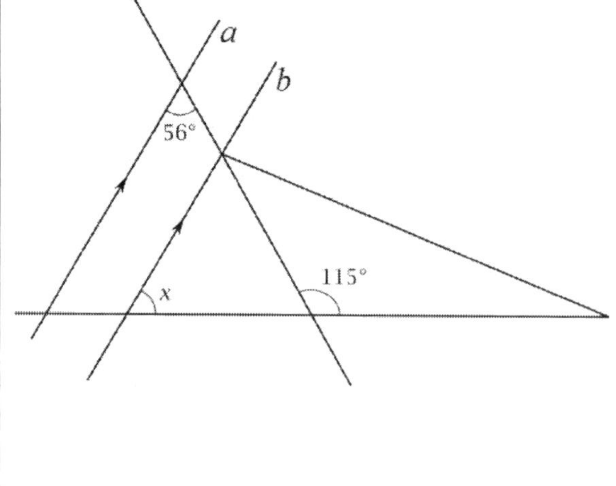

REGENTS QUESTIONS

Multiple Choice

1. In the accompanying diagram, $\overleftrightarrow{AB} \parallel \overleftrightarrow{CD}$. From point E on $\overleftrightarrow{AB}$, transversals $\overleftrightarrow{EF}$ and $\overleftrightarrow{EG}$ are drawn, intersecting $\overleftrightarrow{CD}$ at H and I, respectively.

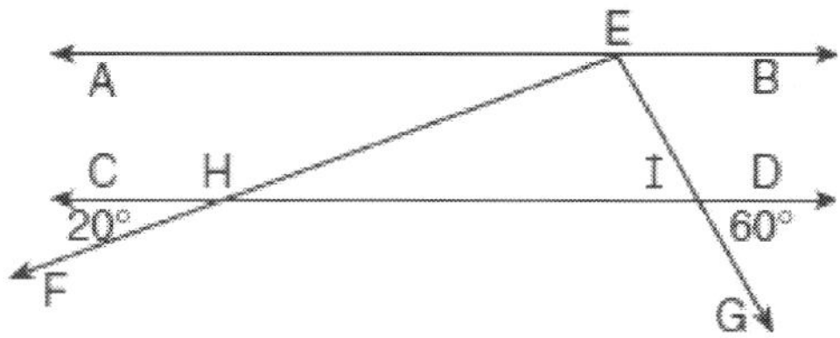

If $m\angle CHF = 20$ and $m\angle DIG = 60$, what is $m\angle HEI$?
(1) 60 (3) 100
(2) 80 (4) 120

2. Juliann plans on drawing $\triangle ABC$, where the measure of $\angle A$ can range from 50° to 60° and the measure of $\angle B$ can range from 90° to 100°. Given these conditions, what is the correct range of measures possible for $\angle C$?
(1) 20° to 40° (3) 80° to 90°
(2) 30° to 50° (4) 120° to 130°

3. In the diagram of $\triangle KLM$ below, $m\angle L = 70$, $m\angle M = 50$, and $\overline{MK}$ is extended through N.

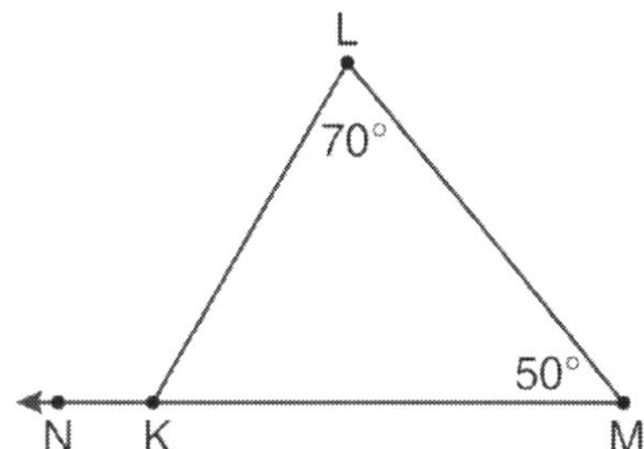

What is the measure of $\angle LKN$?
(1) 60° (3) 180°
(2) 120° (4) 300°

4. In the diagram below of $\triangle BCD$, side $\overline{DB}$ is extended to point A.

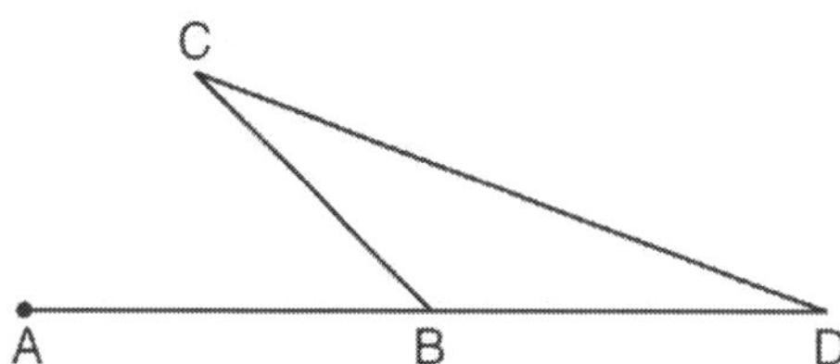

Which statement must be true?

(1) $m\angle C > m\angle D$
(2) $m\angle ABC < m\angle D$
(3) $m\angle ABC > m\angle C$
(4) $m\angle ABC > m\angle C + m\angle D$

5. In $\triangle FGH$, $m\angle F = 42$ and an exterior angle at vertex H has a measure of 104. What is $m\angle G$?

(1) 34
(2) 62
(3) 76
(4) 146

6. In the diagram of $\triangle JEA$ below, $m\angle JEA = 90$ and $m\angle EAJ = 48$. Line segment MS connects points M and S on the triangle, such that $m\angle EMS = 59$.

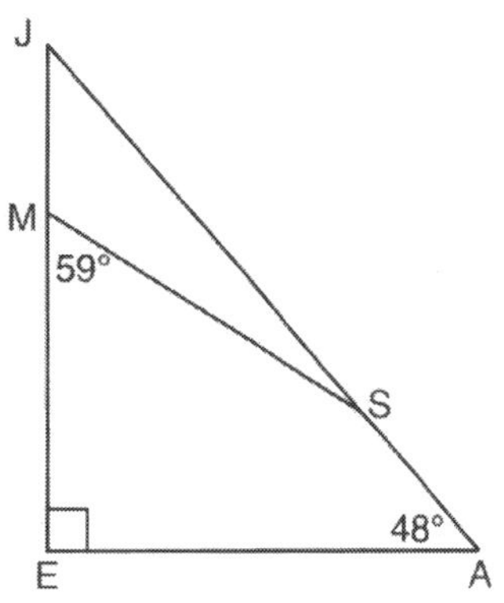

What is $m\angle JSM$?

(1) 163
(2) 121
(3) 42
(4) 17

7. The diagram below shows $\triangle ABD$, with $\overrightarrow{ABC}$, $\overline{BE} \perp \overline{AD}$, and $\angle EBD \cong \angle CBD$.

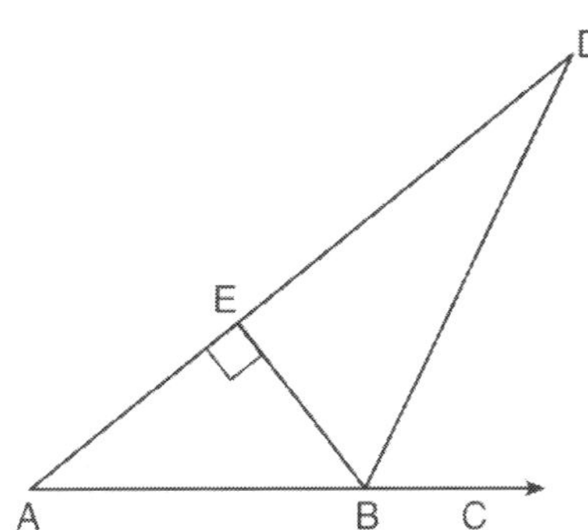

If $m\angle ABE = 52$, what is $m\angle D$?

(1) 26
(2) 38
(3) 52
(4) 64

8. In the diagram below, $\overleftrightarrow{RCBT}$ and $\triangle ABC$ are shown with $m\angle A = 60$ and $m\angle ABT = 125$.

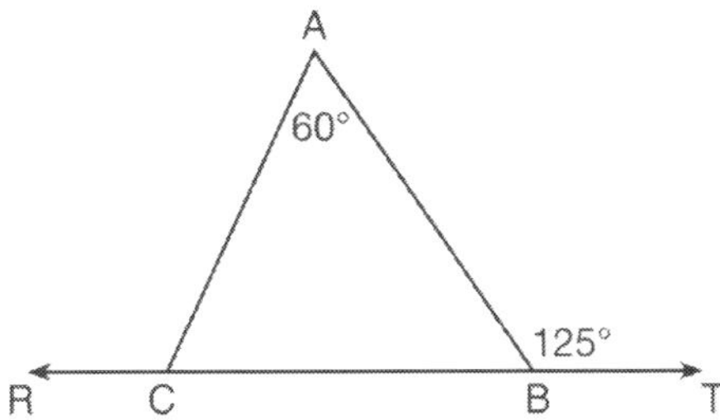

What is $m\angle ACR$?

(1) 125
(2) 115
(3) 65
(4) 55

9. (CC) In the diagram below, $m\angle BDC = 100°$, $m\angle A = 50°$, and $m\angle DBC = 30°$.

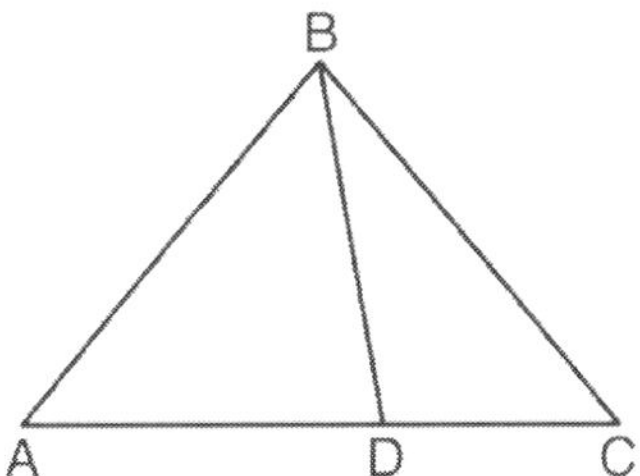

Which statement is true?

(1) $\triangle ABD$ is obtuse.
(2) $\triangle ABC$ is isosceles.
(3) $m\angle ABD = 80°$
(4) $\triangle ABD$ is scalene.

Constructed Response

10. In the diagram below of quadrilateral $ABCD$ with diagonal $\overline{BD}$, $m\angle A = 93$, $m\angle ADB = 43$, $m\angle C = 3x+5$, $m\angle BDC = x+19$, and $m\angle DBC = 2x+6$. Determine if $\overline{AB}$ is parallel to $\overline{DC}$. Explain your reasoning.

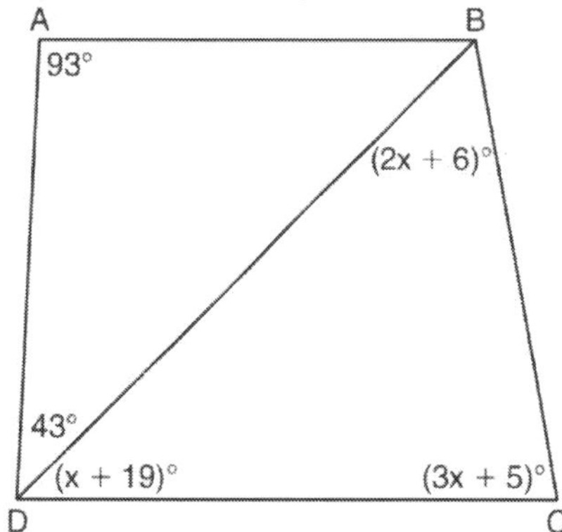

11. In the diagram below, $\ell \parallel m$ and $\overline{QR} \perp \overline{ST}$ at R.

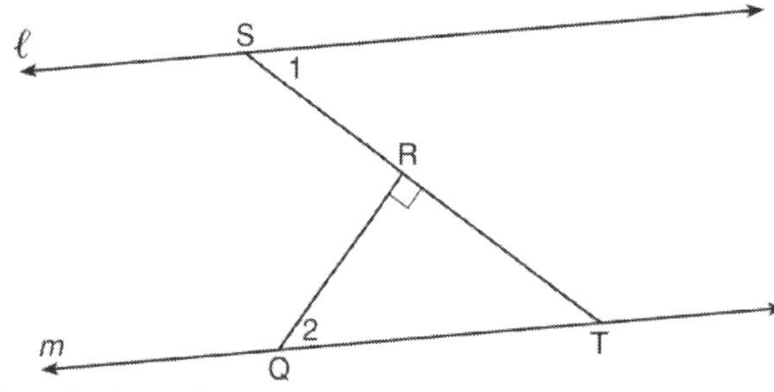

If $m\angle 1 = 63$, find $m\angle 2$.

12. CC Prove the sum of the exterior angles of a triangle is 360°.

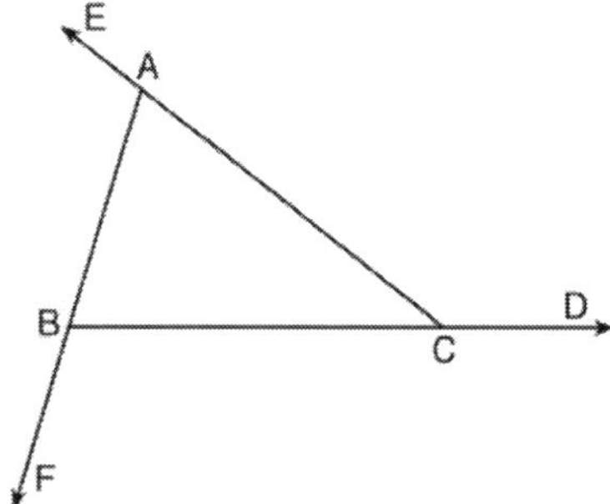

4.2 Medians, Altitudes and Bisectors

Key Terms and Concepts

A **bisector** of a triangle is a segment joining a vertex and the opposite side such that it *bisects the angle* at that vertex.

An **altitude** is a segment joining a vertex and the line containing the opposite side such that it is *perpendicular* to that line.

In a triangle, a **median** is a segment joining a vertex and the *midpoint* of the opposite side.

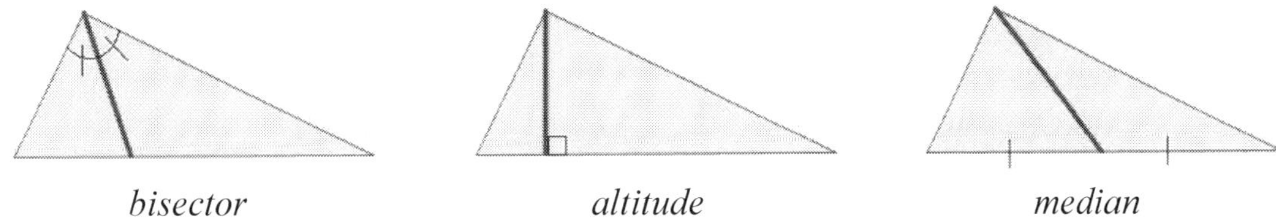

bisector *altitude* *median*

Note the use of **tick marks** in the diagram. If two line segments or two angles are labelled with the same number of tick marks, then they are congruent to each other.

In an *obtuse triangle*, an altitude drawn from a vertex of an acute angle will be perpendicular to the *extension* of the opposite side, as shown below.

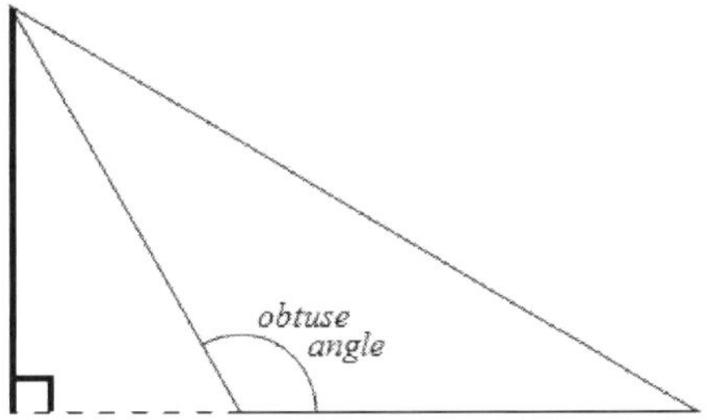

A triangle has three vertices, so it has three medians, three altitudes, and three angle bisectors.

When three or more lines intersect in a single point, they are **concurrent**, and the point of intersection is the **point of concurrency**.

The **incenter** is the point of concurrency of the *three angle bisectors* of a triangle.

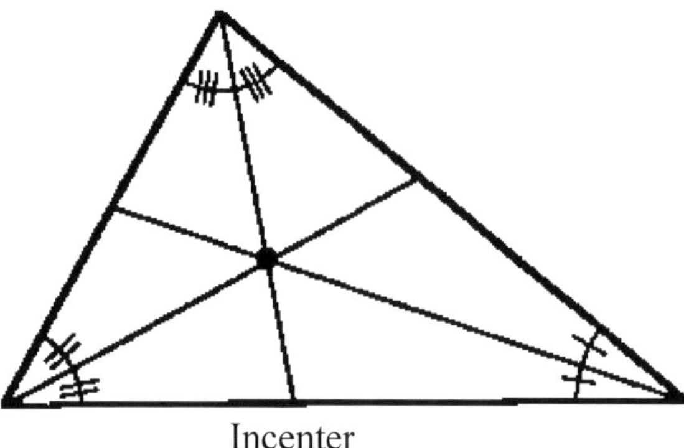

Incenter

The point of concurrency of the *three altitudes* of a triangle is called the **orthocenter**.

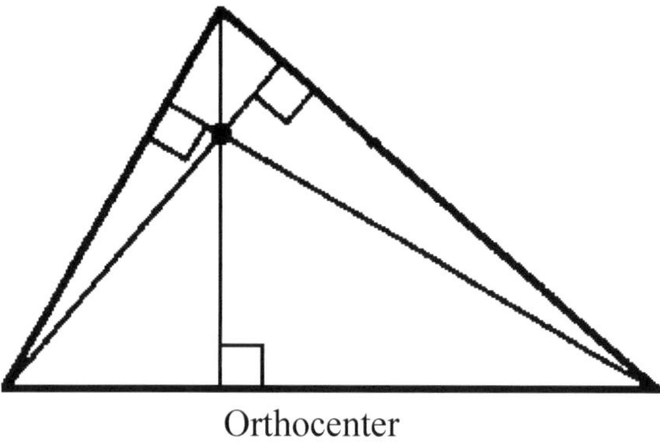

Orthocenter

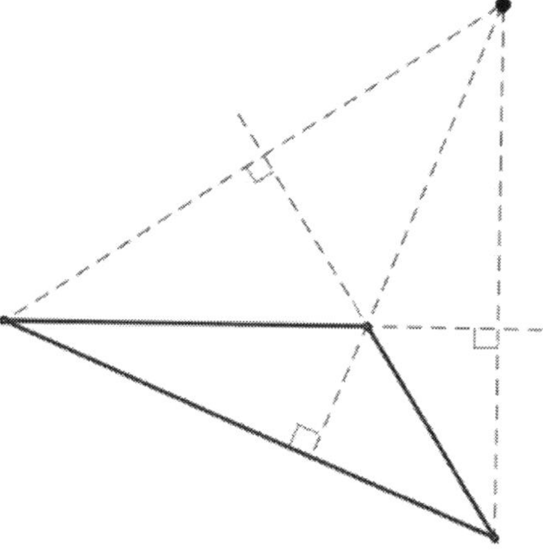

In an obtuse triangle, the orthocenter will lie outside the triangle, as shown to the right. In a right triangle, the orthocenter is the vertex of the right angle.

The point of concurrency of the *three medians* of a triangle is called the **centroid**.

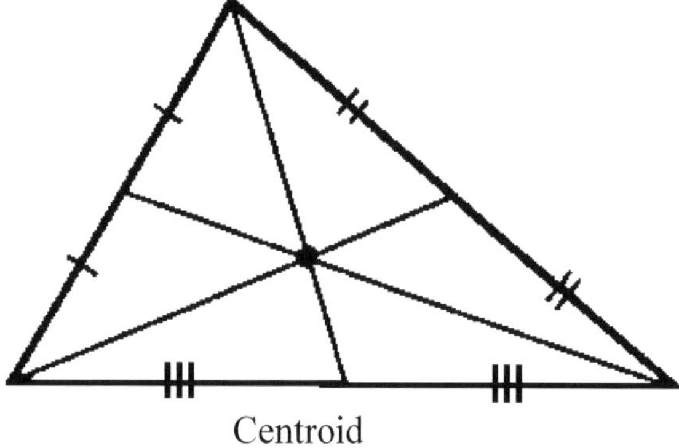

Centroid

The **circumcenter** is the point of concurrency of the *three perpendicular bisectors* of the sides of a triangle.

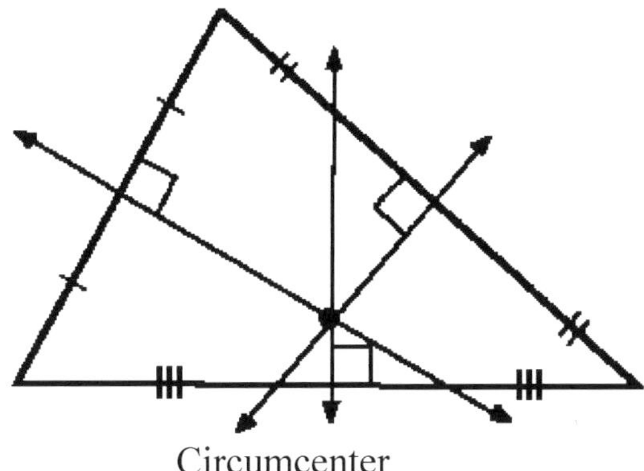

Circumcenter

In an acute triangle, the circumcenter is inside the triangle, but in an obtuse triangle, it falls outside the triangle. In a right triangle, the circumcenter is the midpoint of the hypotenuse.

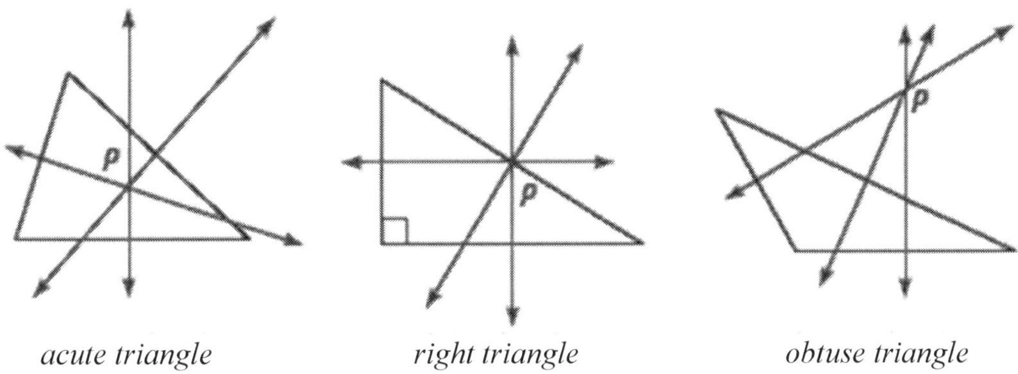

acute triangle *right triangle* *obtuse triangle*

Some important facts about the points of concurrency in a triangle:

- The incenter is equidistant from the three sides of the triangle. In other words, the three segments drawn from the incenter, perpendicular to each side, have the same length.
- The centroid is two-thirds of the distance from each vertex to the midpoint of the opposite side.
- The circumcenter is equidistant to the three vertices of the triangle.

The *circumcenter* is also the center of the circle that **circumscribes** the triangle, and the *incenter* is the center of the circle that is **inscribed** in the triangle. We will explore this more fully in the unit on Constructions.

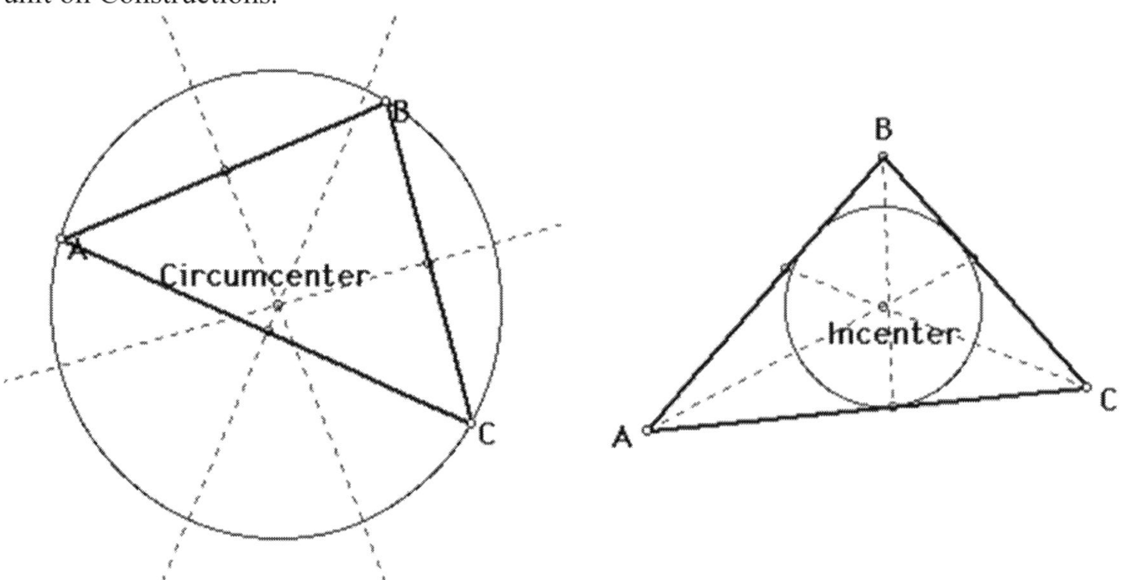

Model Problem:
Where is the orthocenter in a right triangle?

Solution:
It coincides with the vertex at the right angle of the triangle.

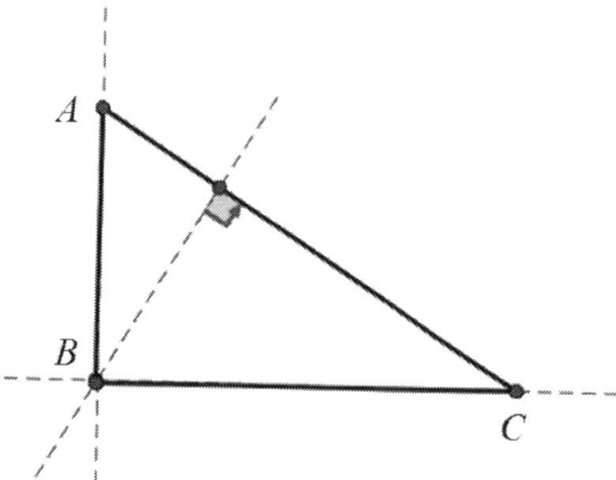

Explanation:
The legs of a right triangle are also altitudes, since they are perpendicular to each other. They both meet at the vertex of the right angle (*B* in the diagram). The third altitude extends from *B* to the hypotenuse. So, all three altitudes intersect at *B*.

Practice Problems

1. In the diagram of $\triangle ABC$ below, name (a) a median, (b) an altitude, and (c) an angle bisector. 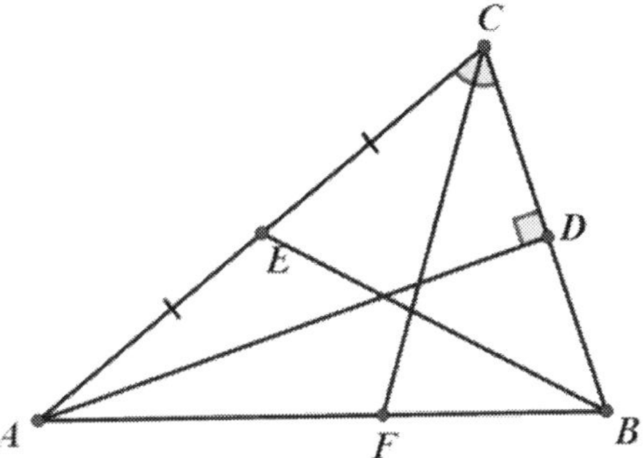	2. Name the point of concurrency for each of the following in a triangle: (a) perpendicular bisectors (b) medians (c) altitudes (d) angle bisectors
3. P is the centroid of $\triangle QRS$, $PT = 8$, and $ST = 24$. Find: (a) QS, (b) PR, and (c) RT. 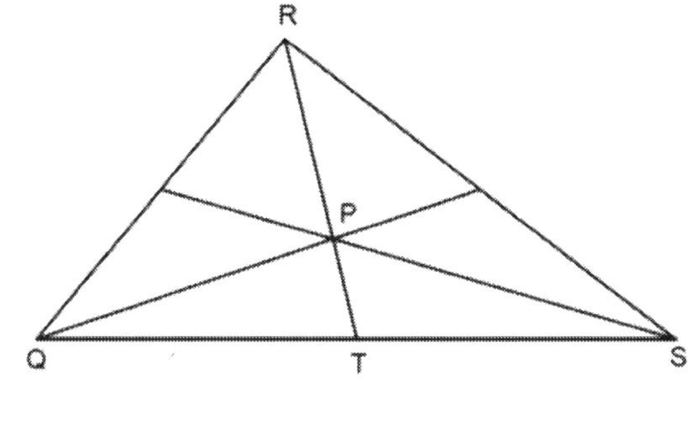	4. In $\triangle DEG$ below, $\overleftrightarrow{FH}$ is the perpendicular bisector of $\overline{DHG}$. (a) If $DH = 2x+3$ and $GH = 7x-47$, find the length of $\overline{DG}$. (b) If $m\angle FHG = (y^2+9)^\circ$ and $m\angle EFH = (12y)^\circ$, find $m\angle EFH$. 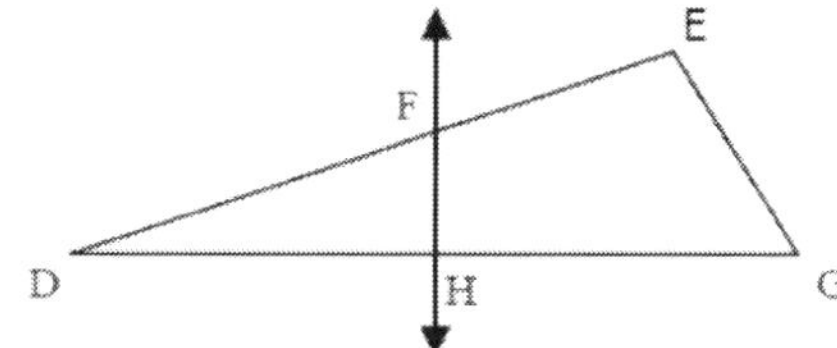

REGENTS QUESTIONS

Multiple Choice

1. In which triangle do the three altitudes intersect outside the triangle?
   (1) a right triangle (3) an obtuse triangle
   (2) an acute triangle (4) an equilateral triangle

2. In the diagram below of $\triangle ABC$, $\overline{CD}$ is the bisector of $\angle BCA$, $\overline{AE}$ is the bisector of $\angle CAB$, and $\overline{BG}$ is drawn.

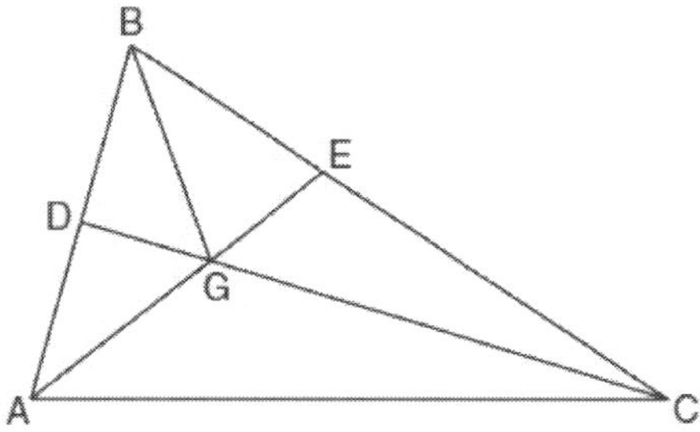

   Which statement must be true?
   (1) DG = EG (3) $\angle AEB \cong \angle AEC$
   (2) AG = BG (4) $\angle DBG \cong \angle EBG$

3. In the diagram below of $\triangle ABC$, $\overline{AE} \cong \overline{BE}$, $\overline{AF} \cong \overline{CF}$, and $\overline{CD} \cong \overline{BD}$.

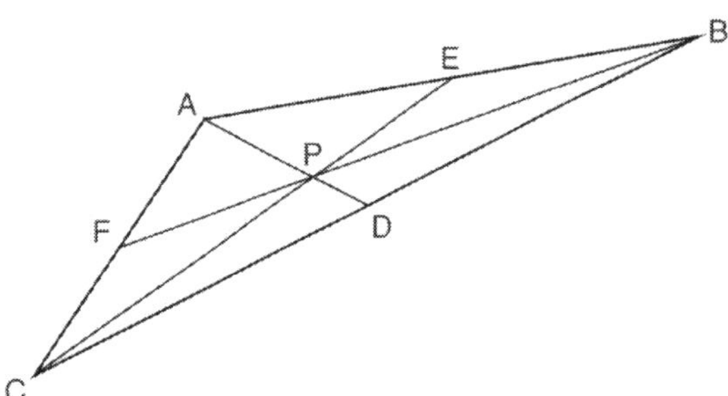

   Point P must be the
   (1) centroid (3) incenter
   (2) circumcenter (4) orthocenter

4. For a triangle, which two points of concurrence could be located outside the triangle?
   (1) incenter and centroid (3) incenter and circumcenter
   (2) centroid and orthocenter (4) circumcenter and orthocenter

Constructed Response

5. In the diagram below, point B is the incenter of $\triangle FEC$, and $\overline{EBR}$, $\overline{CBD}$, and $\overline{FB}$ are drawn.

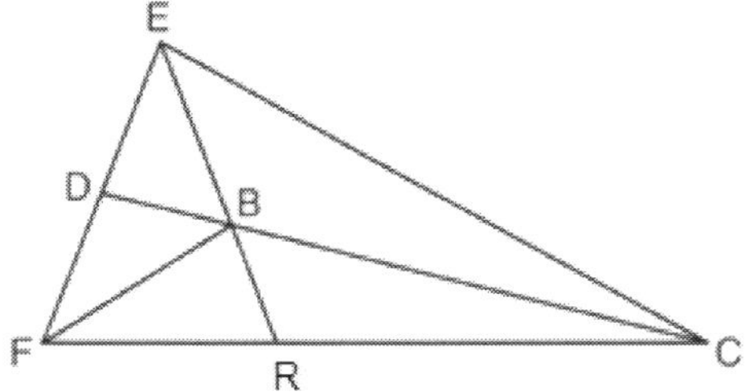

If $m\angle FEC = 84$ and $m\angle ECF = 28$, determine and state $m\angle BRC$.

4.3 Isosceles and Equilateral Triangles

Key Terms and Concepts

Triangles may be classified by how many congruent sides they have. A **scalene triangle** has no congruent sides, an **isosceles triangle** has two congruent sides, and an **equilateral triangle** has three congruent sides.

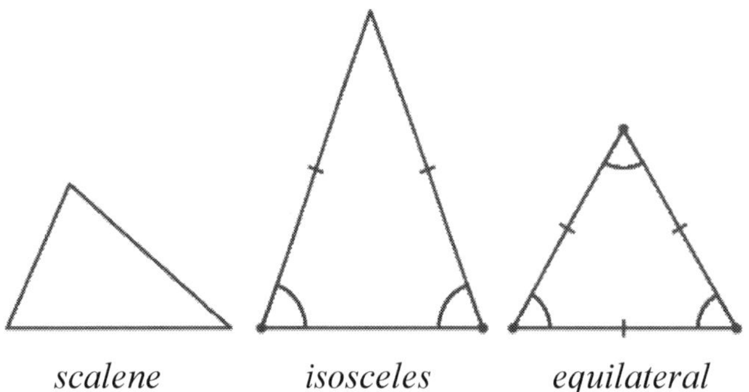

In an isosceles triangle, the congruent sides are called the **legs** and the remaining side is called the **base**. The angle formed by the legs is called the **vertex angle**, and the other two angles, opposite the legs, are called the **base angles**.

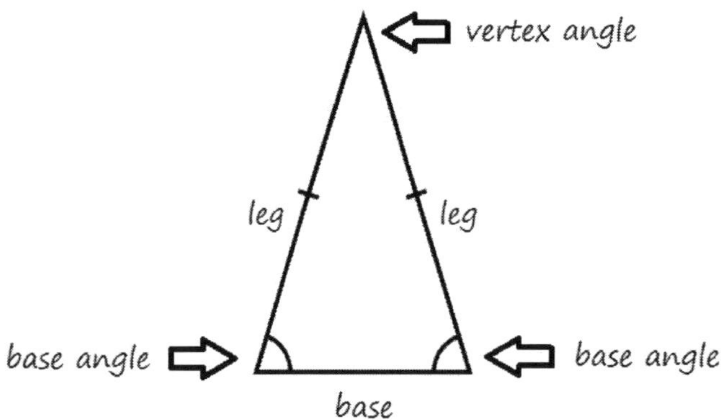

In an isosceles triangle, the base angles are congruent. The converse is also true: if two angles of a triangle are congruent, then the opposite sides are congruent and the triangle is isosceles. The base angles of an isosceles triangle are always acute.

If the measure of the vertex angle, v, is known, then the measure of a base angle, b, of an isosceles triangle can be calculated as

$$b = \frac{180 - v}{2}$$

In an equilateral triangle, all three angles are congruent. Since the sum of the angle measures of a triangle is $180°$, each angle in an equilateral triangle measures $60°$.

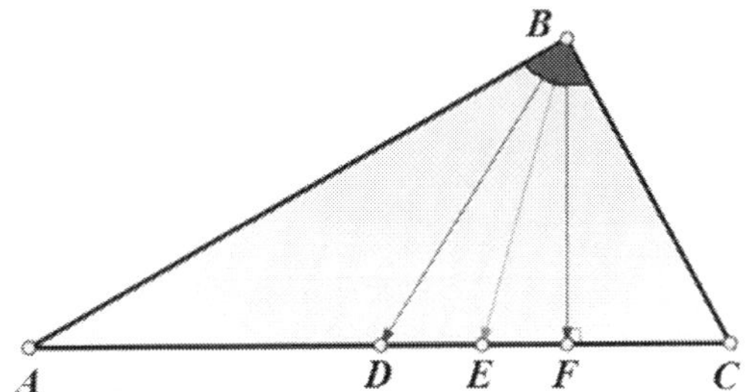

In general, the bisector, median, and altitude drawn from one vertex of a triangle may be three different segments. For example, in the scalene triangle show to the left, $\overline{BD}$ is a median, $\overline{BE}$ is an angle bisector, and $\overline{BF}$ is an altitude.

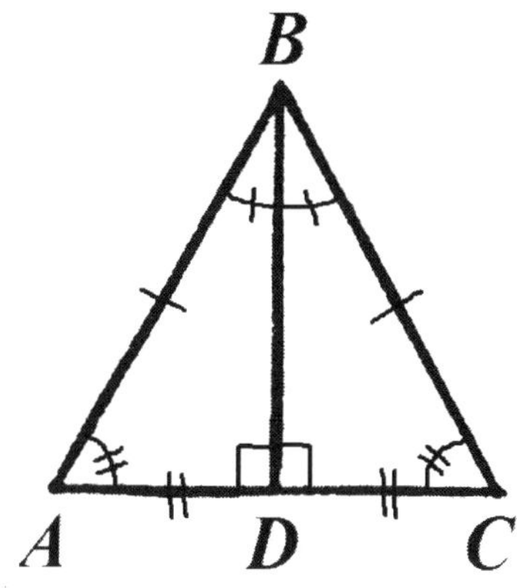

However, in an isosceles triangle, the bisector, median and altitude drawn from the *vertex angle* are all the same segment. This segment is the line of symmetry. The line of symmetry divides the isosceles triangle into two congruent right triangles.

Example: In the diagram of isosceles $\triangle ABC$ to the right, sides $\overline{BA} \cong \overline{BC}$. The line of symmetry, $\overline{BD}$, is drawn. $\overline{BD}$ is a bisector, median, and altitude. Therefore, $\angle ABD \cong \angle CBD$, $\overline{AD} \cong \overline{DC}$, and $\overline{BD} \perp \overline{AC}$.

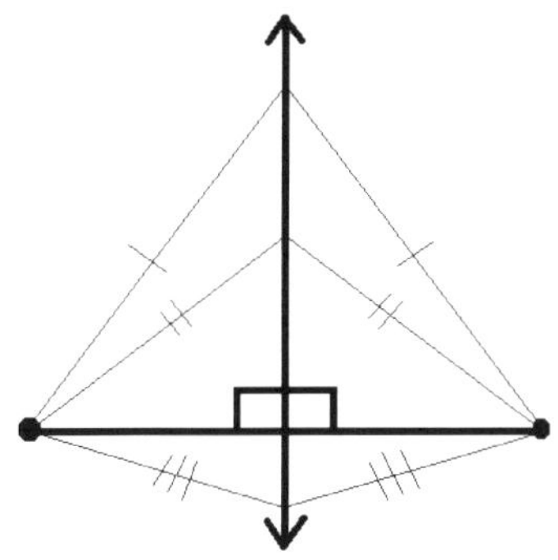

In general, all the points on the perpendicular bisector of a line segment are equidistant to the endpoints of the line segment, as demonstrated in the diagram at the left.

Therefore, if the perpendicular bisector of one side of a triangle passes through the opposite vertex, then the triangle is isosceles, with that side as the base.

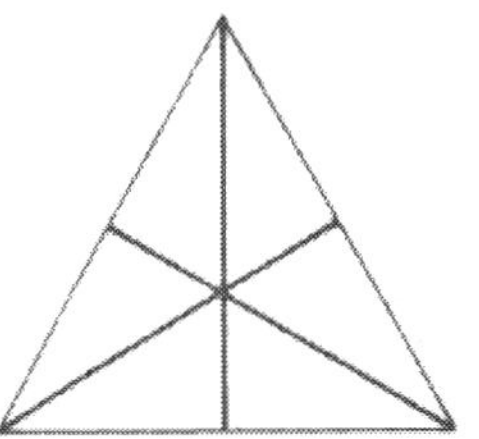

In an equilateral triangle, all three lines of symmetry represent a bisector, median, and altitude.

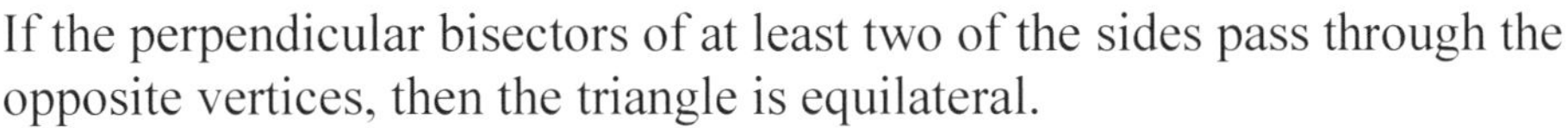

If the perpendicular bisectors of at least two of the sides pass through the opposite vertices, then the triangle is equilateral.

Model Problem:

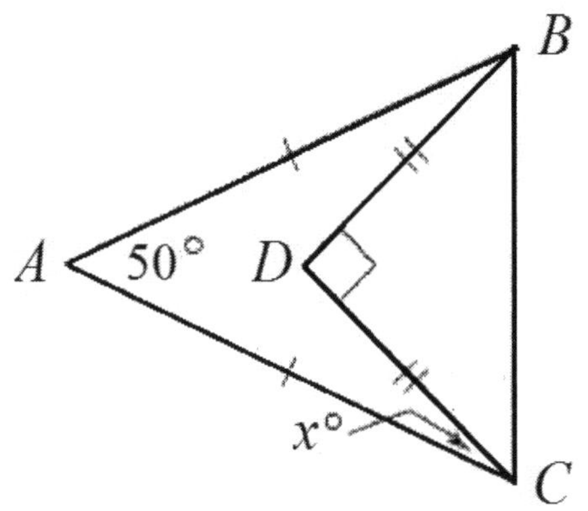

In the diagram at right of $\triangle CAB$ and $\triangle CDB$, $\overline{AC} \cong \overline{AB}$, $\overline{DC} \cong \overline{DB}$, and $\angle CDB$ is a right angle.
Find x, the measure of $\angle ACD$.

Solution:

(A) $m\angle DCB = \dfrac{180-90}{2} = 45°$

(B) $m\angle ACB = \dfrac{180-50}{2} = 65°$

(C) $x = m\angle ACB - m\angle DCB = 65 - 45 = 20°$

Explanation of steps:

(A) Use the formula $b = \dfrac{180-v}{2}$ to calculate the base angle *[of $\triangle CDB$].*

(B) Use the formula $b = \dfrac{180-v}{2}$ to calculate the base angle *[of $\triangle CAB$].*

(C) Use the Partition Postulate: subtract the part [$m\angle DCB$] from the whole [$m\angle ACB$].

Practice Problems

1. For an isosceles triangle whose vertex angle measures 120°, what is the measure of each base angle?

2. In the diagram below, identify (a) a bisector, (b) an altitude, and (c) a median.

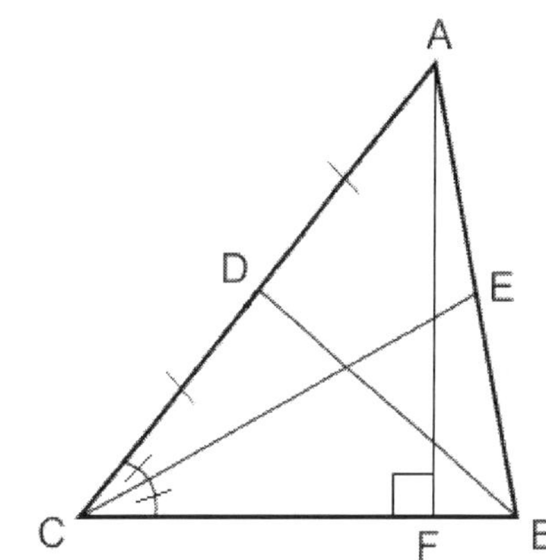

3. For the diagram below, find:
   (a) $m\angle BCA$ (b) $m\angle DCE$ (c) $m\angle BCD$
   (d) $m\angle DEF$ (e) $m\angle BAG$ (f) $m\angle GAH$

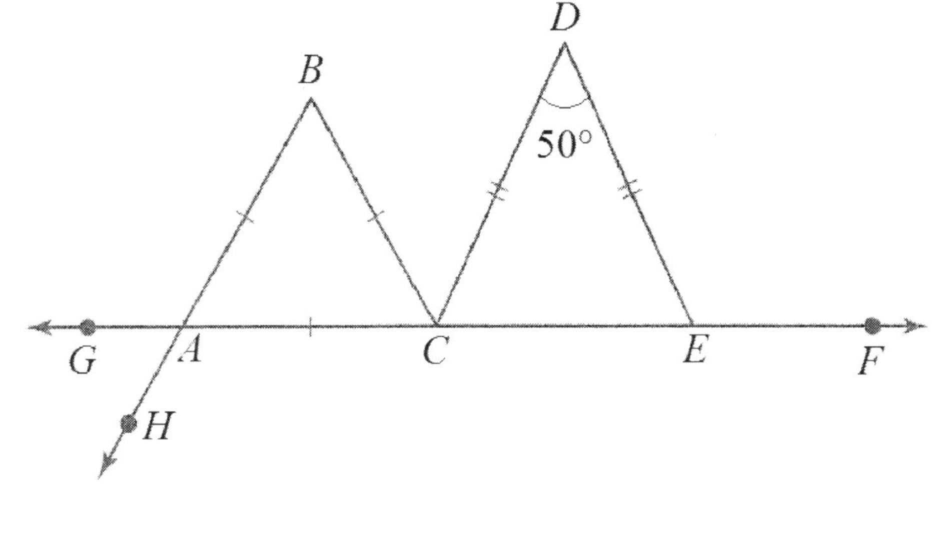

4. For the diagram of triangles PQR and QRS below, where $\overleftrightarrow{PRS}$ is a straight line, find x.

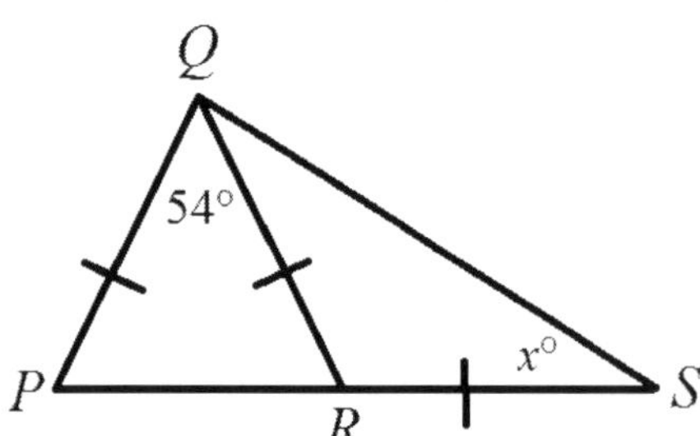

5. For the diagram of triangles DEF and EFG below, find x.

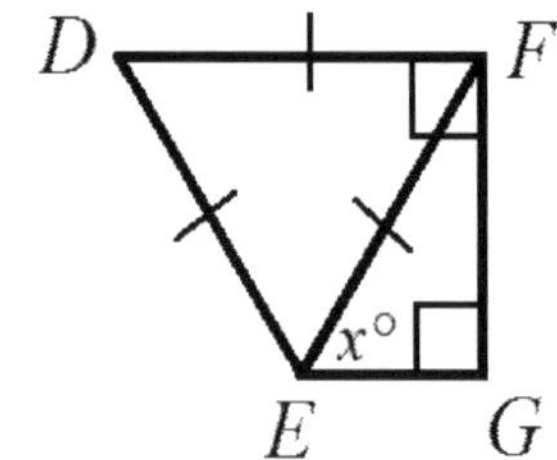

6. Given: $\triangle JKL$, side $\overline{JL}$ extended through L to M, side $\overline{KL}$ extended through L to N, $m\angle K = 70°$, $m\angle MLN = 55°$

Prove: $\triangle JKL$ is isosceles.

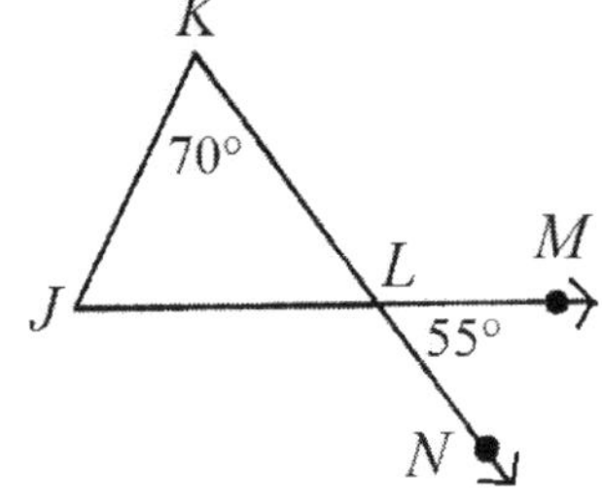

Statements	*Reasons*
$\triangle JKL$, side $\overline{JL}$ extended through L to M, side $\overline{KL}$ extended through L to N, $m\angle K = 70°$, $m\angle MLN = 55°$	Given

REGENTS QUESTIONS

Multiple Choice

1. In the diagram below, $\triangle LMO$ is isosceles with $LO = MO$.

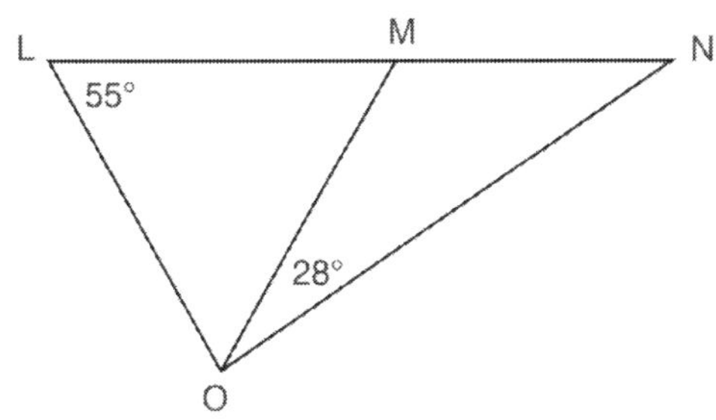

If $m\angle L = 55$ and $m\angle NOM = 28$, what is $m\angle N$?

(1) 27 (3) 42
(2) 28 (4) 70

2. In the diagram below of isosceles $\triangle ABC$, the measure of vertex angle B is 80°. If $\overline{AC}$ extends to point D, what is $m\angle BCD$?

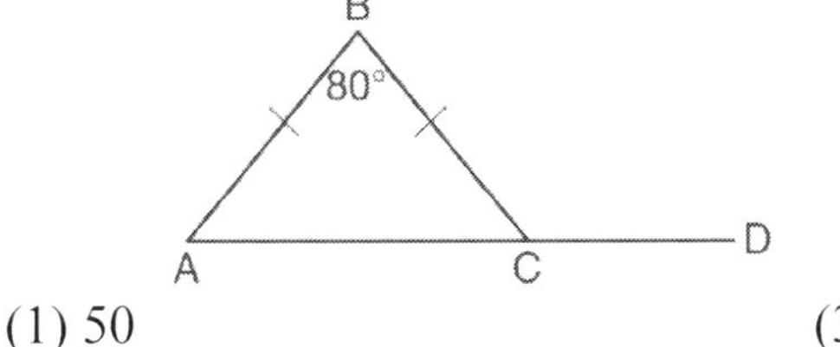

(1) 50 (3) 100
(2) 80 (4) 130

3. In all isosceles triangles, the exterior angle of a base angle must always be

(1) a right angle (3) an obtuse angle
(2) an acute angle (4) equal to the vertex angle

4. In $\triangle JKL$, $\overline{JL} \cong \overline{KL}$. If $m\angle J = 58$, then $m\angle L$ is

(1) 61 (3) 116
(2) 64 (4) 122

5. (CC) Line segment EA is the perpendicular bisector of $\overline{ZT}$, and $\overline{ZE}$ and $\overline{TE}$ are drawn.

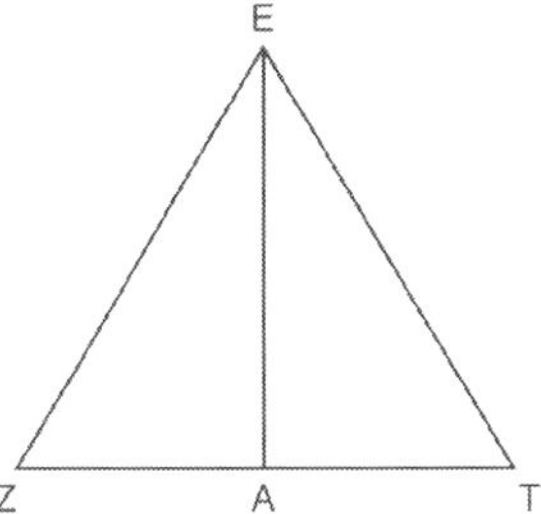

Which conclusion can *not* be proven?

(1) $\overline{EA}$ bisects angle ZET. (3) $\overline{EA}$ is a median of triangle EZT.
(2) Triangle EZT is equilateral. (4) Angle Z is congruent to angle T.

6. (CC) Segment CD is the perpendicular bisector of $\overline{AB}$ at E. Which pair of segments does *not* have to be congruent?
   (1) $\overline{AD}$, $\overline{BD}$
   (2) $\overline{AC}$, $\overline{BC}$
   (3) $\overline{AE}$, $\overline{BE}$
   (4) $\overline{DE}$, $\overline{CE}$

Constructed Response

7. In $\triangle RST$, $m\angle RST = 46$ and $\overline{RS} \cong \overline{ST}$. Find $m\angle STR$.

8. In the diagram below of $\triangle ACD$, B is a point on $\overline{AC}$ such that $\triangle ADB$ is an equilateral triangle, and $\triangle DBC$ is an isosceles triangle with $\overline{DB} \cong \overline{BC}$. Find $m\angle C$.

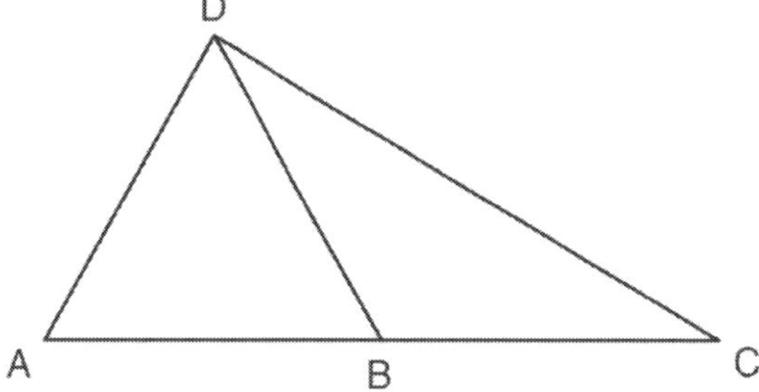

9. In the diagram below of $\triangle GJK$, H is a point on $\overline{GJ}$, $\overline{HJ} \cong \overline{JK}$, $m\angle G = 28$, and $m\angle GJK = 70$. Determine whether $\triangle GHK$ is an isosceles triangle and justify your answer.

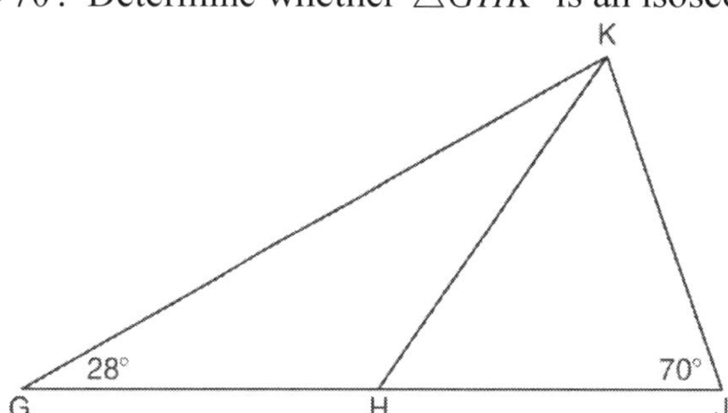

10. (CC) In isosceles $\triangle MNP$, line segment NO bisects vertex $\angle MNP$, as shown below. If $MP = 16$, find the length of $\overline{MO}$ and explain your answer.

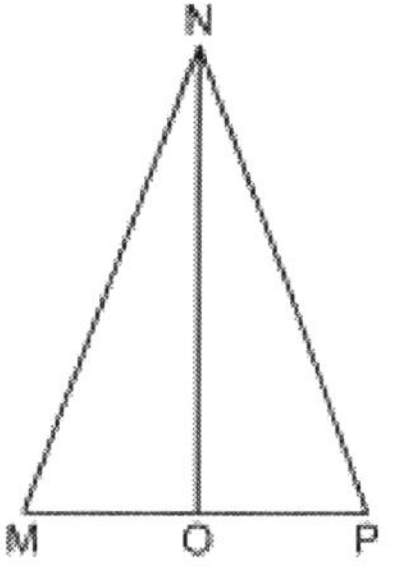

11. (CC) Given: $\triangle XYZ$, $\overline{XY} \cong \overline{ZY}$, and $\overline{YW}$ bisects $\angle XYZ$
Prove that $\angle YWZ$ is a right angle.

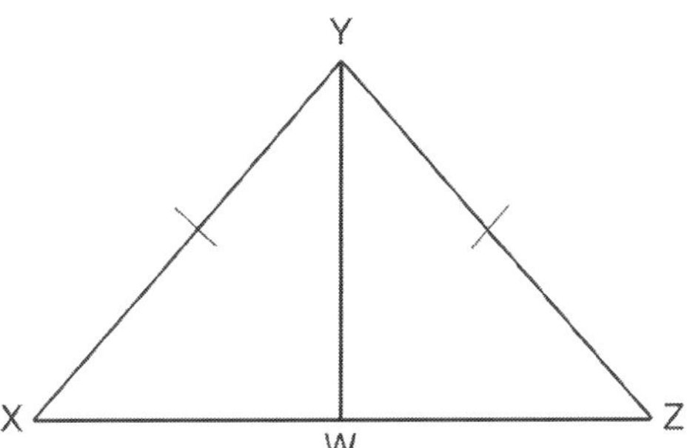

12. (CC) In the diagram below, $\overline{EF}$ intersects $\overline{AB}$ and $\overline{CD}$ at G and H, respectively, and $\overline{GI}$ is drawn such that $\overline{GH} \cong \overline{IH}$.

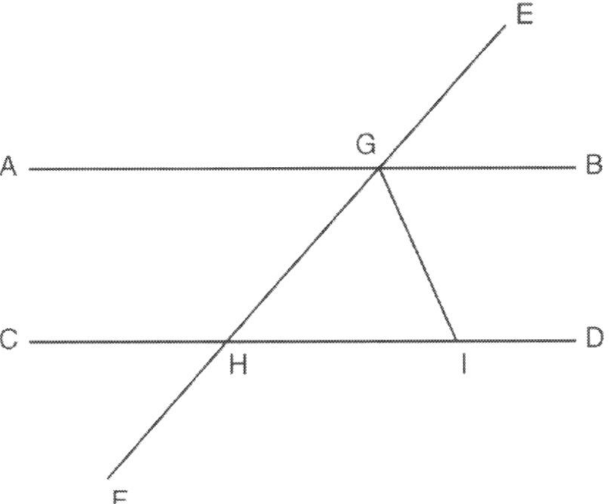

If $m\angle EGB = 50°$ and $m\angle DIG = 115°$, explain why $\overline{AB} \parallel \overline{CD}$.

13. (CC) In the diagram below of isosceles triangle ABC, $\overline{AB} \cong \overline{CB}$ and angle bisectors $\overline{AD}$, $\overline{BF}$, and $\overline{CE}$ are drawn and intersect at X.

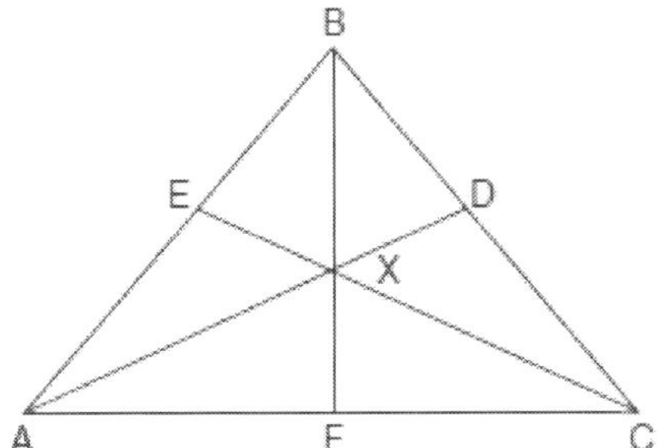

If $m\angle BAC = 50°$, find $m\angle AXC$.

4.4 Congruent Triangles

Key Terms and Concepts

Two figures are congruent if they have the same shape and size. When two triangles are congruent, then all three pairs of **corresponding sides** are congruent and all three pairs of **corresponding angles** are congruent. In the diagram below of congruent triangles *ABC* and *XYZ*, the corresponding parts are marked with the same number of ticks or arcs to show that they are congruent.

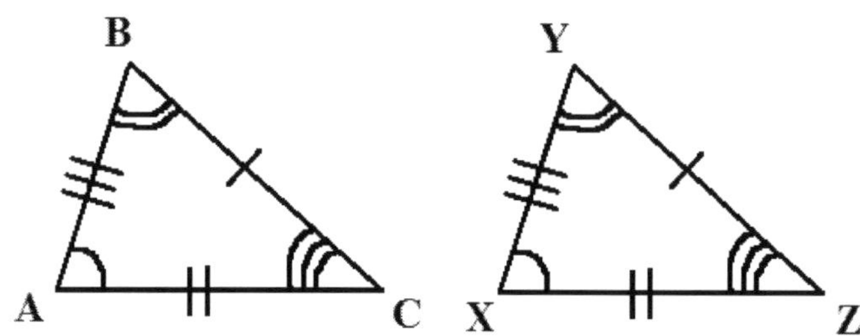

The following are pairs of corresponding angles and corresponding sides:

$\angle A \cong \angle X$	$\overline{BC} \cong \overline{YZ}$
$\angle B \cong \angle Y$	$\overline{AC} \cong \overline{XZ}$
$\angle C \cong \angle Z$	$\overline{AB} \cong \overline{XY}$

Note that corresponding sides must be opposite corresponding angles.

An important statement that we will use in many of our triangle proofs is, "**Corresponding parts of congruent triangles are congruent.**" We will often abbreviate this statement as **CPCTC**. Once we prove that two triangles are congruent, we can use CPCTC as a reason why any pair of corresponding sides or angles are congruent.

We can also refer to the vertices as corresponding, if they are the vertices of the corresponding angles. For example, in the above diagram, *A* corresponds to *X*, *B* to *Y*, and *C* to *Z*.

When we state that the two triangles are congruent, it is necessary to write the vertices in **corresponding order**.

Example: $\triangle NYC \cong \triangle BUF$ means $\angle N \cong \angle B$, $\angle Y \cong \angle U$, and $\angle C \cong \angle F$.
It also means $\overline{NY} \cong \overline{BU}$, $\overline{YC} \cong \overline{UF}$, and $\overline{CN} \cong \overline{FB}$.

Two triangles may be congruent even if they do not have the same orientation.

Example: $\triangle ABC \cong \triangle DEF$, even though you would need to flip one of them to match orientations. The triangle at the right is a mirror image of $\triangle DEF$ – *complete with backward letters as vertex labels for effect!* – and the image now matches the orientation of $\triangle ABC$.

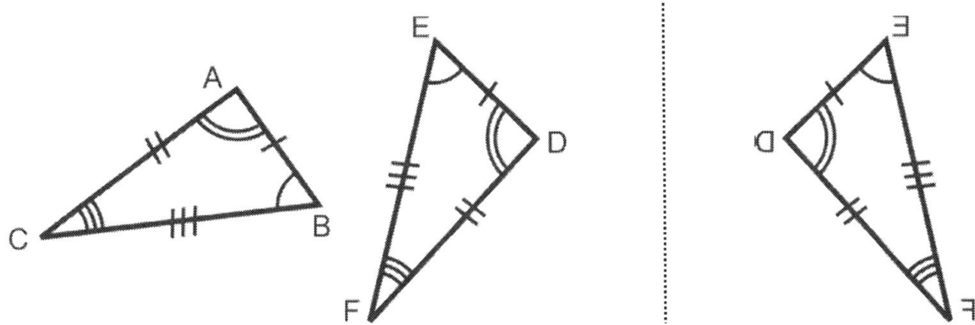

To prove that two triangles are congruent, we don't need to show that all six pairs of corresponding parts are congruent. It is possible to prove congruence using fewer parts.

The following methods are sufficient for proving two triangles are congruent:

- **SSS**: three pairs of corresponding sides are congruent
- **SAS**: two pairs of corresponding sides, and the angles formed by these sides (called the **included angles**), are congruent
- **ASA**: two pairs of corresponding angles, and the sides shared by these angles (called the **included sides**), are congruent
- **AAS**: two pairs of corresponding angles, and a pair of corresponding non-included sides, are congruent

The diagrams below give examples of each of the four valid congruence rules for triangles:

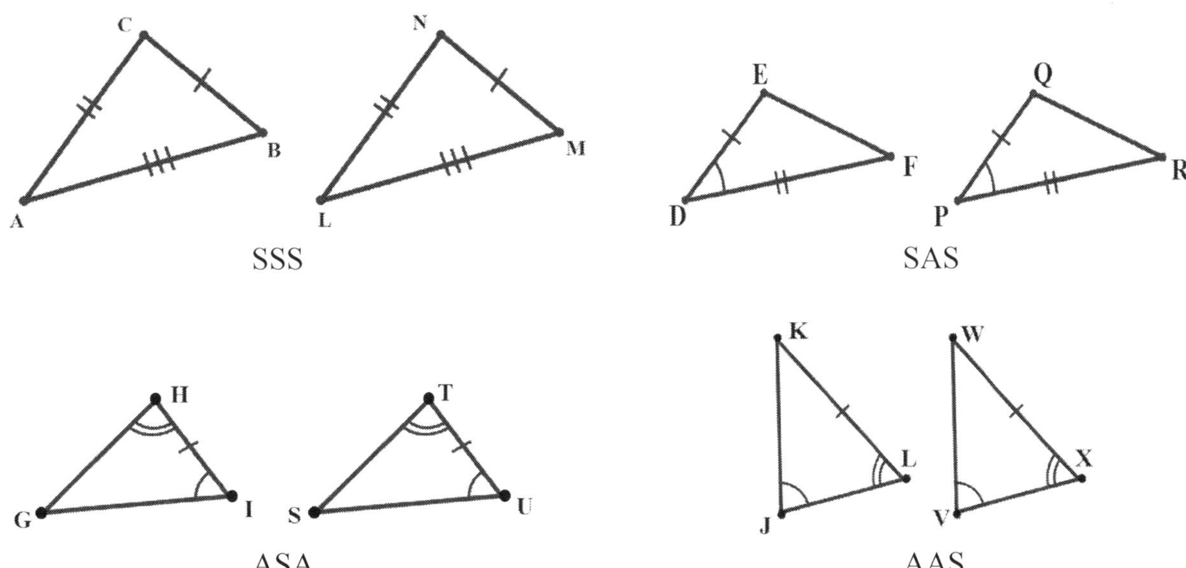

You must always be sure to look at pairs of *corresponding parts only*.

Example: Although the triangles below show two pairs of congruent angles and a pair of congruent sides, the congruent sides are *not corresponding*. Side $\overline{KL}$ is opposite $\angle M$, which has a double-arc marking, so its corresponding side would be $\overline{TU}$, which is opposite $\angle S$. We *cannot* claim these triangles are congruent by AAS.

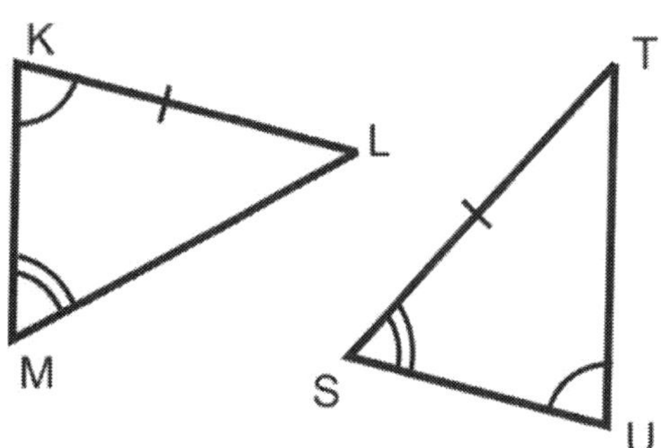

Also, be careful to never use SSA (two sides and a non-included angle) as a method for proving triangles congruent. **SSA is *not* a valid congruence rule.** Why not? Given two pairs of congruent sides and a pair of congruent non-included angles, it is still possible to form two different (non-congruent) triangles, as shown below.

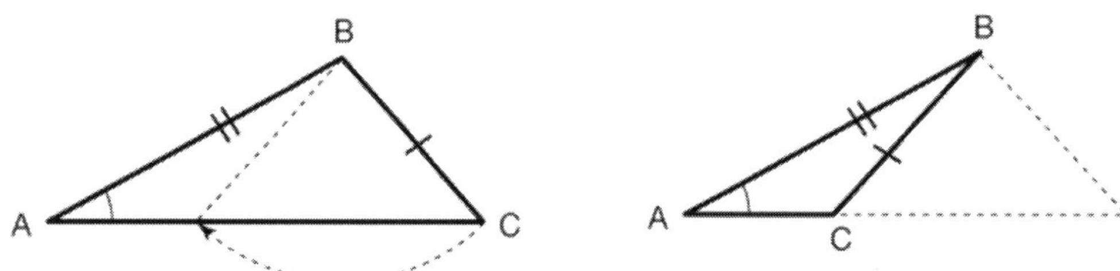

You may be asked to prove two angles or two sides congruent. If they can be shown to be a pair of corresponding parts of triangles, you can first try to prove the triangles congruent. If the triangles are proven congruent, then their corresponding parts are congruent by CPCTC.

In a triangle proof, it is helpful to mark the diagram with ticks or arcs for any pairs of congruent sides or angles. It is also helpful to keep track of which corresponding parts have been proven to be congruent by writing (S) or (A) next to the statements.

Example: Below is a proof that the base angles of an isosceles triangles are congruent.

Given: Isosceles triangle ABC with $\overline{AB} \cong \overline{BC}$. Angle bisector $\overline{BP}$ is drawn.

Prove: $\angle A \cong \angle C$

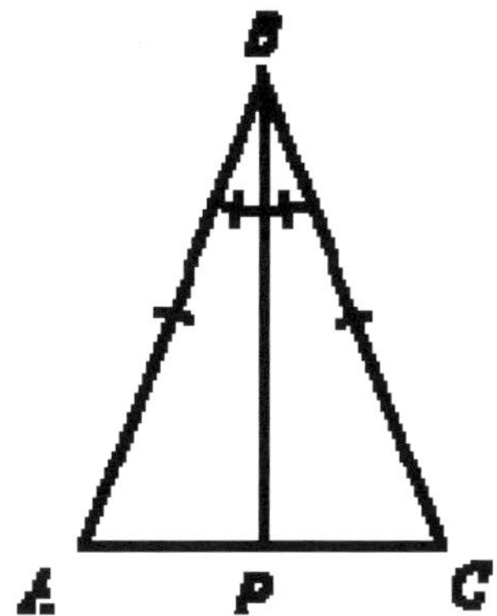

Statements		*Reasons*
$\overline{AB} \cong \overline{BC}$	(S)	Given
$\overline{BP}$ is an angle bisector		Given
$\angle ABP \cong \angle CBP$	(A)	Definition of angle bisector
$\overline{BP} \cong \overline{BP}$	(S)	Reflexive Property
$\triangle ABP \cong \triangle CBP$		SAS
$\angle A \cong \angle C$		CPCTC

If the preceding proof were written in a flowchart format, it may look like this:

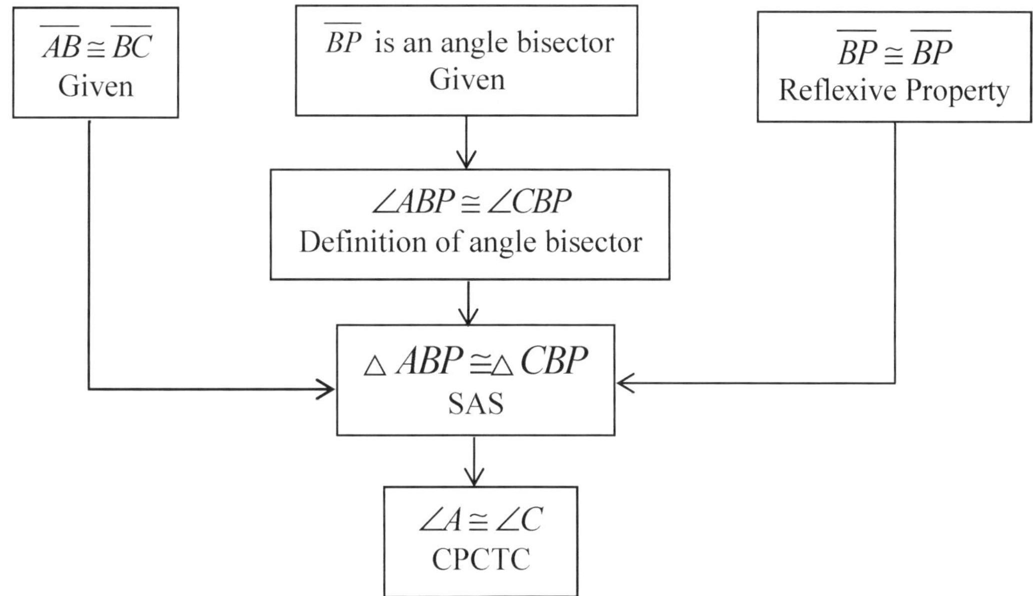

Model Problem:

Given: $\overline{YW}$ is the perpendicular bisector of $\overline{XZ}$.
Prove: $\overline{XY} \cong \overline{ZY}$

Solution:

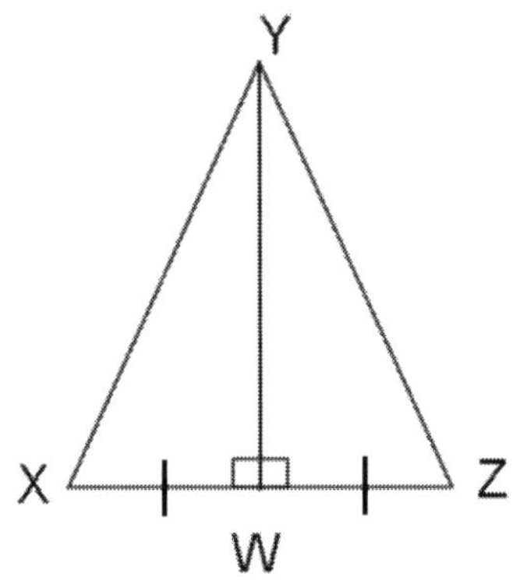

Statements	*Reasons*
$\overline{YW}$ is the perpendicular bisector of $\overline{XZ}$	Given
$\overline{XW} \cong \overline{ZW}$ (S)	Definition of perpendicular bisector
$\angle YWX$ and $\angle YWZ$ are right angles	Definition of perpendicular bisector
$\angle YWX \cong \angle YWZ$ (A)	All right angles are congruent
$\overline{YW} \cong \overline{YW}$ (S)	Reflexive Property
$\triangle XYW \cong \triangle ZYW$	SAS
$\overline{XY} \cong \overline{ZY}$	CPCTC

Explanation of steps:
(A) Include any given statements that will be used in your proof.
(B) Use the given information to deduce any relevant facts, being sure to mark any congruencies or measures on the diagram.
(C) Continue until you get all three parts of a triangle congruence method.
(D) Conclude that the triangles are congruent by that method.
(E) Use CPCTC to show that any other corresponding parts are congruent, as needed.

We can generalize the Model Problem above by allowing point Y to be *any* point on the perpendicular bisector of $\overline{XZ}$. That is, the proof in the Model Problem shows that every point on the perpendicular bisector of a line segment is equidistant from the segment's endpoints.

Practice Problems

1. Which statement properly uses corresponding order?
   (1) $\triangle FET \cong \triangle WRY$
   (2) $\triangle FET \cong \triangle YRW$
   (3) $\triangle EFT \cong \triangle YRW$
   (4) $\triangle EFT \cong \triangle WRY$

   By which congruence method are the two triangles congruent?

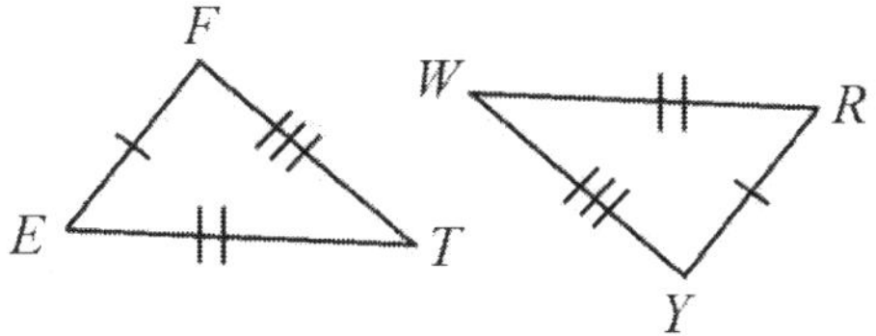

2. Which statement properly uses corresponding order?
   (1) $\triangle UIA \cong \triangle OEY$
   (2) $\triangle UIA \cong \triangle OYE$
   (3) $\triangle AUI \cong \triangle EOY$
   (4) $\triangle IAU \cong \triangle OYE$

   By which congruence method are the two triangles congruent?

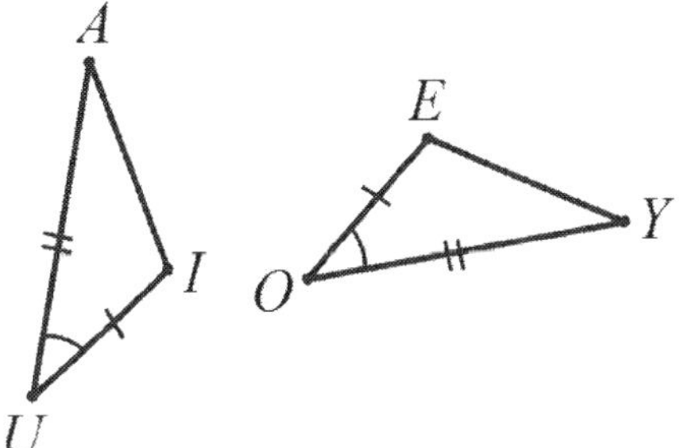

3. In the diagram of $\triangle ABC$ and $\triangle DEF$ below, $\overline{AB} \cong \overline{DE}$, $\angle A \cong \angle D$, and $\angle B \cong \angle E$. Which congruence method can be used to prove $\triangle ABC \cong \triangle DEF$?

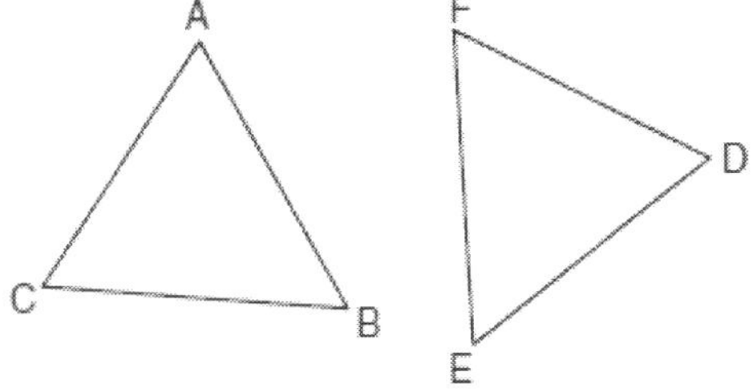

4. In the diagram below of $\triangle DAE$ and $\triangle BCE$, $\overline{AB}$ and $\overline{CD}$ intersect at E, such that $\overline{AE} \cong \overline{CE}$ and $\angle BCE \cong \angle DAE$. Which congruence method can be used to prove $\triangle DAE \cong \triangle BCE$?

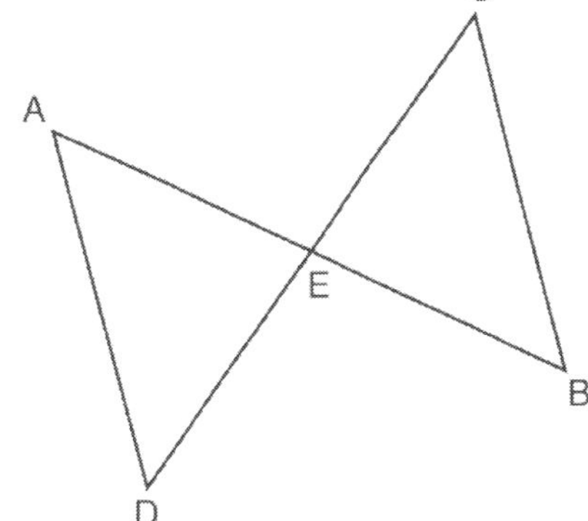

5. In the accompanying diagram, $\overline{HK}$ bisects $\overline{IL}$ and $\angle H \cong \angle K$. What is the most direct congruence method that could be used to prove $\triangle HIJ \cong \triangle KLJ$?

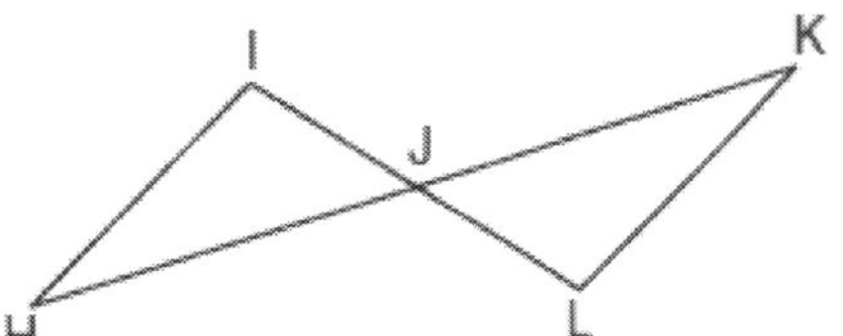

6. In the accompanying diagram, $\overline{AB} \cong \overline{CD}$ and $\angle BAC \cong \angle DCA$. What is the most direct congruence method that could be used to prove $\triangle ABC \cong \triangle CDA$?

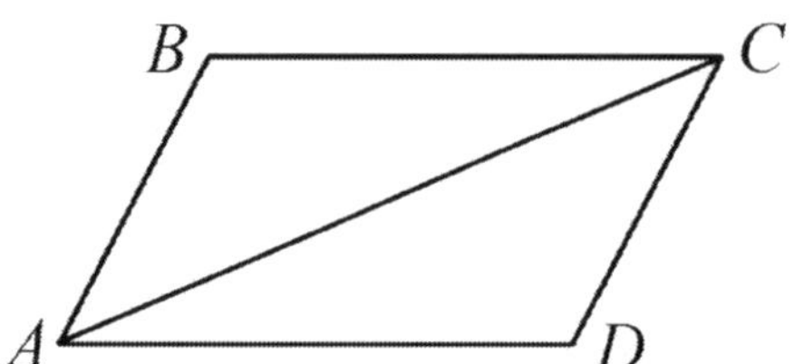

7. Given: $\angle C \cong \angle D$, $\overline{AC} \cong \overline{AD}$
Prove: $\overline{CE} \cong \overline{DB}$

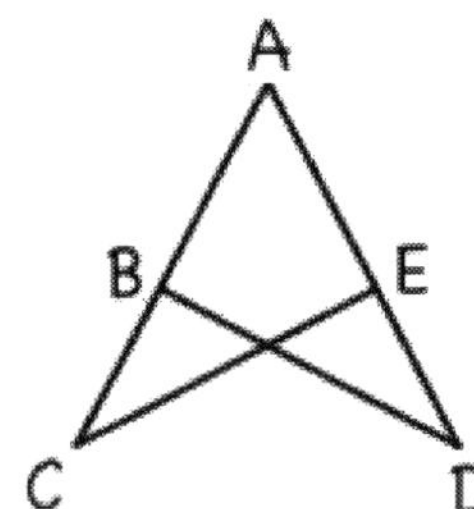

Statements	*Reasons*

REGENTS QUESTIONS

Multiple Choice

1. In the accompanying diagram of triangles BAT and FLU, $\angle B \cong \angle F$ and $\overline{BA} \cong \overline{FL}$.

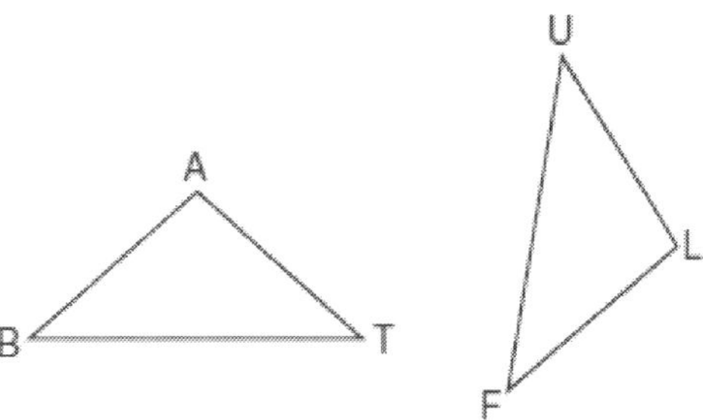

Which statement is needed to prove $\triangle BAT \cong \triangle FLU$?

(1) $\angle A \cong \angle L$
(2) $\overline{AT} \cong \overline{LU}$
(3) $\angle A \cong \angle U$
(4) $\overline{BA} \parallel \overline{FL}$

2. In the diagram below, $\triangle ABC \cong \triangle XYZ$.

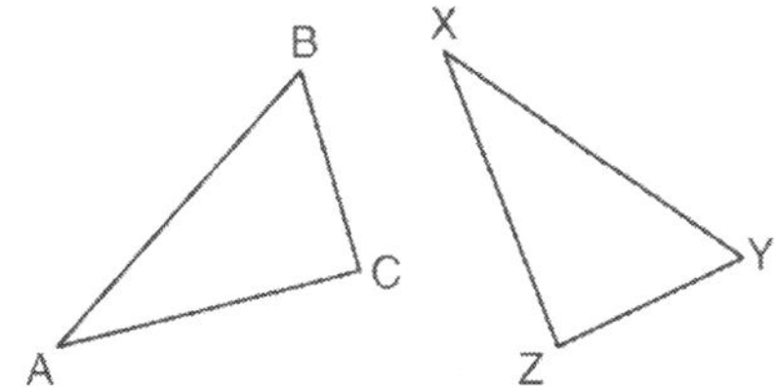

Which two statements identify corresponding congruent parts for these triangles?

(1) $\overline{AB} \cong \overline{XY}$ and $\angle C \cong \angle Y$
(2) $\overline{AB} \cong \overline{YZ}$ and $\angle C \cong \angle X$
(3) $\overline{BC} \cong \overline{XY}$ and $\angle A \cong \angle Y$
(4) $\overline{BC} \cong \overline{YZ}$ and $\angle A \cong \angle X$

3. In the diagram below of $\triangle AGE$ and $\triangle OLD$, $\angle GAE \cong \angle LOD$, and $\overline{AE} \cong \overline{OD}$.

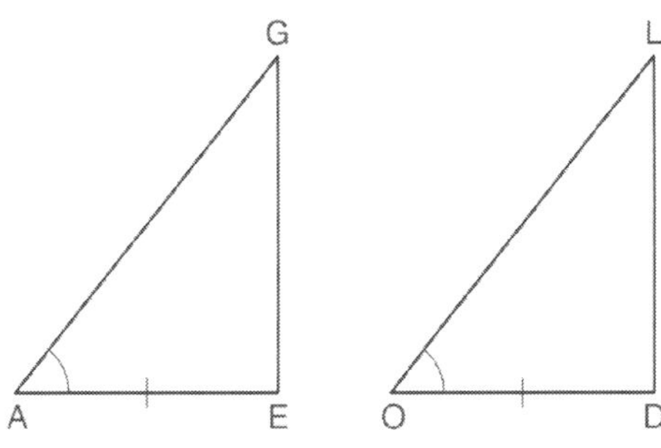

To prove that $\triangle AGE$ and $\triangle OLD$ are congruent by SAS, what other information is needed?

(1) $\overline{GE} \cong \overline{LD}$
(2) $\overline{AG} \cong \overline{OL}$
(3) $\angle AGE \cong \angle OLD$
(4) $\angle AEG \cong \angle ODL$

4. In the diagram of quadrilateral $ABCD$, $\overline{AB} \parallel \overline{CD}$, $\angle ABC \cong \angle CDA$, and diagonal $\overline{AC}$ is drawn.

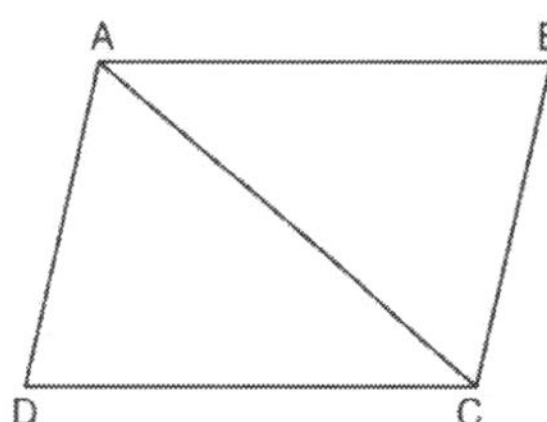

Which method can be used to prove $\triangle ABC$ is congruent to $\triangle CDA$?

(1) AAS
(2) SSA
(3) SAS
(4) SSS

5. If $\triangle JKL \cong \triangle MNO$, which statement is always true?

(1) $\angle KLJ \cong \angle NMO$
(2) $\angle KJL \cong \angle MON$
(3) $\overline{JL} \cong \overline{MO}$
(4) $\overline{JK} \cong \overline{ON}$

6. In the diagram below, $\triangle ABC \cong \triangle XYZ$.

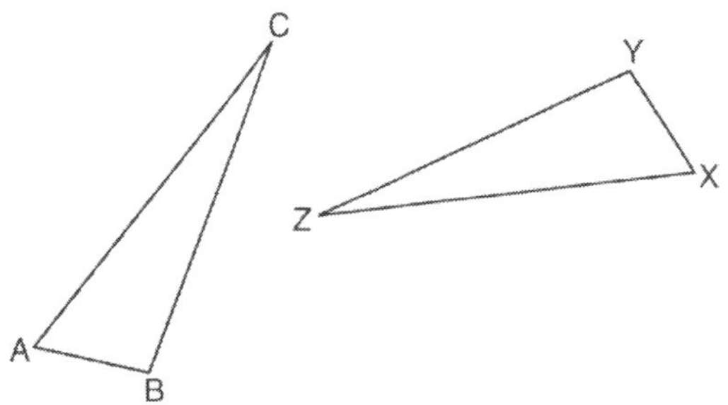

Which statement must be true?

(1) $\angle C \cong \angle Y$
(2) $\angle A \cong \angle X$
(3) $\overline{AC} \cong \overline{YZ}$
(4) $\overline{CB} \cong \overline{XZ}$

7. As shown in the diagram below, $\overline{AC}$ bisects $\angle BAD$ and $\angle B \cong \angle D$.

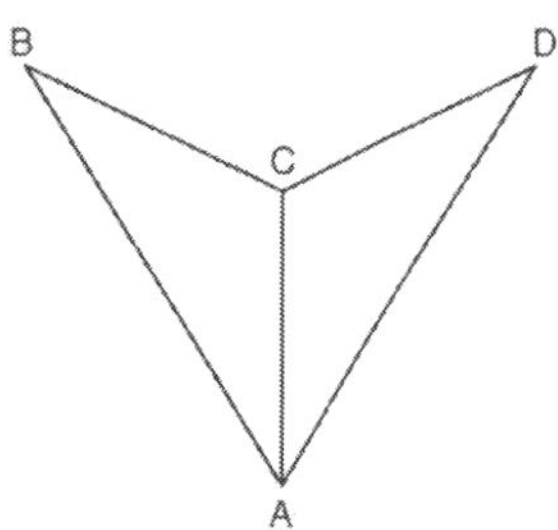

Which method could be used to prove $\triangle ABC \cong \triangle ADC$?

(1) SSS
(2) AAA
(3) SAS
(4) AAS

8. The diagram below shows a pair of congruent triangles, with $\angle ADB \cong \angle CDB$ and $\angle ABD \cong \angle CBD$.

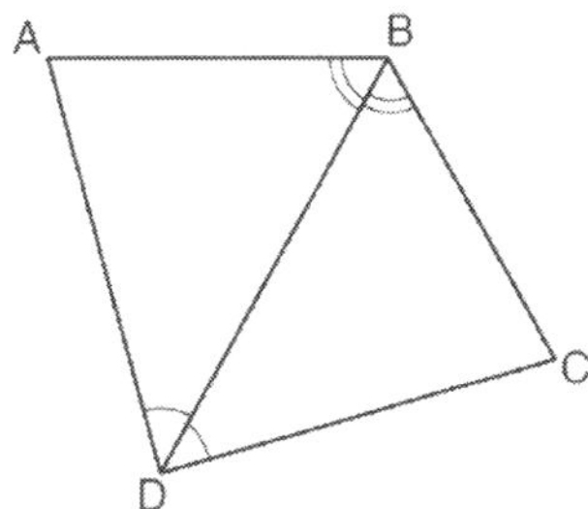

Which statement must be true?

(1) $\angle ADB \cong \angle CBD$
(2) $\angle ABC \cong \angle ADC$
(3) $\overline{AB} \cong \overline{CD}$
(4) $\overline{AD} \cong \overline{CD}$

9. In the diagram below, $\triangle AEC \cong \triangle BED$.

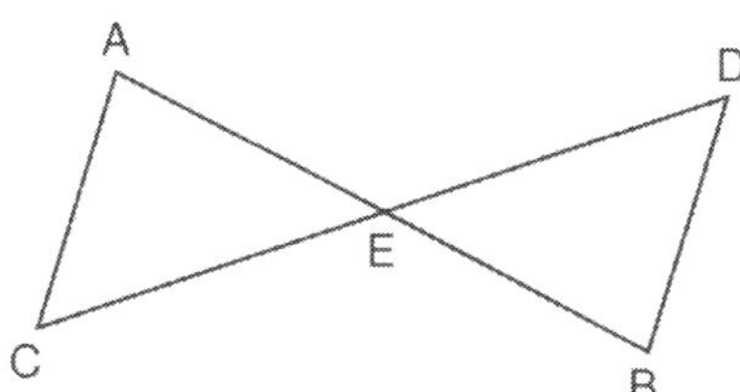

Which statement is *not* always true?

(1) $\overline{AC} \cong \overline{BD}$
(2) $\overline{CE} \cong \overline{DE}$
(3) $\angle EAC \cong \angle EBD$
(4) $\angle ACE \cong \angle DBE$

10. (CC) Given: $\triangle ABE$ and $\triangle CBD$ shown in the diagram below with $\overline{DB} \cong \overline{BE}$

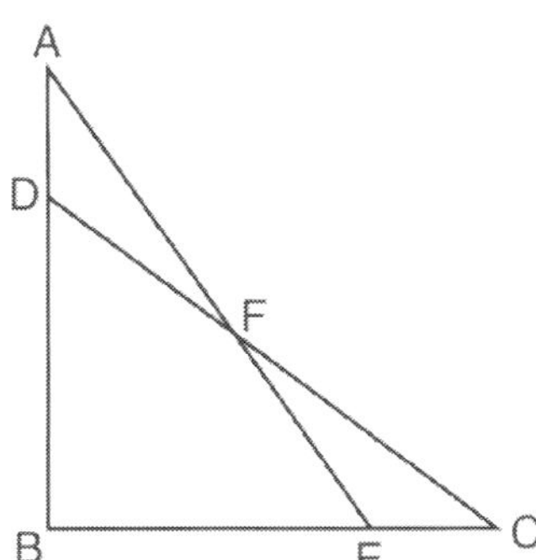

Which statement is needed to prove $\triangle ABE \cong \triangle CBD$ using only SAS ≅ SAS?

(1) $\angle CDB \cong \angle AEB$
(2) $\angle AFD \cong \angle EFC$
(3) $\overline{AD} \cong \overline{CE}$
(4) $\overline{AE} \cong \overline{CD}$

11. (CC) Given $\triangle ABC \cong \triangle DEF$, which statement is *not* always true?

(1) $\overline{BC} \cong \overline{DF}$
(2) $m\angle A = m\angle D$
(3) area of $\triangle ABC$ = area of $\triangle DEF$
(4) perimeter of $\triangle ABC$ = perimeter of $\triangle DEF$

Constructed Response

12. Complete the partial proof below for the accompanying diagram by providing reasons for steps 3, 6, 8, and 9.

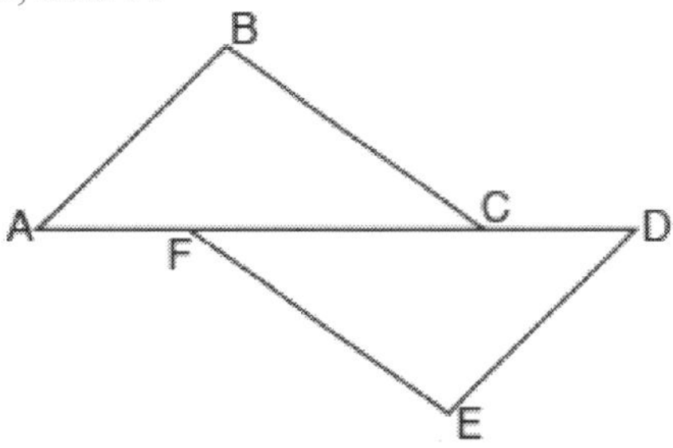

Given: $\overline{AFCD}$, $\overline{AB} \perp \overline{BC}$, $\overline{DE} \perp \overline{EF}$, $\overline{BC} \parallel \overline{FE}$, $\overline{AB} \cong \overline{DE}$

Prove: $\overline{AC} \cong \overline{FD}$

Statements	*Reasons*
1. $\overline{AFCD}$	1. Given
2. $\overline{AB} \perp \overline{BC}$, $\overline{DE} \perp \overline{EF}$	2. Given
3. $\angle B$ and $\angle E$ are right angles	3. ________________
4. $\angle B \cong \angle E$	4. All right angles are congruent
5. $\overline{BC} \parallel \overline{FE}$	5. Given
6. $\angle BCA \cong \angle EFD$	6. ________________
7. $\overline{AB} \cong \overline{DE}$	7. Given
8. $\triangle ABC \cong \triangle DEF$	8. ________________
9. $\overline{AC} \cong \overline{FD}$	9. ________________

13. Given: $\overline{AD}$ bisects $\overline{BC}$ at E, $\overline{AB} \perp \overline{BC}$, $\overline{DC} \perp \overline{BC}$

Prove: $\overline{AB} \cong \overline{DC}$

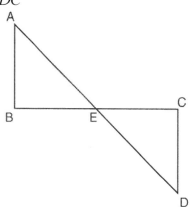

14. Given: $\triangle ABC$, $\overline{BD}$ bisects $\angle ABC$, $\overline{BD} \perp \overline{AC}$

Prove: 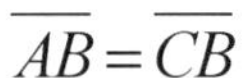 $\overline{AB} = \overline{CB}$

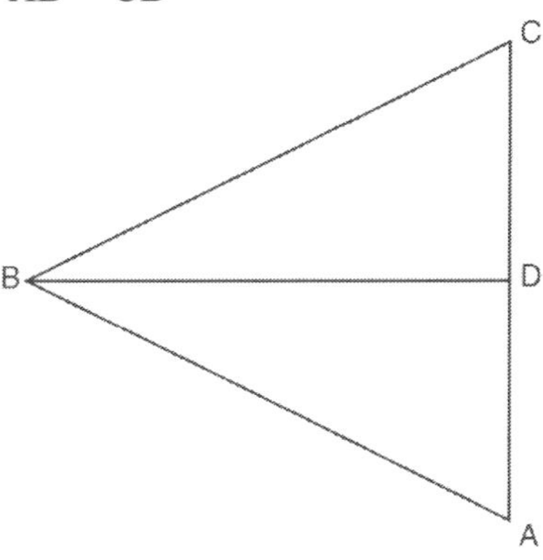

15. Given: $\overline{BE}$ and $\overline{AD}$ intersect at point C, $\overline{BC} \cong \overline{EC}$, $\overline{AC} \cong \overline{DC}$, $\overline{AB}$ and $\overline{DE}$ are drawn

Prove: $\triangle ABC \cong \triangle DEF$

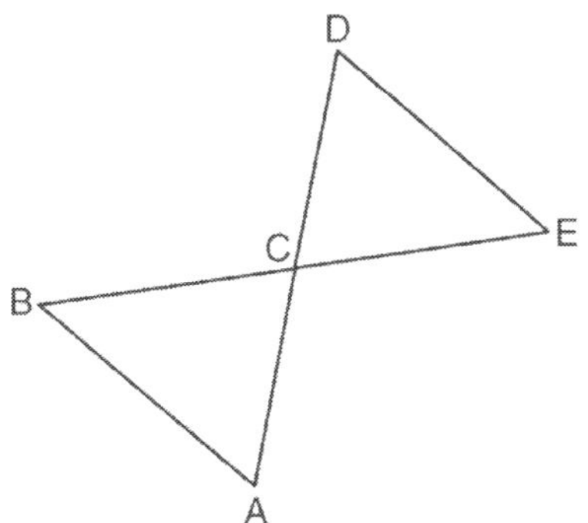

4.5 Similar Triangles

Key Terms and Concepts

Whereas congruent figures have the same size and shape, **similar figures** have the same shape but not necessarily the same size. The ~ symbol is used for similarity; $\triangle ABC \sim \triangle DEF$ means that these two triangles are similar.

Because similar polygons have the same shape, we know that *corresponding angles of similar polygons are congruent.* If two similar polygons are not congruent, their sides will not be congruent. However, the lengths of their corresponding sides will be *in proportion.* This is called the Side Proportionality theorem.

Side Proportionality: *Corresponding sides of similar polygons are proportional*

A triangle is a polygon, so all of these facts about similar polygons apply to similar triangles.

Example: For the similar triangles shown below, $\frac{5}{10} = \frac{10}{20} = \frac{12}{24}$ (or $\frac{10}{5} = \frac{20}{10} = \frac{24}{12}$).

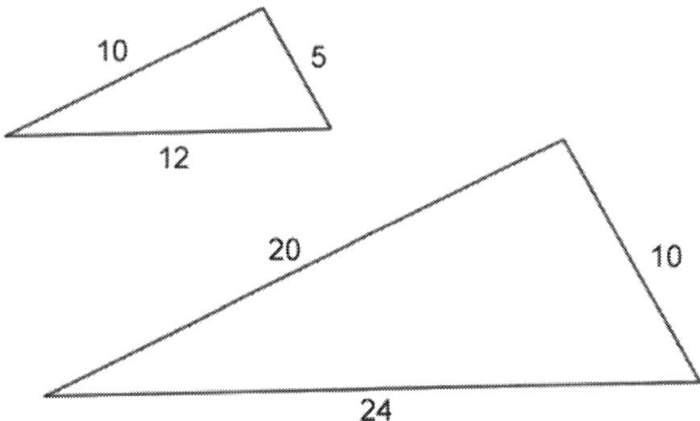

The **scale factor** is the ratio between the lengths of a pair of corresponding sides of two similar polygons. As an extension of this fact, the ratio of the perimeters of similar polygons is also proportional to the scale factor.

Example: The triangles above have a scale factor of $\frac{1}{2}$ (or 2).

The perimeters are $5 + 10 + 12 = 27$ and $10 + 20 + 24 = 54$, so the perimeters are also in the same proportion.

There are several valid methods of proving two triangles are similar:

AA Similarity: two pairs of corresponding angles are congruent
SSS Similarity: three pairs of corresponding side lengths are proportional
SAS Similarity: two pairs of corresponding side lengths are proportional, and their included angles are congruent

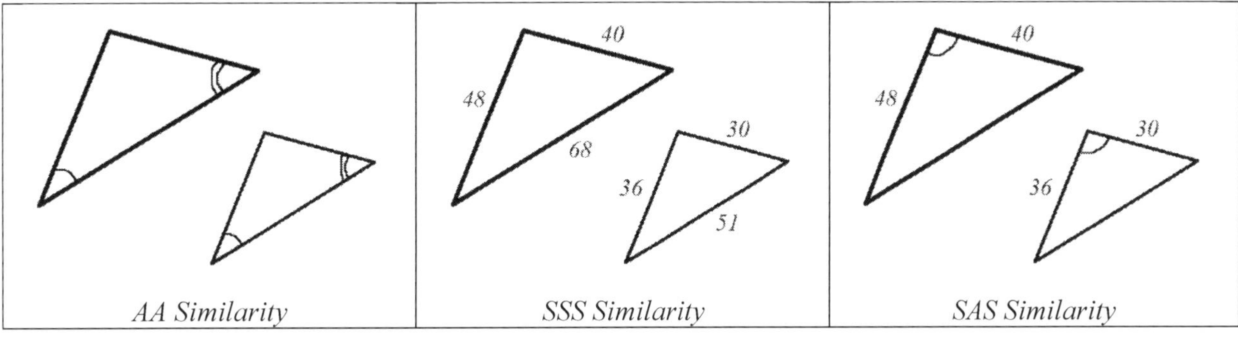

AA Similarity *SSS Similarity* *SAS Similarity*

Model Problem:

Given: $\overline{AB} \parallel \overline{DE}$, $\overline{AE}$ and $\overline{BD}$ intersect at C, $AB = 20$, $BC = 15$, and $DE = 12$. Find CD.

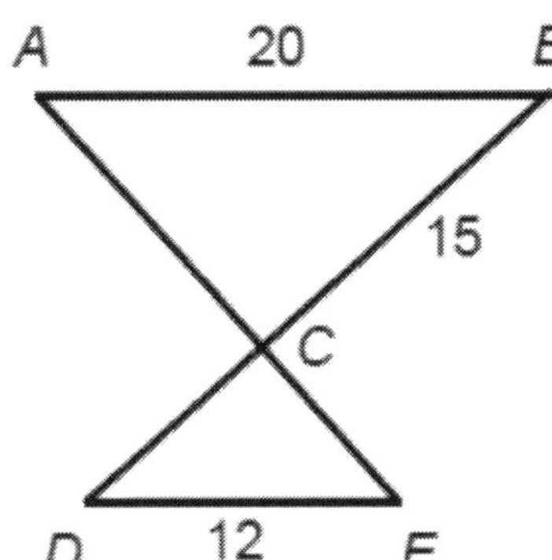

Solution:

(A) $\angle A \cong \angle E$ and $\angle B \cong \angle D$
(B) $\triangle ABC \sim \triangle DEC$
(C) $\frac{CD}{15} = \frac{12}{20}$
(D) $CD = 9$

Explanation of steps:

(A) Parallel lines cut by a transversal form congruent pairs of alternate interior angles. *[It is sufficient to show that two pairs of corresponding angles are congruent, but it is also true that $\angle ACB \cong \angle ECD$ since they are a pair of vertical angles.]*
(B) Triangles are congruent by AA Similarity.
(C) Corresponding sides of similar triangles are in proportion.
(D) Solve for CD.

Practice Problems

1. $\triangle ABC \sim \triangle XYZ$. Find AC.

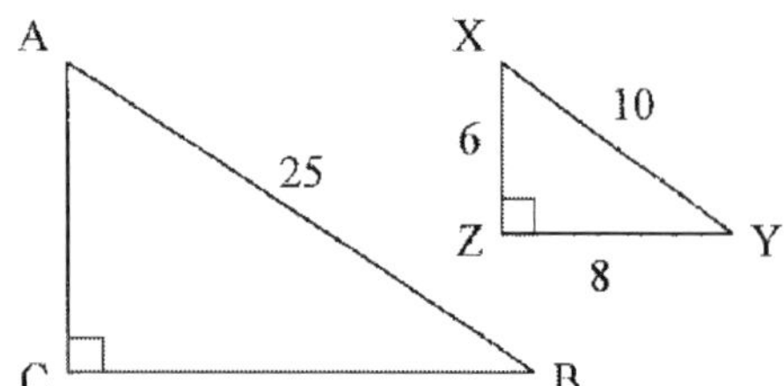

2. $\triangle ABC \sim \triangle XYZ$. Find XY.

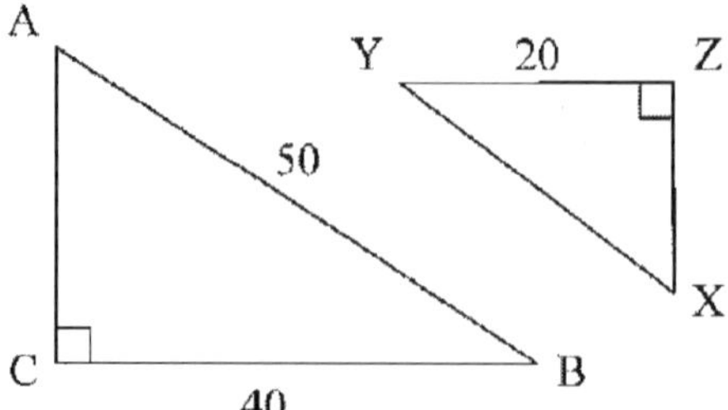

3. $\triangle ABC \sim \triangle PQR$. Find $m\angle Q$.

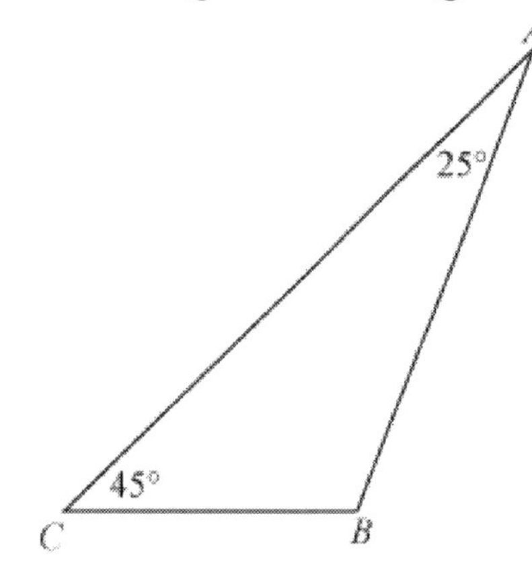

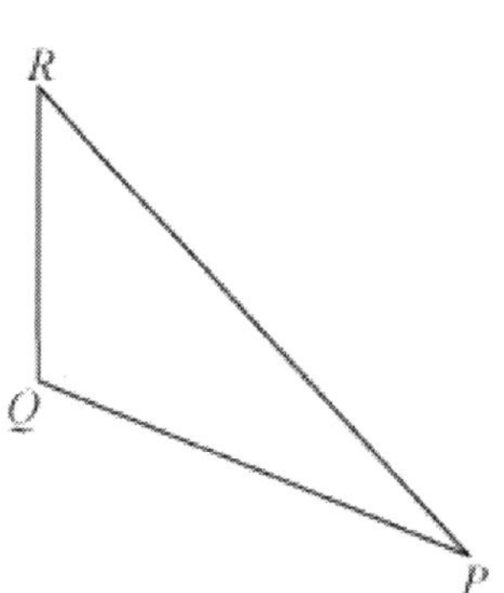

4. In the diagram below, $\overline{QR} \parallel \overline{TU}$. Name the pair of similar triangles, using corresponding order. Why are the triangles similar?

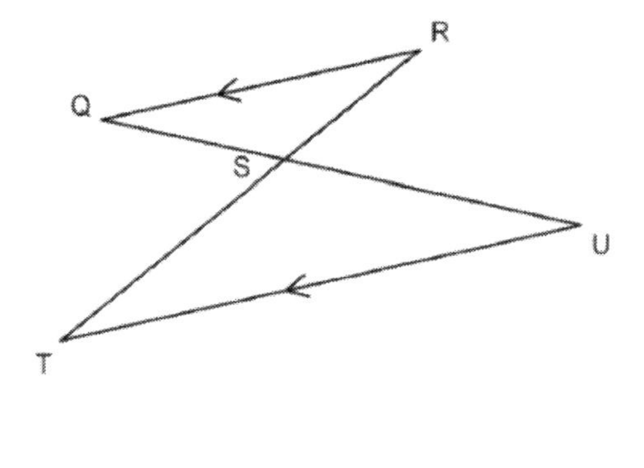

5. In the diagram below, $\overline{AB} \parallel \overline{CD}$. Name the pair of similar triangles, using corresponding order. Why are the triangles similar?

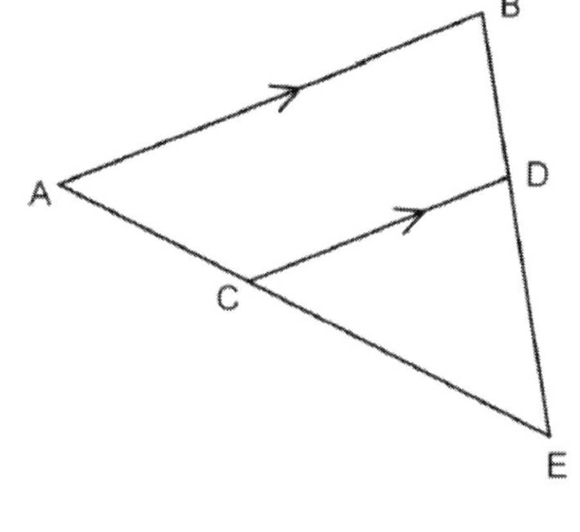

REGENTS QUESTIONS

Multiple Choice

1. In the diagram below, $\overline{SQ}$ and $\overline{PR}$ intersect at T, $\overline{PQ}$ is drawn, and $\overline{PS} \parallel \overline{QR}$.

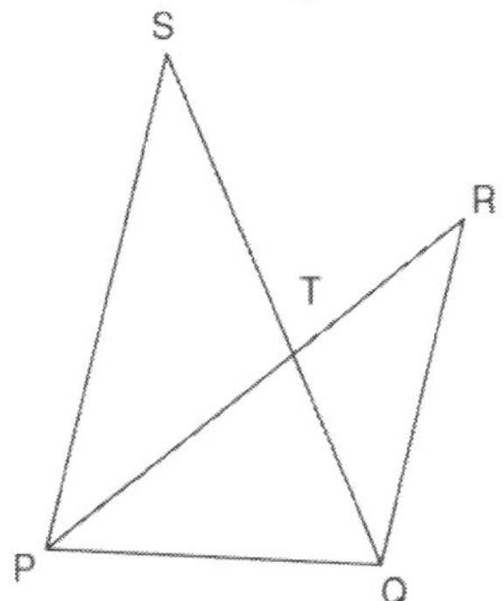

What technique can be used to prove that $\triangle PST \sim \triangle RQT$?

(1) SAS (3) ASA
(2) SSS (4) AA

2. If $\triangle ABC \sim \triangle ZXY$, $m\angle A = 50$, and $m\angle C = 30$, what is $m\angle X$?

(1) 30 (3) 80
(2) 50 (4) 100

3. In $\triangle ABC$ and $\triangle DEF$, $\frac{AC}{DF} = \frac{CB}{FE}$. Which additional information would prove $\triangle ABC \sim \triangle DEF$?

(1) $AC = DF$ (3) $\angle ACB \cong \angle DFE$
(2) $CB = FE$ (4) $\angle BAC \cong \angle EDF$

4. Triangle ABC is similar to triangle DEF. The lengths of the sides of $\triangle ABC$ are 5, 8, and 11. What is the length of the shortest side of $\triangle DEF$ if its perimeter is 60?

(1) 10 (3) 20
(2) 12.5 (4) 27.5

5. In triangles ABC and DEF, $AB = 4$, $AC = 5$, $DE = 8$, $DF = 10$, and $\angle A \cong \angle D$. Which method could be used to prove $\triangle ABC \sim \triangle DEF$?

(1) AA (3) SSS
(2) SAS (4) ASA

6. The sides of a triangle are 8, 12, and 15. The longest side of a similar triangle is 18. What is the ratio of the perimeter of the smaller triangle to the perimeter of the larger triangle?

(1) 2:3 (3) 5:6
(2) 4:9 (4) 25:36

7. For which diagram is the statement $\triangle ABC \sim \triangle ADE$ *not* always true?

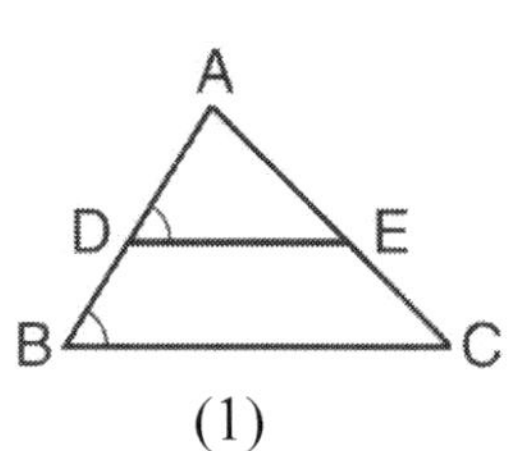

(1)

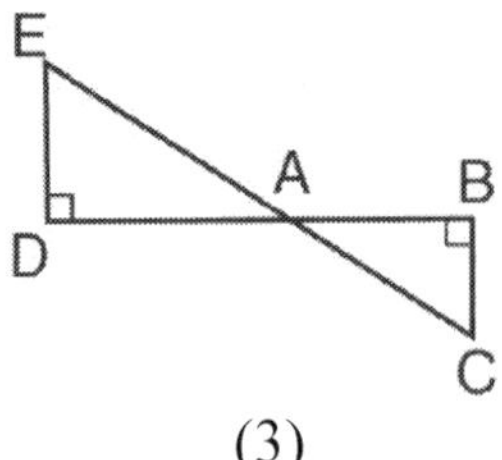

(3)

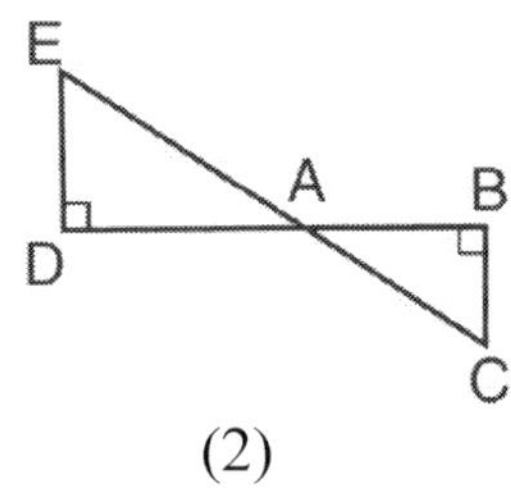

(2)

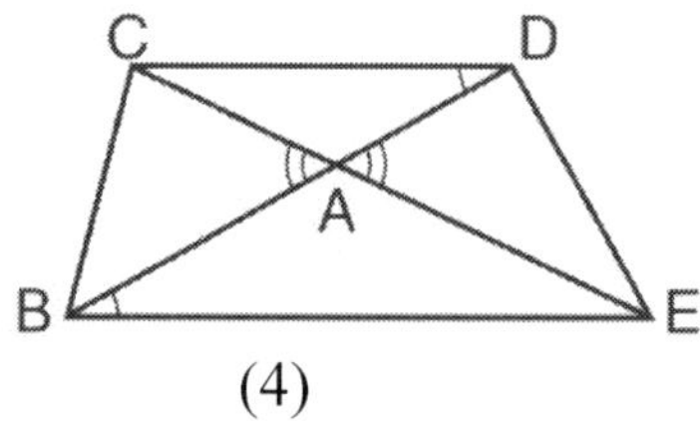

(4)

8. Ⓒ Triangles ABC and DEF are drawn below.

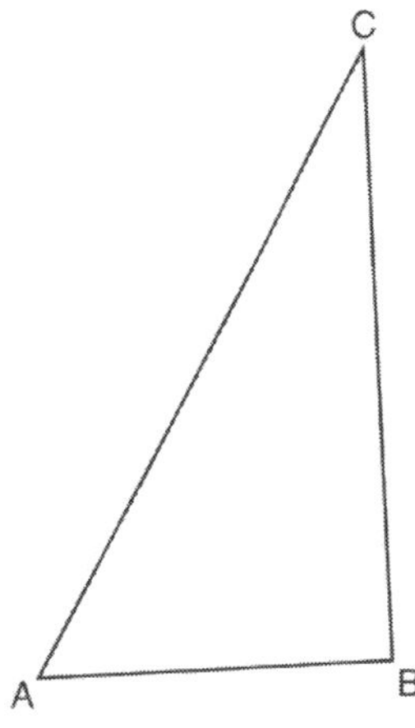

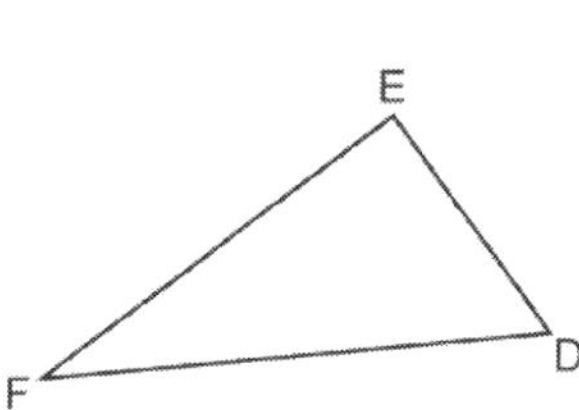

If $AB = 9$, $BC = 15$, $DE = 6$, $EF = 10$, and $\angle B \cong \angle E$, which statement is true?

(1) $\angle CAB \cong \angle DEF$

(2) $\dfrac{AB}{CB} = \dfrac{FE}{DE}$

(3) $\triangle ABC \sim \triangle DEF$

(4) $\dfrac{AB}{DE} = \dfrac{FE}{CB}$

9. (CC) In the diagram below, $\triangle ABC \sim \triangle DEC$.

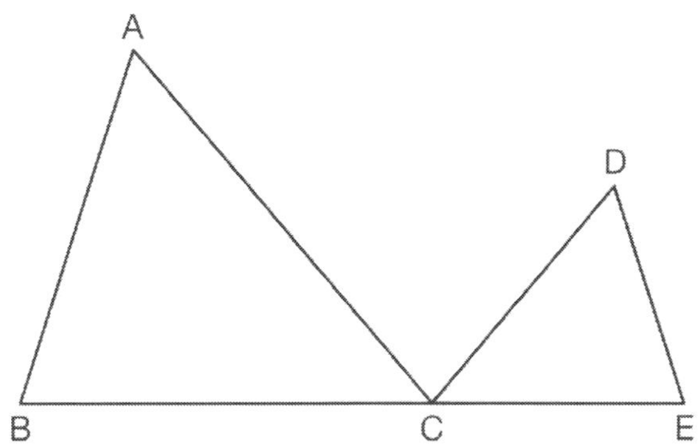

If $AE = 12$, $DC = 7$, $DE = 5$, and the perimeter of $\triangle ABC$ is 30, what is the perimeter of $\triangle DEC$?

(1) 12.5
(2) 14.0
(3) 14.8
(4) 17.5

10. (CC) As shown in the diagram below, $\overline{AB}$ and $\overline{CD}$ intersect at E, and $\overline{AC} \parallel \overline{BD}$.

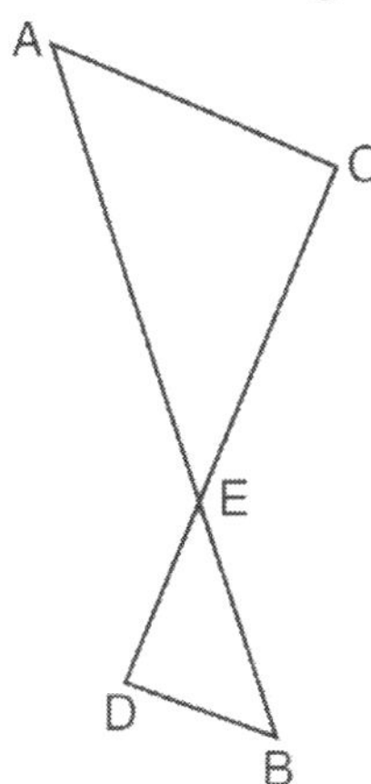

Given $\triangle AEC \sim \triangle BED$, which equation is true?

(1) $\frac{CE}{DE} = \frac{EB}{EA}$
(2) $\frac{AE}{BE} = \frac{AC}{BD}$
(3) $\frac{EC}{AE} = \frac{BE}{ED}$
(4) $\frac{ED}{EC} = \frac{AC}{BD}$

11. (CC) In the diagram below, $\triangle ABC \sim \triangle DEF$.

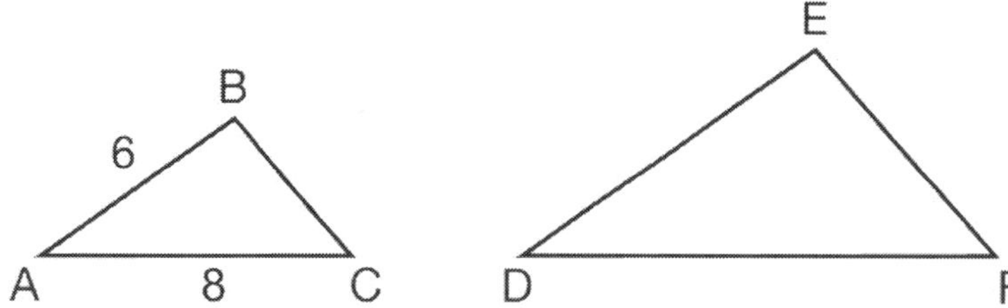

If $AB = 6$ and $AC = 8$, which statement will justify similarity by SAS?
(1) $DE = 9$, $DF = 12$, and $\angle A \cong \angle D$
(2) $DE = 8$, $DF = 10$, and $\angle A \cong \angle D$
(3) $DE = 36$, $DF = 64$, and $\angle C \cong \angle F$
(4) $DE = 15$, $DF = 20$, and $\angle C \cong \angle F$

12. (CC) The ratio of similarity of $\triangle BOY$ to $\triangle GRL$ is $1:2$. If $BO = x+3$ and $GR = 3x-1$, then the length of $\overline{GR}$ is
(1) 5
(2) 7
(3) 10
(4) 20

13. (CC) In $\triangle SCU$ shown below, points T and O are on $\overline{SU}$ and $\overline{CU}$, respectively. Segment OT is drawn so that $\angle C \cong \angle OTU$.

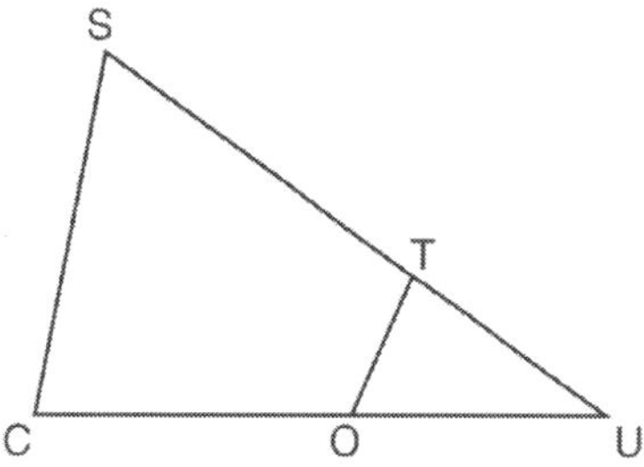

If $TU = 4$, $OU = 5$, and $OC = 7$, what is the length of $\overline{ST}$?
(1) 5.6
(2) 8.75
(3) 11
(4) 15

14. (CC) Using the information given below, which set of triangles can *not* be proven similar?

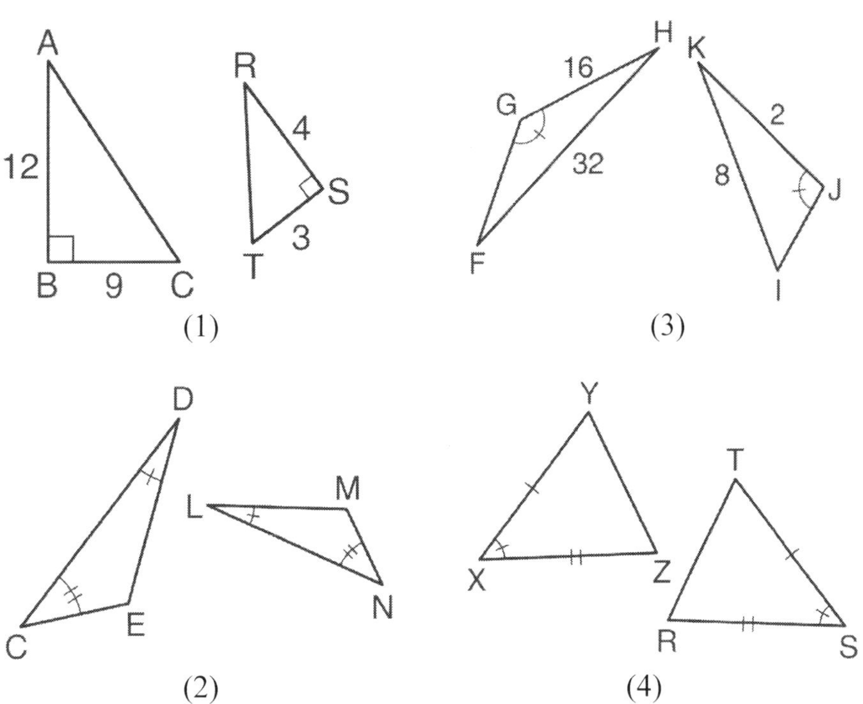

15. (CC) In the diagram below, $\overline{DB}$ and $\overline{AF}$ intersect at point C, and $\overline{AD}$ and $\overline{FBE}$ are drawn.

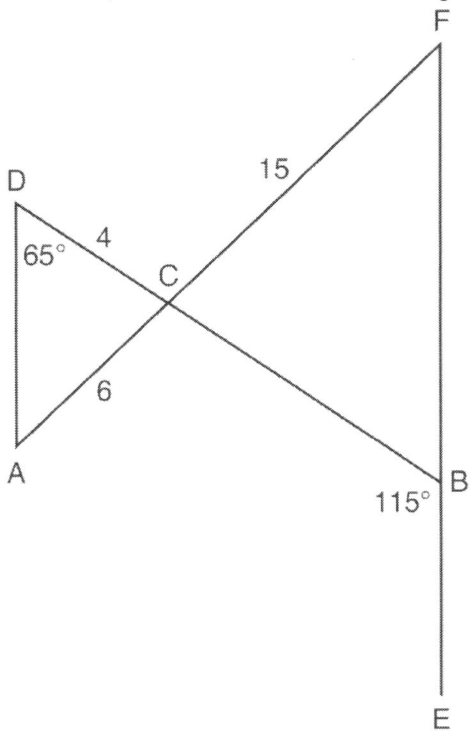

If $AC = 6$, $DC = 4$, $FC = 15$, $m\angle D = 65°$, and $m\angle CBE = 115°$, what is the length of $\overline{CB}$?

(1) 10
(2) 12
(3) 17
(4) 22.5

16. (CC) In triangle *CHR*, *O* is on $\overline{HR}$, and *D* is on $\overline{CR}$ so that $\angle H \cong \angle RDO$.

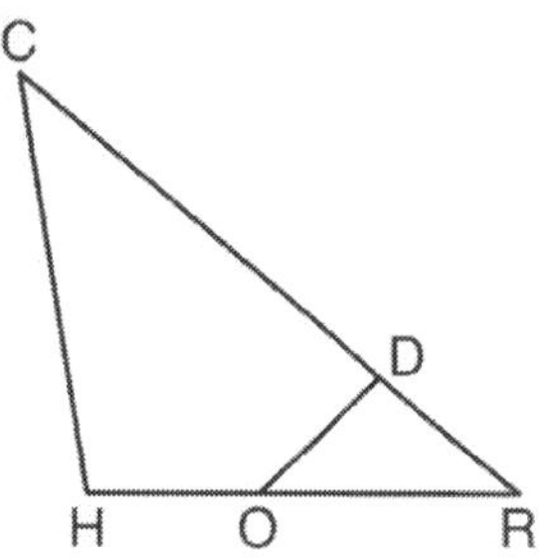

If RD = 4, RO = 6, and OH = 4, what is the length of $\overline{CD}$?

(1) $2\frac{2}{3}$ (3) 11

(2) $6\frac{2}{3}$ (4) 15

Constructed Response

17. In the accompanying diagram, $\triangle QRS$ is similar to $\triangle LMN$, $RQ = 30$, $QS = 21$, $SR = 27$, and $LN = 7$. What is the length of $\overline{ML}$?

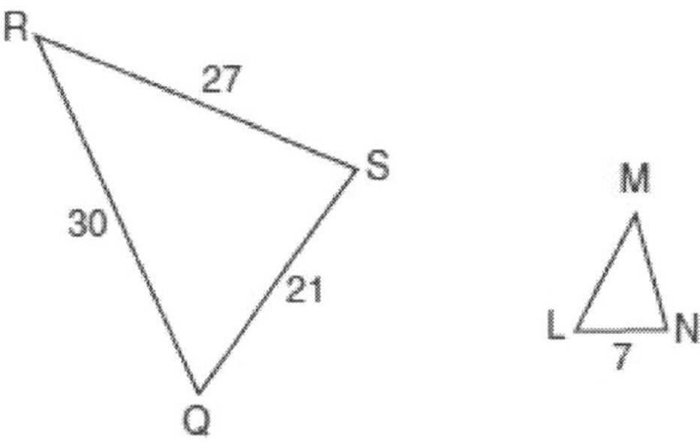

18. In the diagram below, $\overline{BFCE}$, $\overline{AB} \perp \overline{BE}$, $\overline{DE} \perp \overline{BE}$, and $\angle BFD \cong \angle ECA$. Prove that $\triangle ABC \sim \triangle DEF$.

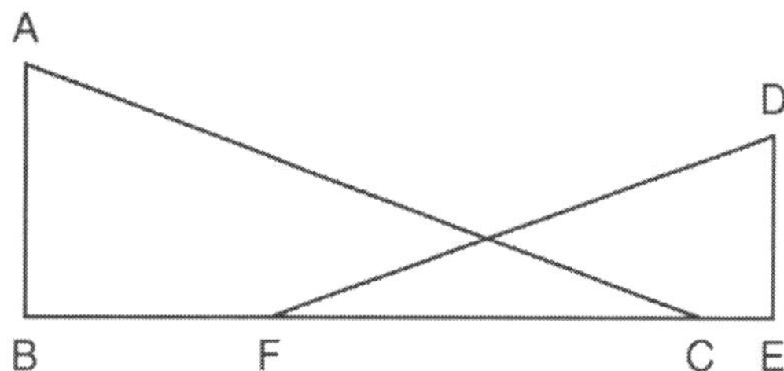

19. The diagram below shows $\triangle ABC$, with $\overline{AEB}$, $\overline{ADC}$, and $\angle ACB \cong \angle AED$. Prove that $\triangle ABC$ is similar to $\triangle ADE$.

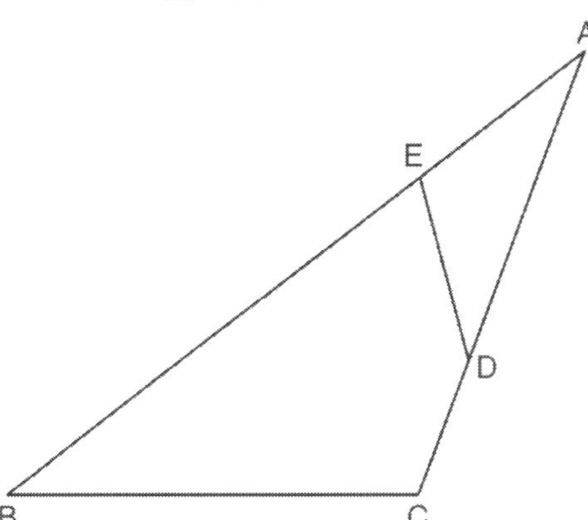

20. The sides of a triangle measure 7, 4, and 9. If the longest side of a similar triangle measures 36, determine and state the length of the shortest side of this triangle.

21. (CC) Triangles *RST* and *XYZ* are drawn below. If $RS = 6$, $ST = 14$, $XY = 9$, $YZ = 21$, and $\angle S \cong \angle Y$, is $\triangle RST$ similar to $\triangle XYZ$? Justify your answer.

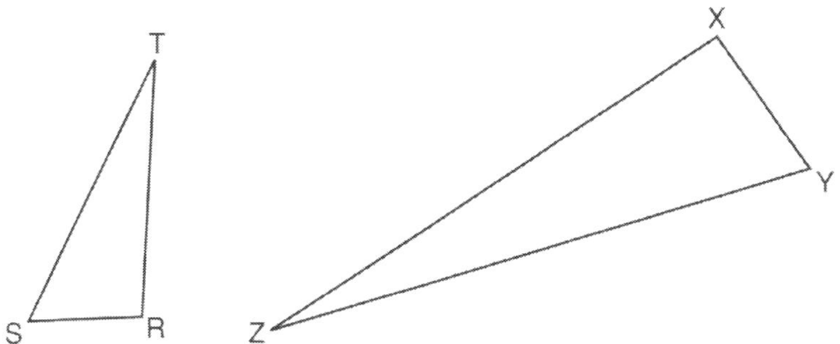

22. (CC) In the diagram below, $\overline{GI}$ is parallel to $\overline{NT}$, and $\overline{IN}$ intersects $\overline{GT}$ at *A*.

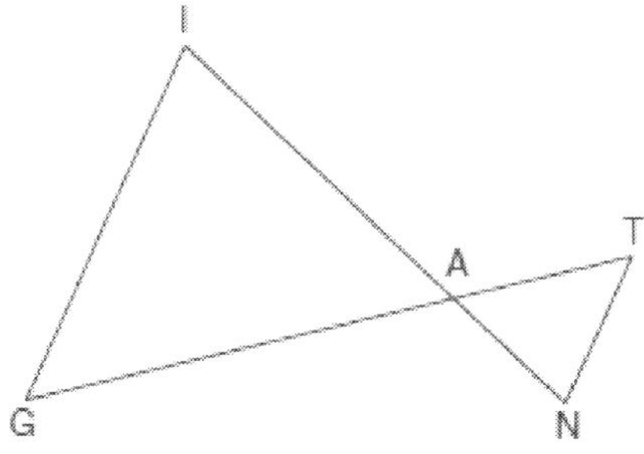

Prove: $\triangle GIA \sim \triangle TNA$

4.6 Similar Triangle Theorems

Key Terms and Concepts

Angle Bisector Theorem: The bisector of an angle of a triangle splits the opposite side into segments that are proportional to the adjacent sides.

Example: In the diagram below, $\overline{AD}$ bisects $\angle A$. Therefore, $\frac{BD}{CD} = \frac{BA}{CA}$.

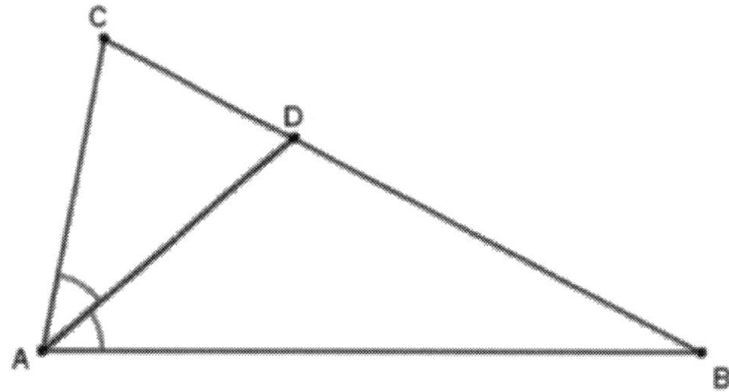

Side Splitter Theorem (*also known as the Triangle Proportionality Theorem*):
A line parallel to one side of a triangle divides the other two sides proportionally.

Example: In the diagram below, $\overline{BE} \parallel \overline{CD}$. Therefore, $\frac{AB}{BC} = \frac{AE}{ED}$.

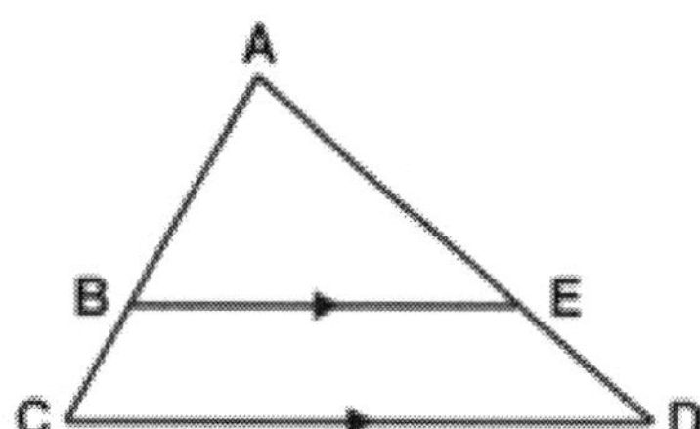

Triangle Midsegment Theorem: The segment joining midpoints of two sides of a triangle is parallel to the third side and half its length.

Example: In the diagram below, D and E are the midpoints of $\overline{AB}$ and $\overline{BC}$, respectively. Therefore, $\overline{DE} \parallel \overline{AC}$ and $DE = \frac{1}{2}AC$.

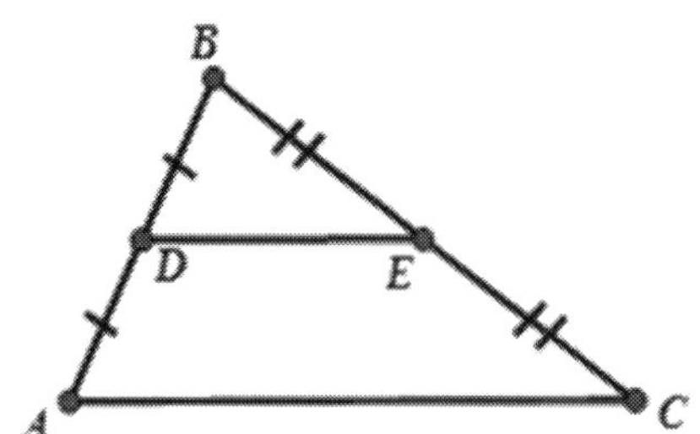

Important Proofs

Proof of the Angle Bisector Theorem:

Given: $\triangle ABC$, $\overline{AD}$ bisects $\angle A$

Prove: $\frac{BD}{CD} = \frac{BA}{CA}$

To prove this, we'll extend $\overleftrightarrow{AD}$ and draw $\overleftrightarrow{CE} \parallel \overline{AB}$, where E is the point of intersection with $\overleftrightarrow{AD}$, and prove $\triangle BAD \sim \triangle CED$.

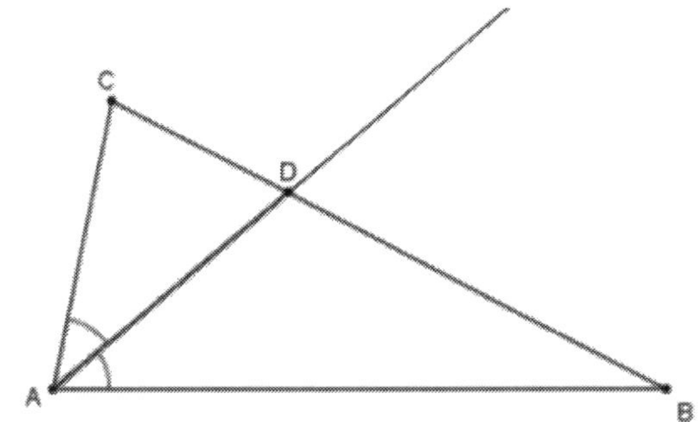

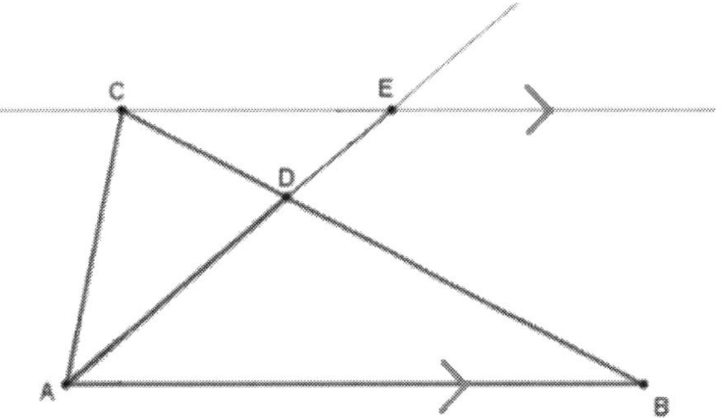

Statements	*Reasons*
$\overline{AD}$ bisects $\angle A$	Given
$\overleftrightarrow{CE} \parallel \overline{AB}$	Through a point not on a given line, a parallel line may be drawn
$\angle ECD \cong \angle ABD$ (A) $\angle CED \cong \angle BAD$ (A)	Parallel lines cut by a transversal form congruent alternate interior angles
$\triangle BAD \sim \triangle CED$	AA Similarity
$\frac{BD}{CD} = \frac{BA}{CE}$	Side Proportionality
$\angle BAD \cong \angle CAD$	Definition of angle bisector
$\angle CED \cong \angle CAD$	Transitive Property
$CE = CA$	If two angles of a triangle are congruent, the opposite sides are congruent
$\frac{BD}{CD} = \frac{BA}{CA}$	Substitution Property

Proof of the Side Splitter Theorem:

Given: $\triangle ACD$, $\overleftrightarrow{ABC}$, $\overleftrightarrow{AED}$, $\overline{BE} \parallel \overline{CD}$.

Prove: $\frac{AB}{BC} = \frac{AE}{ED}$.

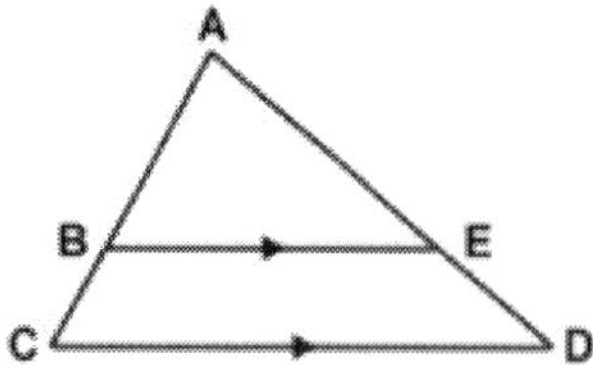

We can prove this theorem by first showing that $\triangle ABE \sim \triangle ACD$.

Statements	*Reasons*
$\overline{BE} \parallel \overline{CD}$	Given
$\angle A \cong \angle A$ (A)	Reflexive Property
$\angle ABE \cong \angle ACD$ (A) *(or,* $\angle AEB \cong \angle ADC$ *)*	Parallel lines cut by a transversal form congruent corresponding angles.
$\triangle ABE \sim \triangle ACD$	AA Similarity
$\frac{AB}{AC} = \frac{AE}{AD}$	Side Proportionality
$AC = AB + BC$ $AD = AE + ED$	Segment Addition Postulate
$\frac{AB}{AB+BC} = \frac{AE}{AE+ED}$	Substitution
$AB(AE+ED) = AE(AB+BC)$	In a proportion, the product of the means = the product of the extremes
$AB \cdot AE + AB \cdot ED = AE \cdot AB + AE \cdot BC$	Distributive Property
$AB \cdot AE = AE \cdot AB$	Commutative Property
$AB \cdot ED = AE \cdot BC$	Subtraction Property
$\frac{AB}{BC} = \frac{AE}{ED}$	In a proportion, the product of the means = the product of the extremes

Proof of the Triangle Midsegment Theorem:.

Given: $\triangle ABC$, D and E are the midpoints of $\overline{AB}$ and $\overline{BC}$, respectively

Prove: $\overline{DE} \parallel \overline{AC}$ and $DE = \frac{1}{2}AC$

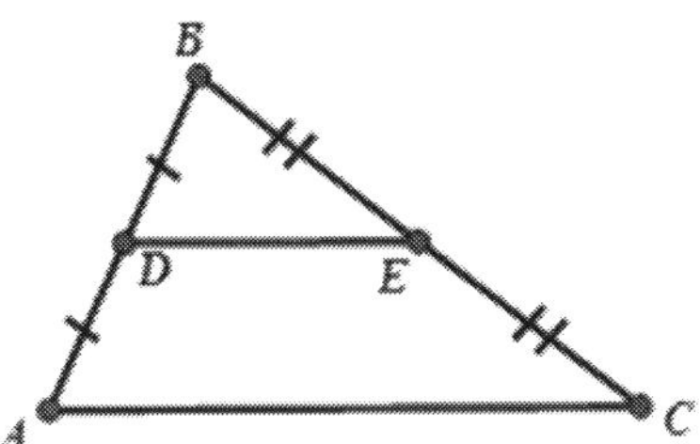

We can prove this theorem by first showing that $\triangle ABC \sim \triangle DBE$.

Statements	*Reasons*
D is the midpoint of $\overline{AB}$ E is the midpoint of $\overline{BC}$	Given
$\overline{DB} \cong \overline{AD}$, $\overline{EB} \cong \overline{CE}$	Definition of midpoint
$DB = AD$, $EB = CE$	Definition of congruence
$DB + AD = AB$	Segment Addition Postulate
$DB + DB = AB$, or $2DB = AB$	Substitution Property, and simplify
$\frac{AB}{DB} = 2$ (S)	Division Property
$\angle B \cong \angle B$ (A)	Reflexive Property
$EB + CE = CB$	Segment Addition Postulate
$EB + EB = CB$, or $2EB = CB$	Substitution Property, and simplfy
$\frac{CB}{AB} = 2$ (S)	Division Property
$\triangle ABC \sim \triangle DBE$	SAS Similarity
$\angle BDE \cong \angle BAC$	Corresponding angles of similar triangles are congruent
$\overline{DE} \parallel \overline{AC}$	If a transversal intersecting two lines forms congruent corresponding angles, then the lines are parallel
$\frac{AC}{DE} = 2$	Side Proportionality
$DE = \frac{1}{2}AC$	Solve for DE

Model Problem 1: *angle bisector theorem*

In the diagram below, $\overline{LK}$ bisects $\angle L$. Find x.

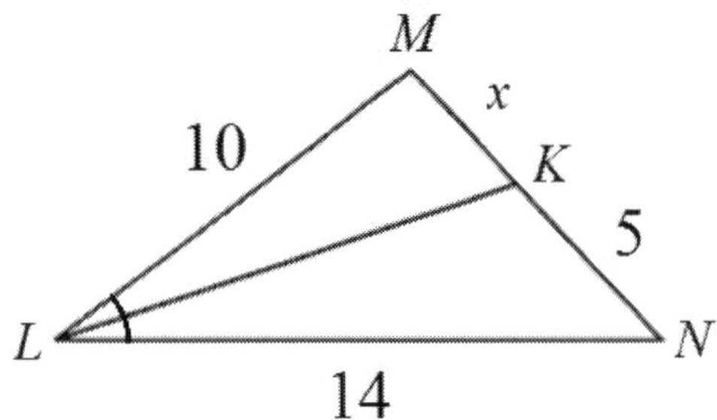

Solution:

$$\frac{x}{5} = \frac{10}{14}$$
$$14x = 50$$
$$x \approx 3.57$$

Explanation of steps:

The bisector of an angle of a triangle splits the opposite side into segments that are proportional to the adjacent sides.

$$\left[\frac{MK}{NK} = \frac{ML}{NL}\right]$$

Model Problem 2: *side-splitter theorem*

In the diagram below, $\overline{PQ} \parallel \overline{TR}$. Find x.

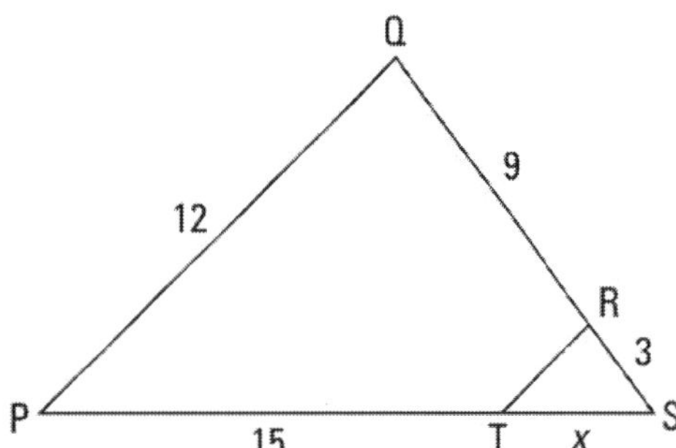

Solution:

$$\frac{x}{15} = \frac{3}{9}$$
$$9x = 45$$
$$x = 5$$

Explanation of steps:

A line parallel to one side of a triangle divides the other two sides proportionally.

$$\left[\frac{ST}{TP} = \frac{SR}{RQ}\right]$$

Practice Problems

1. In the diagram below, an angle bisector of the triangle is shown. Find x.

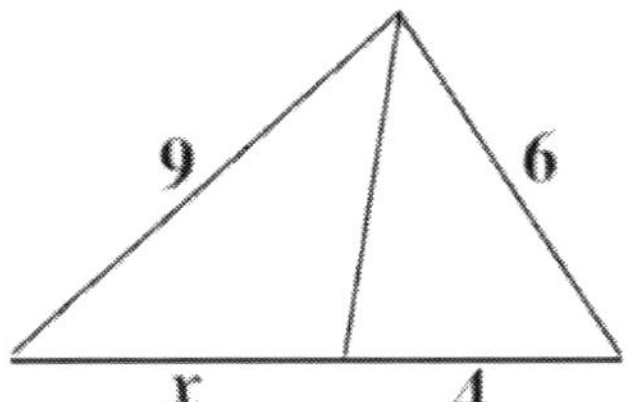

2. In the diagram below, an angle bisector of the triangle is shown. Find x.

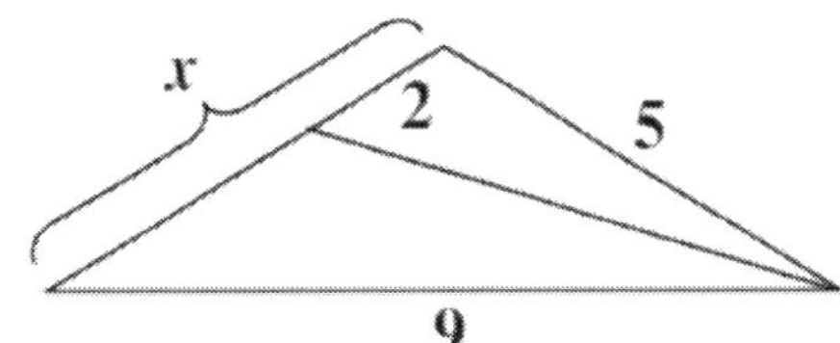

3. Find x.

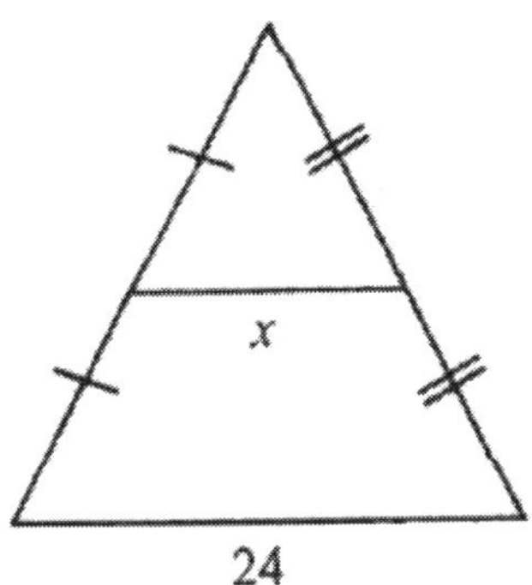

4. In the diagram below, $\overline{AE} \parallel \overline{BD}$. Find x.

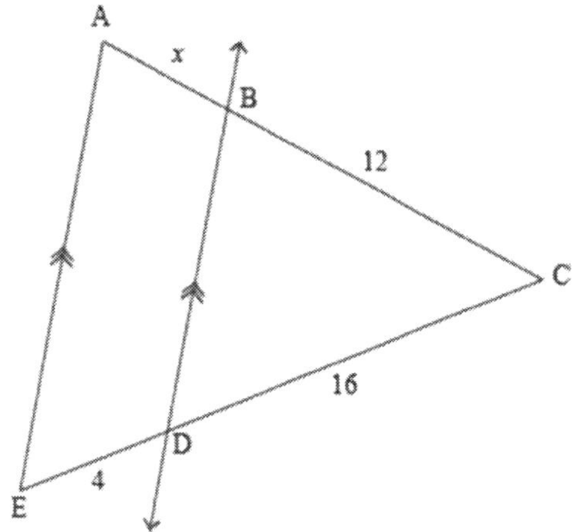

5. In the diagram, $\overline{NR} \parallel \overline{PQ}$.
$MQ = 42$, $MN = 13$, and $NP = 8$.
Find RQ and MR.

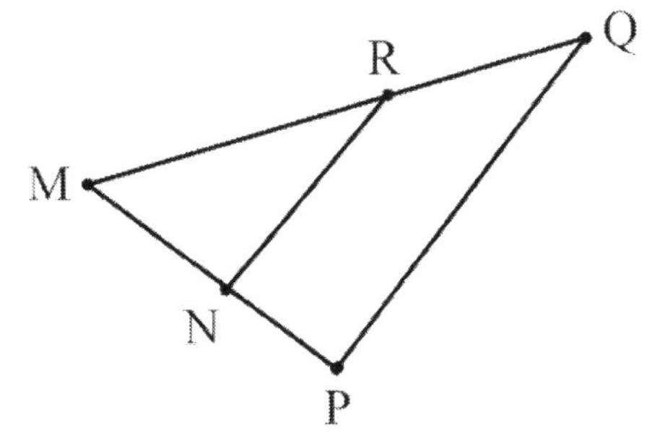

6. In triangle BWL:
$\overline{WE}$ is the bisector of $\angle W$, with E on $\overline{BL}$
$\overline{EO}$ is parallel to $\overline{BW}$, with O on $\overline{WL}$
$WO = 6$ and $OL = 9$
Find BW.

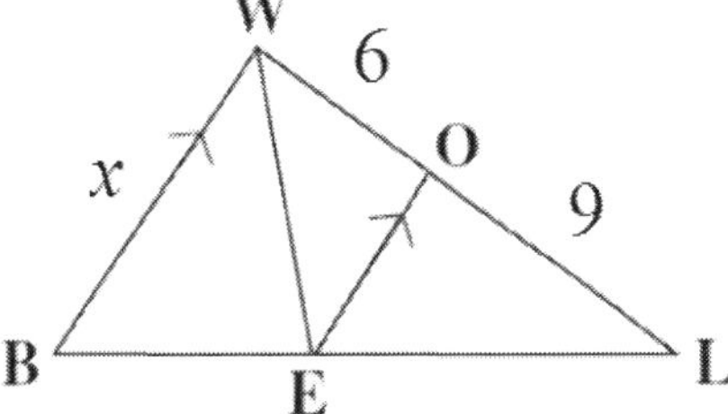

REGENTS QUESTIONS

Multiple Choice

1. In the diagram below, the vertices of $\triangle DEF$ are the midpoints of the sides of equilateral triangle *ABC*, and the perimeter of $\triangle ABC$ is 36 cm.

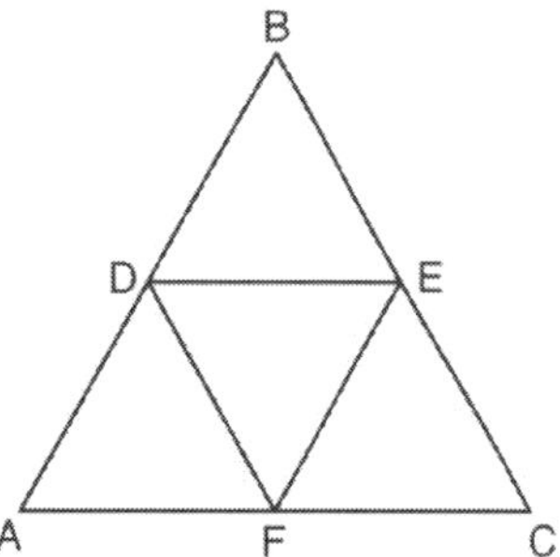

What is the length, in centimeters, of $\overline{EF}$?

(1) 6
(2) 12
(3) 18
(4) 4

2. In the diagram below of $\triangle ACT$, $\overleftrightarrow{BE} \parallel \overline{AT}$.

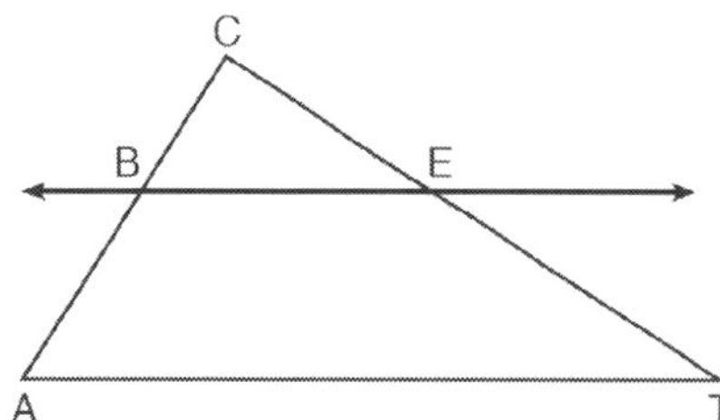

If $CB = 3$, $CA = 10$, and $CE = 6$, what is the length of $\overline{ET}$?

(1) 5
(2) 14
(3) 20
(4) 26

3. In the diagram of $\triangle ABC$ shown below, *D* is the midpoint of $\overline{AB}$, *E* is the midpoint of $\overline{BC}$, and *F* is the midpoint of $\overline{AC}$.

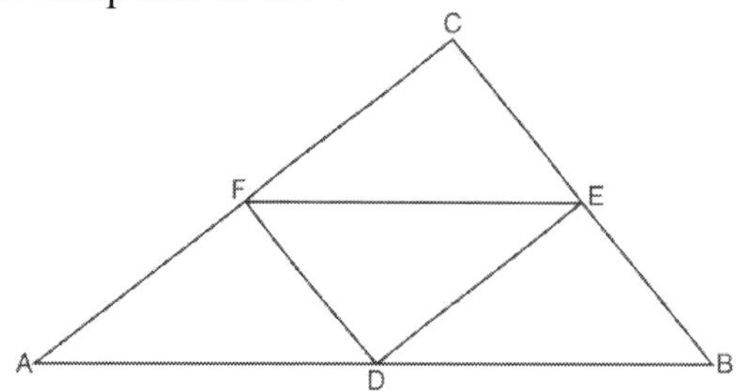

If AB = 20, BC = 12, and AC = 16, what is the perimeter of trapezoid *ABEF*?

(1) 24
(2) 36
(3) 40
(4) 44

4. In the diagram of $\triangle ABC$ shown below, $\overline{DE} \parallel \overline{BC}$.

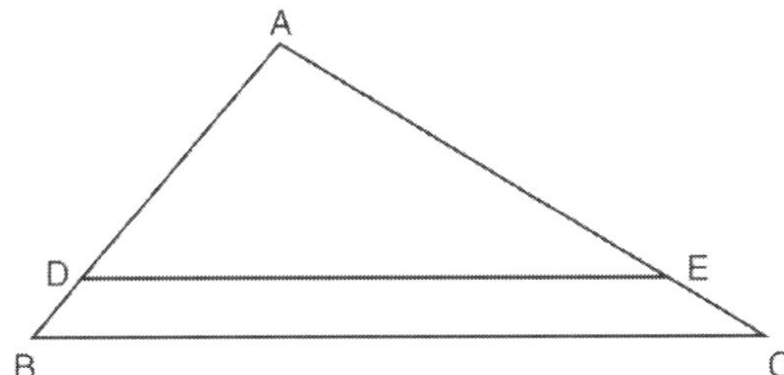

If AB = 10, AD = 8, and AE = 12, what is the length of $\overline{EC}$?

(1) 6
(2) 2
(3) 3
(4) 15

5. Triangle ABC is shown in the diagram below.

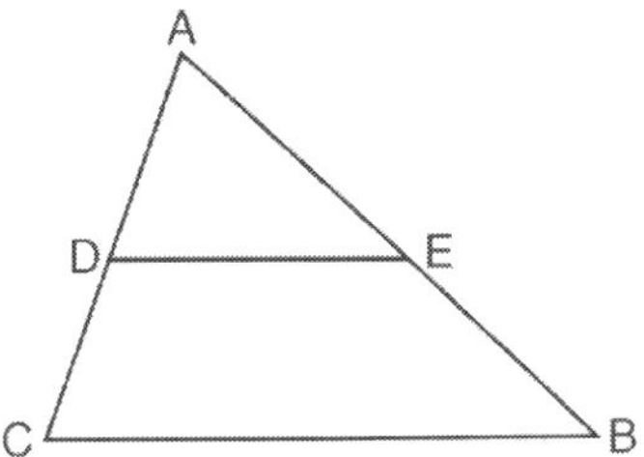

If $\overline{DE}$ joins the midpoints of $\overline{ADC}$ and $\overline{AEB}$, which statement is *not* true?

(1) $DE = \frac{1}{2}CB$
(2) $\overline{DE} \parallel \overline{CB}$
(3) $\frac{AD}{DC} = \frac{DE}{CB}$
(4) $\triangle ABC \sim \triangle AED$

6. In the diagram of $\triangle ABC$ below, $\overline{DE} \parallel \overline{BC}$, $AD = 3$, $DB = 2$, and $DE = 6$.

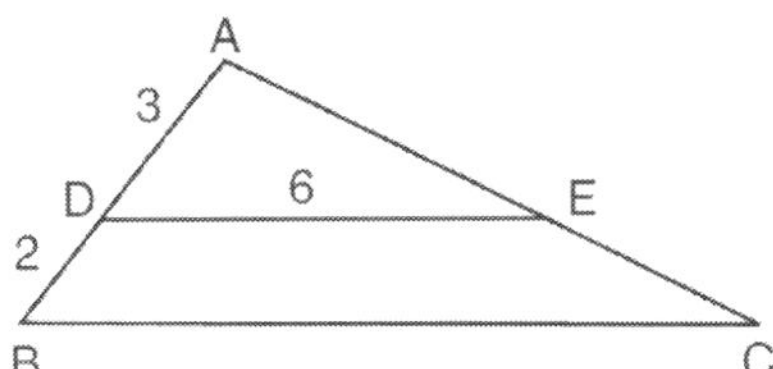

What is the length of $\overline{BC}$?

(1) 12
(2) 10
(3) 8
(4) 4

7. In the diagram of $\triangle ABC$ below, $\overline{DE} \parallel \overline{AB}$.

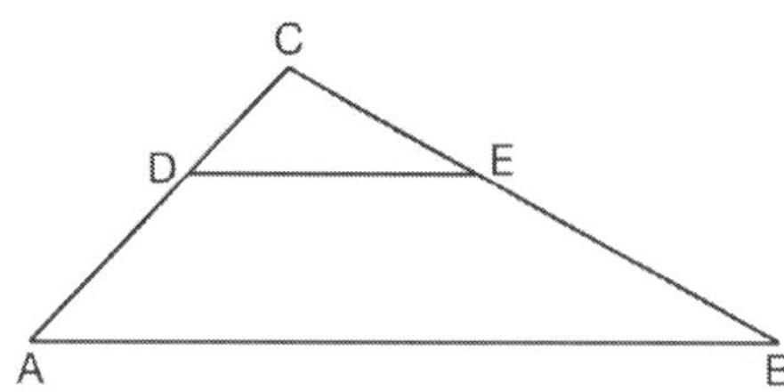

If CD = 4, CA = 10, $CE = x+2$, and $EB = 4x-7$, what is the length of $\overline{CE}$?

(1) 10 (3) 6

(2) 8 (4) 4

8. In $\triangle ABC$ shown below, L is the midpoint of $\overline{BC}$, M is the midpoint of $\overline{AB}$, and N is the midpoint of $\overline{AC}$.

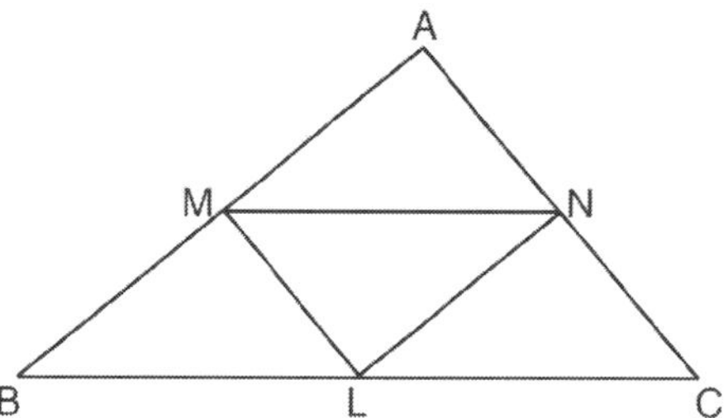

If MN = 8, ML = 5, and NL = 6, the perimeter of trapezoid *BMNC* is

(1) 26 (3) 30

(2) 28 (4) 35

9. In the diagram below of $\triangle ABC$, with $\overline{CDEA}$ and $\overline{BGFA}$, $\overline{EF} \parallel \overline{DG} \parallel \overline{CB}$.

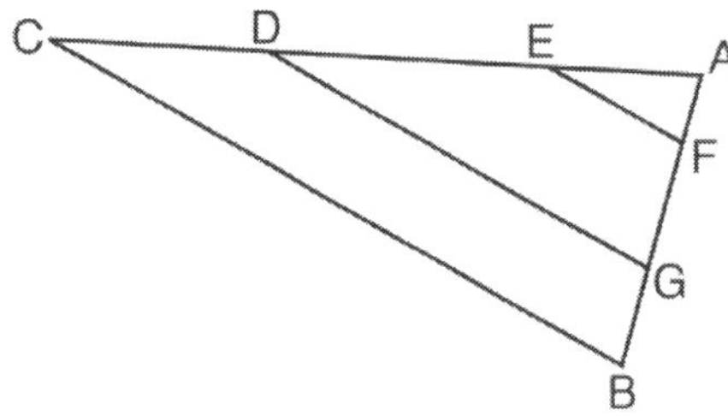

Which statement is *false*?

(1) $\frac{AC}{AD} = \frac{AB}{AG}$ (3) $\frac{AE}{AD} = \frac{EC}{AC}$

(2) $\frac{AE}{AF} = \frac{AC}{AB}$ (4) $\frac{BG}{BA} = \frac{CD}{CA}$

10. CC In the diagram of $\triangle ADC$ below, $\overline{EB} \parallel \overline{DC}$, $AE = 9$, $ED = 5$, and $AB = 9.2$.

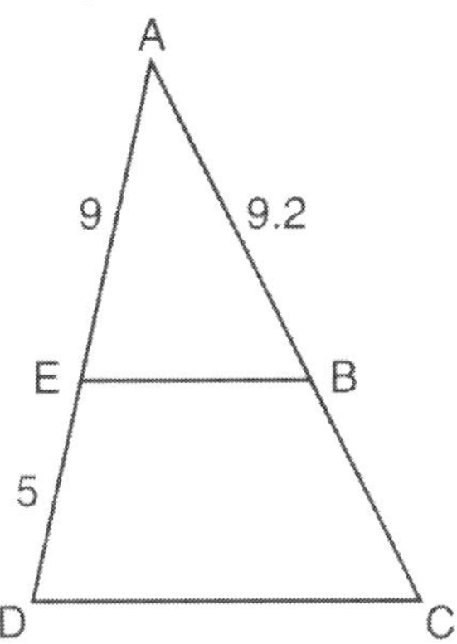

What is the length of $\overline{AC}$, to the *nearest tenth*?

(1) 5.1
(2) 5.2
(3) 14.3
(4) 14.4

11. CC In the diagram below, $\triangle ABC \sim \triangle ADE$.

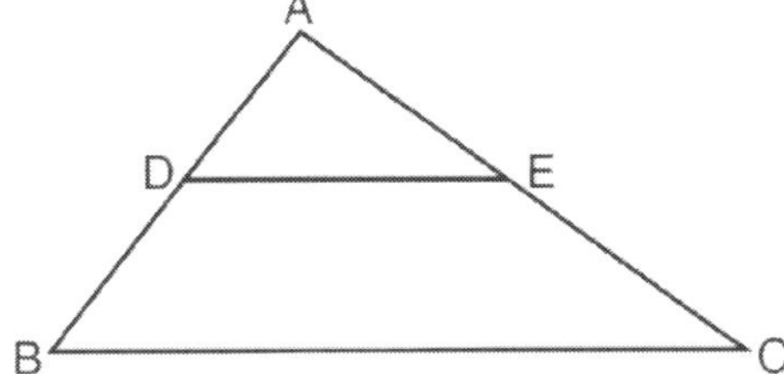

Which measurements are justified by this similarity?

(1) $AD = 3$, $AB = 6$, $AE = 4$, and $AC = 12$
(2) $AD = 5$, $AB = 8$, $AE = 7$, and $AC = 10$
(3) $AD = 3$, $AB = 9$, $AE = 5$, and $AC = 10$
(4) $AD = 2$, $AB = 6$, $AE = 5$, and $AC = 15$

12. CC In the diagram of $\triangle ABC$, points D and E are on $\overline{AB}$ and $\overline{CB}$, respectively, such that $\overline{AC} \parallel \overline{DE}$.

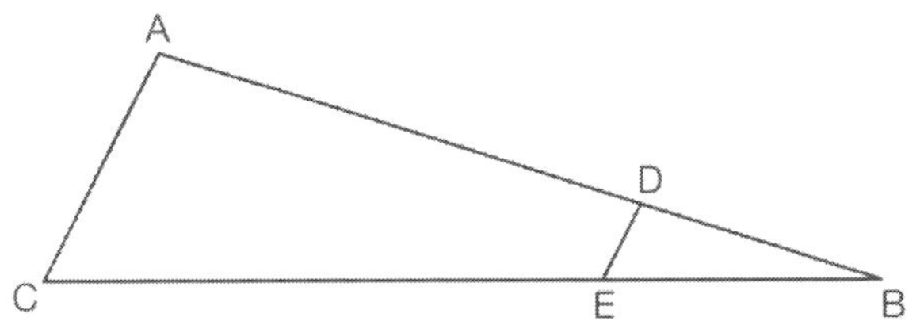

If $AD = 24$, $DB = 12$, and $DE = 4$, what is the length of $\overline{AC}$?

(1) 8
(2) 12
(3) 16
(4) 72

13. (CC) In the diagram below, $\overline{DE}$, $\overline{DF}$, and $\overline{EF}$ are midsegments of $\triangle ABC$.

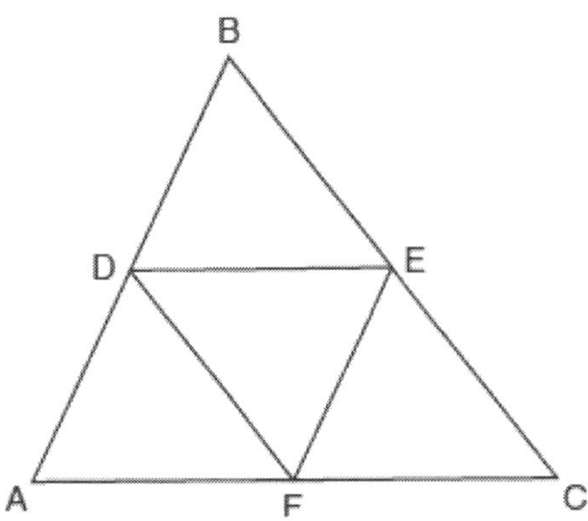

The perimeter of quadrilateral $ADEF$ is equivalent to

(1) $AB + BC + AC$

(2) $\frac{1}{2}AB + \frac{1}{2}AC$

(3) $2AB + 2AC$

(4) $AB + AC$

Constructed Response

14. The accompanying diagram shows a section of the city of Tacoma. High Road, State Street, and Main Street are parallel and 5 miles apart. Ridge Road is perpendicular to the three parallel streets. The distance between the intersection of Ridge Road and State Street and where the railroad tracks cross State Street is 12 miles. What is the distance between the intersection of Ridge Road and Main Street and where the railroad tracks cross Main Street?

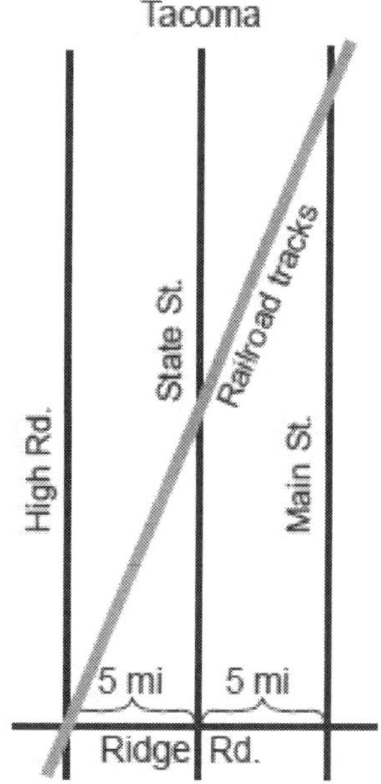

15. In the diagram below of $\triangle ACD$, E is a point on $\overline{AD}$ and B is a point on $\overline{AC}$, such that $\overline{EB} \parallel \overline{DC}$. If $AE = 3$, $ED = 6$, and $DC = 15$, find the length of $\overline{EB}$.

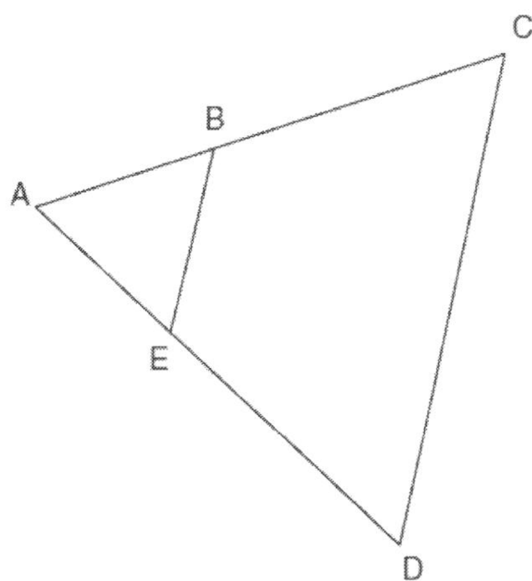

16. In the diagram below of $\triangle ADE$, B is a point on $\overline{AE}$ and C is a point on $\overline{AD}$ such that $\overline{BC} \parallel \overline{ED}$, $AC = x - 3$, $BE = 20$, $AB = 16$, and $AD = 2x + 2$. Find the length of $\overline{AC}$.

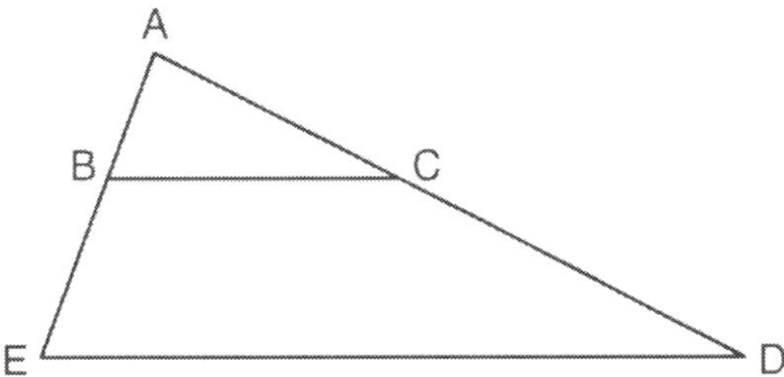

17. In the diagram below of $\triangle ABC$, D is a point on $\overline{AB}$, E is a point on $\overline{BC}$, $\overline{AC} \parallel \overline{DE}$, $CE =$ 25 inches, $AD = 18$ inches, and $DB = 12$ inches. Find, to the *nearest tenth of an inch*, the length of $\overline{EB}$.

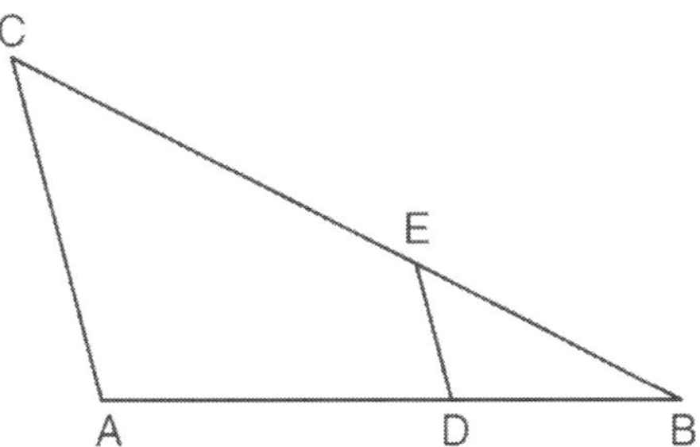

18. In isosceles triangle RST shown below, $\overline{RS} \cong \overline{RT}$, M and N are midpoints of $\overline{RS}$ and $\overline{RT}$, respectively, and $\overline{MN}$ is drawn. If $MN = 3.5$ and the perimeter of $\triangle RST$ is 25, determine and state the length of $\overline{NT}$.

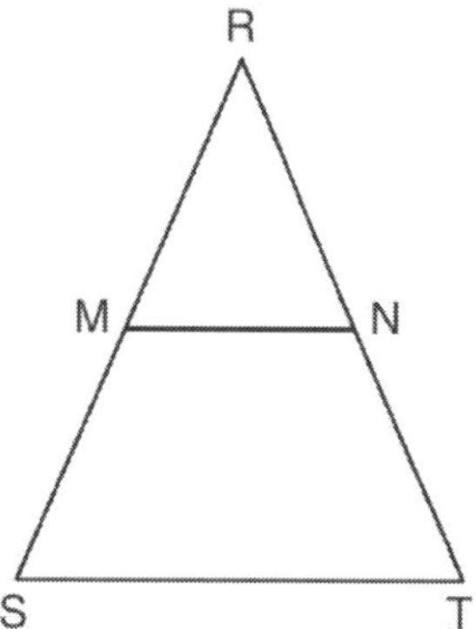

19. (CC) A flagpole casts a shadow 16.60 meters long. Tim stands at a distance of 12.45 meters from the base of the flagpole, such that the end of Tim's shadow meets the end of the flagpole's shadow. If Tim is 1.65 meters tall, determine and state the height of the flagpole to the *nearest tenth of a meter*.

20. (CC) To find the distance across a pond from point B to point C, a surveyor drew the diagram below. The measurements he made are indicated on his diagram.

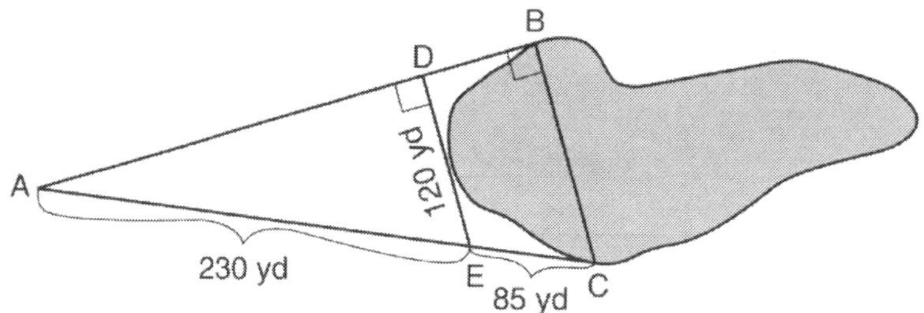

Use the surveyor's information to determine and state the distance from point B to point C, to the *nearest yard*.

21. (CC) In $\triangle CED$ as shown below, points A and B are located on sides $\overline{CE}$ and $\overline{ED}$, respectively. Line segment AB is drawn such that $AE = 3.75$, $AC = 5$, $EB = 4.5$, and $BD = 6$.

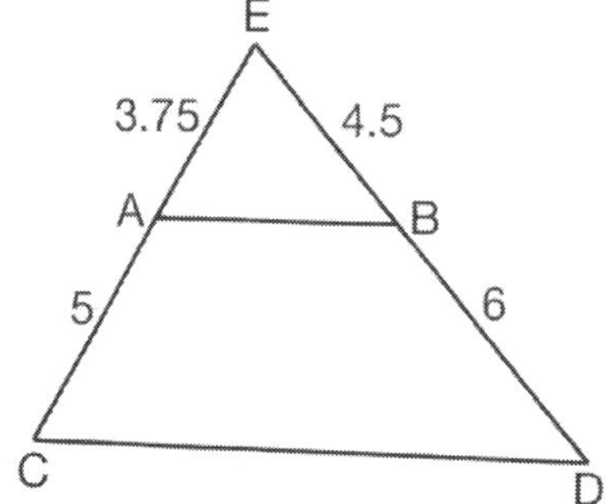

Explain why $\overline{AB}$ is parallel to $\overline{CD}$.

Chapter 5. Right Triangles and Trigonometry

5.1 Pythagorean Theorem

Key Terms and Concepts

The **legs** of a right triangle are the two shortest sides. The legs are perpendicular; they form the right angle. The **hypotenuse** is the longest side.

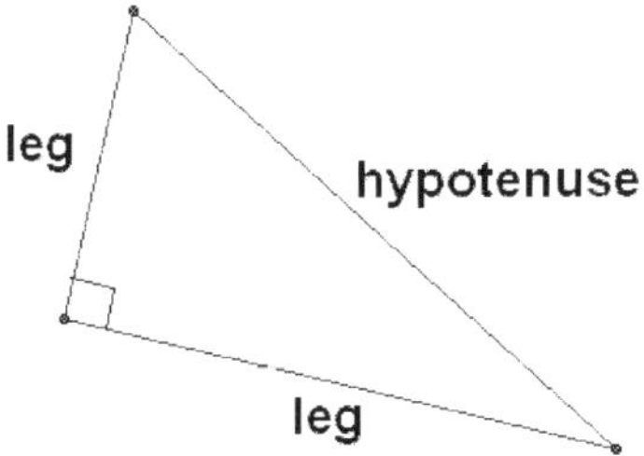

The **Pythagorean Theorem** states a relationship among the sides of any right triangle. If a and b represent the lengths of the legs, and c represents the length of the hypotenuse,

$$a^2 + b^2 = c^2$$

A set of three positive integers that can satisfy this equation is called a **Pythagorean triple.**
Examples: {3,4,5} {5,12,13} {8,15,17}

Informal Proof of the Pythagorean Theorem:

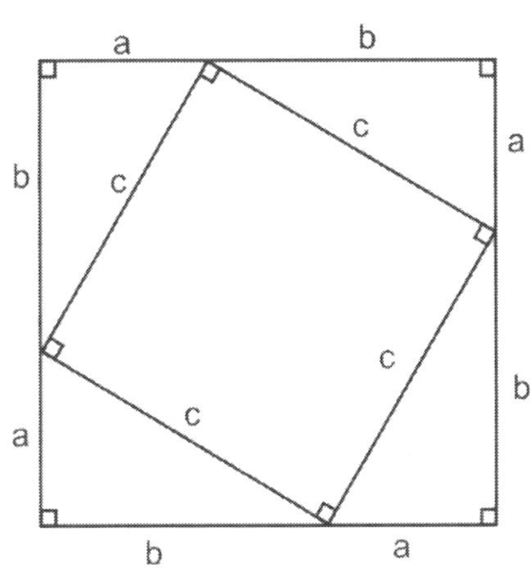

One method of proving this theorem, given a right triangle with legs of a and b and hypotenuse c, is to make four copies of the triangle and arrange them so that their legs form an outer square and their hypotenuses form an inner square, as shown to the left.

The area of the outer square minus the area of the four triangles equals the area of the inner square, so:

$$(a+b)^2 - 4\left(\tfrac{1}{2}ab\right) = c^2$$

Simplifying gives us the Pythagorean Theorem:

$$a^2 + 2ab + b^2 - 2ab = c^2$$
$$a^2 + b^2 = c^2$$

Model Problem:

A wall is supported by a brace 10 feet long, as shown in the diagram below. If one end of the brace is placed 6 feet from the base of the wall, how many feet up the wall does the brace reach?

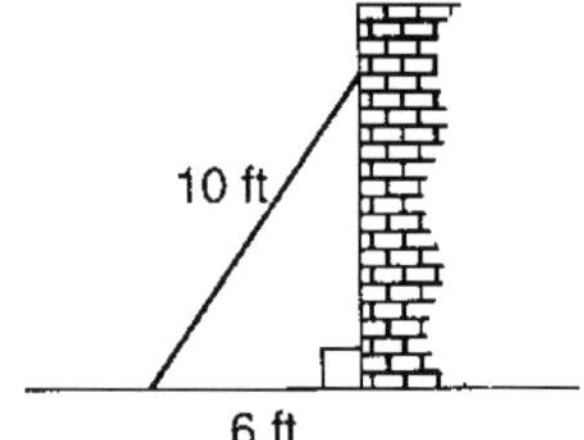

Solution:

(A) $a^2 + b^2 = c^2$

(B) $6^2 + b^2 = 10^2$

(C) $36 + b^2 = 100$

$b^2 = 64$

$b = \sqrt{64} = 8\ ft$

Explanation of steps:

(A) Given two sides of a right triangle, use the Pythagorean Theorem to find the third side.

(B) Substitute given legs as a and b (in either order), and substitute the hypotenuse, if given, as c. *[leg a = 6 and hypotenuse c = 10]*

(C) Solve for the remaining variable, and simplify the radical if possible. When taking the square root of both sides, ignore the negative square root since the length of a side must be positive.

Practice Problems

1. If the lengths of the legs of a right triangle are 5 and 7, what is the length of the hypotenuse?	2. The hypotenuse of a right triangle is 26 centimeters and one leg is 24 centimeters. Find the number of centimeters in the second leg.

3. A 10-foot ladder is placed against the side of a building as shown in Figure 1 below. The bottom of the ladder is 8 feet from the base of the building. In order to increase the reach of the ladder against the building, it is moved 4 feet closer to the base of the building as shown in Figure 2. To the *nearest tenth of a foot*, how much further up the building does the ladder now reach?

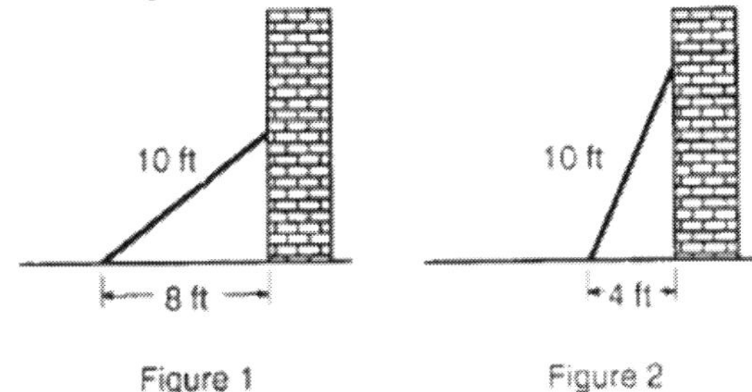

4. A rectangular garden is going to be planted in a person's rectangular backyard, as shown in the accompanying diagram. Some dimensions of the backyard and the width of the garden are given. Find the area of the garden to the *nearest square foot.*

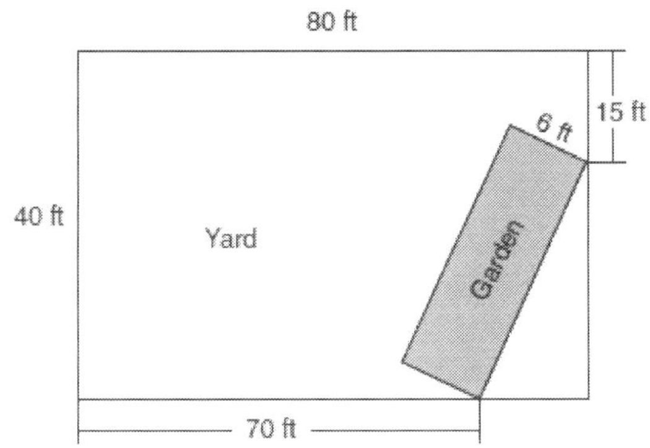

5. In $\triangle DEF$, $\overline{DG}$ bisects $\angle D$, $m\angle E = 90°$, $DE = 6$ and $EF = 8$. Find DG.

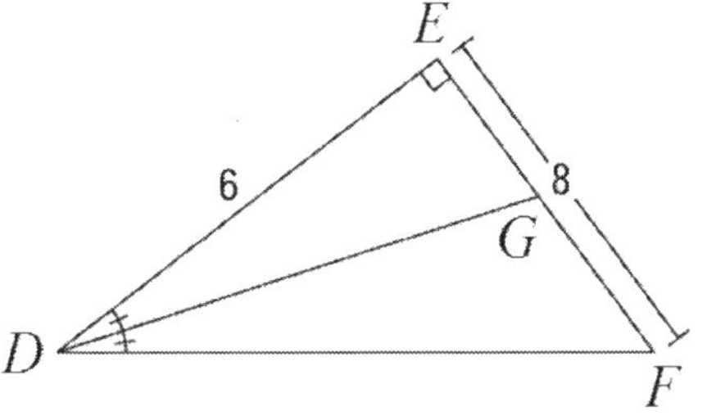

REGENTS QUESTIONS

Multiple Choice

1. Tanya runs diagonally across a rectangular field that has a length of 40 yards and a width of 30 yards, as shown in the diagram below.

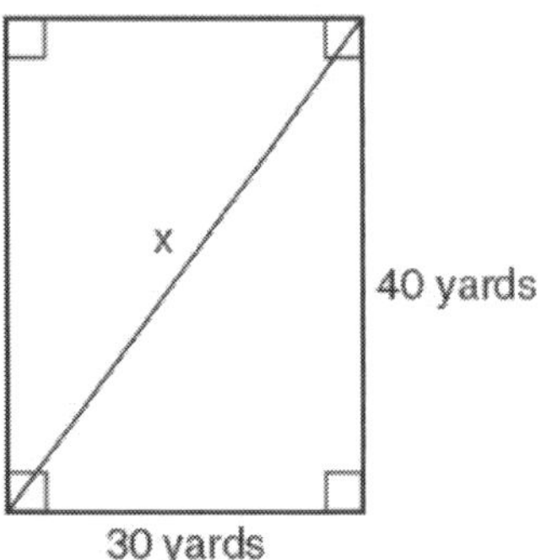

What is the length of the diagonal, in yards, that Tanya runs?

(1) 50 (3) 70
(2) 60 (4) 80

2. Don placed a ladder against the side of his house as shown in the diagram below.

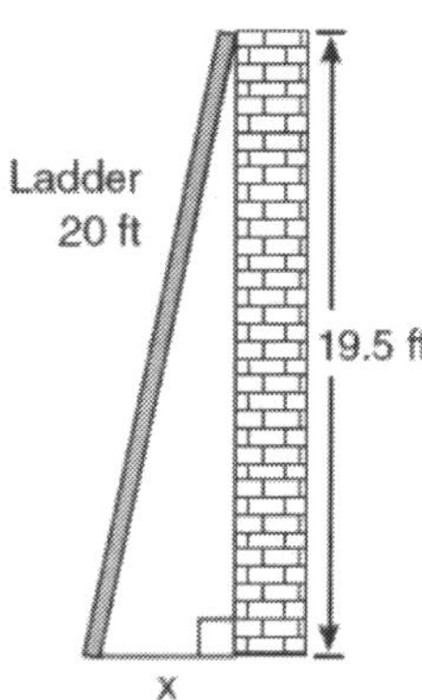

Which equation could be used to find the distance, x, from the foot of the ladder to the base of the house?

(1) $x = 20 - 19.5$ (3) $x = \sqrt{20^2 - 19.5^2}$
(2) $x = 20^2 - 19.5^2$ (4) $x = \sqrt{20^2 + 19.5^2}$

3. The length of the hypotenuse of a right triangle is 34 inches and the length of one of its legs is 16 inches. What is the length, in inches, of the other leg of this right triangle?

(1) 16 (3) 25
(2) 18 (4) 30

4. What is the value of x, in inches, in the right triangle below?

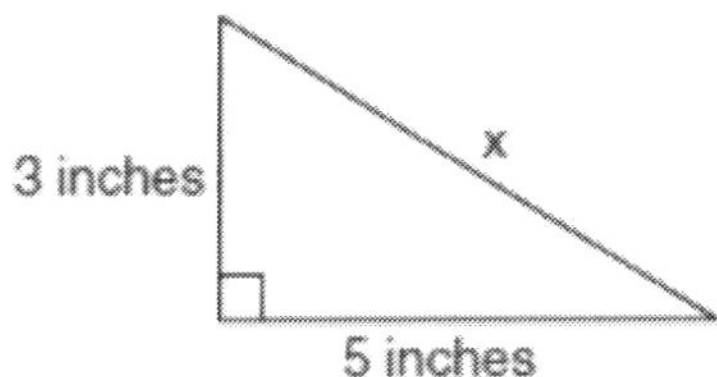

(1) $\sqrt{15}$
(2) 8
(3) $\sqrt{34}$
(4) 4

5. Nancy's rectangular garden is represented in the diagram below.

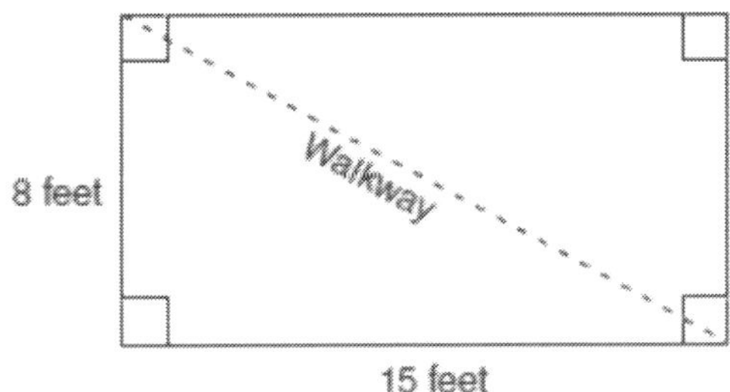

If a diagonal walkway crosses her garden, what is its length, in feet?

(1) 17
(2) 22
(3) $\sqrt{161}$
(4) $\sqrt{529}$

6. The diagram below shows a pennant in the shape of an isosceles triangle. The equal sides each measure 13, the altitude is $x+7$, and the base is $2x$.

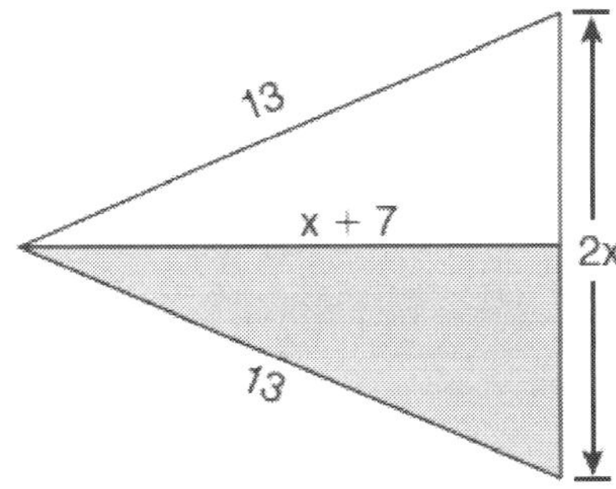

What is the length of the base?

(1) 5
(2) 10
(3) 12
(4) 24

7. The end of a dog's leash is attached to the top of a 5-foot-tall fence post, as shown in the diagram below. The dog is 7 feet away from the base of the fence post.

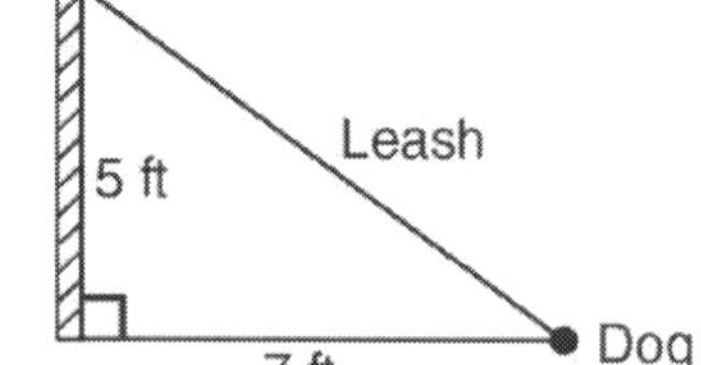

How long is the leash, to the *nearest tenth of a foot*?

(1) 4.9
(2) 8.6
(3) 9.0
(4) 12.0

8. The rectangle shown below has a diagonal of 18.4 cm and a width of 7 cm.

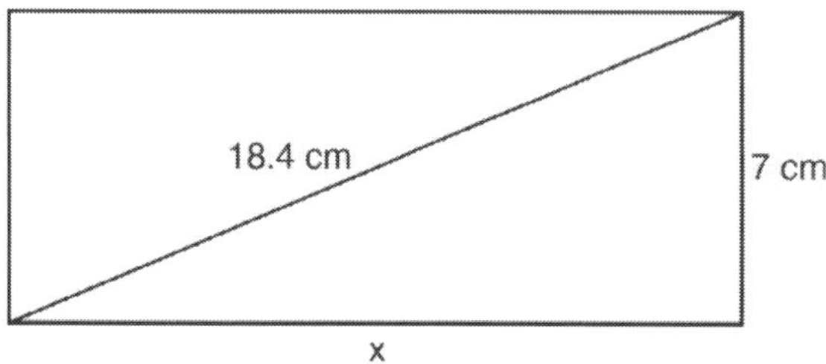

To the *nearest centimeter*, what is the length, x, of the rectangle?

(1) 11
(2) 17
(3) 20
(4) 25

9. The legs of an isosceles right triangle each measure 10 inches. What is the length of the hypotenuse of this triangle, to the *nearest tenth of an inch*?

(1) 6.3
(2) 7.1
(3) 14.1
(4) 17.1

10. As shown in the diagram below, a kite needs a vertical and a horizontal support bar attached at opposite corners. The upper edges of the kite are 7 inches, the side edges are x inches, and the vertical support bar is $(x + 1)$ inches.

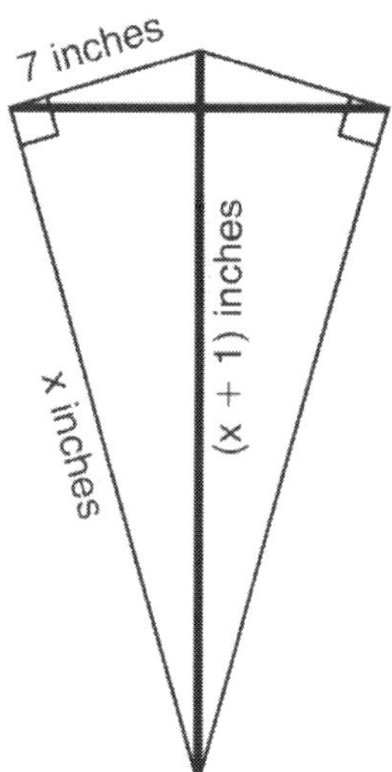

What is the measure, in inches, of the vertical support bar?

(1) 23
(2) 24
(3) 25
(4) 26

11. Campsite A and campsite B are located directly opposite each other on the shores of Lake Omega, as shown in the diagram below. The two campsites form a right triangle with Sam's position, S. The distance from campsite B to Sam's position is 1,300 yards, and campsite A is 1,700 yards from his position.

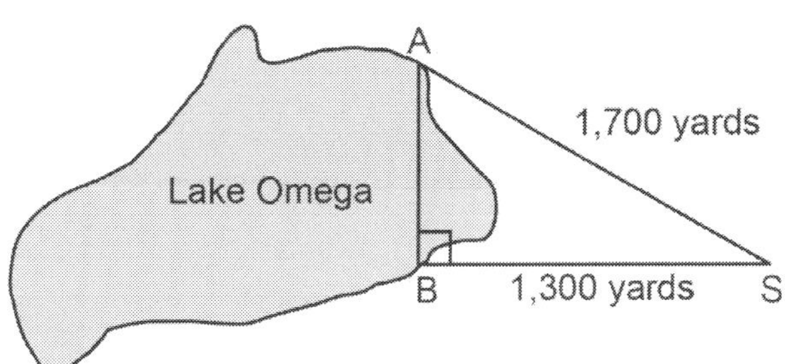

What is the distance from campsite A to campsite B, to the *nearest yard*?

(1) 1,095	(3) 2,140
(2) 1,096	(4) 2,141

12. The length of one side of a square is 13 feet. What is the length, to the *nearest foot*, of a diagonal of the square?

(1) 13	(3) 19
(2) 18	(4) 26

13. In triangle RST, angle R is a right angle. If $TR = 6$ and $TS = 8$, what is the length of $\overline{RS}$?

(1) 10	(3) $2\sqrt{7}$
(2) 2	(4) $7\sqrt{2}$

14. In right triangle ABC, $m\angle C = 90$, $AC = 7$, and $AB = 13$. What is the length of $\overline{BC}$?

(1) 6	(3) $\sqrt{120}$
(2) 20	(4) $\sqrt{218}$

15. (CC) Linda is designing a circular piece of stained glass with a diameter of 7 inches. She is going to sketch a square inside the circular region. To the *nearest tenth of an inch*, the largest possible length of a side of the square is

(1) 3.5	(3) 5.0
(2) 4.9	(4) 6.9

16. (CC) An equilateral triangle has sides of length 20. To the nearest tenth, what is the height of the equilateral triangle?

(1) 10.0	(3) 17.3
(2) 11.5	(4) 23.1

Constructed Response

17. An 18-foot ladder leans against the wall of a building. The base of the ladder is 9 feet from the building on level ground. How many feet up the wall, to the *nearest tenth of a foot*, is the top of the ladder?

18. The "Little People" day care center has a rectangular, fenced play area behind its building. The play area is 30 meters long and 20 meters wide. Find, to the *nearest meter*, the length of a pathway that runs along the diagonal of the play area.

19. (CC) The aspect ratio (the ratio of screen width to height) of a rectangular flat-screen television is $16:9$. The length of the diagonal of the screen is the television's screen size. Determine and state, to the *nearest inch*, the screen size (diagonal) of this flat-screen television with a screen height of 20.6 inches.

5.2 Congruent Right Triangles

Key Terms and Concepts

In addition to the methods we've already seen for proving triangles congruent, we may also prove *two right triangles* congruent by using the **hypotenuse-leg** method (abbreviated HL):

HL: corresponding hypotenuses and a pair of corresponding legs are congruent

Example: $\triangle ABC \cong \triangle XYZ$

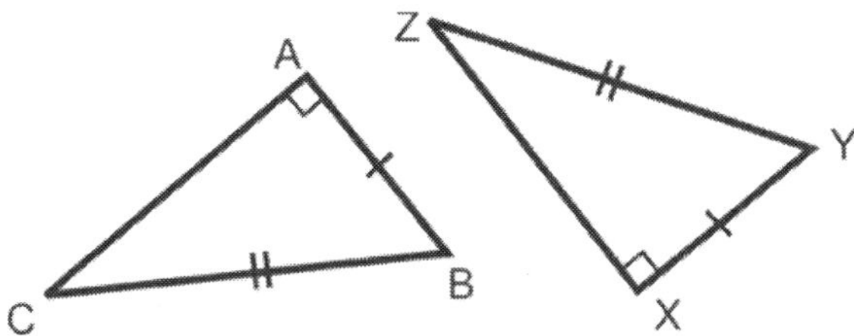

This leads us to the fact that, if *any two pairs* of corresponding sides of right triangles are congruent, then the triangles are congruent. We already know that if two pairs of corresponding legs are congruent, then the right triangles are congruent by SAS. The angles shared by the legs are congruent because they are both right angles.

Example: $\triangle JKL \cong \triangle OMN$ by SAS.

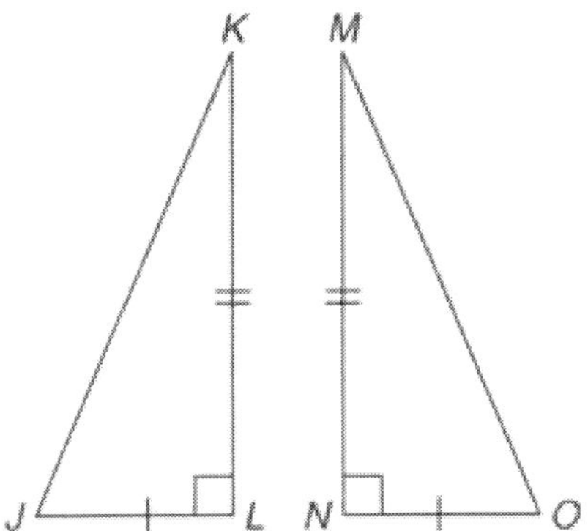

Important Proofs

Proof of the Hypotenuse-Leg Theorem

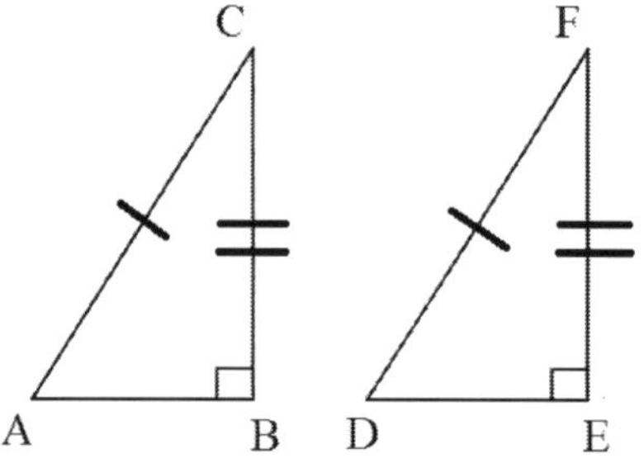

Given: Right triangles ABC and DEF, hypotenuses $\overline{AC} \cong \overline{DF}$, and legs $\overline{BC} \cong \overline{EF}$

Prove: $\triangle ABC \cong \triangle DEF$

To prove these triangles congruent, extend $\overleftrightarrow{DE}$ to point G such that $\overline{AB} \cong \overline{EG}$ and draw $\overline{FG}$. Below, we prove $\triangle ABC \cong \triangle GEF$ (by SAS), then we prove $\triangle GEF \cong \triangle DEF$ (by AAS), which gives us $\triangle ABC \cong \triangle DEF$ by the Transitive Property.

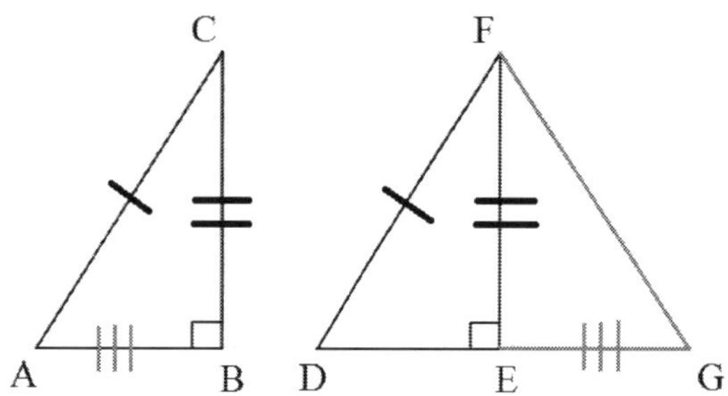

Statements	*Reasons*
$\angle B$ and $\angle E$ are right angles, $\overline{AC} \cong \overline{DF}$, $\overline{BC} \cong \overline{EF}$ (S)	Given
$\overline{FE} \perp \overline{DG}$	Definition of perpendicular
$\angle FEG$ is a right angle	Definition of perpendicular
$\angle ABC \cong \angle GEF$ (A)	All right angles are congruent
$\overline{AB} \cong \overline{EG}$ (S)	A line may be extended indefinitely or by any length
$\triangle ABC \cong \triangle GEF$	SAS
$\overline{AC} \cong \overline{FG}$	CPCTC
$\overline{DF} \cong \overline{FG}$ (S)	Transitive Property
$\triangle DFG$ is isosceles	Definition of isosceles
$\angle D \cong \angle G$ (A)	Base angles of isosceles triangles are congruent
$\angle DEF \cong \angle GEF$ (A)	All right angles are congruent
$\triangle GEF \cong \triangle DEF$	AAS
$\triangle ABC \cong \triangle DEF$	Transitive Property

Model Problem:

Given: Triangles ABD and ACD with right angles at B and C, $\overline{AB} \cong \overline{CD}$.

Prove: $\overline{AC} \cong \overline{BD}$.

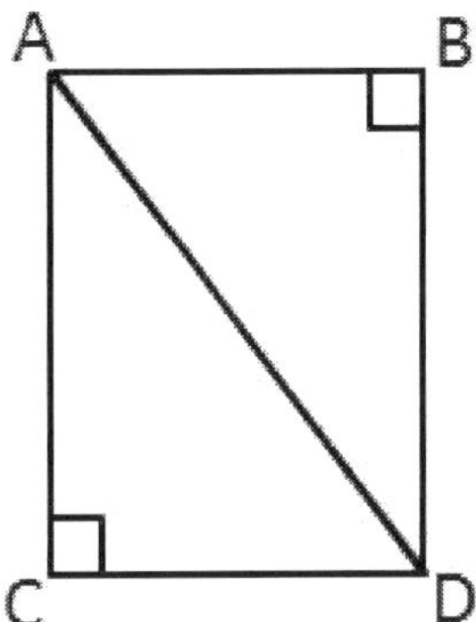

Solution:

1. $\angle B$ and $\angle C$ are right angles (Given)
2. $\overline{AD} \cong \overline{AD}$ (Reflexive Property)
3. $\overline{AB} \cong \overline{CD}$ (Given)
4. $\triangle ABD \cong \triangle DCA$ (HL)
5. $\overline{AC} \cong \overline{BD}$ (CPCTC)

Practice Problems

1. For these triangles, select the triangle congruence statement and the congruence method that supports it.

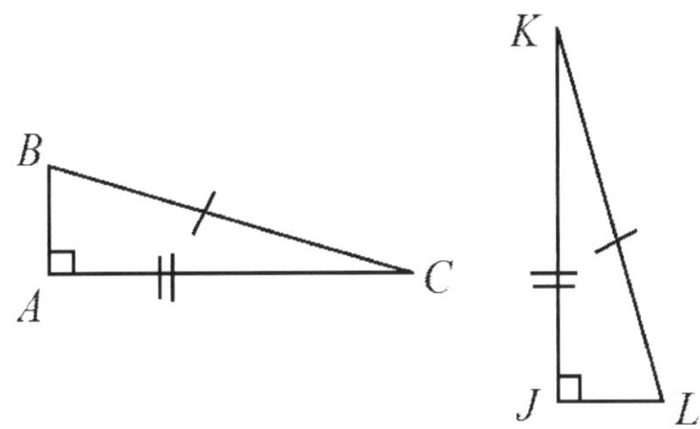

(1) $\triangle ABC \cong \triangle JLK$, HL
(2) $\triangle ABC \cong \triangle JLK$, SAS
(3) $\triangle ABC \cong \triangle JKL$, HL
(4) $\triangle ABC \cong \triangle JKL$, SAS

2. Given: Right triangles MAT and HTM, $\overline{MT} \cong \overline{AH}$.
Prove: $\angle M \cong \angle H$

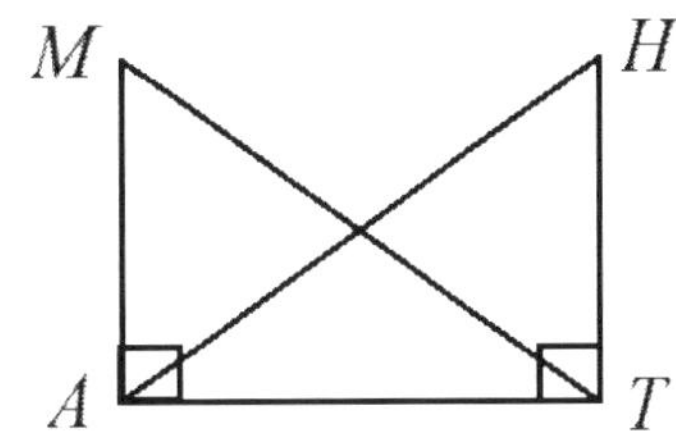

REGENTS QUESTIONS

Multiple Choice

1. In the accompanying diagram, $\overline{CA} \perp \overline{AB}$, $\overline{ED} \perp \overline{DF}$, $\overline{ED} \parallel \overline{AB}$, $\overline{CE} \cong \overline{BF}$, $\overline{AB} \cong \overline{ED}$, and $m\angle CAB = m\angle FDE = 90$.

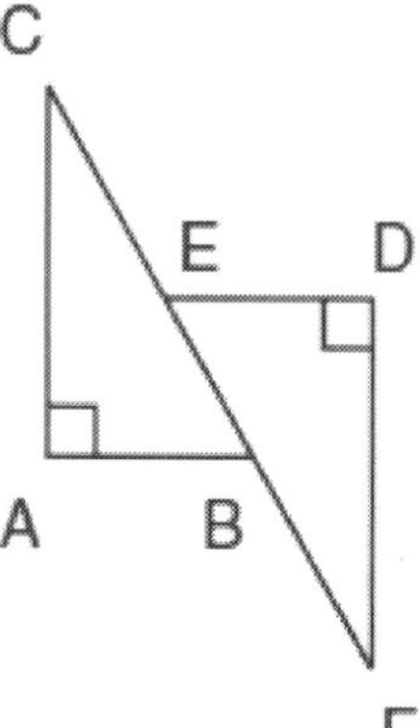

Which statement would *not* be used to prove $\triangle ABC \cong \triangle DEF$?

(1) SSS ≅ SSS (3) AAS ≅ AAS
(2) SAS ≅ SAS (4) HL ≅ HL

2. In the diagram below, four pairs of triangles are shown. Congruent corresponding parts are labeled in each pair.

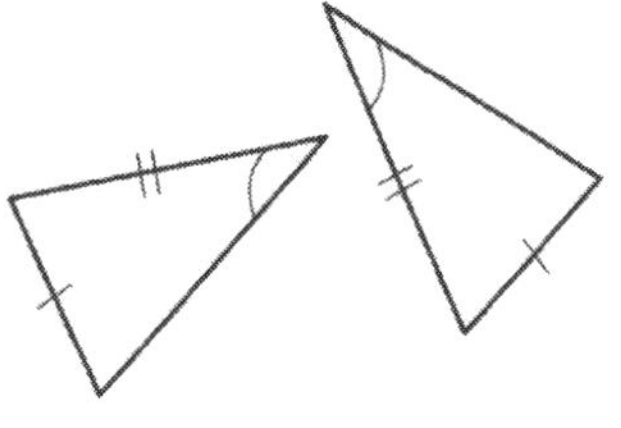

A

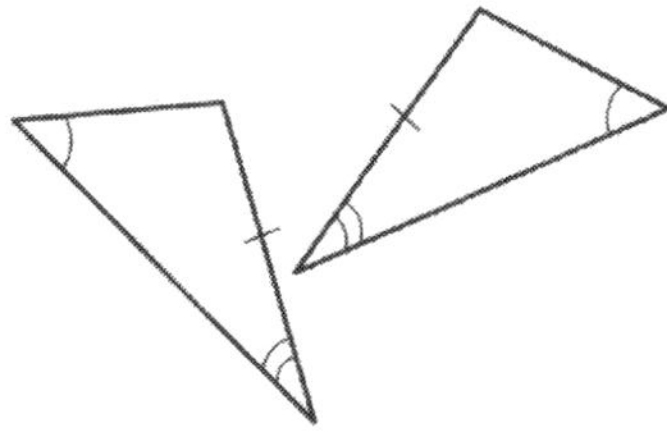

C

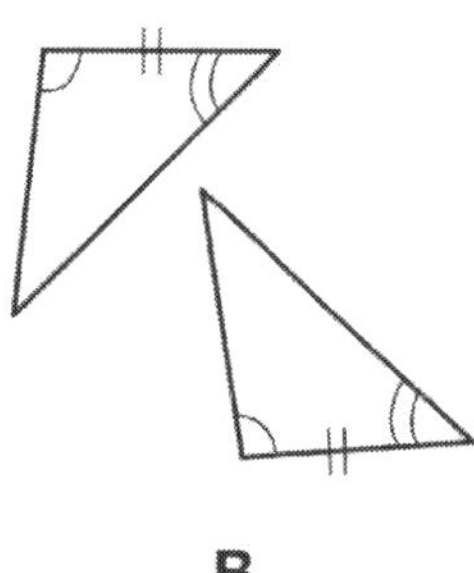

B

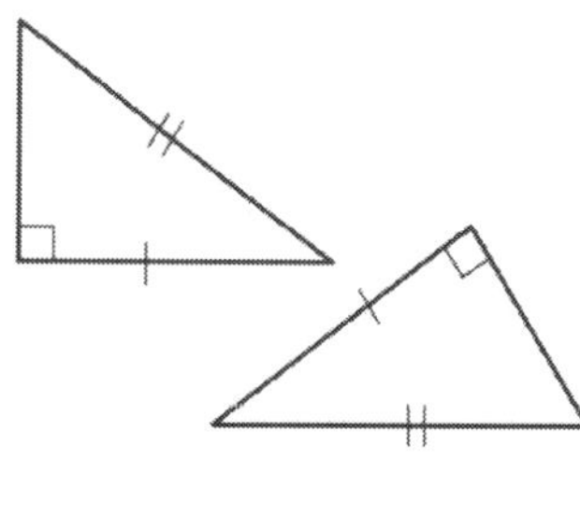

D

Using only the information given in the diagrams, which pair of triangles can *not* be proven congruent?

(1) A (3) C
(2) B (4) D

3. (CC) Two right triangles must be congruent if
(1) an acute angle in each triangle is congruent
(2) the lengths of the hypotenuses are equal
(3) the corresponding legs are congruent
(4) the areas are equal

5.3 Trigonometric Ratios

Key Terms and Concepts

In a right triangle, the ratio of the lengths of two sides can be expressed as the **sine**, **cosine**, or **tangent** (abbreviated **sin**, **cos**, or **tan**) of one of the acute angles of the triangle.

In relation to one of the acute angles, the **adjacent leg** (*adj*) is the leg that has the vertex of the angle as one of its endpoints, while the **opposite leg** (*opp*) is the leg that does not. In other words, the *adjacent* leg is *next to* the angle and the *opposite* leg is *across from* the angle. The **hypotenuse** (*hyp*) is always the longest side, across from the right angle.

For example: In the right triangle below, when referring to $\angle A$, side b is the adjacent leg and side a is the opposite leg. But when referring to $\angle B$, side a is the adjacent leg and side b is the opposite leg. In either case, side c is the hypotenuse.

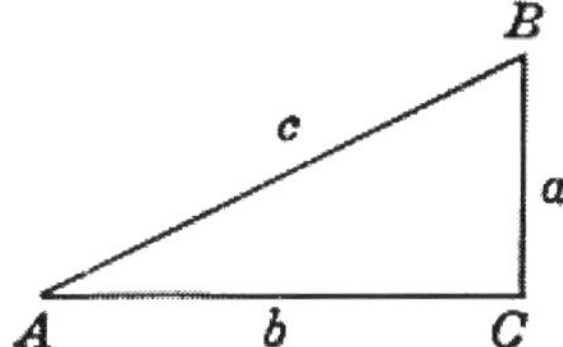

The **trigonometric ratios** are defined as follows, where A is one of the acute angles:

$$\sin A = \frac{opposite}{hypotenuse} \qquad \cos A = \frac{adjacent}{hypotenuse} \qquad \tan A = \frac{opposite}{adjacent}$$

To remember these ratios, the acronym **SOH-CAH-TOA** is often used as a mnemonic.

Model Problem:

For the right triangle shown below, find, to the *nearest thousandth*, sin A and tan B.

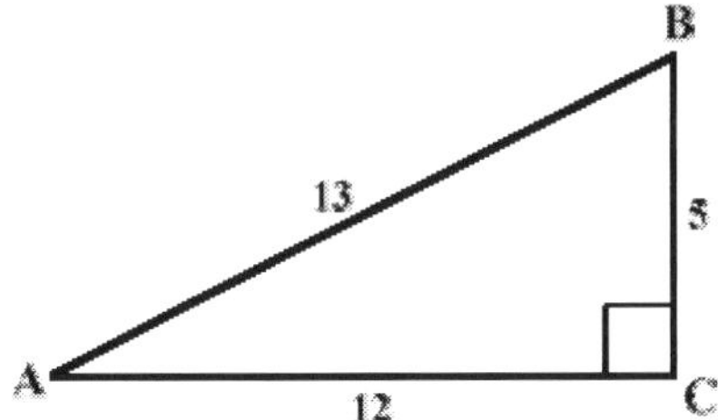

Solution:

$$\sin A \overset{(A)}{=} \frac{opp}{hyp} \overset{(B)}{=} \frac{5}{13} \overset{(C)}{\approx} 0.385 \qquad \tan B \overset{(A)}{=} \frac{opp}{adj} \overset{(B)}{=} \frac{12}{5} \overset{(C)}{=} 2.4$$

Explanation of steps:

(A) Write the correct trigonometric ratio.

(B) Substitute the lengths of the appropriate sides *[5 is opposite to A, but it is adjacent to B; 12 is opposite to B; 13 is the hypotenuse regardless of the angle].*

(C) Divide, and round if necessary.

Practice Problems

1. For the right triangle below, find sin A, cos A, and tan A. Leave in fraction form.

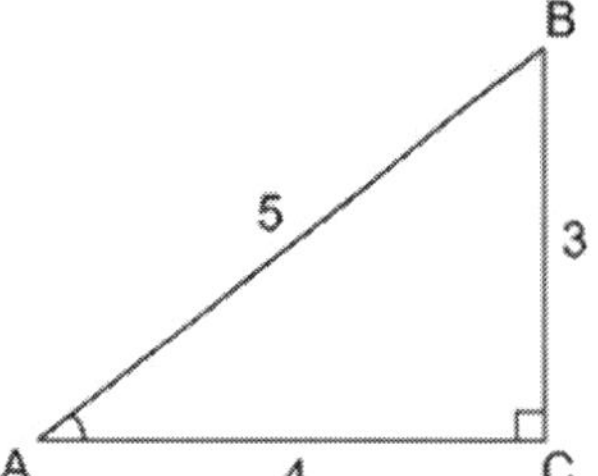

2. For the right triangle below, find sin B, cos B, and tan B. Leave in fraction form.

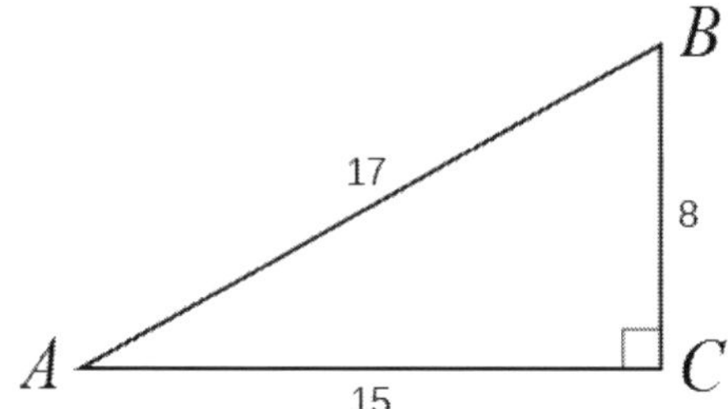

3. For the right triangle below, state two trigonometric functions that are equal to the ratio, $\dfrac{40}{41}$.

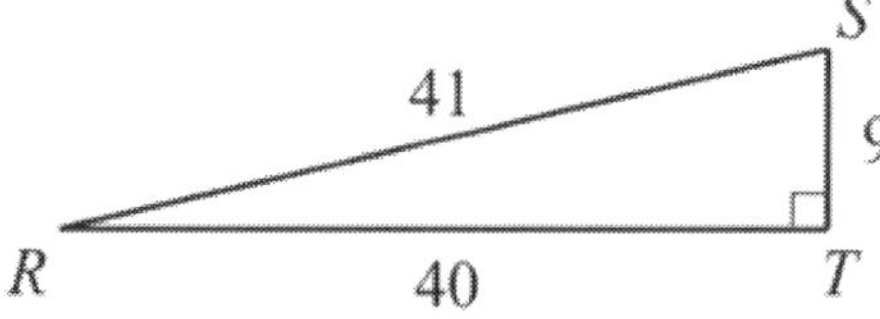

4. In $\triangle LMN$, $\angle M = 90°$, $LM = 7$, $MN = 24$, and $LN = 25$. What is the sine of $\angle N$ written as a ratio?

5. In $\triangle PQR$, $\angle PQR = 90°$, $PQ = 20$, $QR = 21$, and $PR = 29$. What is the tangent of $\angle QPR$ written as a ratio?

6. In $\triangle ABC$, shown below, $\angle C = 90°$. What is the sine of $\angle B$ written as a ratio in lowest terms?

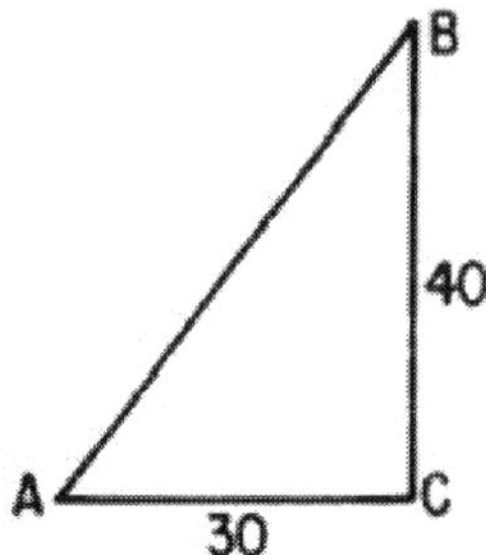

REGENTS QUESTIONS

Multiple Choice

1. Which equation shows a correct trigonometric ratio for angle A in the right triangle below?

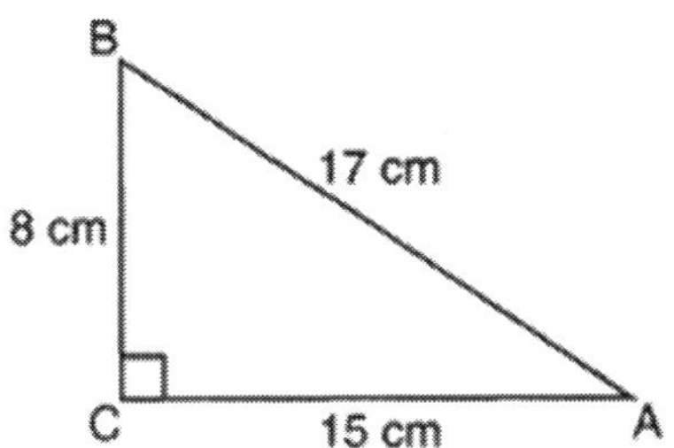

(1) $\sin A = \frac{15}{17}$ (3) $\cos A = \frac{15}{17}$

(2) $\tan A = \frac{8}{17}$ (4) $\tan A = \frac{15}{8}$

2. In $\triangle ABC$, the measure of $\angle B = 90°$, $AC = 50$, $AB = 48$, and $BC = 14$. Which ratio represents the tangent of $\angle A$?

(1) $\frac{14}{50}$ (3) $\frac{48}{50}$

(2) $\frac{14}{48}$ (4) $\frac{48}{14}$

3. Right triangle ABC has legs of 8 and 15 and a hypotenuse of 17, as shown in the diagram below.

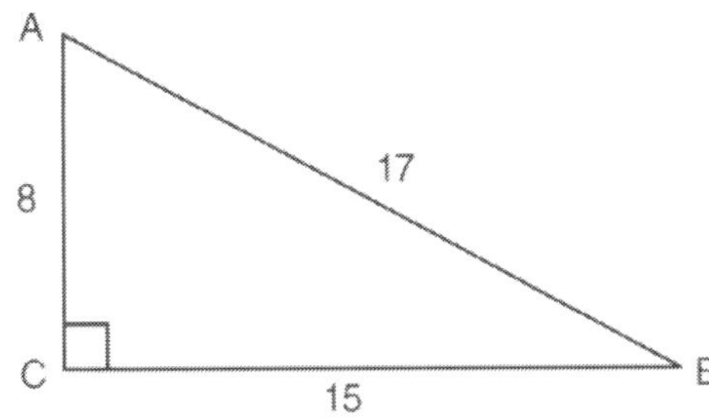

The value of the tangent of $\angle B$, to the *nearest ten-thousandth*, is

(1) 0.4706 (3) 0.8824

(2) 0.5333 (4) 1.8750

4. Which ratio represents $\sin x$ in the right triangle shown below?

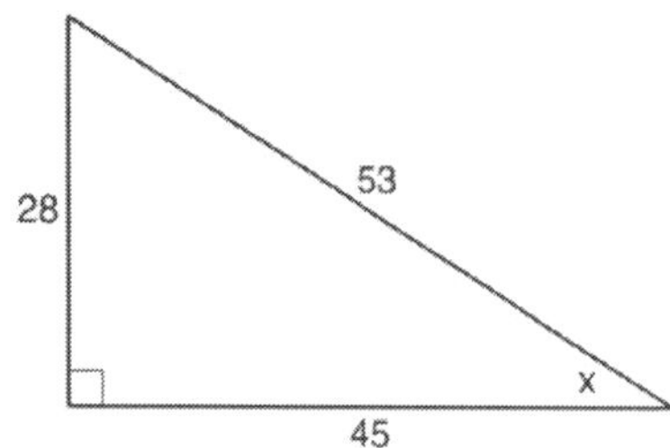

(1) $\frac{28}{53}$ (3) $\frac{45}{53}$

(2) $\frac{28}{45}$ (4) $\frac{53}{28}$

5. The diagram below shows right triangle ABC. Which ratio represents the tangent of $\angle ABC$?

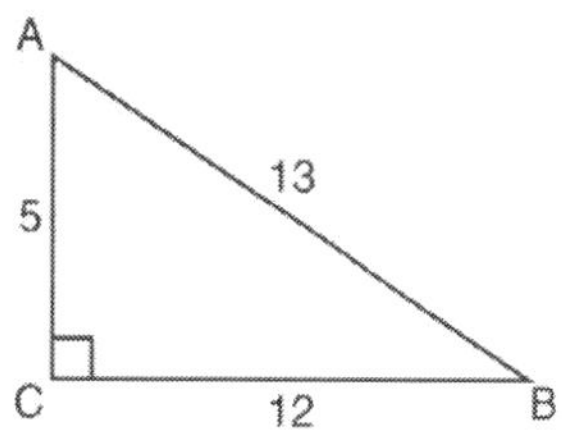

(1) $\frac{5}{13}$ (3) $\frac{12}{13}$

(2) $\frac{5}{12}$ (4) $\frac{12}{5}$

6. The diagram below shows right triangle LMP.

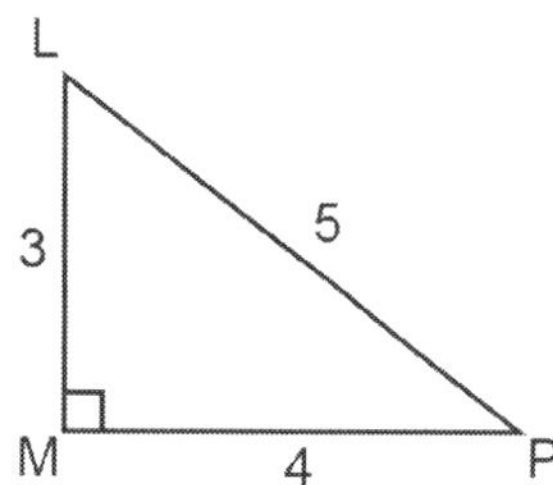

Which ratio represents the tangent of $\angle PLM$?

(1) $\frac{3}{4}$ (3) $\frac{4}{3}$

(2) $\frac{3}{5}$ (4) $\frac{5}{4}$

7. In $\triangle ABC$, $m\angle C = 90$. If $AB = 5$ and $AC = 4$, which statement is *not* true?

(1) $\cos A = \frac{4}{5}$

(2) $\tan A = \frac{3}{4}$

(3) $\sin B = \frac{4}{5}$

(4) $\tan B = \frac{5}{3}$

8. In right triangle ABC shown below, what is the value of $\cos A$?

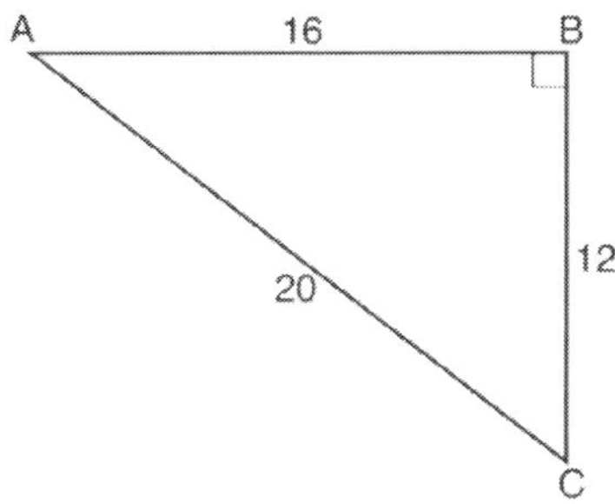

(1) $\frac{12}{20}$

(2) $\frac{16}{20}$

(3) $\frac{20}{12}$

(4) $\frac{20}{16}$

9. Which equation could be used to find the measure of angle D in the right triangle shown in the diagram below?

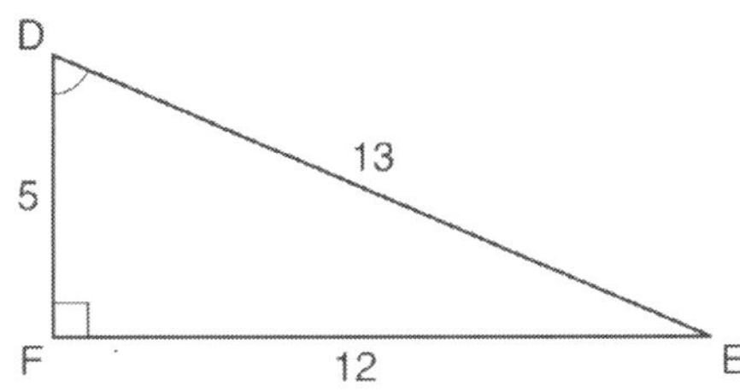

(1) $\cos D = \frac{12}{13}$

(2) $\cos D = \frac{13}{12}$

(3) $\sin D = \frac{5}{13}$

(4) $\sin D = \frac{12}{13}$

10. Which ratio represents the cosine of angle A in the right triangle below?

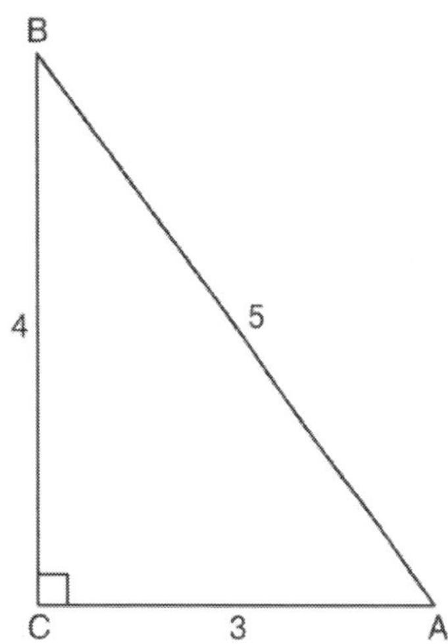

(1) $\frac{3}{5}$ (3) $\frac{4}{5}$

(2) $\frac{5}{3}$ (4) $\frac{4}{3}$

11. (CC) In the diagram below, $\triangle ERM \sim \triangle JTM$.

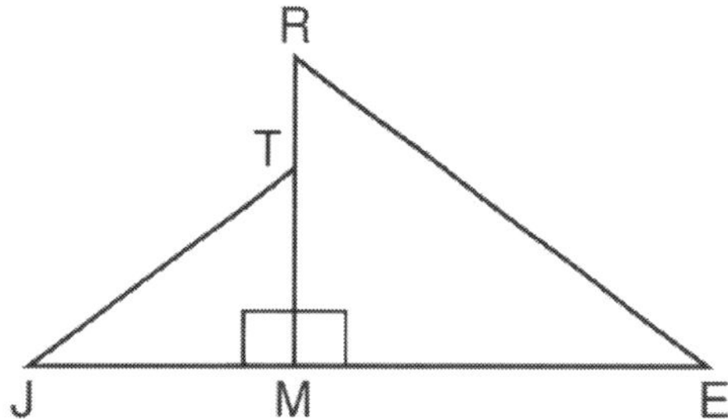

Which statement is always true?

(1) $\cos J = \frac{RM}{RE}$ (3) $\tan T = \frac{RM}{EM}$

(2) $\cos R = \frac{JM}{JT}$ (4) $\tan E = \frac{TM}{JM}$

12. (CC) In the diagram of right triangle ADE below, $\overline{BC} \parallel \overline{DE}$.

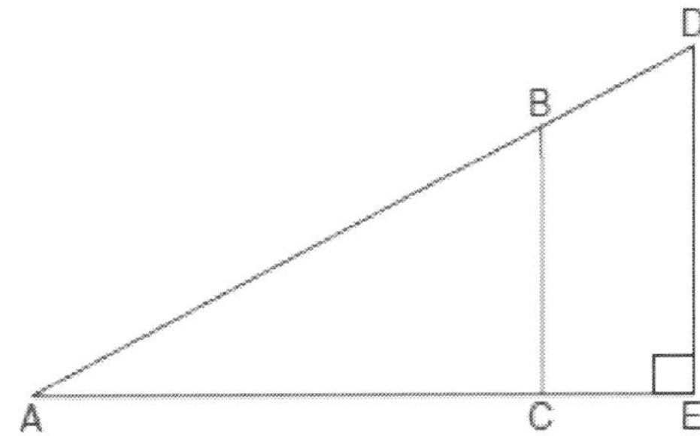

Which ratio is always equivalent to the sine of $\angle A$?

(1) $\dfrac{AD}{DE}$

(2) $\dfrac{AE}{AD}$

(3) $\dfrac{BC}{AB}$

(4) $\dfrac{AB}{AC}$

5.4 Use Trigonometry to Find a Side

Key Terms and Concepts

When given one side of a right triangle and an acute angle, it is possible to find the measure of another side by using an **appropriate trigonometric ratio** with a variable for the needed side.

You will need to use the [SIN], [COS] or [TAN] key on your calculator.

Note: Press [MODE] *on your calculator to check that it is set for* [Degree] *and not* [Radian].

Steps for using trigonometry to find a side:

1. Select the appropriate trig ratio for the given and needed measures.
2. Substitute the given measures and the variable for the unknown.
3. Solve for (isolate) the variable.
4. Use a calculator to find the answer.

Example: If we can determine that $\sin 30° = \frac{x}{10}$, where x is the length of a side, then $x = 10\sin 30°$. On the calculator, 10 [SIN] (30) gives us 5, so $x = 5$.

Some trigonometric models will refer to an angle of elevation or depression. These terms are used to describe an angle where the vertex is the observer's eye looking at an object. However, the model will not necessarily have an observer with a literal "eye." It may simply be a **vantage point** from which a line may be drawn to the object.

The **line of sight** is the line from the vantage point (vertex) to the object. The **horizontal** is a horizontal line (the ground, or a line parallel to the ground) that passes through the vantage point.

The **angle of elevation or depression** is the angle between the line of sight and the horizontal. It is called an angle of *elevation* when the object is *higher* than the vantage point or an angle of *depression* when the object is *lower* than the vantage point.

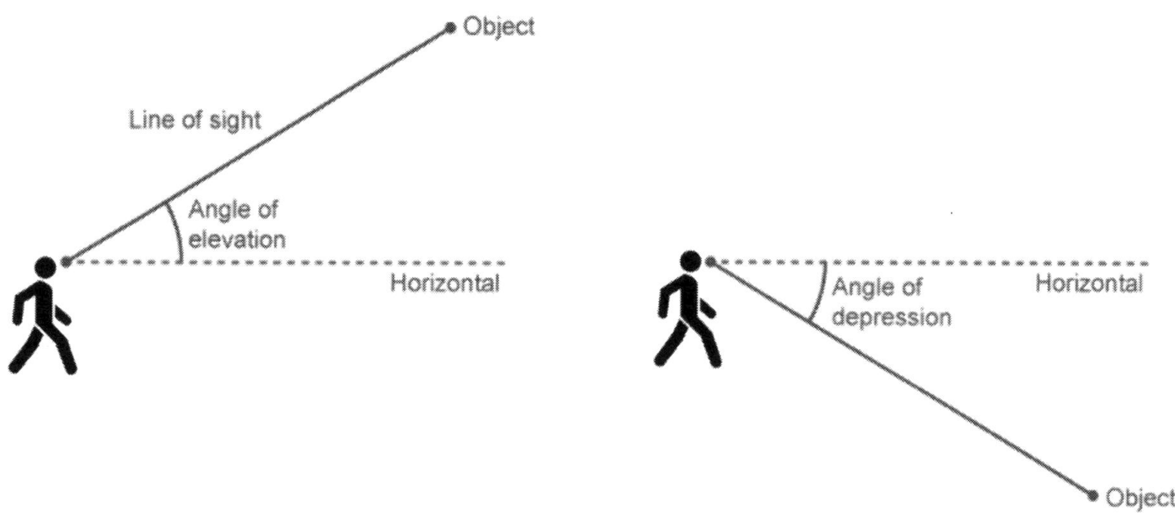

Model Problem:

Given the measures shown in the diagram below, find a to the *nearest tenth of a foot*.

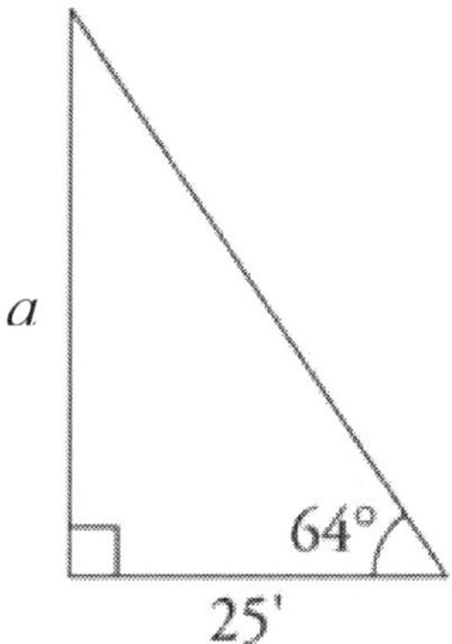

Solution:

(A) $\tan A = \dfrac{opposite}{adjacent}$

(B) $\tan 64° = \dfrac{a}{25}$

(C) $25 \tan 64° = a$

(D) $51.3 = a$

Explanation of steps:

(A) Select the appropriate trig ratio for the given and needed measures *[since we are given the adjacent leg and need to find the opposite leg, the tangent ratio should be used]*.

(B) Substitute the given measures and the variable for the unknown.

(C) Solve for (isolate) the variable *[multiply both sides by 25]*.

(D) Use a calculator to find the answer *[enter* $25\tan(64)$ *using the* TAN *key]*.

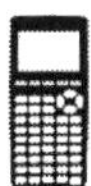

Practice Problems

1. The accompanying diagram shows a ramp 30 feet long leaning against a wall at a construction site. If the ramp forms an angle of 32° with the ground, how high above the ground, to the *nearest tenth of a foot*, is the top of the ramp?

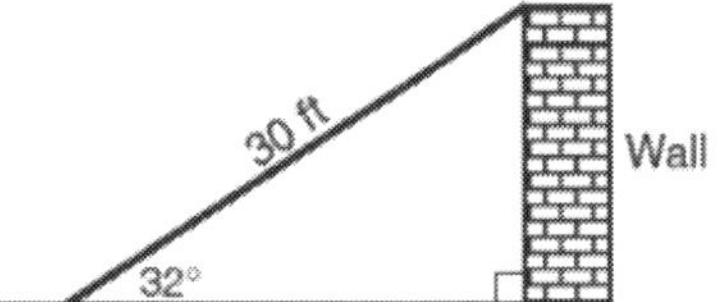

2. Find, to the *nearest tenth of a foot*, the height of the tree represented below.

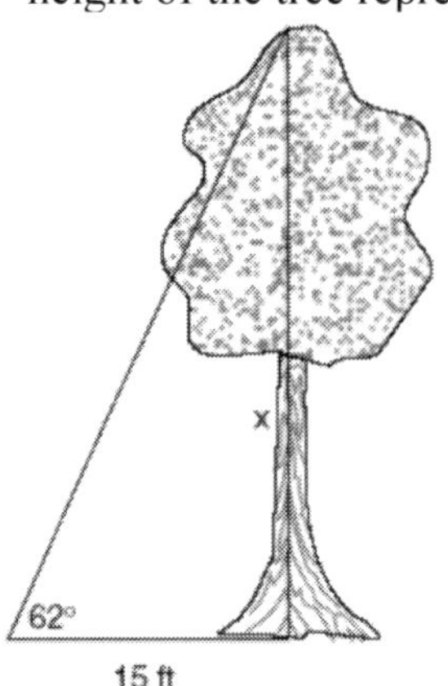

3. An airplane is climbing at an angle of 11° with the ground. Find, to the *nearest foot*, the ground distance the airplane has traveled when it has attained an altitude of 400 feet.

4. Find the area of the triangle below, to the *nearest tenth of a square foot.*

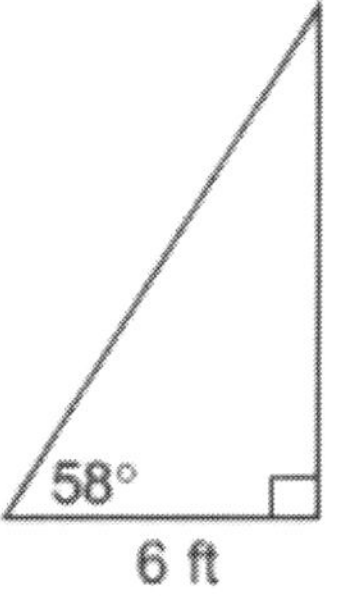

5. A 10-foot ladder is to be placed against the side of a building. The base of the ladder must be placed at an angle of 72° with the level ground for a secure footing. Find, to the *nearest inch*, how far the base of the ladder should be from the side of the building *and* how far up the side of the building the ladder will reach.

6. A lighthouse is built on the edge of a cliff near the ocean, as shown in the diagram. From a boat located 200 feet from the base of the cliff, the angle of elevation to the top of the cliff is 18° and the angle of elevation to the top of the lighthouse is 28°. What is the height of the lighthouse, x, to the *nearest tenth of a foot*?

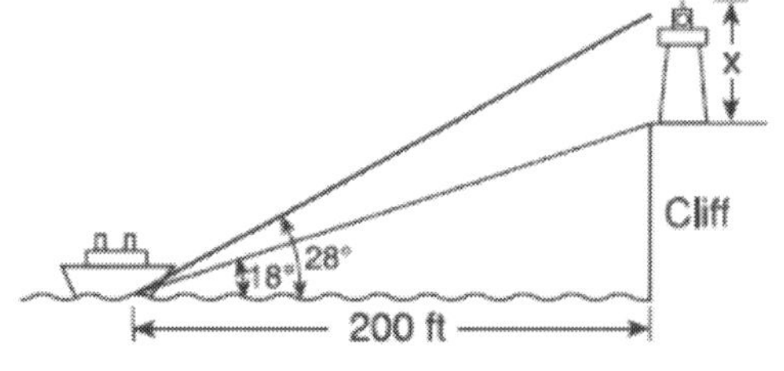

REGENTS QUESTIONS

Multiple Choice

1. As shown in the accompanying diagram, the diagonal of the rectangle is 10 inches long. The diagonal makes a 15° angle with the longer side of the rectangle. What is the width, w, of the rectangle to the *nearest tenth of an inch*?

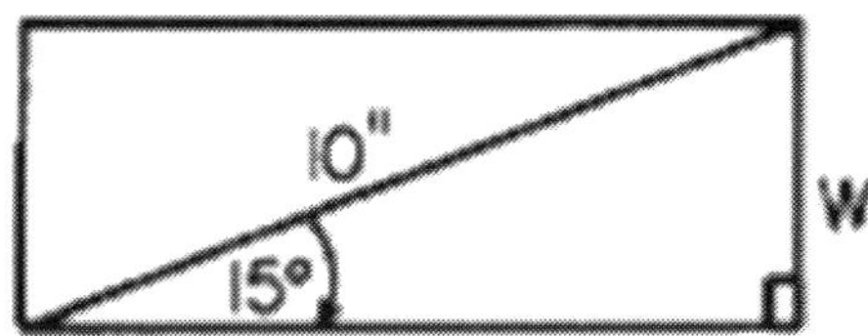

(1) 2.5
(2) 2.6
(3) 3.9
(4) 9.7

2. The hypotenuse of a right triangle measures 10 inches and one angle measures 31°. Which equation could be used to find the length of the side opposite the 31° angle?

(1) $\sin 31° = \frac{10}{x}$
(2) $\cos 31° = \frac{x}{10}$
(3) $\sin 31° = \frac{x}{10}$
(4) $\tan 31° = \frac{x}{10}$

3. A tree casts a 25-foot shadow on a sunny day, as shown in the diagram below.

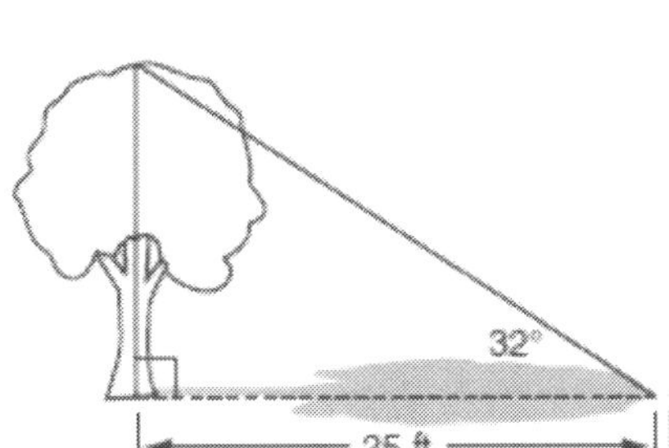

If the angle of elevation from the tip of the shadow to the top of the tree is 32°, what is the height of the tree to the *nearest tenth of a foot*?

(1) 13.2
(2) 15.6
(3) 21.2
(4) 40.0

4. An 8-foot rope is tied from the top of a pole to a stake in the ground, as shown in the diagram below.

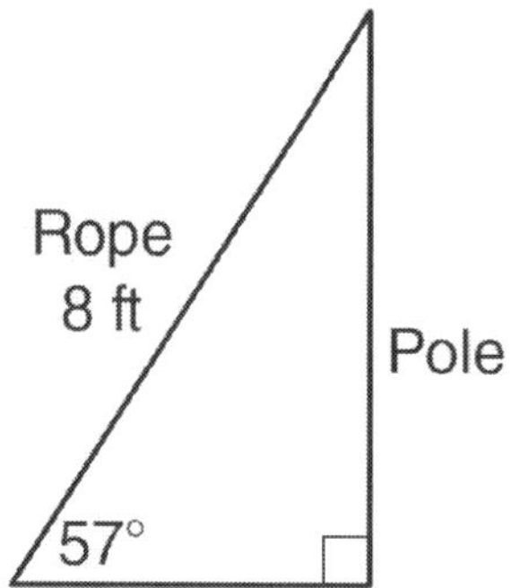

If the rope forms a 57° angle with the ground, what is the height of the pole, to the *nearest tenth of a foot*?

(1) 4.4
(2) 6.7
(3) 9.5
(4) 12.3

5. A right triangle contains a 38° angle whose adjacent side measures 10 centimeters. What is the length of the hypotenuse, to the *nearest hundredth of a centimeter*?

(1) 7.88
(2) 12.69
(3) 12.80
(4) 16.24

6. (CC) As shown in the diagram below, the angle of elevation from a point on the ground to the top of the tree is 34°.

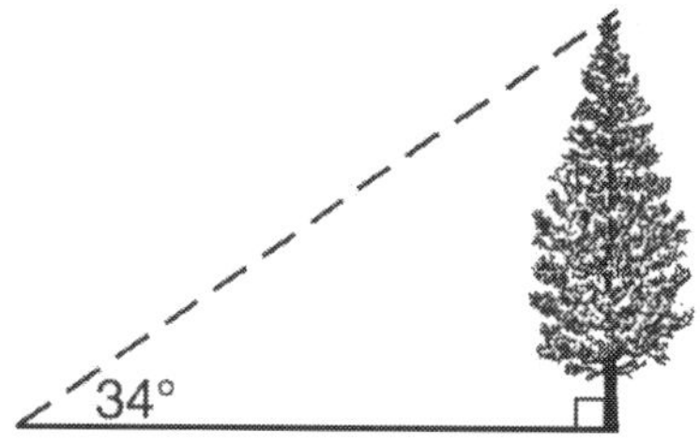

If the point is 20 feet from the base of the tree, what is the height of the tree, to the *nearest tenth of a foot*?

(1) 29.7
(2) 16.6
(3) 13.5
(4) 11.2

7. (CC) A 20-foot support post leans against a wall, making a 70° angle with the ground. To the *nearest tenth of a foot*, how far up the wall will the support post reach?

(1) 6.8
(2) 6.9
(3) 18.7
(4) 18.8

8. (CC) The diagram below shows two similar triangles.

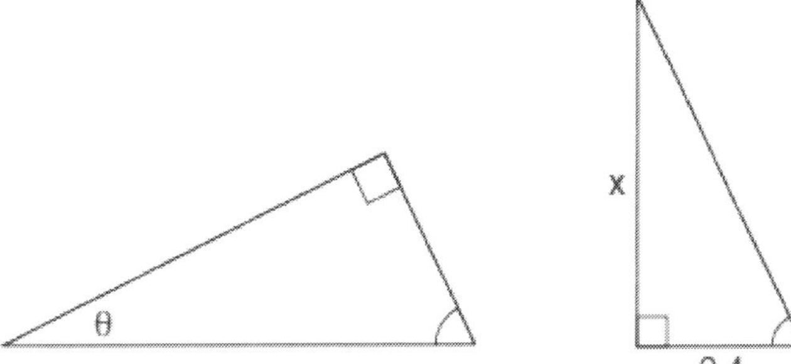

If $\tan\theta = \frac{3}{7}$, what is the value of x, to the *nearest tenth*?

(1) 1.2
(2) 5.6
(3) 7.6
(4) 8.8

9. (CC) Given the right triangle in the diagram below, what is the value of x, to the *nearest foot*?

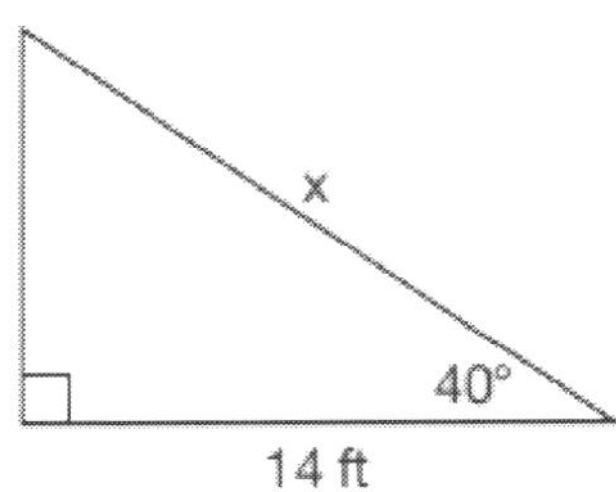

(1) 11
(2) 17
(3) 18
(4) 22

Constructed Response

10. A stake is to be driven into the ground away from the base of a 50-foot pole, as shown in the diagram below. A wire from the stake on the ground to the top of the pole is to be installed at an angle of elevation of 52°.

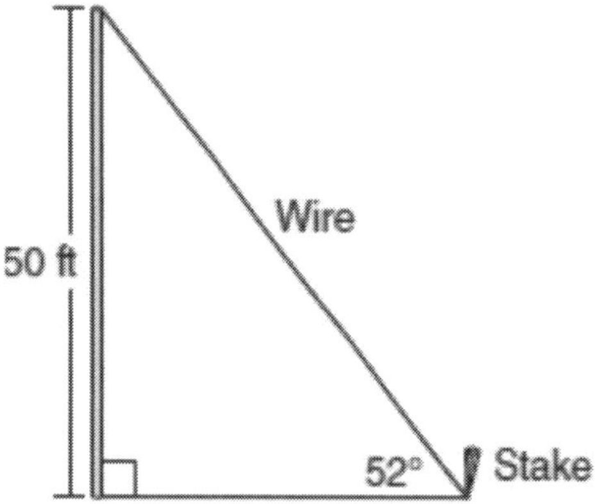

How far away from the base of the pole should the stake be driven in, to the *nearest foot*? What will be the length of the wire from the stake to the top of the pole, to the *nearest foot*?

11. A hot-air balloon is tied to the ground with two taut (straight) ropes, as shown in the diagram below. One rope is directly under the balloon and makes a right angle with the ground. The other rope forms an angle of 50° with the ground.

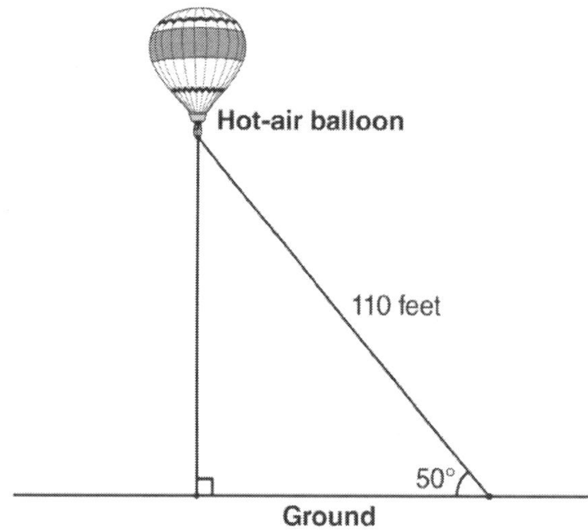

Determine the height, to the *nearest foot*, of the balloon directly above the ground. Determine the distance, to the *nearest foot*, on the ground between the two ropes.

12. As shown in the diagram below, a ladder 5 feet long leans against a wall and makes an angle of 65° with the ground. Find, to the *nearest tenth of a foot*, the distance from the wall to the base of the ladder.

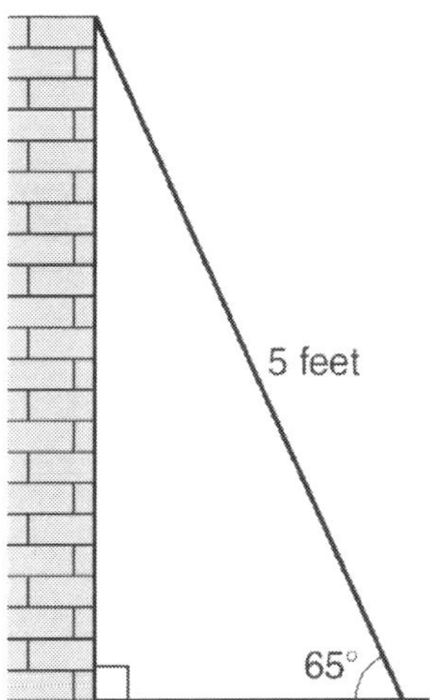

13. A metal pipe is used to hold up a 9-foot fence, as shown in the diagram below. The pipe makes an angle of 48° with the ground.

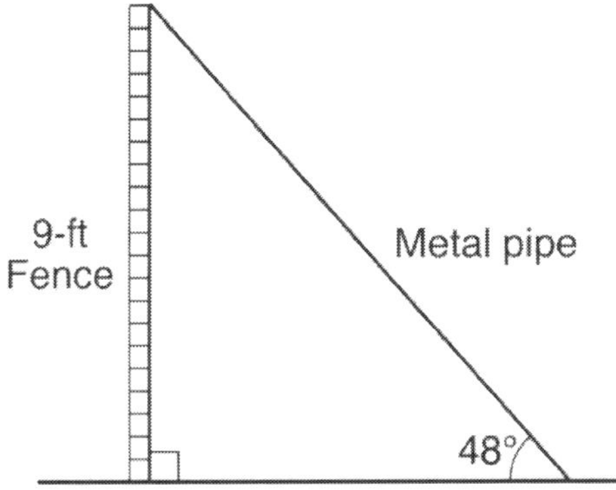

Determine, to the *nearest foot*, how far the bottom of the pipe is from the base of the fence. Determine, to the *nearest foot*, the length of the metal pipe.

14. (CC) The map below shows the three tallest mountain peaks in New York State: Mount Marcy, Algonquin Peak, and Mount Haystack. Mount Haystack, the shortest peak, is 4960 feet tall. Surveyors have determined the horizontal distance between Mount Haystack and Mount Marcy is 6336 feet and the horizontal distance between Mount Marcy and Algonquin Peak is 20,493 feet.

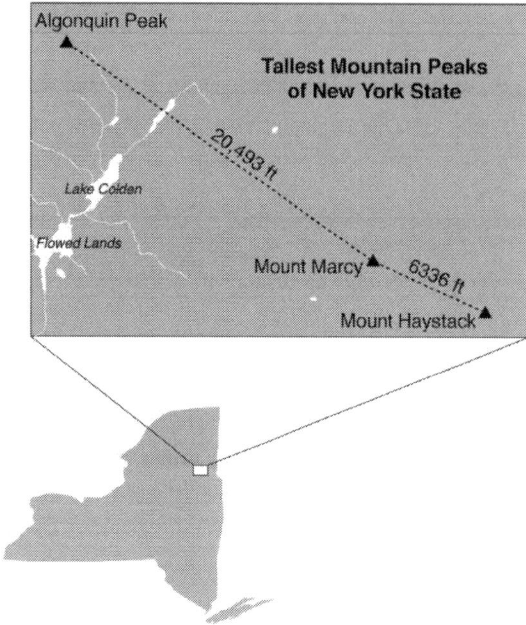

The angle of depression from the peak of Mount Marcy to the peak of Mount Haystack is 3.47 degrees. The angle of elevation from the peak of Algonquin Peak to the peak of Mount Marcy is 0.64 degrees. What are the heights, to the *nearest foot*, of Mount Marcy and Algonquin Peak? Justify your answer.

15. (CC) As shown below, a canoe is approaching a lighthouse on the coastline of a lake. The front of the canoe is 1.5 feet above the water and an observer in the lighthouse is 112 feet above the water.

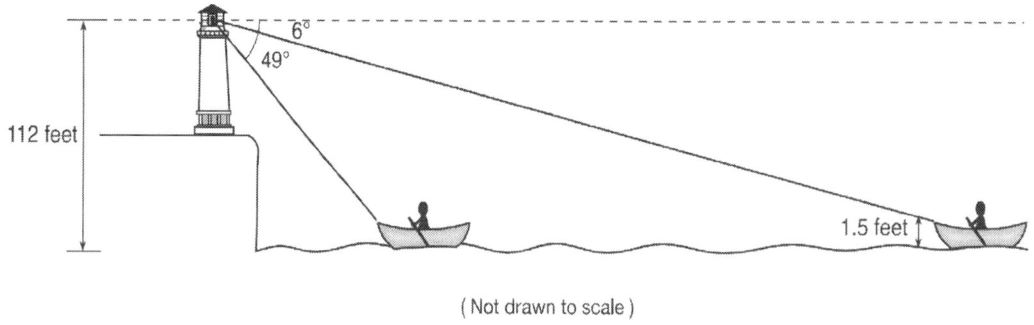

(Not drawn to scale)

At 5:00, the observer in the lighthouse measured the angle of depression to the front of the canoe to be 6°. Five minutes later, the observer measured and saw the angle of depression to the front of the canoe had increased by 49°. Determine and state, to the *nearest foot per minute*, the average speed at which the canoe traveled toward the lighthouse.

16. (CC) As shown in the diagram below, a ship is heading directly toward a lighthouse whose beacon is 125 feet above sea level. At the first sighting, point A, the angle of elevation from the ship to the light was 7°. A short time later, at point D, the angle of elevation was 16°.

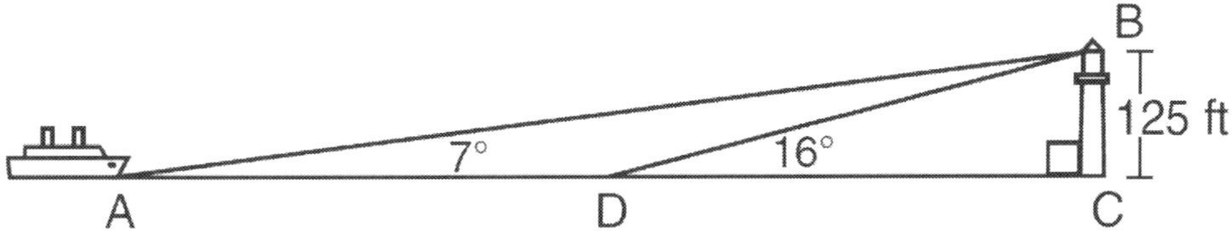

To the *nearest foot*, determine and state how far the ship traveled from point A to point D.

17. (CC) A carpenter leans an extension ladder against a house to reach the bottom of a window 30 feet above the ground. As shown in the diagram below, the ladder makes a 70° angle with the ground. To the *nearest foot*, determine and state the length of the ladder.

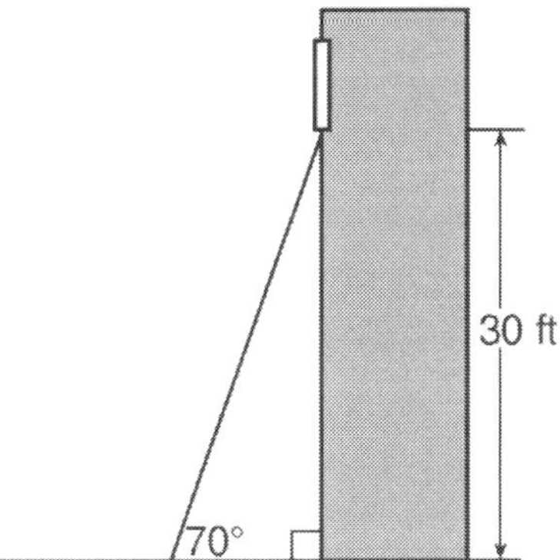

18. (CC) Cathy wants to determine the height of the flagpole shown in the diagram below. She uses a survey instrument to measure the angle of elevation to the top of the flagpole, and determines it to be 34.9°. She walks 8 meters closer and determines the new measure of the angle of elevation to be 52.8°. At each measurement, the survey instrument is 1.7 meters above the ground.

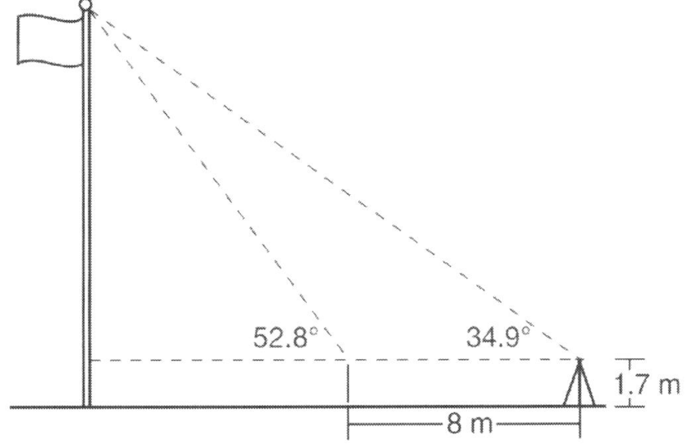

Determine and state, to the *nearest tenth of a meter*, the height of the flagpole.

19. (CC) In the diagram below, a window of a house is 15 feet above the ground. A ladder is placed against the house with its base at an angle of 75° with the ground. Determine and state the length of the ladder to the *nearest tenth of a foot*.

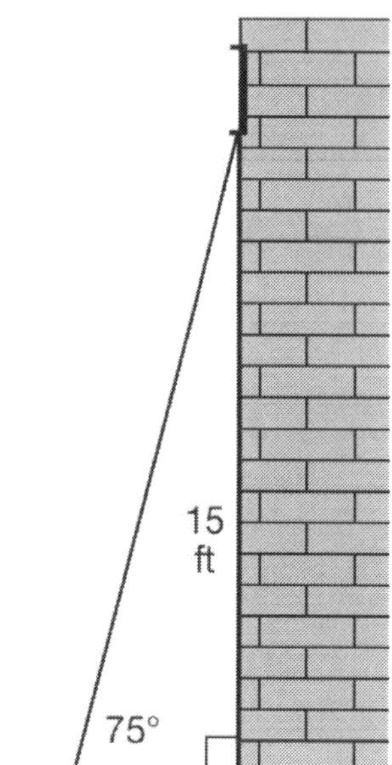

5.5 Use Trigonometry to Find an Angle

Key Terms and Concepts

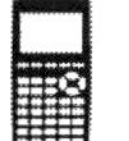

Note: Press [MODE] *on your calculator to check that it is set for* [Degree] *and not* [Radian].

If given two sides of a right triangle, the measure of an acute angle can be found by using the appropriate trigonometric ratio.

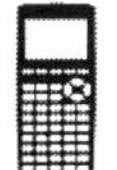

To find an angle measure when you know its sine, cosine, or tangent, you will need to use the corresponding **inverse trig function** key: [SIN⁻¹], [COS⁻¹] or [TAN⁻¹]. The inverse functions are entered by pressing [2nd] followed by the appropriate trig function key on the calculator.

For example: If $\sin A = \frac{1}{\sqrt{2}}$, enter $\sin^{-1}(1/\sqrt{2})$ using [2nd] [SIN⁻¹].

The result is 45, so the measure of angle A is 45°.

These inverse functions are often called the **arcsine**, **arccosine**, and **arctangent**. Therefore, you may see $\sin^{-1}$ written as **arcsin**, $\cos^{-1}$ written as **arccos**, and $\tan^{-1}$ written as **arctan**.

Model Problem:

In the right triangle below, the lengths of the legs are 9 and 10 as shown. Find the measure of angle x to the *nearest degree*.

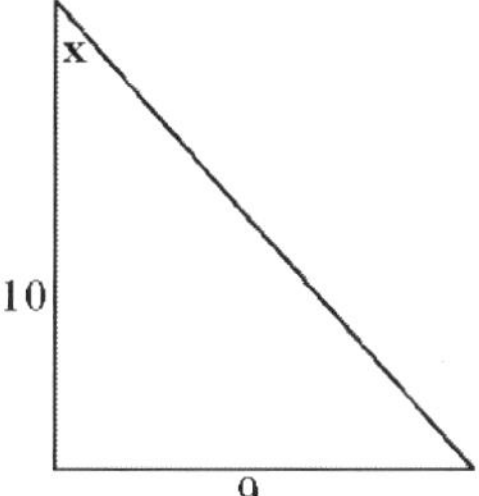

Solution:

(A) $\tan x = \frac{opposite}{adjacent}$

(B) $\tan x = \frac{9}{10}$

(C) $x = \tan^{-1}(9/10)$

(D) $x \approx 42°$

Explanation of steps:

(A) Select the appropriate trig ratio *[since the opposite and adjacent legs are given, use tan]*.

(B) Substitute the given measures.

(C) Since the angle is the unknown, isolate the variable by taking the inverse trig function of both sides. Think of a trig function and its inverse as "canceling out."

(D) Use the inverse trig function on the calculator. *[Press* [2nd] [TAN⁻¹] 9 ÷ 10 [)] [ENTER]*.]* Round your answer as specified.

Practice Problems

1. What is the measure of the angle, x, to the *nearest tenth of a degree*?

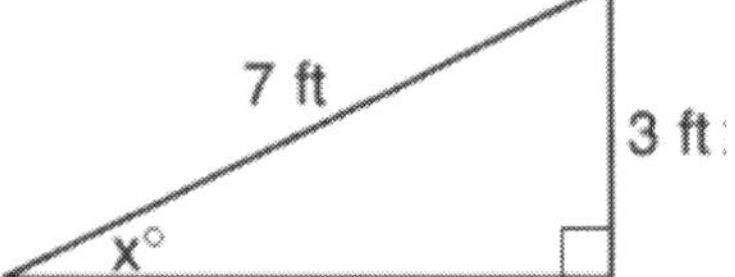

2. A person standing on level ground is 2,000 feet away from the foot of a 420-foot-tall building, as shown in the diagram. To the *nearest degree*, what is the value of x?

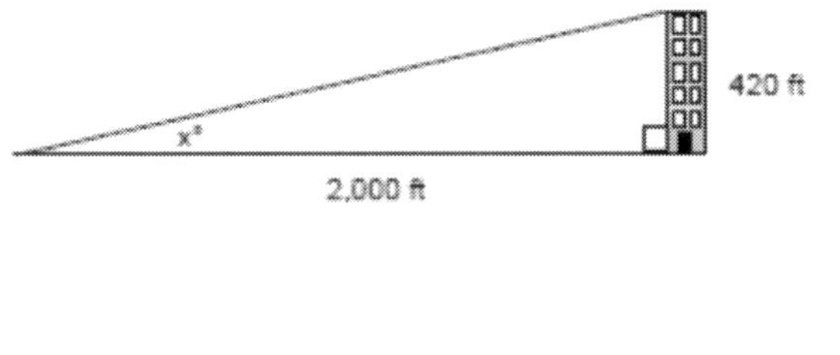

3. The diagram shows a flagpole that stands on level ground. Two cables, r and s, are attached to the pole at a point 16 feet above the ground. The combined length of the two cables is 50 feet. If cable r is attached to the ground 12 feet from the base of the pole, what is the measure of the angle, x, to the *nearest degree*, that cable s makes with the ground?

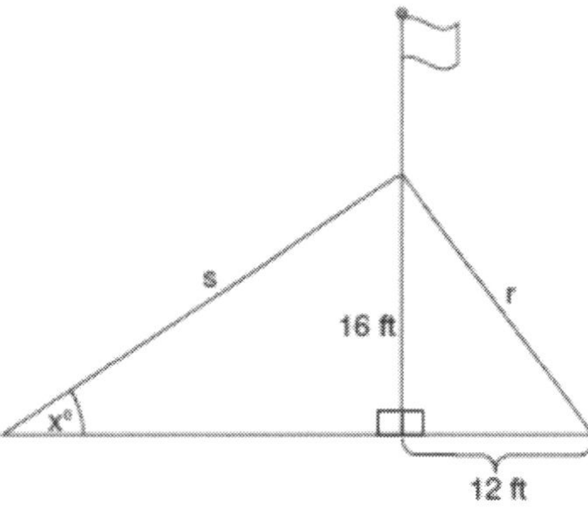

4. The base of a 15-foot ladder rests on the ground 4 feet from a 6-foot fence. If the ladder touches the top of the fence and the side of a building, what angle, to the *nearest degree*, does the ladder make with the ground? How far up the side of the building does the top of the ladder reach, to the *nearest foot*?

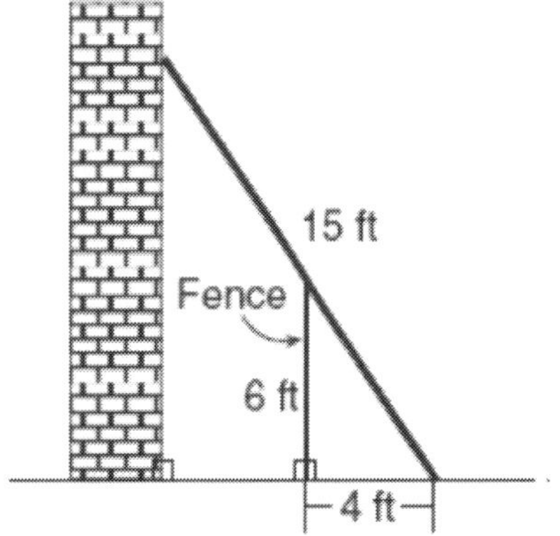

REGENTS QUESTIONS

Multiple Choice

1. The center pole of a tent is 8 feet long, and a side of the tent is 12 feet long as shown in the diagram below.

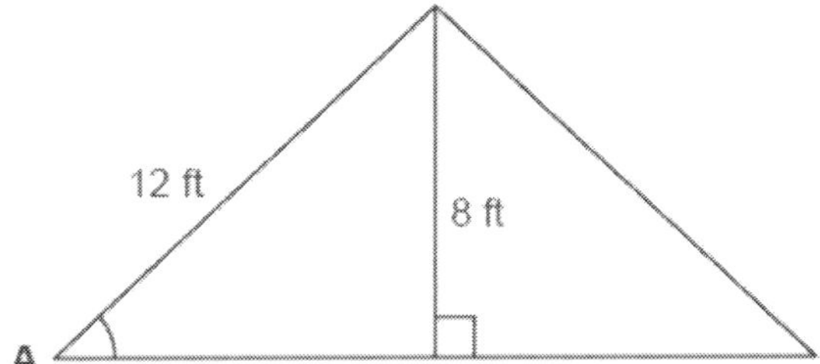

If a right angle is formed where the center pole meets the ground, what is the measure of angle A to the nearest degree?

(1) 34
(2) 42
(3) 48
(4) 56

2. In the diagram of $\triangle ABC$ shown below, $BC = 10$ and $AB = 16$.

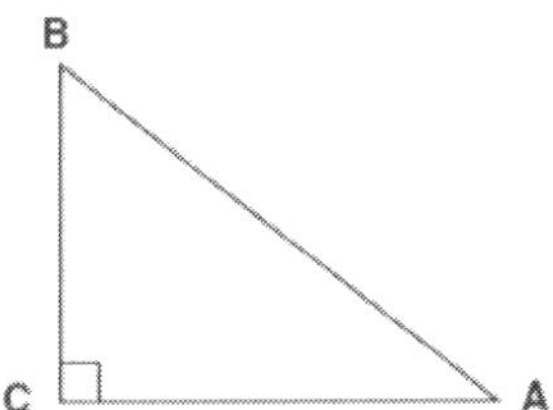

To the *nearest tenth of a degree*, what is the measure of the largest acute angle in the triangle?

(1) 32.0
(2) 38.7
(3) 51.3
(4) 90.0

3. In right triangle ABC shown below, $AB = 18.3$ and $BC = 11.2$.

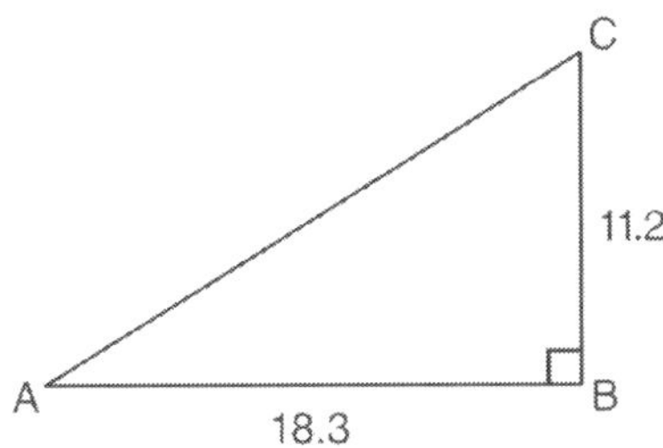

What is the measure of $\angle A$, to the *nearest tenth of a degree*?

(1) 31.5
(2) 37.7
(3) 52.3
(4) 58.5

4. CC A man who is 5 feet 9 inches tall casts a shadow of 8 feet 6 inches. Assuming that the man is standing perpendicular to the ground, what is the angle of elevation from the end of the shadow to the top of the man's head, to the *nearest tenth of a degree*?

   (1) 34.1 (3) 42.6
   (2) 34.5 (4) 55.9

5. CC In the diagram of right triangle ABC shown below, $AB = 14$ and $AC = 9$.

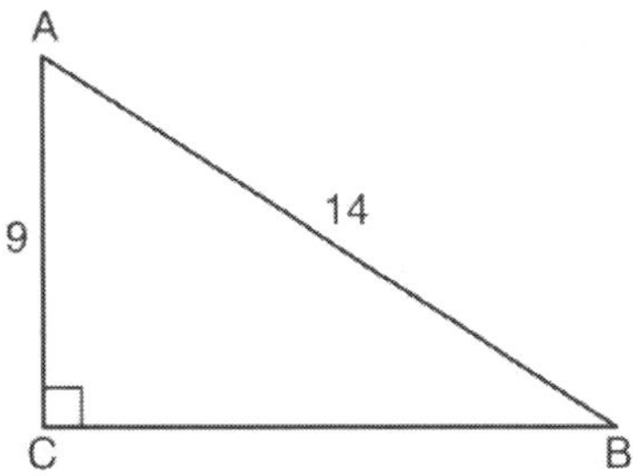

   What is the measure of $\angle A$, to the *nearest degree*?

   (1) 33 (3) 50
   (2) 40 (4) 57

Constructed Response

6. In right triangle ABC, $AB = 20$, $AC = 12$, $BC = 16$, and $m\angle C = 90$. Find, to the *nearest degree*, the measure of $\angle A$.

7. A communications company is building a 30-foot antenna to carry cell phone transmissions. As shown in the diagram below, a 50-foot wire from the top of the antenna to the ground is used to stabilize the antenna.

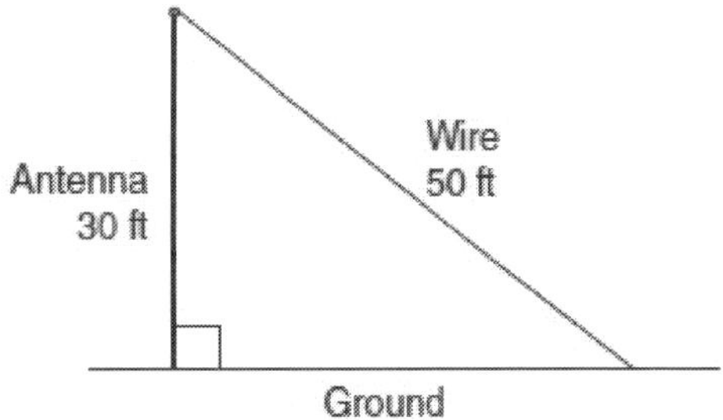

   Find, to the *nearest degree,* the measure of the angle that the wire makes with the ground.

8. A trapezoid is shown below.

   28 feet
   x
   12 feet
   36 feet

   Calculate the measure of angle x, to the *nearest tenth of a degree*.

9. A 28-foot ladder is leaning against a house. The bottom of the ladder is 6 feet from the base of the house. Find the measure of the angle formed by the ladder and the ground, to the *nearest degree*.

10. In right triangle ABC shown below, $AC = 29$ inches, $AB = 17$ inches, and m$\angle ABC = 90$. Find the number of degrees in the measure of angle BAC, to the *nearest degree*.

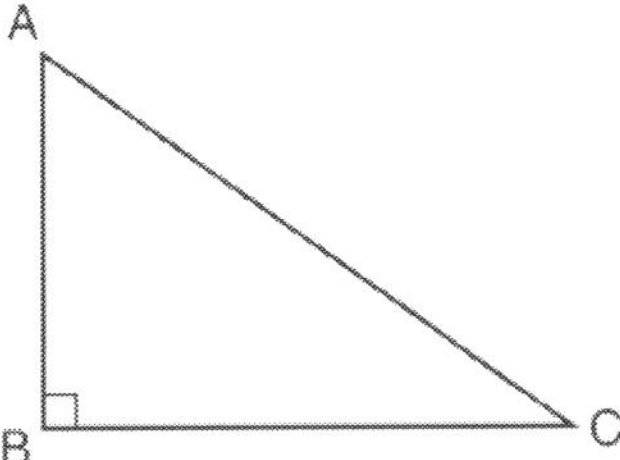

Find the length of $\overline{BC}$ to the *nearest inch*.

11. A man standing on level ground is 1000 feet away from the base of a 350-foot-tall building. Find, to the *nearest degree*, the measure of the angle of elevation to the top of the building from the point on the ground where the man is standing.

12. (CC) The diagram below shows a ramp connecting the ground to a loading platform 4.5 feet above the ground. The ramp measures 11.75 feet from the ground to the top of the loading platform.

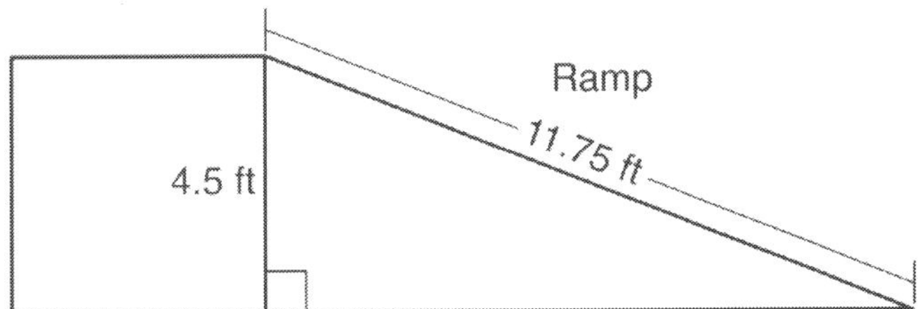

Determine and state, to the *nearest degree*, the angle of elevation formed by the ramp and the ground.

13. (CC) A ladder leans against a building. The top of the ladder touches the building 10 feet above the ground. The foot of the ladder is 4 feet from the building. Find, to the *nearest degree*, the angle that the ladder makes with the level ground.

14. (CC) As modeled below, a movie is projected onto a large outdoor screen. The bottom of the 60-foot-tall screen is 12 feet off the ground. The projector sits on the ground at a horizontal distance of 75 feet from the screen.

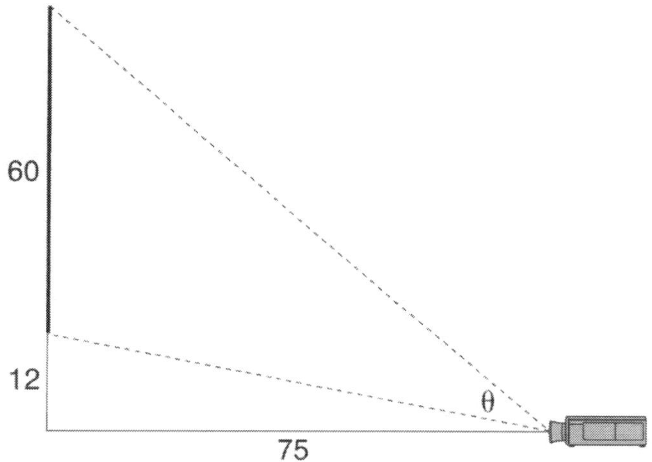

Determine and state, to the *nearest tenth of a degree*, the measure of θ, the projection angle.

5.6 Special Triangles (+)

*In the Geometry Draft of Revised Standards for 2018-19, NYS has recommended **adding** this topic.*

Key Terms and Concepts

There are two right triangles that are given "special" prominence: the 30-60-90 triangle and the 45-45-90 triangle.

In the **30-60-90 triangle**, the legs have a ratio of $1:\sqrt{3}$, with the shorter leg across from the 30° angle, and the ratio of the shorter leg to the hypotenuse is $1:2$.

In the **45-45-90 triangle**, the legs are congruent, making it an *isosceles right triangle*, and the ratio of a leg to the hypotenuse is $1:\sqrt{2}$.

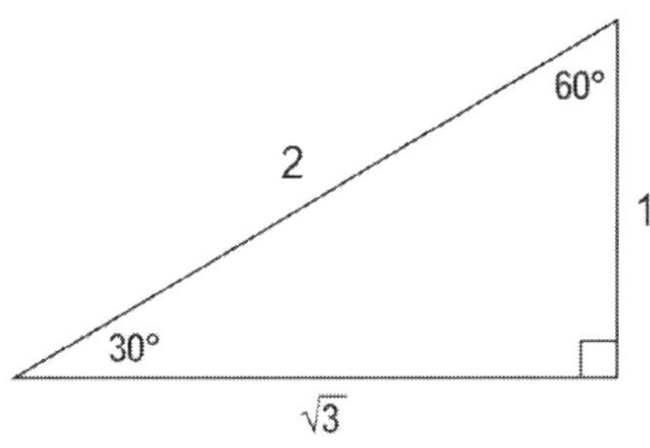

30-60-90 triangle

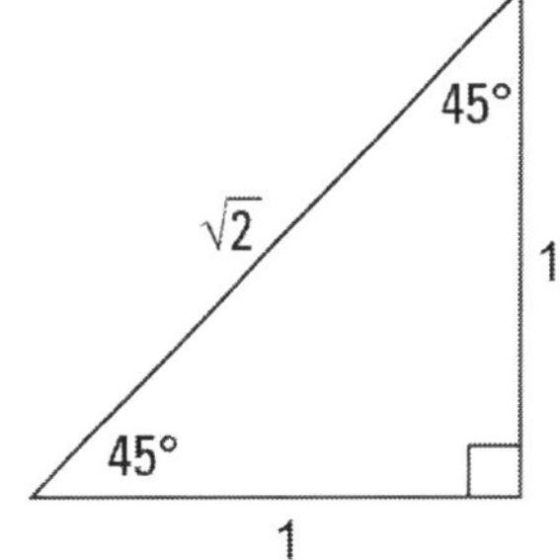

45-45-90 triangle

These triangles are important enough that their trigonometric ratios are often memorized by students. One advantage of knowing them is that they are exact values, rather than the decimal approximations, of trigonometric functions. A table of these special ratios is given below. You should be able to derive each of these ratios from the triangles themselves.

	30°	45°	60°
sin	$\frac{1}{2}$	$\frac{1}{\sqrt{2}}$	$\frac{\sqrt{3}}{2}$
cos	$\frac{\sqrt{3}}{2}$	$\frac{1}{\sqrt{2}}$	$\frac{1}{2}$
tan	$\frac{1}{\sqrt{3}}$	1	$\sqrt{3}$

Of course, we can shrink or enlarge either special triangle by any factor, and the sides would shrink or grow by that same factor. The more general 30-60-90 and 45-45-90 triangles, with scale factors of x, are shown below.

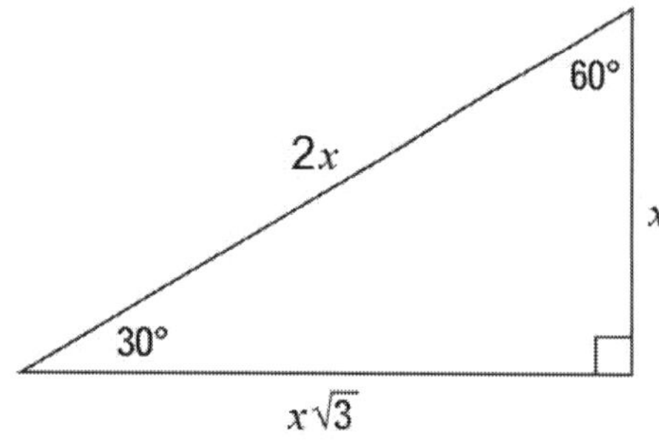

30-60-90 general triangle

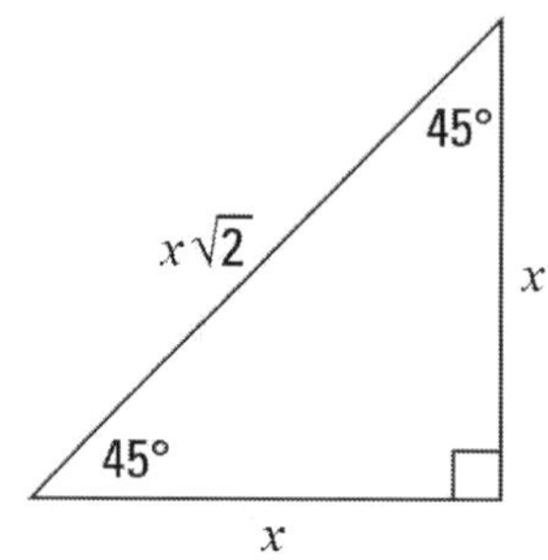

45-45-90 general triangle

When we take trigonometric ratios for these triangles, the factor cancels out, so *the ratios are the same regardless of the size of the triangle*. However, knowing the factor can help to determine the missing sides of a special triangle.

Example: A 30-60-90 triangle has a hypotenuse that is 20 units long. What are the lengths of the legs of the triangle? *Answer:* Since $2x = 20$, the factor x is 10. So the legs are 10 and $10\sqrt{3}$ units long.

Model Problem:

A straight post leaning against a wall makes a 45° angle with the ground. The base of the post is 3.25 meters from the building. What is the length of the post, to the *nearest tenth of a meter*?

Solution:

$$3.25 \times \sqrt{2} \approx 4.6 \text{ meters}$$

Explanation:

The post is the hypotenuse of a 45-45-90 triangle as it leans against the building. Therefore, if one leg, x, is 3.25 meters, then the hypotenuse can be calculated as $x\sqrt{2} = 3.25 \times \sqrt{2}$.

Practice Problems

1. The perimeter of a square is 20 inches. Find the length of a diagonal.	2. The altitude of an equilateral triangle is $2\sqrt{3}$ units. Find the length of a side.

REGENTS QUESTIONS

Multiple Choice

1. In the right triangle shown in the diagram below, what is the value of x to the *nearest whole number*?

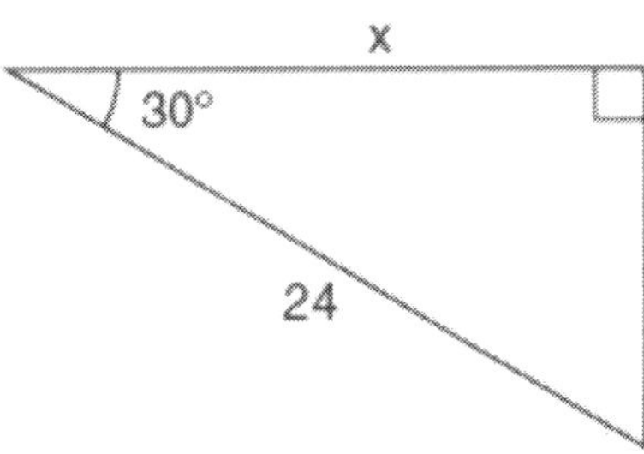

(1) 12
(2) 14
(3) 21
(4) 28

2. (CC) The diagram shows rectangle *ABCD*, with diagonal $\overline{BD}$.

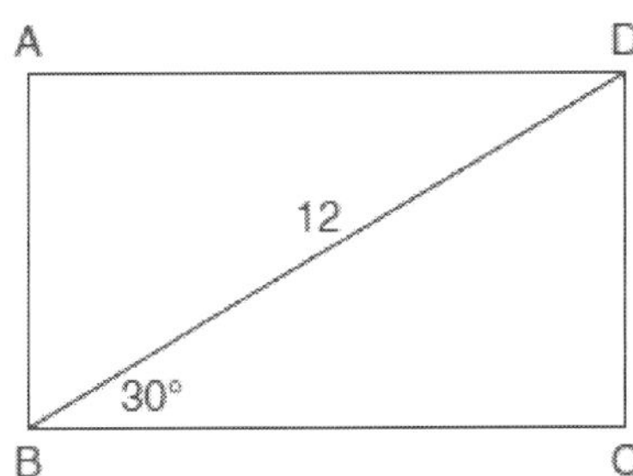

What is the perimeter of rectangle *ABCD*, to the *nearest tenth*?
(1) 28.4
(2) 32.8
(3) 48.0
(4) 62.4

5.7 Cofunctions

Key Terms and Concepts

In a right triangle, the sine of one acute angle is equal to the cosine of the other acute angle. For this reason, we say that sin and cos are **cofunctions**.

Example: In the right triangle below, $\sin A = \frac{a}{c}$ and $\cos B = \frac{a}{c}$, so $\sin A = \cos B$.

Also, $\sin B = \frac{b}{h}$ and $\cos A = \frac{b}{c}$, so $\sin B = \cos A$.

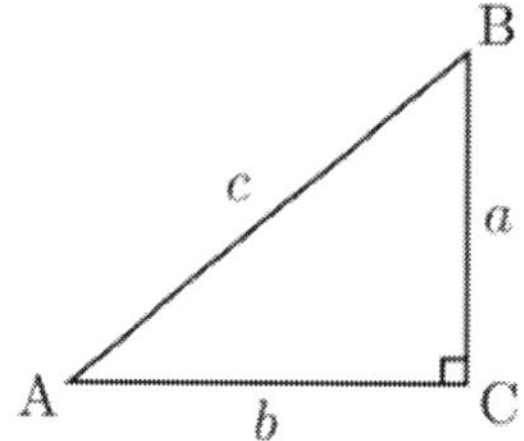

The two acute angles of a right triangle are always complementary (their measures add to 90°). Therefore, the following statements are true for any acute angle ($0° < x < 90°$):

$$\sin(x) = \cos(90 - x)$$
$$\cos(x) = \sin(90 - x)$$

Model Problem:

If $\sin 2x = \cos 3x$, find the value of x.

Solution:

$2x + 3x = 90$
$5x = 90$
$x = 18$

Explanation:
Cofunctions of complementary angles are equal. If two angles are complementary, their measures add to 90°.

Practice Problems

1. If $\sin 25° = \cos B$, find $m\angle B$.	2. If $\sin(x+15)° = \cos(x-5)°$, find x.

REGENTS QUESTIONS

Multiple Choice

1. If $\cos(2x-1)^\circ = \sin(3x+6)^\circ$, then the value of x is
   (1) –7
   (2) 17
   (3) 35
   (4) 71

2. If $\sin(x+20^\circ) = \cos x$, the value of x is
   (1) 35°
   (2) 45°
   (3) 55°
   (4) 70°

3. If $\sin(x-3)^\circ = \cos(2x+6)^\circ$, then the value of x is
   (1) –9
   (2) 26
   (3) 29
   (4) 64

4. CC In scalene triangle ABC shown in the diagram below, $m\angle C = 90^\circ$.

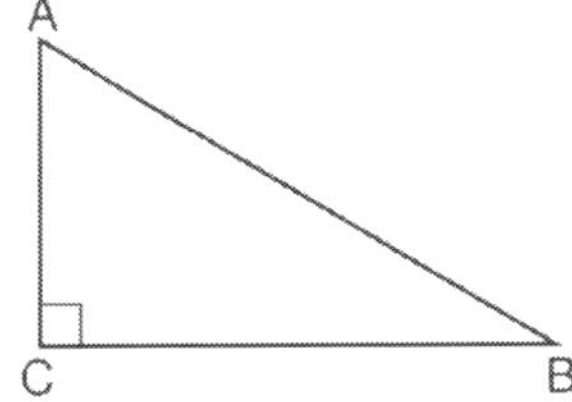

   Which equation is always true?
   (1) $\sin A = \sin B$
   (2) $\cos A = \cos B$
   (3) $\cos A = \sin C$
   (4) $\sin A = \cos B$

5. CC Which expression is always equivalent to $\sin x$ when $0^\circ < x < 90^\circ$?
   (1) $\cos(90^\circ - x)$
   (2) $\cos(45^\circ - x)$
   (3) $\cos(2x)$
   (4) $\cos x$

6. CC In $\triangle ABC$, the complement of $\angle B$ is $\angle A$. Which statement is always true?
   (1) $\tan\angle A = \tan\angle B$
   (2) $\sin\angle A = \sin\angle B$
   (3) $\cos\angle A = \tan\angle B$
   (4) $\sin\angle A = \cos\angle B$

7. CC In $\triangle ABC$, where $\angle C$ is a right angle, $\cos A = \frac{\sqrt{21}}{5}$. What is $\sin B$?
   (1) $\frac{\sqrt{21}}{5}$
   (2) $\frac{\sqrt{21}}{2}$
   (3) $\frac{2}{5}$
   (4) $\frac{5}{\sqrt{21}}$

Constructed Response

8. If $\cos 72° = \sin x$, find the number of degrees in the measure of acute angle x.

9. If $\cos(2x-25)° = \sin 55°$, find the value of x.

10. If $\sin(2x+20)° = \cos 40°$, find x.

11. If $3x$ is the measure of a positive acute angle and $\cos 3x = \sin 60°$, find the value of x.

12. (CC) In right triangle ABC with the right angle at C, $\sin A = 2x + 0.1$ and $\cos B = 4x - 0.7$. Determine and state the value of x. Explain your answer.

13. (CC) Explain why $\cos(x) = \sin(90-x)$ for x such that $0 < x < 90$.

14. (CC) Find the value of R that will make the equation $\sin 73° = \cos R$ true when $0° < R < 90°$. Explain your answer.

15. (CC) When instructed to find the length of $\overline{HJ}$ in right triangle HJG, Alex wrote the equation $\sin 28° = \frac{HJ}{20}$ while Marlene wrote $\cos 62° = \frac{HJ}{20}$. Are both students' equations correct? Explain why.

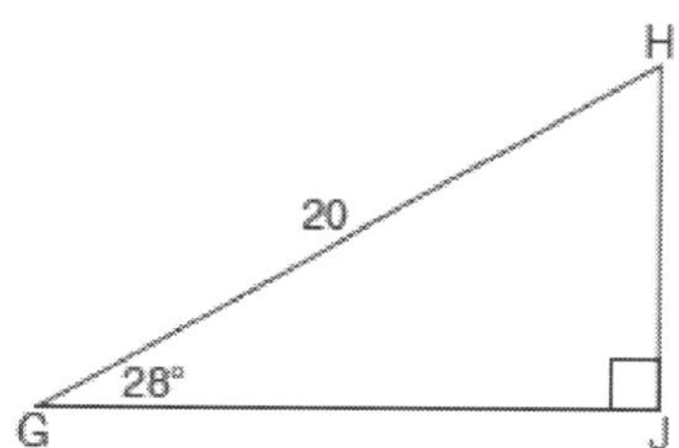

5.8 Mean Proportional

Key Terms and Concepts

When the altitude is drawn from the right angle vertex of a right triangle, a special relationship of three similar triangles is formed.

Example: The altitude $\overline{CP}$ of right triangle ABC splits the triangle into two smaller triangles, APC and CPB, shown below. Not only are these two smaller triangles similar to each other, but they are both also similar to the original larger triangle, ACB.

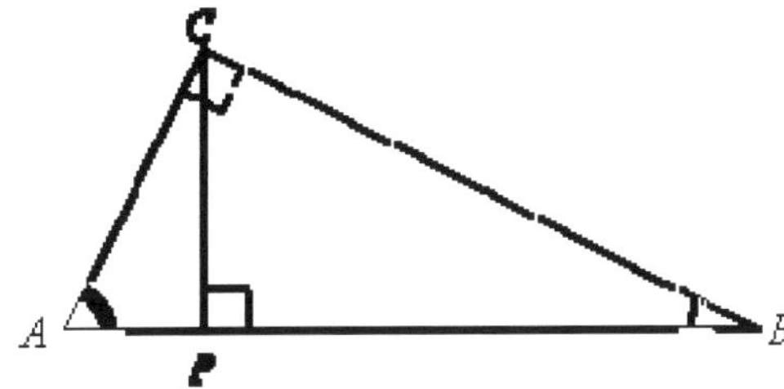

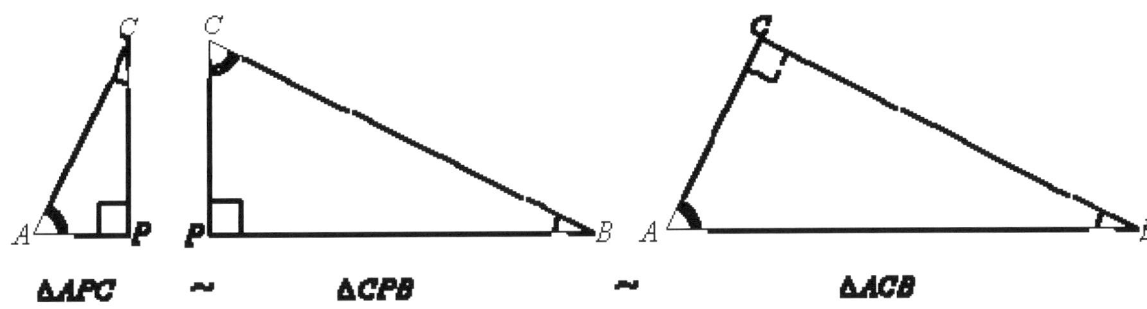

This relationship allows us to set up proportions to find a missing side. In these proportions, either the *altitude* or a *leg* of the original triangle is used as the *means*.

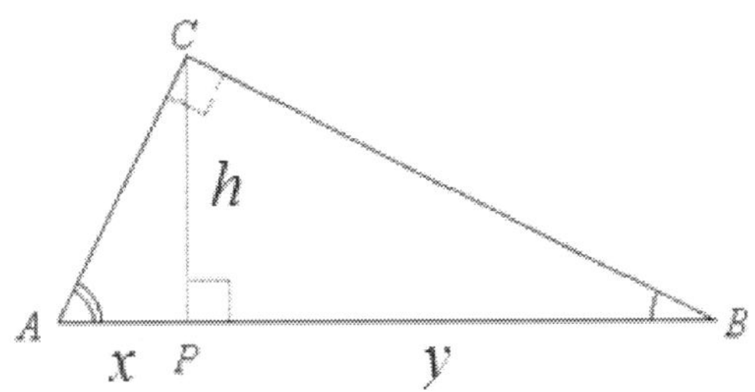

Diagram A – Altitude Rule: $h^2 = xy$

In *Diagram A* at left, the altitude of length h separates the hypotenuse into two segments, with lengths of x and y. In our first proportion, h is the mean between x and y. In other words:

$\frac{x}{h} = \frac{h}{y}$, which gives us the equation, $h^2 = xy$.

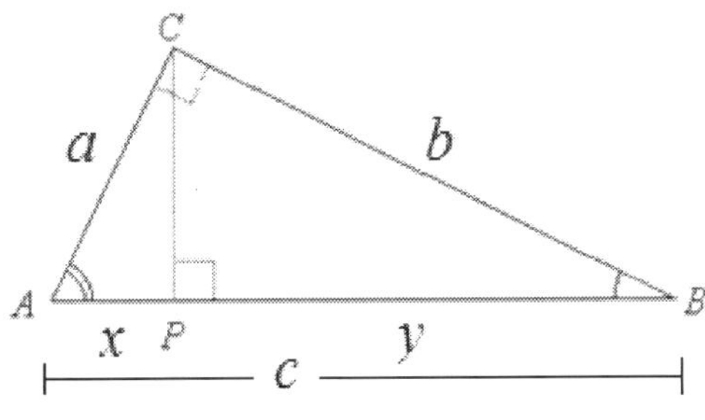

Diagram B – Leg Rules: $a^2 = cx$ *and* $b^2 = cy$

In *Diagram B*, we label the legs a and b, and the hypotenuse, c. The length of each leg will be the mean between the length of c and the length of the segment of the hypotenuse (x or y) that is adjacent to that leg. This is represented by two proportions:

a) $\frac{c}{a} = \frac{a}{x}$, which gives us $a^2 = cx$, and

b) $\frac{c}{b} = \frac{b}{y}$, which gives us $b^2 = cy$.

The Leg Rules can be used to prove the Pythagorean Theorem.

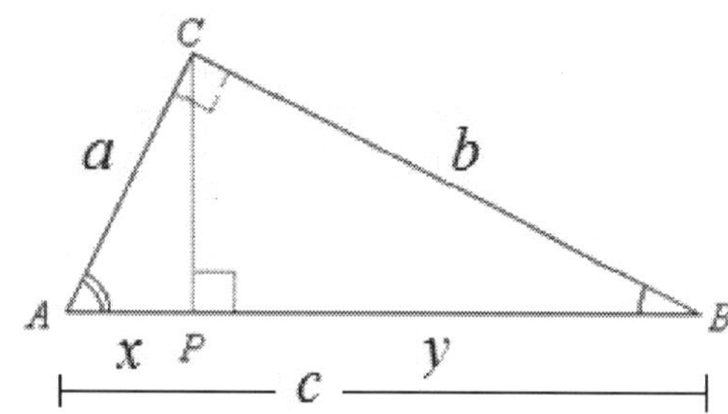

We've already developed the two equations, $a^2 = cx$ and $b^2 = cy$, from the similarity relationships of these triangles. We also know the segments x and y add to the length of c. Therefore,

$$\begin{aligned} a^2 + b^2 &= cx + cy && \textit{(add the two equations)} \\ &= c(x+y) && \textit{(factor)} \\ &= c^2 && \textit{(since } x+y=c\textit{)} \end{aligned}$$

In the diagrams above, $\overline{AP}$ and $\overline{BP}$ (with lengths of x and y) are often called the **projections** of the legs onto the hypotenuse.

Model Problem:

In the diagram below, $\overline{BE}$ bisects $\angle ABC$.
(a) Find DC. (b) Find BC. (c) Find DE to the *nearest tenth*.

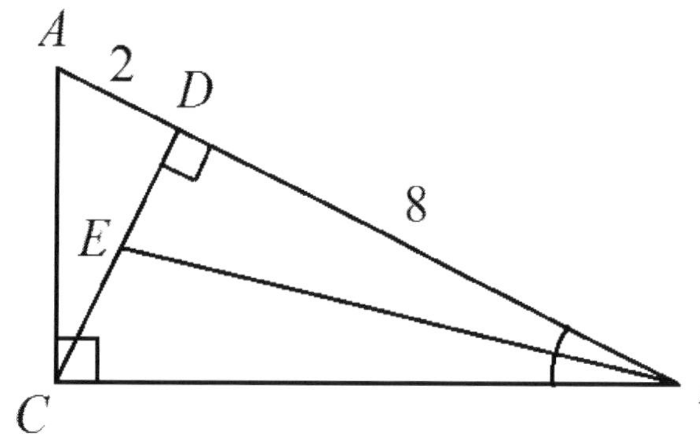

Solution:

(A) By the Altitude Rule,

$$(DC)^2 = 2 \cdot 8$$
$$(DC)^2 = 16$$
$$DC = 4$$

(B) By the Legs Rule,

$$(BC)^2 = (8+2) \cdot 8$$
$$(BC)^2 = 80$$
$$BC = \sqrt{80} = 4\sqrt{5}$$

(C) By the Angle Bisector Theorem,

$$\frac{DE}{EC} = \frac{BD}{BC} \qquad \frac{x}{4-x} = \frac{8}{4\sqrt{5}}$$

$$4\sqrt{5} \cdot x = 8(4-x)$$
$$4\sqrt{5} \cdot x = 32 - 8x$$
$$4\sqrt{5} \cdot x + 8x = 32$$

$$(4\sqrt{5}+8)x = 32$$
$$x = \frac{32}{4\sqrt{5}+8} \approx 1.9$$

Explanation of steps:

(A) $\overline{DC}$ is the altitude of $\triangle ACB$, so it is the mean proportional to AD and BD. $[(DC)^2 = AD \cdot BD]$

(B) $\overline{BC}$ is a leg of $\triangle ACB$, so it is the mean proportional to and the hypotenuse, AB, and its projection onto the hypotenuse, BD. $[(BC)^2 = AB \cdot BD]$
(Note: alternatively, the Pythagorean Theorem could have been used on $\triangle CDB$*.)*

(C) The bisector of an angle of a triangle splits the opposite side into segments that are proportional to the adjacent sides.

Practice Problems

1. If $AD = 3$ and $DB = 9$, find CD. 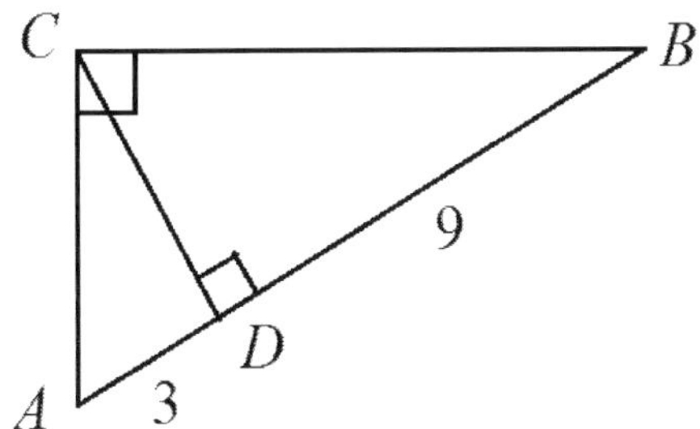	2. If $AD = 3$ and $DB = 9$, find AC. 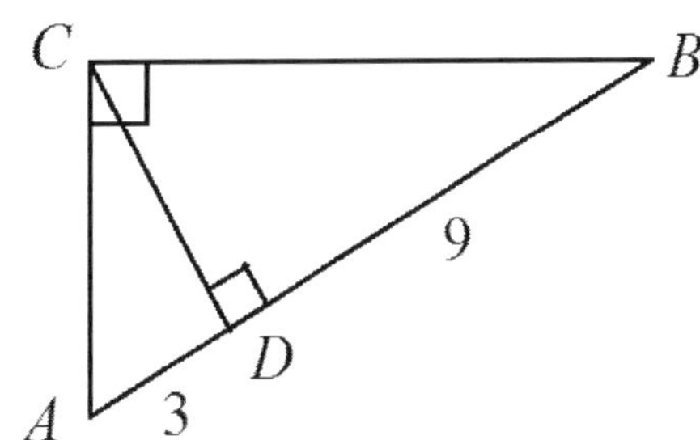
3. If $AD = 21$ and $BC = 10$, find BD. 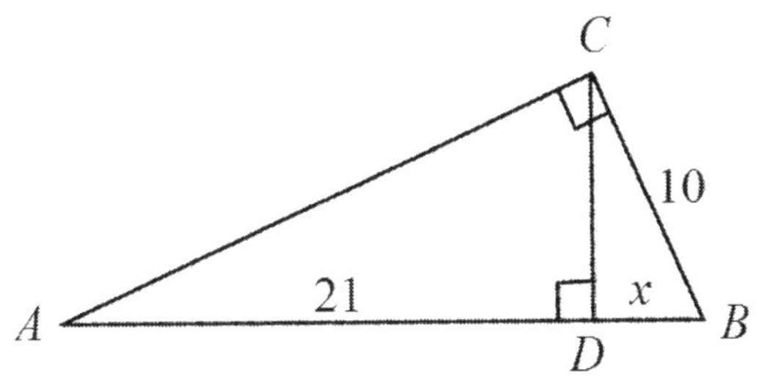	4. Find x, y, and z. 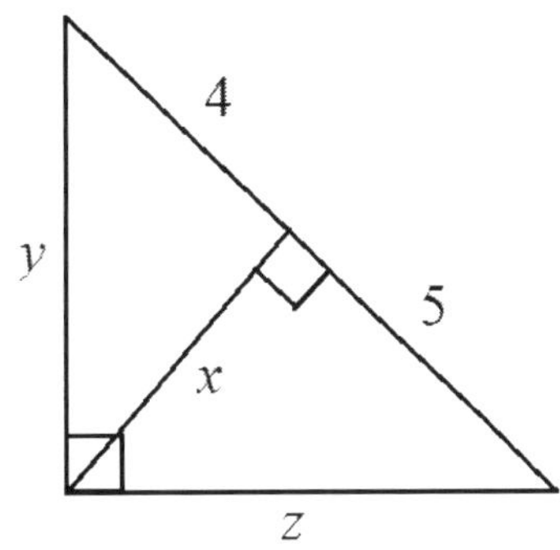
5. Find x, y, and z. 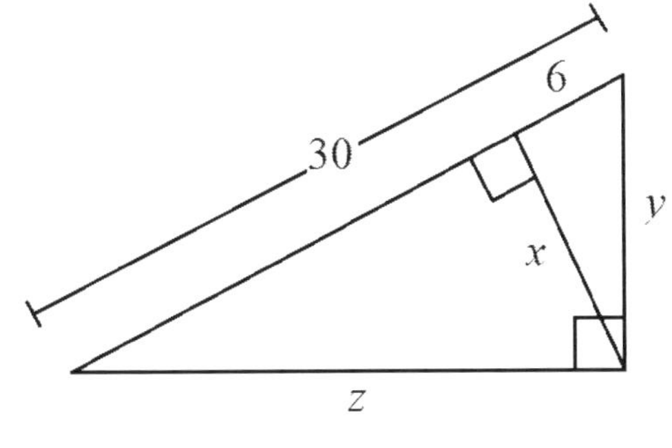	6. In the 30-60-90 triangle below, an altitude is drawn to the hypotenuse. What is the length of this altitude? 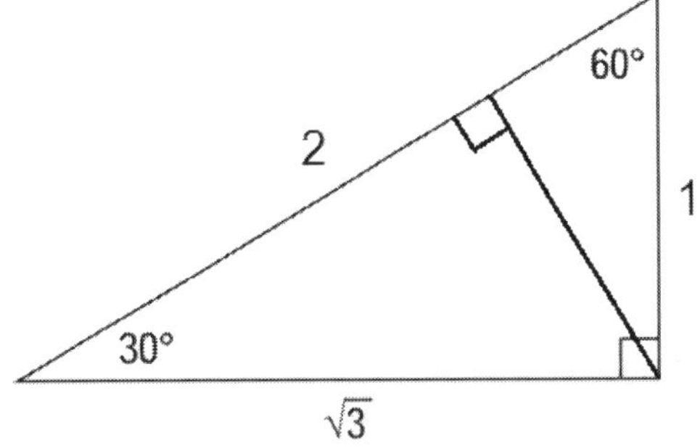

REGENTS QUESTIONS

Multiple Choice

1. In the diagram below of right triangle ABC, $\overline{CD}$ is the altitude to hypotenuse $\overline{AB}$, $CB = 6$, and $AD = 5$.

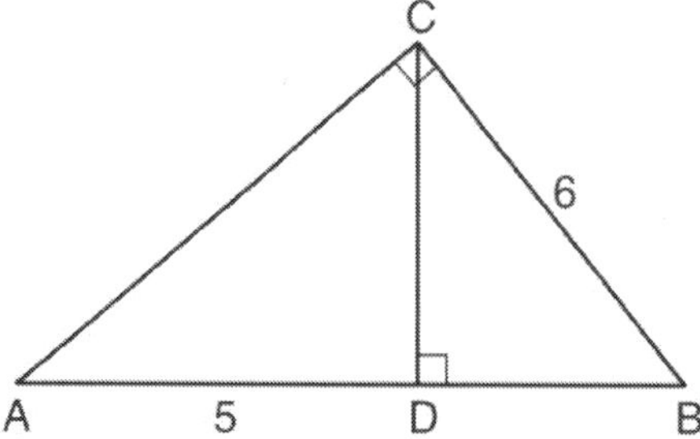

What is the length of $\overline{BD}$?

(1) 5
(2) 9
(3) 3
(4) 4

2. In the diagram below of right triangle ABC, altitude $\overline{BD}$ is drawn to hypotenuse $\overline{AC}$, $AC = 16$, and $CD = 7$.

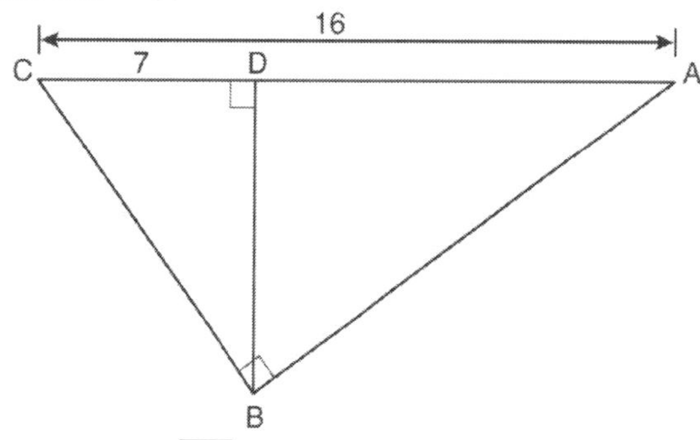

What is the length of $\overline{BD}$?

(1) $3\sqrt{7}$
(2) $4\sqrt{7}$
(3) $7\sqrt{3}$
(4) 12

3. In the diagram below of right triangle ABC, altitude $\overline{CD}$ is drawn to hypotenuse $\overline{AB}$.

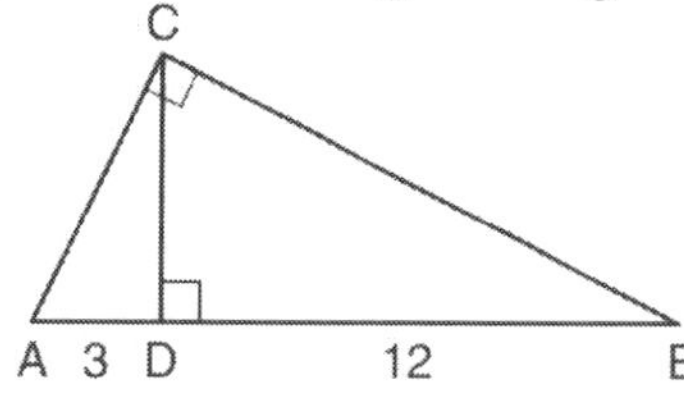

If $AD = 3$ and $DB = 12$, what is the length of altitude $\overline{CD}$?

(1) 6
(2) $6\sqrt{5}$
(3) 3
(4) $3\sqrt{5}$

4. In right triangle ABC shown in the diagram below, altitude $\overline{BD}$ is drawn to hypotenuse $\overline{AC}$, $CD = 12$, and $AD = 3$.

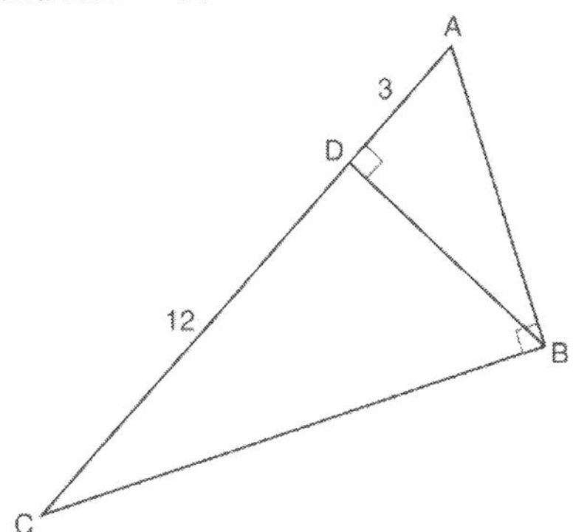

What is the length of $\overline{AB}$?

(1) $5\sqrt{3}$ (3) $3\sqrt{5}$
(2) 6 (4) 9

5. Triangle ABC shown below is a right triangle with altitude $\overline{AD}$ drawn to the hypotenuse $\overline{BC}$.

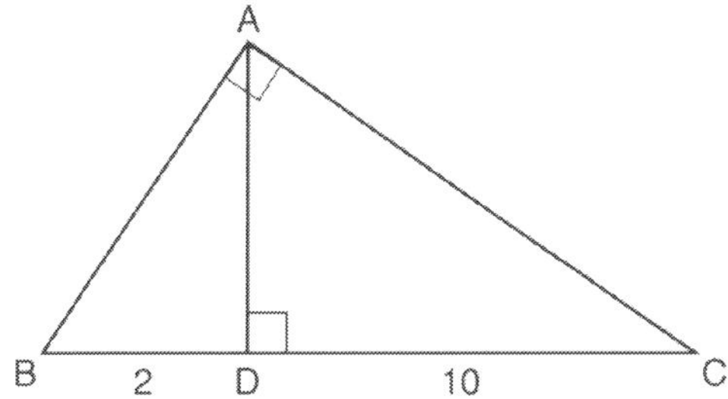

If $BD = 2$ and $DC = 10$, what is the length of $\overline{AB}$?

(1) $2\sqrt{2}$ (3) $2\sqrt{6}$
(2) $2\sqrt{5}$ (4) $2\sqrt{30}$

6. In the diagram below of right triangle ABC, an altitude is drawn to the hypotenuse $\overline{AB}$.

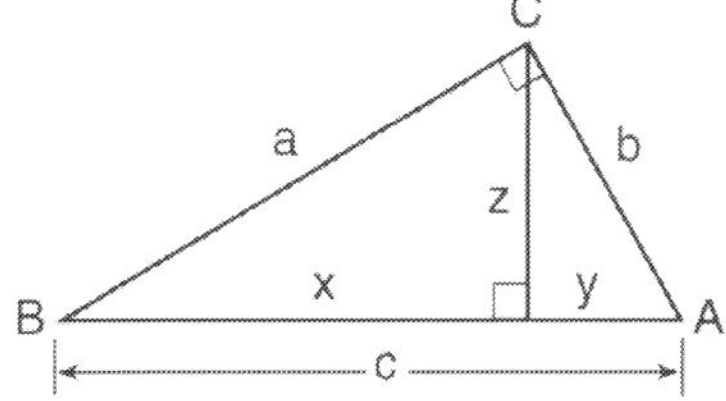

Which proportion would always represent a correct relationship of the segments?

(1) $\frac{c}{z} = \frac{z}{y}$ (3) $\frac{x}{z} = \frac{z}{y}$

(2) $\frac{c}{a} = \frac{a}{y}$ (4) $\frac{y}{b} = \frac{b}{x}$

7. In the diagram below, $\overline{QM}$ is an altitude of right triangle PQR, $PM = 8$, and $RM = 18$.

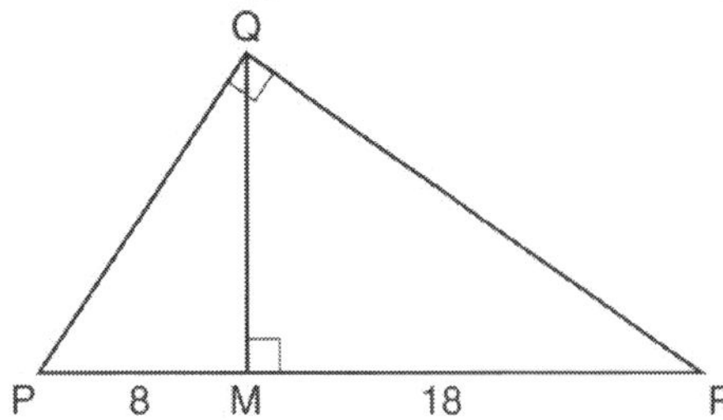

What is the length of $\overline{QM}$?

(1) 20
(2) 16
(3) 12
(4) 10

8. In the diagram below of right triangle ABC, $\overline{CD}$ is the altitude to hypotenuse $\overline{AB}$, $AD = 3$, and $DB = 4$.

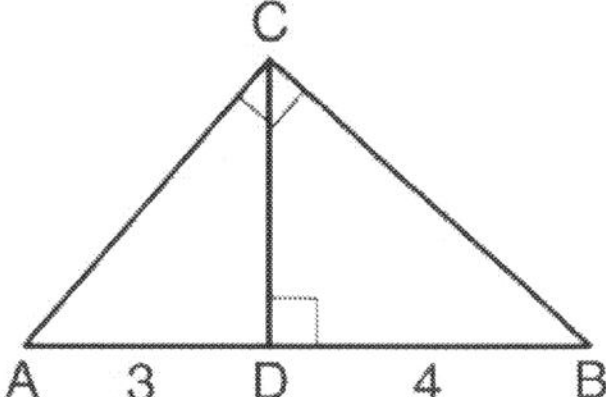

What is the length of $\overline{CB}$?

(1) $2\sqrt{3}$
(2) $\sqrt{21}$
(3) $2\sqrt{7}$
(4) $4\sqrt{3}$

9. (CC) In the diagram below, $\overline{CD}$ is the altitude drawn to the hypotenuse $\overline{AB}$ of right triangle ABC.

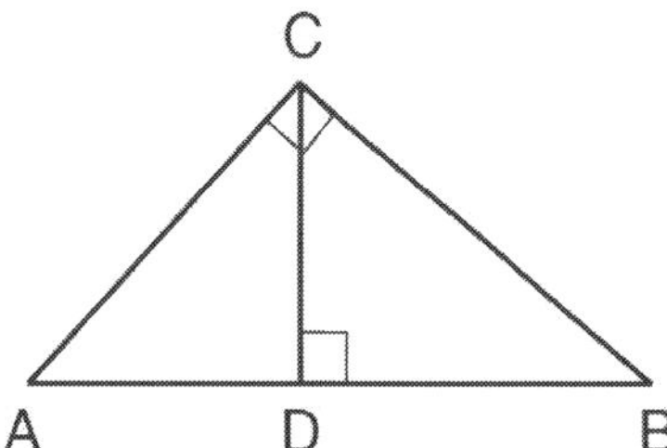

Which lengths would not produce an altitude that measures $6\sqrt{2}$?

(1) $AD = 2$ and $DB = 36$
(2) $AD = 3$ and $AB = 24$
(3) $AD = 6$ and $DB = 12$
(4) $AD = 8$ and $AB = 17$

10. (CC) In $\triangle RST$ shown below, altitude $\overline{SU}$ is drawn to $\overline{RT}$ at U.

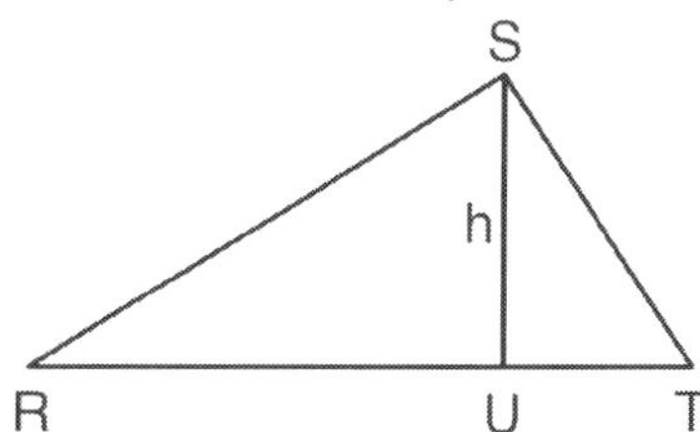

If $SU = h$, $UT = 12$, and $RT = 42$, which value of h will make $\triangle RST$ a right triangle with $\angle RST$ as a right angle?

(1) $6\sqrt{3}$
(2) $6\sqrt{10}$
(3) $6\sqrt{14}$
(4) $6\sqrt{35}$

11. (CC) In the diagram of right triangle ABC, $\overline{CD}$ intersects hypotenuse $\overline{AB}$ at D.

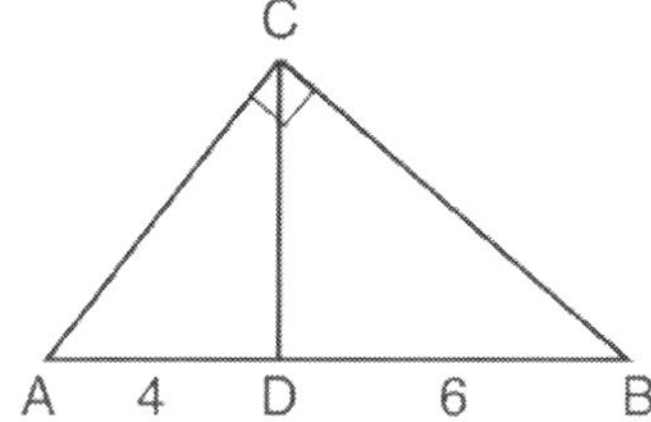

If $AD = 4$ and $DB = 6$, which length of $\overline{AC}$ makes $\overline{CD} \perp \overline{AB}$?

(1) $2\sqrt{6}$
(2) $2\sqrt{10}$
(3) $2\sqrt{15}$
(4) $4\sqrt{2}$

Constructed Response

12. Four streets in a town are illustrated in the accompanying diagram. If the distance on Poplar Street from F to P is 12 miles and the distance on Maple Street from E to M is 10 miles, find the distance on Maple Street, in miles, from M to P.

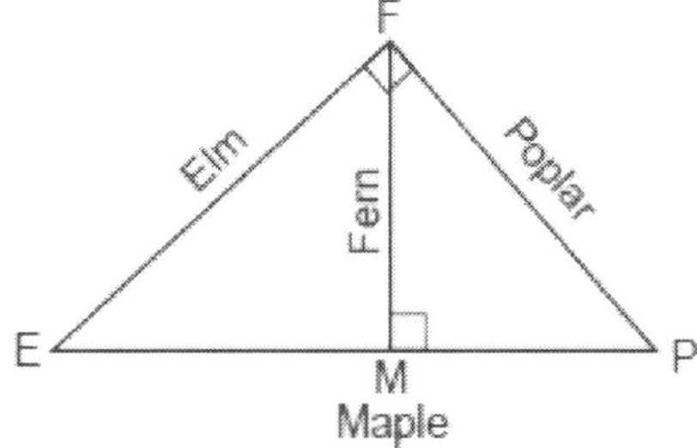

13. In the diagram below of right triangle ACB, altitude $\overline{CD}$ intersects $\overline{AB}$ at D. If $AD = 3$ and $DB = 4$, find the length of $\overline{CD}$ in simplest radical form.

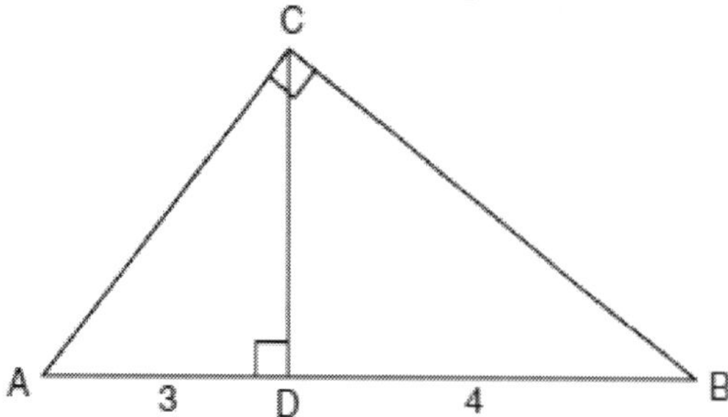

14. The drawing for a right triangular roof truss, represented by $\triangle ABC$, is shown in the accompanying diagram. If $\angle ABC$ is a right angle, altitude $BD = 4$ meters, and $\overline{DC}$ is 6 meters longer than $\overline{AD}$, find the length of base $\overline{AC}$ in meters.

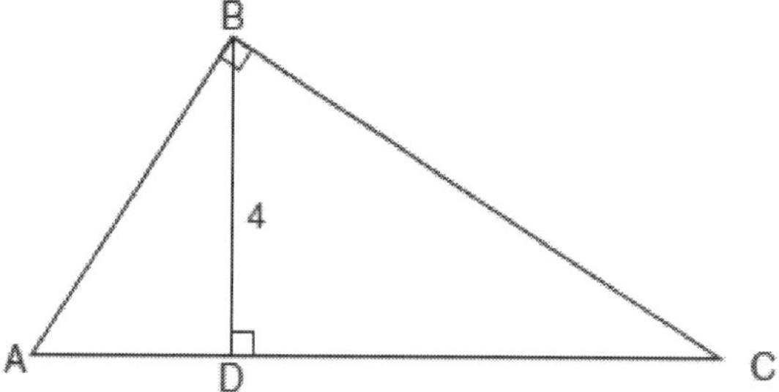

15. In right triangle ABC shown below, altitude $\overline{BD}$ is drawn to hypotenuse $\overline{AC}$.

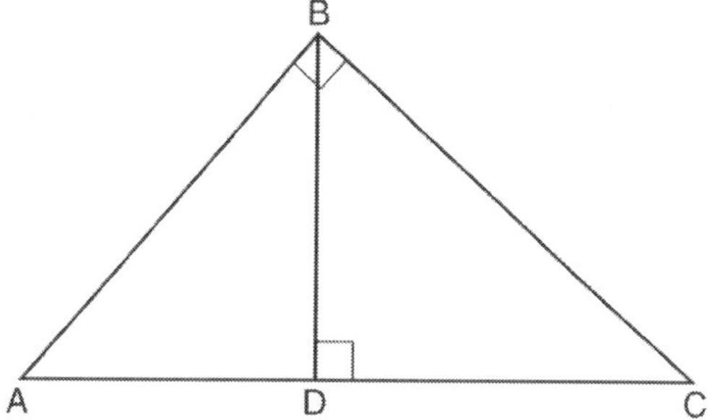

If AD = 8 and DC = 10, determine and state the length of $\overline{AB}$.

16. (CC) In the diagram below, the line of sight from the park ranger station, P, to the lifeguard chair, L, on the beach of a lake is perpendicular to the path joining the campground, C, and the first aid station, F. The campground is 0.25 mile from the lifeguard chair. The straight paths from both the campground and first aid station to the park ranger station are perpendicular.

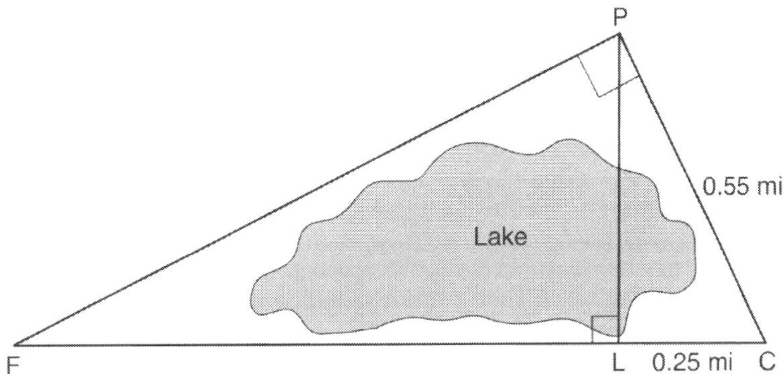

If the path from the park ranger station to the campground is 0.55 mile, determine and state, to the *nearest hundredth of a mile*, the distance between the park ranger station and the lifeguard chair.

Gerald believes the distance from the first aid station to the campground is at least 1.5 miles. Is Gerald correct? Justify your answer.

Chapter 6. Oblique Triangles

6.1 Trigonometric Ratios of Obtuse Angles (+)

*In the Geometry Draft of Revised Standards for 2018-19, NYS has recommended **adding** this topic.*

Key Terms and Concepts

An **oblique triangle** is a triangle without a right angle. The previous unit focused on right triangles only, but the formulas and concepts in this unit will work on any triangle – either right or oblique.

In an oblique triangle, there may be an obtuse angle. An **obtuse angle** is larger than a right angle but smaller than a straight angle ($90° < x < 180°$).

We can still use the trigonometric functions – sine, cosine, and tangent – on obtuse angles. Their values will be equal to, or the negation of, the values of their reference angles. We can find the **reference angle** of an obtuse angle by subtracting from 180°.

Example: For a 150° angle, the reference angle is 180° – 150° = 30°
For a 135° angle, the reference angle is 180° – 135° = 45°

The sine of an obtuse angle is equal to the sine of its reference angle.

Example: $\sin 150° = \sin 30° = \frac{1}{2} = 0.5$

$\sin 135° = \sin 45° = \frac{1}{\sqrt{2}} \approx 0.7071$

The cosine or tangent of an obtuse angle is equal to the *negation* of its reference angle.

Example: $\cos 150° = -\cos 30° = -\frac{\sqrt{3}}{2} \approx -0.8660$

$\tan 150° = -\tan 30° = -\frac{1}{\sqrt{3}} \approx -0.5774$

Model Problem:

$\cos x = -\sin 70°$. Find x.

Solution:

$-\sin 70° = -\cos 20° = \cos 160°$, so $x = 160°$.

Explanation:

Remember that sin and cos are cofunctions, and cofunctions of complementary angles are equal [$\sin 70° = \cos 20°$]. The cosine of an obtuse angle equals the negation of the cosine of its reference angle [$180° - 160° = 20°$, so $\cos 160° = -\cos 20°$].

Practice Problems

1. What is the reference angle for a 100° angle?	2. Name an obtuse angle whose reference angle is 40°.
3. $\sin 175°$ is equal to (1) $\sin 5°$ (3) $\cos 5°$ (2) $-\sin 5°$ (4) $\cos 175°$	4. $\cos 120°$ is equal to (1) $\frac{\sqrt{3}}{2}$ (3) $\frac{1}{2}$ (2) $-\frac{\sqrt{3}}{2}$ (4) $-\frac{1}{2}$

REGENTS QUESTIONS

Multiple Choice

1. Which expression is equivalent to $\cos 120°$?
   (1) $\cos 60°$
   (2) $\cos 30°$
   (3) $-\sin 60°$
   (4) $-\sin 30°$

2. Two straight roads intersect at an angle whose measure is 125°. Which expression is equivalent to the cosine of this angle?
   (1) $\cos 35°$
   (2) $-\cos 35°$
   (3) $\cos 55°$
   (4) $-\cos 55°$

6.2 SAS Sine Formula for Area of a Triangle (+)

*In the Geometry Draft of Revised Standards for 2018-19, NYS has recommended **adding** this topic.*

Key Terms and Concepts

We can use trigonometry to find the area of *any* triangle (either right or oblique) where we are given the lengths of two sides and the measure of the included angle (SAS). The formula is:

$Area = \frac{1}{2}ab\sin C$, where C is the known angle and a and b are the lengths of the adjacent sides.

Note that if C is a right angle, this is equivalent to $Area = \frac{1}{2}ab$, since $\sin 90° = 1$.

The SAS Sine Formula is derived as follows:
Given triangle ABC, we can draw an altitude from vertex B, with length h, as shown.

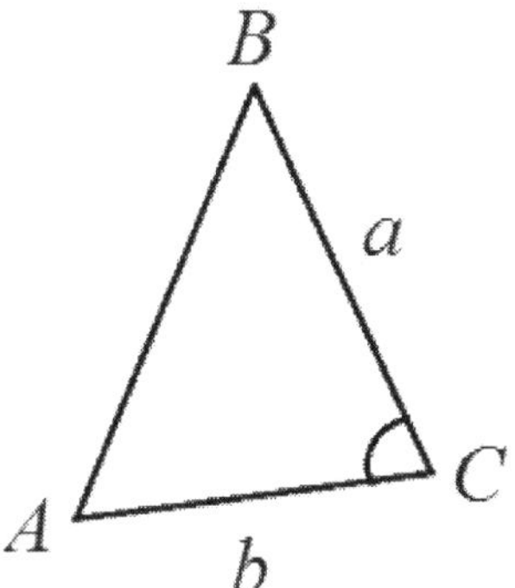

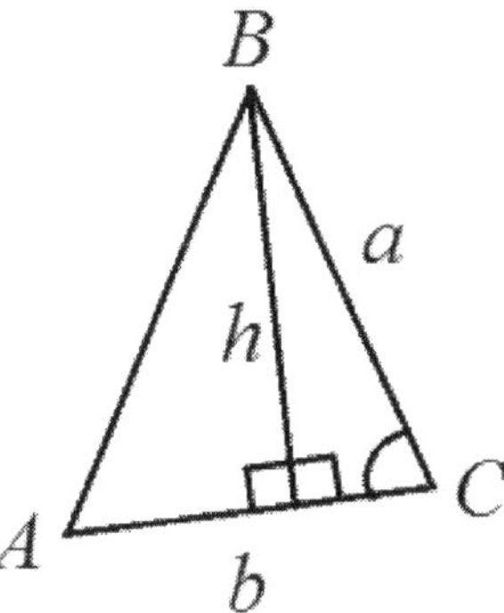

We know a formula for the area of the triangle is $Area = \frac{1}{2}bh$, but the length of h is not given. However, by drawing an altitude, we've created a right triangle, shaded below.

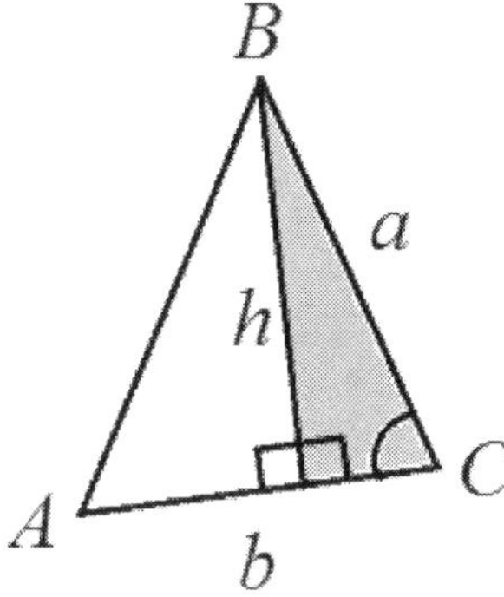

In the shaded right triangle, there is the relationship, $\sin C = \frac{h}{a}$.

Solving for h, this gives us $h = a\sin C$.

We can now substitute for h in the area formula, $Area = \frac{1}{2}bh$, giving us $Area = \frac{1}{2}b \cdot a\sin C$, or more simply written as $Area = \frac{1}{2}ab\sin C$.

Model Problem:

Find the area of triangle *PQR*, shown below, in square units.

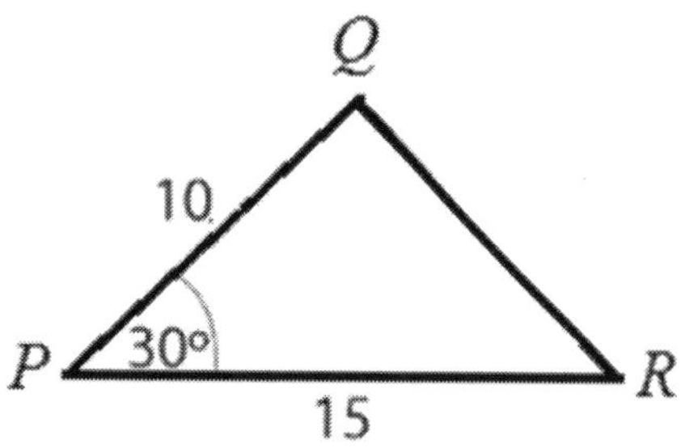

Solution:

(A) $Area = \frac{1}{2}ab\sin C$

(B) $= \frac{1}{2}(10)(15)\sin 30°$

(C) $= 37.5$

Explanation of steps:

(A) Write the formula $Area = \frac{1}{2}ab\sin C$

(B) Substitute the measures of the known angle *[30°]* for *C* and the adjacent sides *[10 and 15]* for *a* and *b*.

(C) Simplify, using the calculator to find the sine of the angle.

Practice Problems

1. In $\triangle ABC$, $a = 6$, $b = 8$, and $\sin C = \frac{1}{4}$. Find the area of $\triangle ABC$.	2. In $\triangle ABC$, $a = 12$, $b = 15$, and $m\angle C = 150°$. Find the area of $\triangle ABC$.
3. Find the area of $\triangle ABC$ to the nearest tenth of a square unit. 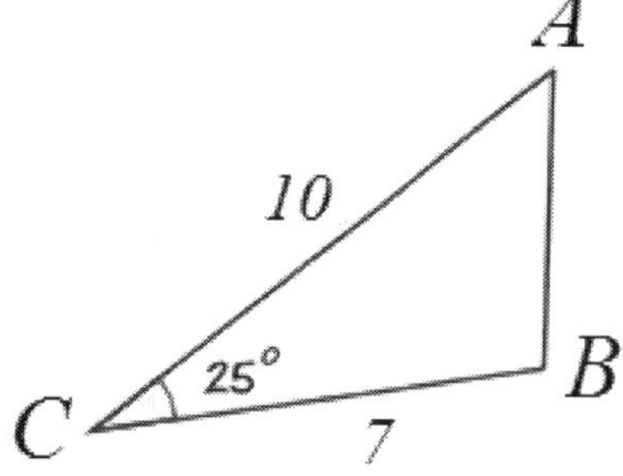	4. Find the area of $\triangle ABC$, shown below. 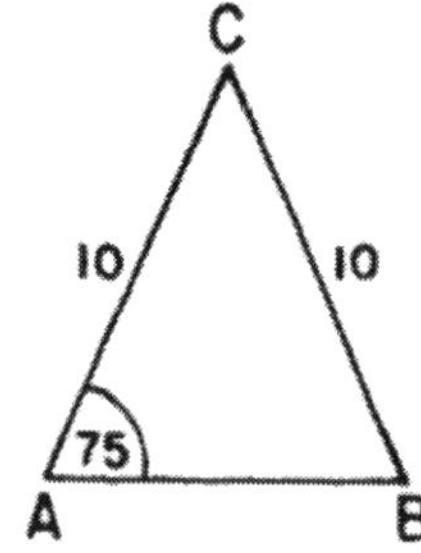

REGENTS QUESTIONS

Multiple Choice

1. If the vertex angle of an isosceles triangle measures 30° and each leg measures 4, the area of the triangle is
   (1) $8\sqrt{3}$ (3) $4\sqrt{3}$
   (2) 8 (4) 4

2. In $\triangle ABC$, $m\angle C = 30$ and $a = 8$. If the area of the triangle is 12, what is the length of side b?
   (1) 6 (3) 3
   (2) 8 (4) 4

3. What is the best approximation for the area of a triangle with consecutive sides of 4 and 5 and an included angle of 59°?
   (1) 5.0 (3) 10.0
   (2) 8.6 (4) 17.1

4. Jack is planting a triangular rose garden. The lengths of two sides of the plot are 8 feet and 12 feet, and the angle between them is 87°. Which expression could be used to find the area of this garden?
   (1) $8 \cdot 12 \cdot \sin 87°$ (3) $\frac{1}{2} \cdot 8 \cdot 12 \cdot \cos 87°$
   (2) $8 \cdot 12 \cdot \cos 87°$ (4) $\frac{1}{2} \cdot 8 \cdot 12 \cdot \sin 87°$

5. In $\triangle ABC$, $m\angle A = 120$, $b = 10$, and $c = 18$. What is the area of $\triangle ABC$ to *the nearest square inch*?
   (1) 52 (3) 90
   (2) 78 (4) 156

6. The area of triangle ABC is 42. If $AB = 8$ and $m\angle B = 61$, the length of $\overline{BC}$ is approximately
   (1) 5.1 (3) 12.0
   (2) 9.2 (4) 21.7

7. Two sides of a triangular-shaped sandbox measure 22 feet and 13 feet. If the angle between these two sides measures 55°, what is the area of the sandbox, to the *nearest square foot*?
   (1) 82 (3) 143
   (2) 117 (4) 234

8. In $\triangle RST$, $m\angle S = 135$, $r = 27$, and $t = 19$. What is the area of $\triangle RST$ to the *nearest tenth of a square unit*?
   (1) 90.7 (3) 256.5
   (2) 181.4 (4) 362.7

Constructed Response

9. In $\triangle ABC$, $m\angle C = 30$ and $a = 24$. If the area of the triangle is 42, what is the length of side b?

10. The accompanying diagram shows the floor plan for a kitchen. The owners plan to carpet all of the kitchen except the "work space," which is represented by scalene triangle ABC. Find the area of this work space to the *nearest tenth of a square foot*.

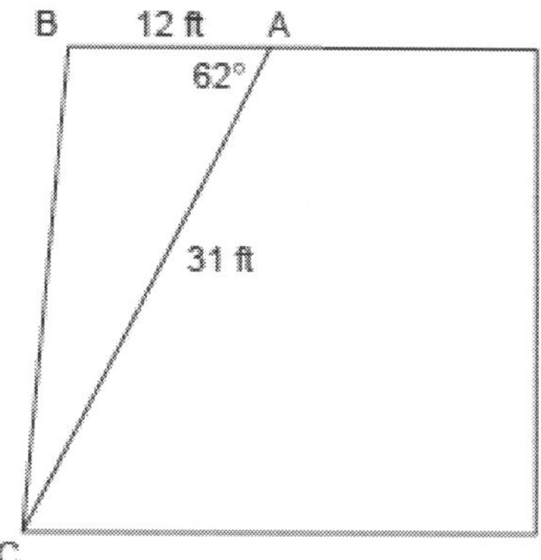

11. In $\triangle DEF$, $m\angle D = 40$, $DE = 12$ meters, and $DF = 8$ meters. Find the area of $\triangle DEF$ to the *nearest tenth of a square meter*.

12. Two sides of a triangular-shaped pool measure 16 feet and 21 feet, and the included angle measures 58°. What is the area, to the *nearest tenth of a square foot*, of a nylon cover that would exactly cover the surface of the pool?

13. In $\triangle ABC$, $m\angle B = 30$ and side $a = 6$. If the area of the triangle is 12, what is the length of side c?

14. The triangular top of a table has two sides of 14 inches and 16 inches, and the angle between the sides is 30°. Find the area of the tabletop, in square inches.

15. If $m\angle A = 30$, side $b = 8$, and side $c = 4$, find the area of $\triangle ABC$.

16. A landscape architect is designing a triangular garden to fit in the corner of a lot. The corner of the lot forms an angle of 70°, and the sides of the garden including this angle are to be 11 feet and 13 feet, respectively. Find, to the *nearest integer*, the number of square feet in the area of the garden.

17. The accompanying diagram shows the peak of a roof that is in the shape of an isosceles triangle. A base angle of the triangle is 50° and each side of the roof is 20.4 feet. Determine, to the *nearest tenth of a square foot*, the area of this triangular region.

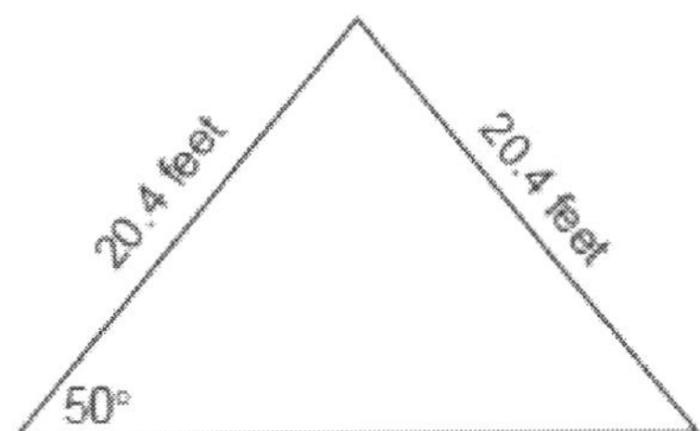

18. In $\triangle ABC$, $a = 12$, $b = 20.5$, and $m\angle C = 73$. Find the area of $\triangle ABC$, to the *nearest tenth.*

19. Determine the area, to the *nearest integer*, of $\triangle SRO$ shown below.

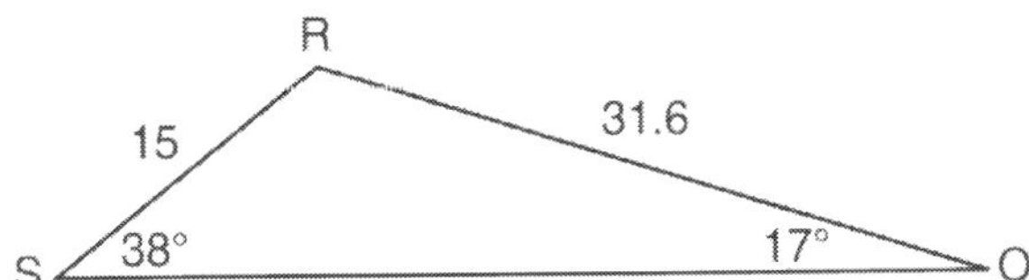

6.3 Law of Sines (+)

*In the Geometry Draft of Revised Standards for 2018-19, NYS has recommended **adding** this topic.*

Key Terms and Concepts

We can use the Law of Sines to find an unknown side or angle in an oblique triangle. (Although this method also works for right triangles, we would prefer to use the more direct methods from the previous unit in the case of a right triangle.)

The **Law of Sines** states that, in any triangle, the ratios of each side to the sine of its opposite angle are in proportion. The formula may be stated as:

$$\frac{a}{\sin A} = \frac{b}{\sin B} = \frac{c}{\sin C}$$, where side a is opposite angle A, etc.

We can derive this formula by drawing altitudes in the triangle. For example, in the triangle below, an altitude is drawn from vertex A, dividing the triangle into two right triangles.

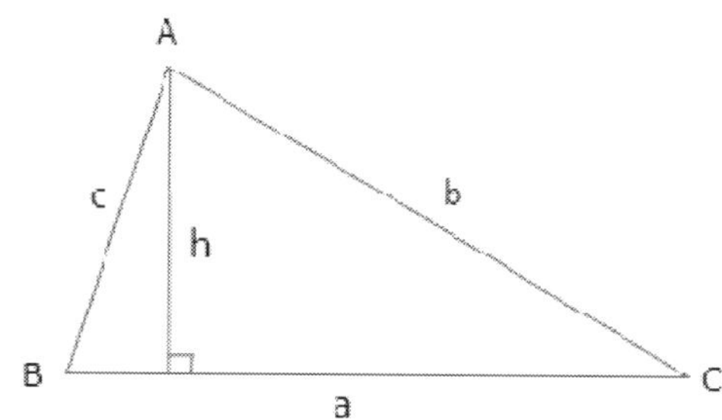

This determines two sine ratios:
$\sin B = \frac{h}{c}$ and $\sin C = \frac{h}{b}$.

Solving both equations for h gives us $h = c \sin B$ and $h = b \sin C$.

Therefore, $c \sin B = b \sin C$.

Dividing through by sin B and then by sin C gives us $\frac{c}{\sin C} = \frac{b}{\sin B}$.

Repeating this process by drawing an altitude from vertex B would establish that $\frac{c}{\sin C} = \frac{a}{\sin A}$.

To use the Law of Sines:

1. Write the Law of Sines: $\frac{a}{\sin A} = \frac{b}{\sin B} = \frac{c}{\sin C}$
2. Substitute the known values into the formula.
3. Remove the fraction that is unhelpful.
4. Solve the remaining equation.

Note to instructors: The ambiguous case for the Law of Sines is not addressed in this course.

Model Problem 1: *finding an unknown side*
Find the length of side a.

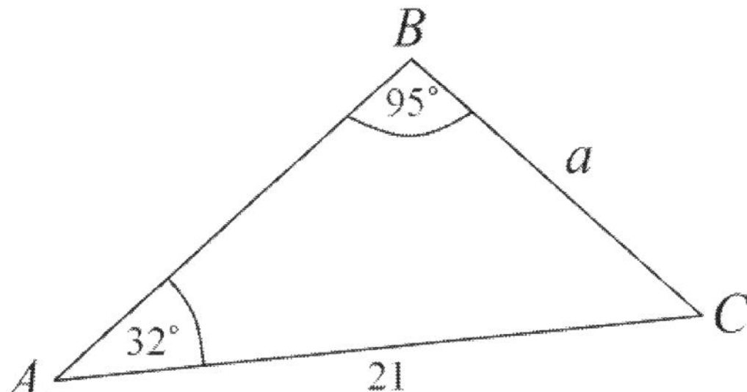

Solution:

(A) $\dfrac{a}{\sin A} = \dfrac{b}{\sin B} = \dfrac{c}{\sin C}$

(B) $\dfrac{a}{\sin 32°} = \dfrac{21}{\sin 95°} = \cancel{\dfrac{c}{\sin C}}$

(C) $\dfrac{a}{\sin 32°} = \dfrac{21}{\sin 95°}$

(D) $a = \dfrac{21}{\sin 95°} \cdot \sin 32°$
$a \approx 11.17$

Explanation of steps:

(A) Write the Law of Sines.
(B) Substitute the known values into the formula.
[b = 21, A = 32°, and B = 95°]
(C) Remove the fraction that is unhelpful.
[Since we are given neither c nor C, the last fraction may be removed.]
(D) Solve the remaining equation. *[Solve for a.]*

Practice Problems

1. In $\triangle FUN$, $f = 4$, $m\angle F = 26$, and $m\angle N = 67$. Find the value of n to the *nearest integer*.	2. In $\triangle ABC$, $\sin A = \frac{1}{4}$, $\sin B = \frac{1}{8}$, and $b = 20$. What is the length of a?

3. A ship at sea heads directly toward a cliff on the shoreline. The accompanying diagram shows the top of the cliff, D, sighted from two locations, A and B, separated by distance S. If $m\angle DAC = 30$, $m\angle DBC = 45$, and $S = 30$ feet, what is the height of the cliff, to the *nearest foot*?

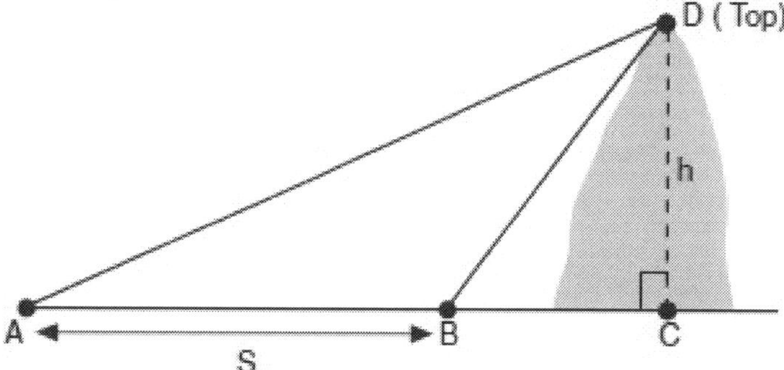

Model Problem 2: *finding an unknown angle*
Find the measure of angle B.

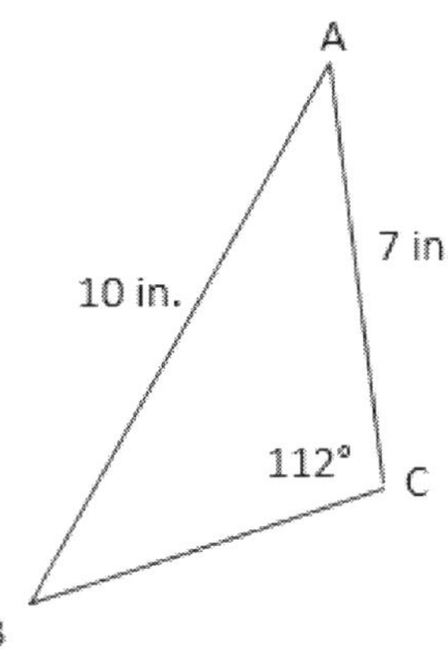

Solution:

(E) $\dfrac{a}{\sin A} = \dfrac{b}{\sin B} = \dfrac{c}{\sin C}$

(F) $\dfrac{a}{\sin A} = \dfrac{7}{\sin B} = \dfrac{10}{\sin 112°}$

(G) $\dfrac{7}{\sin B} = \dfrac{10}{\sin 112°}$

$7(\sin 112°) = 10(\sin B)$

(H) $\dfrac{7(\sin 112°)}{10} = \sin B$

$\sin B \approx 0.6490$

(I) $m\angle B \approx \sin^{-1} 0.6490$

$m\angle B \approx 40.5°$

Explanation of steps:

(E) Write the Law of Sines.

(F) Substitute the known values into the formula.
[b = 7, c = 10, and C = 112°]

(G) Remove the fraction that is unhelpful.
[Since we are given neither a nor A, the first fraction may be removed.]

(H) Solve the remaining equation.
[Cross multiply, then express in terms of sin B. If rounding, go to at least 4 decimal places.]

(I) To find the angle measure, you'll need to use the arcsin function, [SIN⁻¹], on your calculator.

Practice Problems

4. In acute triangle ABC, side $a = 10$, side $b = 12$, and $m\angle A = 42$. Find $m\angle B$ to the *nearest degree.*	5. In $\triangle ABC$, side $a = 3$, side $c = 3\sqrt{2}$, and $m\angle A = 45$. Find $m\angle C$.

REGENTS QUESTIONS

Multiple Choice

1. In $\triangle ABC$, $m\angle A = 33$, $a = 12$, and $b = 15$. What is $m\angle B$ to the *nearest degree*?
   (1) 41 (3) 44
   (2) 43 (4) 48

2. In $\triangle ABC$, $m\angle A = 30$, $a = 12$, and $b = 10$. Which type of triangle is $\triangle ABC$?
   (1) acute (3) obtuse
   (2) isosceles (4) right

3. In the accompanying diagram of $\triangle ABC$, $m\angle A = 30$, $m\angle C = 50$, and $AC = 13$.

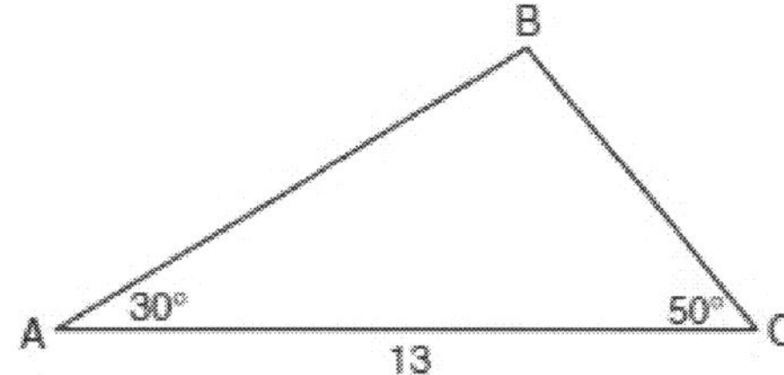

   What is the length of side $\overline{AB}$ to the *nearest tenth*?
   (1) 6.6 (3) 11.5
   (2) 10.1 (4) 12.0

Constructed Response

4. In the accompanying diagram of $\triangle ABC$, $m\angle A = 65$, $m\angle B = 70$, and the side opposite vertex B is 7. Find the length of the side opposite vertex A, and find the area of $\triangle ABC$.

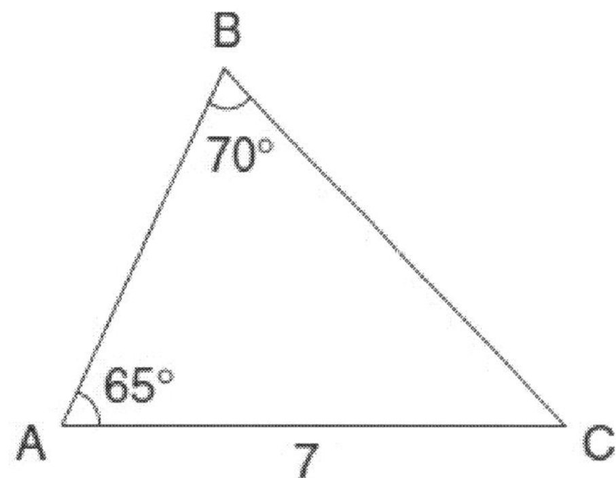

5. A ski lift begins at ground level 0.75 mile from the base of a mountain whose face has a 50° angle of elevation, as shown in the accompanying diagram. The ski lift ascends in a straight line at an angle of 20°. Find the length of the ski lift from the beginning of the ski lift to the top of the mountain, to the *nearest hundredth of a mile*.

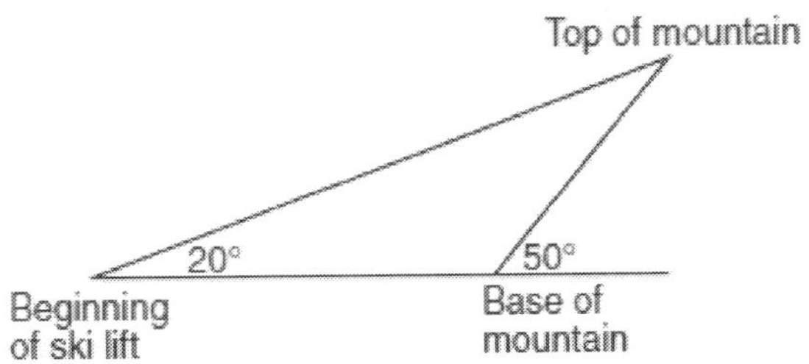

6. As shown in the accompanying diagram, two tracking stations, A and B, are on an east-west line 110 miles apart. A forest fire is located at F, on a bearing 42° northeast of station A and 15° northeast of station B. How far, to the *nearest mile*, is the fire from station A?

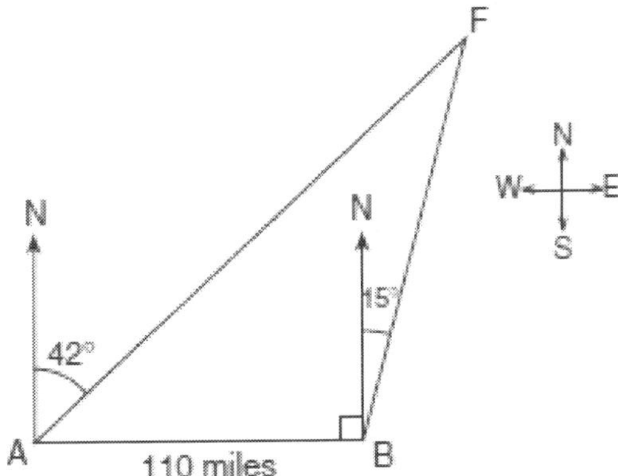

7. The Vietnam Veterans Memorial in Washington, D.C., is made up of two walls, each 246.75 feet long, that meet at an angle of 125.2°. Find, to the *nearest foot*, the distance between the ends of the walls that do not meet.

8. In $\triangle ABC$, $m\angle A = 53$, $m\angle B = 14$, and $a = 10$. Find b to the *nearest integer*.

9. In the accompanying diagram of a streetlight, the light is attached to a pole at R and supported by a brace, $\overline{PQ}$, $RQ = 10$ feet, $RP = 6$ feet, $\angle PRQ$ is an obtuse angle, and $m\angle PQR = 30$. Find the length of the brace, $\overline{PQ}$, to the *nearest foot*.

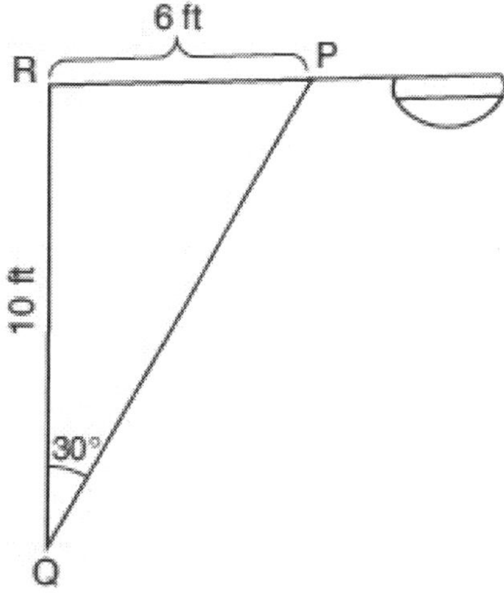

10. In $\triangle ABC$, $\sin A = 0.6$, $a = 10$, and $b = 7$. Find $\sin B$.

11. In the accompanying diagram of $\triangle RST$, $RS = 30$ centimeters, $m\angle T = 105$, and $m\angle R = 40$. Find the area of $\triangle RST$, to the *nearest square centimeter*.

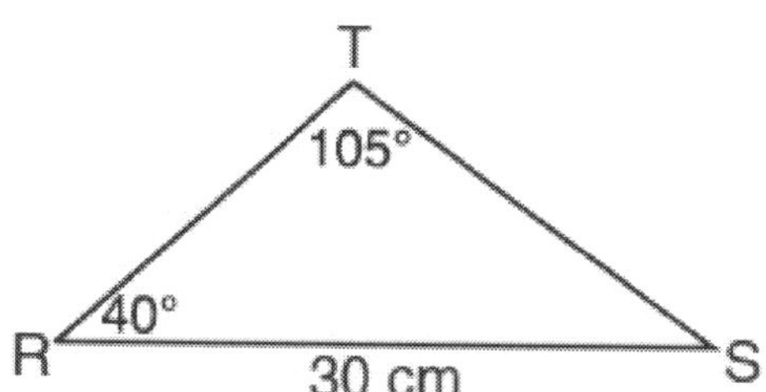

12. In $\triangle ABC$, $m\angle A = 32$, $a = 12$, and $b = 10$. Find the measures of the missing angles and side of $\triangle ABC$. Round each measure to the *nearest tenth*.

13. The diagram below shows the plans for a cell phone tower. A guy wire attached to the top of the tower makes an angle of 65 degrees with the ground. From a point on the ground 100 feet from the end of the guy wire, the angle of elevation to the top of the tower is 32 degrees. Find the height of the tower, to the *nearest foot*.

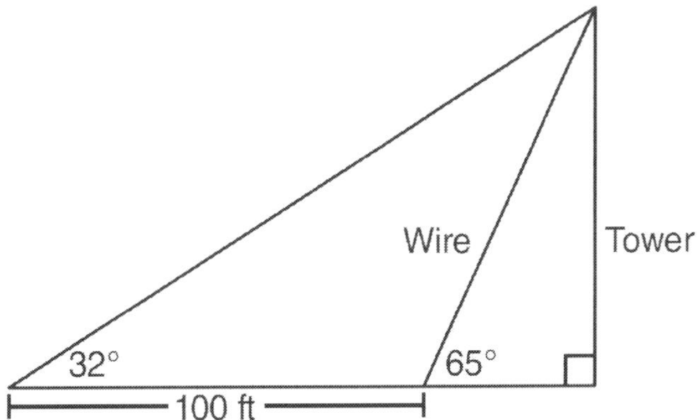

14. As shown in the diagram below, fire-tracking station A is 100 miles due west of fire-tracking station B. A forest fire is spotted at F, on a bearing 47° northeast of station A and 15° northeast of station B. Determine, to the *nearest tenth of a mile*, the distance the fire is from *both* station A and station B. [N represents due north.]

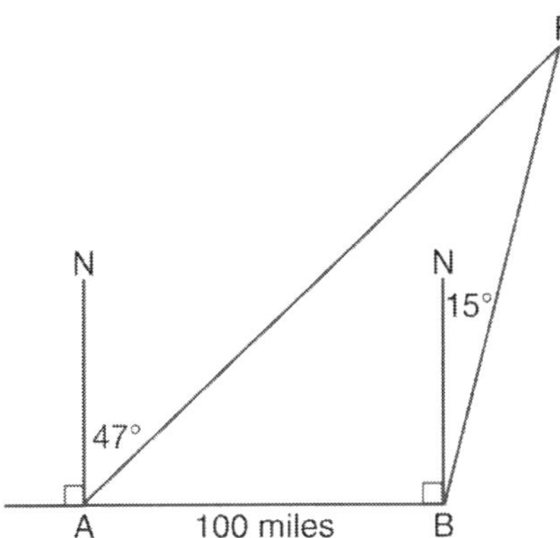

15. A ranch in the Australian Outback is shaped like triangle ACE, with $m\angle A = 42$, $m\angle E = 103$, and $AC = 15$ miles. Find the area of the ranch, to the *nearest square mile*.

6.4 Law of Cosines (+)

*In the Geometry Draft of Revised Standards for 2018-19, NYS has recommended **adding** this topic.*

Key Terms and Concepts

The Law of Sines depends on being given an angle and its opposite side. However, if this is not given, the Law of Cosines may be used instead.

The **Law of Cosines** states:

$$a^2 = b^2 + c^2 - 2bc \cos A$$

To derive the formula, we can draw an altitude from vertex *C* to point *D* as shown below, dividing *c* into two parts labelled *x* and *c* – *x*.

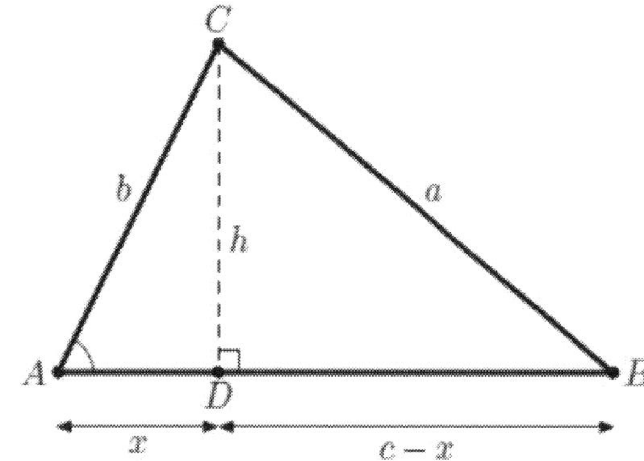

Now, in right triangle ADC, we know $x^2 + h^2 = b^2$ by the Pythagorean Theorem. We also know that $\cos A = \frac{x}{b}$, so $x = b\cos A$.

In right triangle BDC, we know $a^2 = (c - x)^2 + h^2$ by the Pythagorean Theorem. By squaring the binomial in parentheses, this can be expanded to $a^2 = c^2 - 2cx + x^2 + h^2$.

Since $x^2 + h^2 = b^2$, we can substitute b^2 for $x^2 + h^2$ in the equation, $a^2 = c^2 - 2cx + \underline{x^2 + h^2}$, which gives us $a^2 = c^2 - 2cx + b^2$. Using the commutative property to rearrange terms, we can rewrite this equation as $a^2 = b^2 + c^2 - 2cx$.

Finally, since $x = b\cos A$, we can substitute for *x*, which gives us $a^2 = b^2 + c^2 - 2bc \cos A$.

To use the Law of Cosines:

1. Write the Law of Cosines: $a^2 = b^2 + c^2 - 2bc \cos A$
2. Substitute the known values into the formula, making sure that the side (*a*) on the left side of the equation is opposite the angle (*A*) on the right side of the equation.
3. Solve for the missing side or angle.

To determine which technique is best to use in a given situation, you can follow this chart:

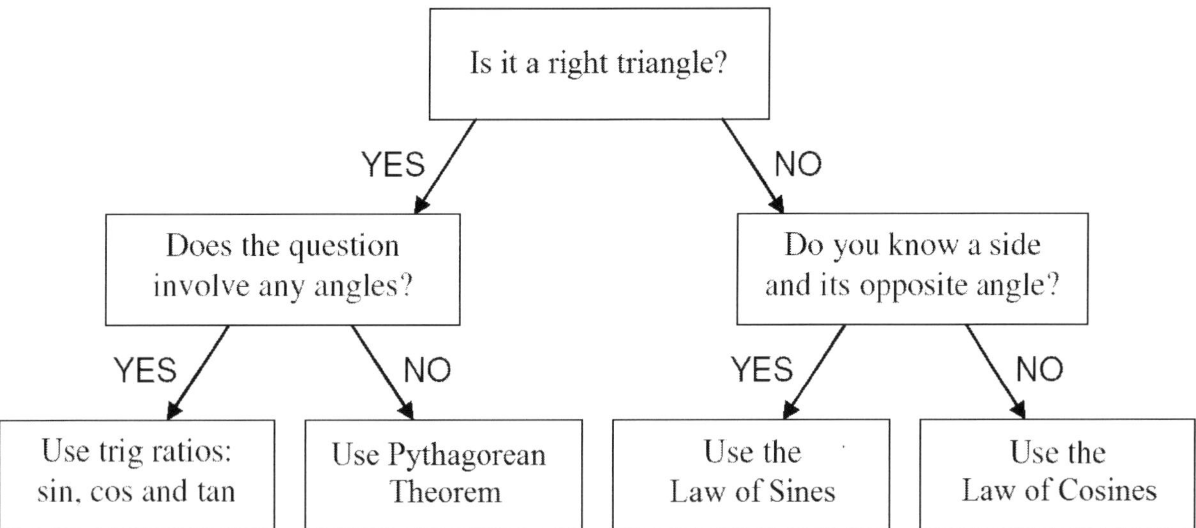

Model Problem 1: *finding a side*

Find a, the length of $\overline{BC}$, in triangle ABC shown to the right.

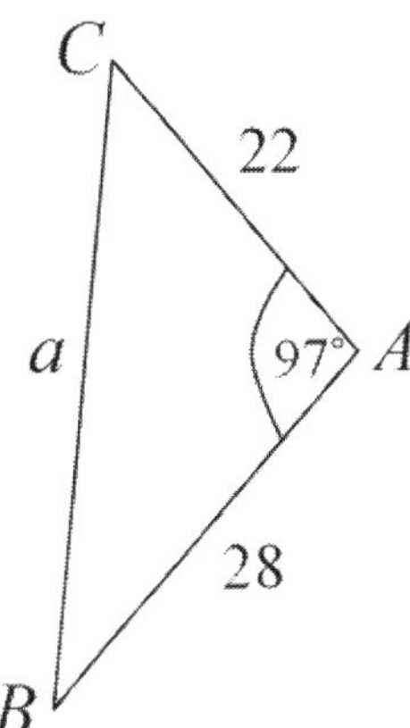

Solution:

(A) $a^2 = b^2 + c^2 - 2bc\cos A$

(B) $a^2 = 22^2 + 28^2 - 2(22)(28)\cos 97°$

(C) $a^2 \approx 1{,}418.143$

(D) $a \approx 37.7$

Explanation of steps:

(A) Write the Law of Cosines.

(B) Substitute the known values into the formula, making sure that the side (a) on the left side of the equation is opposite the angle (A) on the right side of the equation. *[b = 22, c = 28, and A = 97°]*

(C) Evaluate the right side.

(D) Take the square root of both sides (ignoring the negative root).

Practice Problems

1. In $\triangle CAT$, $a = 4$, $c = 5$, and $\cos T = \frac{1}{8}$. What is the length of t to the *nearest tenth*?	2. In triangle ABC, $a = 5$, $b = 8$, and $m\angle C = 60$. Find the length of side c.

Model Problem 2: *finding an angle*

In $\triangle ABC$, if $a = 5$, $b = 6$, and $c = 8$, find $m\angle A$ to the *nearest degree.*

Solution:

(A) $a^2 = b^2 + c^2 - 2bc\cos A$

(B) $5^2 = 6^2 + 8^2 - 2(6)(8)\cos A$

(C) $25 = 100 - 96\cos A$

$-75 = -96\cos A$

$0.78125 = \cos A$

(D) $m\angle A = \cos^{-1} 0.78125 \approx 39$

Explanation of steps:

(A) Write the Law of Cosines.

(B) Substitute the known values into the formula, making sure that the side (a) on the left side of the equation is opposite the angle (A) on the right side of the equation.

(C) Solve the equation for $\cos A$.

(D) Use the inverse function, $\cos^{-1}$ (called the arccosine), to find the angle measure.

Practice Problems

3. In $\triangle ABC$, if $a = 4$, $b = 3$, and $c = 3$, find $m\angle A$ to the *nearest degree*.	4. The sides of a triangle measure 6, 7, and 9. What is the measure of the largest angle?

5. Firefighters dug three trenches in the shape of a triangle to prevent a fire from completely destroying a forest. The lengths of the trenches were 250 feet, 312 feet, and 490 feet. Find, to the *nearest thousandth of a degree*, the smallest angle formed by the trenches.

Find the area of the plot of land within the trenches, to the *nearest square foot*.

REGENTS QUESTIONS

Multiple Choice

1. To the *nearest degree*, what is the measure of the largest angle in a triangle with sides measuring 10, 12, and 18 centimeters?
   (1) 109 (3) 71
   (2) 81 (4) 32

2. Al is standing 50 yards from a maple tree and 30 yards from an oak tree in the park. His position is shown in the accompanying diagram. If he is looking at the maple tree, he needs to turn his head 120° to look at the oak tree.

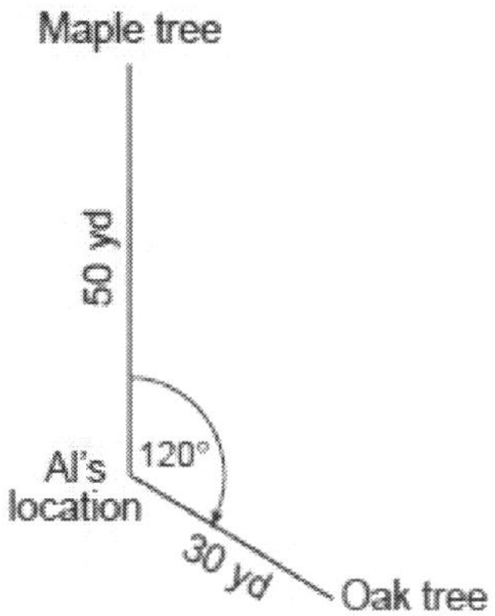

   How many yards apart are the two trees?
   (1) 58.3 (3) 70
   (2) 65.2 (4) 75

3. In $\triangle ABC$, $a = 3$, $b = 5$, and $c = 7$. What is $m\angle C$?
   (1) 22 (3) 60
   (2) 38 (4) 120

4. In $\triangle ABC$, $a = 15$, $b = 14$, and $c = 13$, as shown in the diagram below. What is the $m\angle C$, to the *nearest degree*?

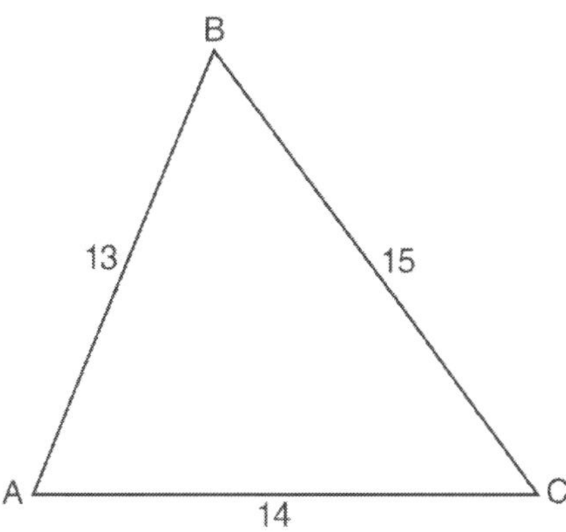

(1) 53
(2) 59
(3) 67
(4) 127

5. In $\triangle FGH$, f = 6, g = 9, and $m\angle H = 57$. Which statement can be used to determine the numerical value of *h*?

(1) $h^2 = 6^2 + 9^2 - 2(9)(h)\cos 57°$
(2) $h^2 = 6^2 + 9^2 - 2(6)(9)\cos 57°$
(3) $6^2 = 9^2 + h^2 - 2(9)(h)\cos 57°$
(4) $9^2 = 6^2 + h^2 - 2(6)(h)\cos 57°$

Constructed Response

6. Hersch says if a triangle is an obtuse triangle, then it cannot also be an isosceles triangle. Using a diagram, show that Hersch is incorrect, and indicate the measures of all the angles and sides to justify your answer.

7. The playground at a day-care center has a triangular-shaped sandbox. Two of the sides measure 20 feet and 14.5 feet and form an included angle of 45. Find the length of the third side of the sandbox to the *nearest tenth of a foot*.

8. Two straight roads, Elm Street and Pine Street, intersect creating a 40° angle, as shown in the accompanying diagram. John's house (*J*) is on Elm Street and is 3.2 miles from the point of intersection. Mary's house (*M*) is on Pine Street and is 5.6 miles from the intersection. Find, to the *nearest tenth of a mile*, the direct distance between the two houses.

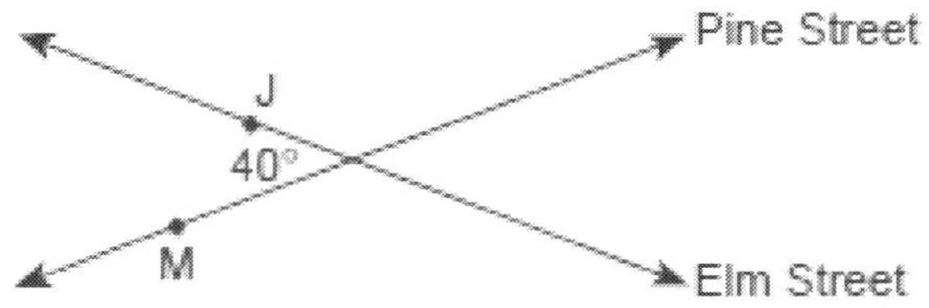

9. In $\triangle DEF$, if side $d = 14$, side $e = 10$, and side $f = 12$, find $m\angle F$ to the *nearest degree*.

10. A ship at sea is 70 miles from one radio transmitter and 130 miles from another. The angle between the signals sent to the ship by the transmitters is 117.4°. Find the distance between the two transmitters, to the *nearest mile*.

11. As shown in the accompanying diagram, a ship at sea is 75 miles from radio transmitter A and 120 miles from radio transmitter B. The angle between the signals sent to the ship by the two transmitters measures 135.

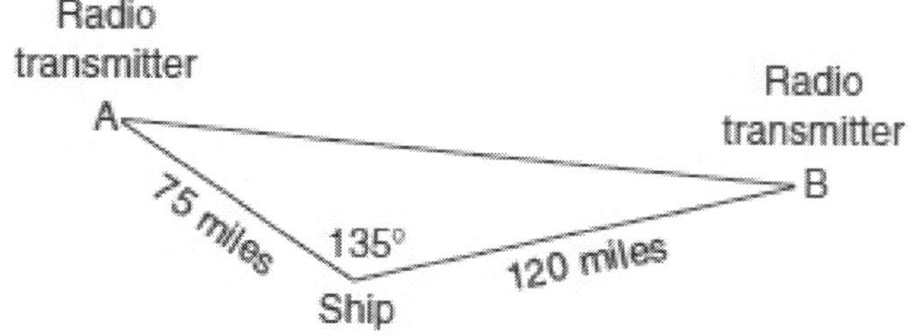

Find the distance between the transmitters to the *nearest mile*. Using this answer, find the measure of angle B to the *nearest degree*.

12. A surveyor is mapping a triangular plot of land. He measures two of the sides and the angle formed by these two sides and finds that the lengths are 400 yards and 200 yards and the included angle is 50°. What is the measure of the third side of the plot of land, to the *nearest yard*? What is the area of this plot of land, to the *nearest square yard*?

13. To measure the distance through a mountain for a proposed tunnel, surveyors chose points A and B at each end of the proposed tunnel and a point C near the mountain. They determined that AC = 3,800 meters, BC = 2,900 meters, and $m\angle ACB = 110$. Draw a diagram to illustrate this situation and find the length of the tunnel, to the *nearest meter*.

14. In $\triangle ABC$, $a = 24$, $b = 36$, and $c = 30$. Find $m\angle A$ to the *nearest tenth of a degree*.

15. In a triangle, two sides that measure 6 cm and 10 cm form an angle that measures 80°. Find, to the *nearest degree*, the measure of the smallest angle in the triangle.

16. Find the measure of the smallest angle, to the *nearest degree*, of a triangle whose sides measure 28, 47, and 34.

17. In a triangle, two sides that measure 8 centimeters and 11 centimeters form an angle that measures 82°. To the *nearest tenth of a degree*, determine the measure of the smallest angle in the triangle.

18. The lengths of the sides of a triangle are 6 cm, 11 cm, and 7 cm. Determine, to the *nearest tenth of a degree*, the measure of the largest angle of the triangle.

Chapter 7. Quadrilaterals

7.1 Properties of Quadrilaterals

Key Terms and Concepts

The following tree diagram shows the classification of various types of quadrilaterals. Each figure inherits the properties of its parent.

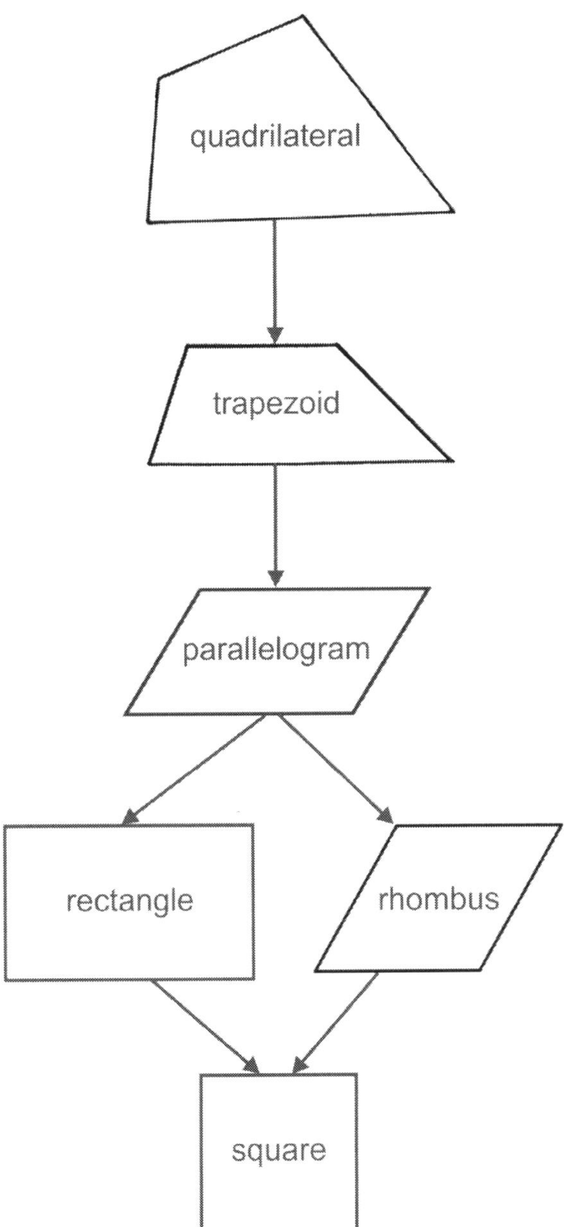

A **quadrilateral** is a four-sided polygon.

A **trapezoid** is a quadrilateral with at least one pair of parallel sides. A trapezoid has:

- at least one pair of parallel sides

In an **isosceles trapezoid**, at least one pair of opposite sides are congruent.

A **parallelogram** is a trapezoid with two pairs of parallel sides.
Parallelograms have the following properties:

- both pairs of opposite sides are parallel
- both pairs of opposite sides are congruent
- both pairs of opposite angles are congruent
- consecutive angles are supplementary
- diagonals bisect each other

A **rectangle** is an equiangular parallelogram.
It has all the properties of a parallelogram plus:

- all four angles are right angles
- diagonals are congruent

A **rhombus** is an equilateral parallelogram.
It has all the properties of a parallelogram plus:

- all four sides are congruent
- diagonals are perpendicular
- diagonals are angle bisectors

A **square** is both equiangular and equilateral, so it has all the properties of a rectangle and a rhombus.

Model Problem:

Find the measures of the numbered angles in the parallelogram.

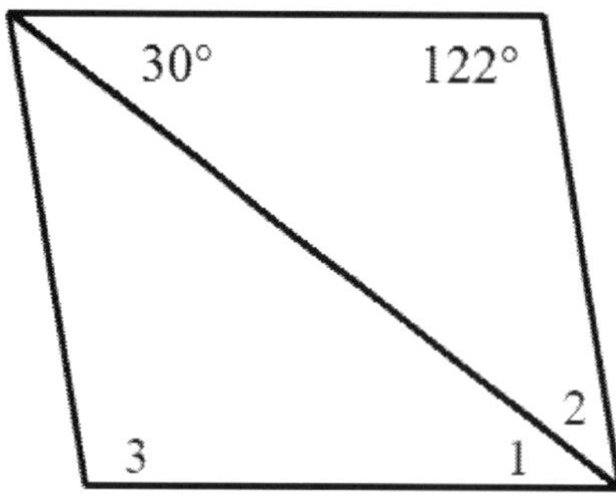

Solution:

(A) $m\angle 1 = 30$
(B) $m\angle 2 = 28$
(C) $m\angle 3 = 122$

Explanation of steps:

(A) In a parallelogram, opposite sides are parallel. Therefore, a diagonal acts as a transversal forming alternate interior angles that are congruent *[$\angle 1$ and the 30° angle are alternate interior angles]*.

(B) In a parallelogram, consecutive angles are supplementary. *[The whole angle consisting of $\angle 1 + \angle 2$ is supplementary to the 122° angle. Since $m\angle 1 = 30$, $m\angle 2 = (180 - 122) - 30 = 28$.]*

(C) In a parallelogram, opposite angles are congruent. *[$\angle 3$ is opposite the 122° angle.]*

Practice Problems

1. Which figure can serve as a counterexample to the statement, "If a quadrilateral has a pair of parallel sides and a pair of congruent sides, then the quadrilateral is a parallelogram."

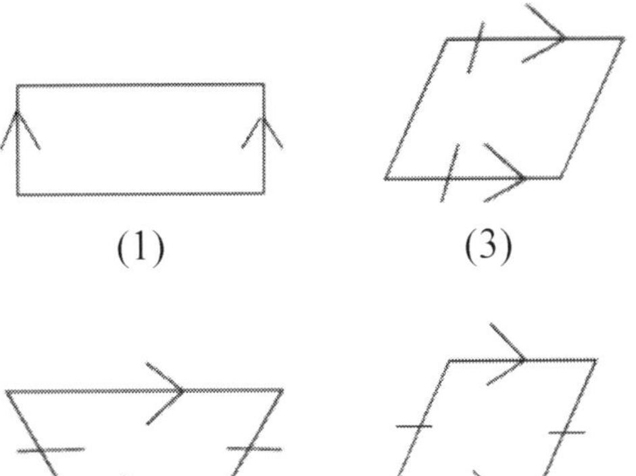

2. In parallelogram $MNOP$, the diagonals intersect at A and $AO = 10$. Find AM.

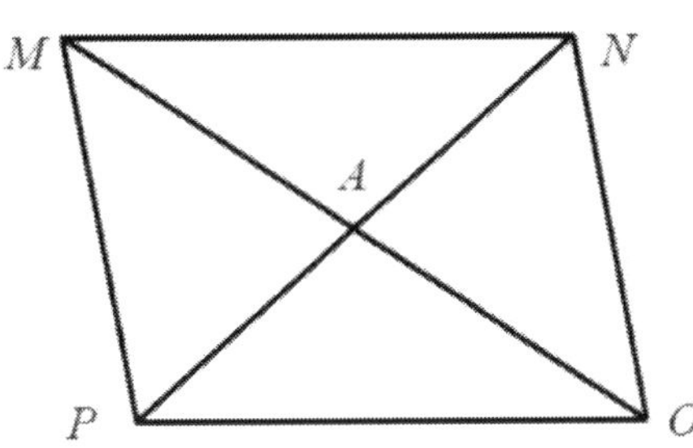

3. $ABCD$ is a parallelogram. $m\angle A = 6x - 30$ and $m\angle C = 4x + 10$. Show that $ABCD$ is a rectangle.

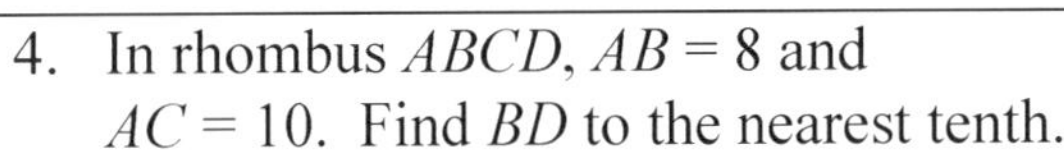

4. In rhombus $ABCD$, $AB = 8$ and $AC = 10$. Find BD to the nearest tenth.

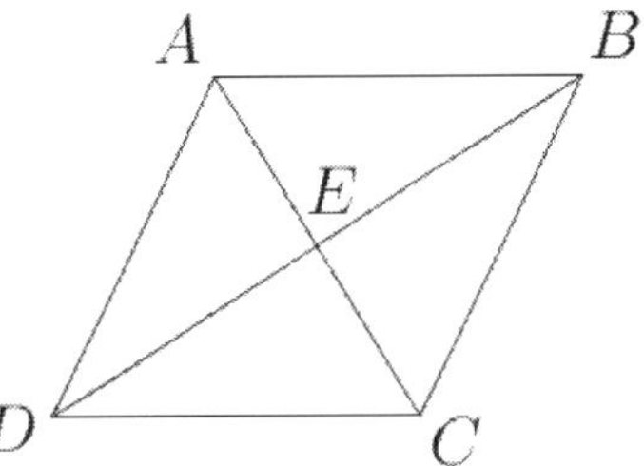

5. Given $ABCD$ is a rhombus, $m\angle BAC = 30$, and $AD = 24$. Find DE.

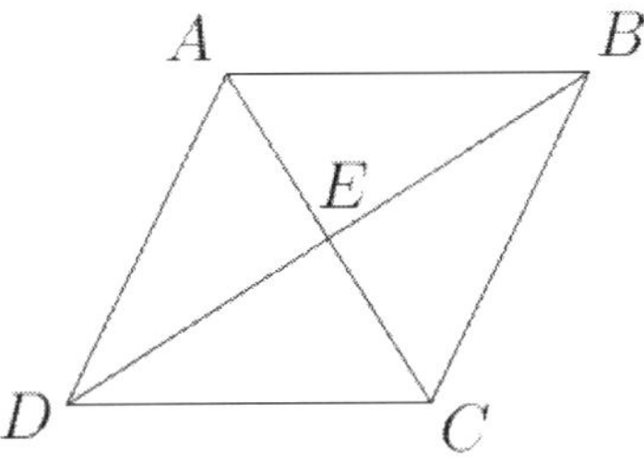

6. Given $ABCD$ is a rhombus, $m\angle DCB = 60$, and $EB = 18$. Find DC.

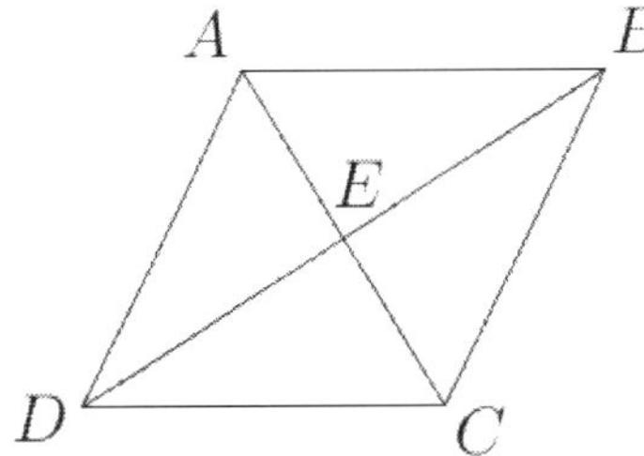

REGENTS QUESTIONS

Multiple Choice

1. Which statement is *not* always true about a parallelogram?
   (1) The diagonals are congruent.
   (2) The opposite sides are congruent.
   (3) The opposite angles are congruent.
   (4) The opposite sides are parallel.

2. In a certain quadrilateral, two opposite sides are parallel, and the other two opposite sides are not congruent. This quadrilateral could be a
   (1) rhombus
   (2) parallelogram
   (3) square
   (4) trapezoid

3. Which statement is *false*?
   (1) All parallelograms are quadrilaterals.
   (2) All rectangles are parallelograms.
   (3) All squares are rhombuses.
   (4) All rectangles are squares.

4. In the diagram below of parallelogram $ABCD$ with diagonals $\overline{AC}$ and $\overline{BD}$, $m\angle 1 = 45$ and $m\angle DCB = 120$.

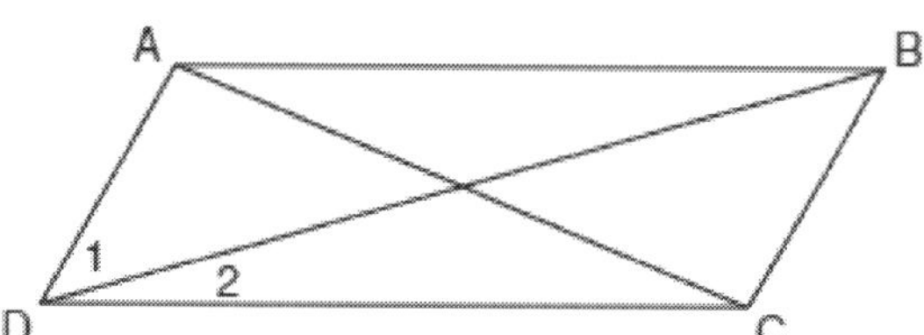

   What is the measure of $\angle 2$?
   (1) 15°
   (2) 30°
   (3) 45°
   (4) 60°

5. If the diagonals of a quadrilateral do not bisect each other, then the quadrilateral could be a
   (1) rectangle
   (2) rhombus
   (3) square
   (4) trapezoid

6. Which statement is true about every parallelogram?
   (1) All four sides are congruent.
   (2) The interior angles are all congruent.
   (3) Two pairs of opposite sides are congruent.
   (4) The diagonals are perpendicular to each other.

7. In the diagram below of rhombus $ABCD$, $m\angle C = 100$.

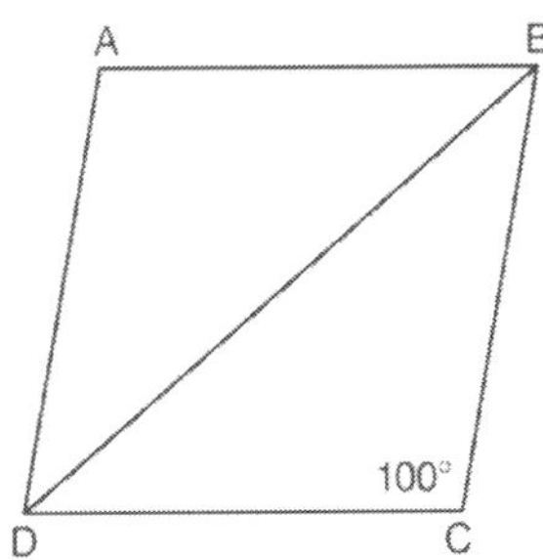

What is $m\angle DBC$?

(1) 40
(2) 45
(3) 50
(4) 80

8. In the diagram below, parallelogram $ABCD$ has diagonals $\overline{AC}$ and $\overline{BD}$ that intersect at point E.

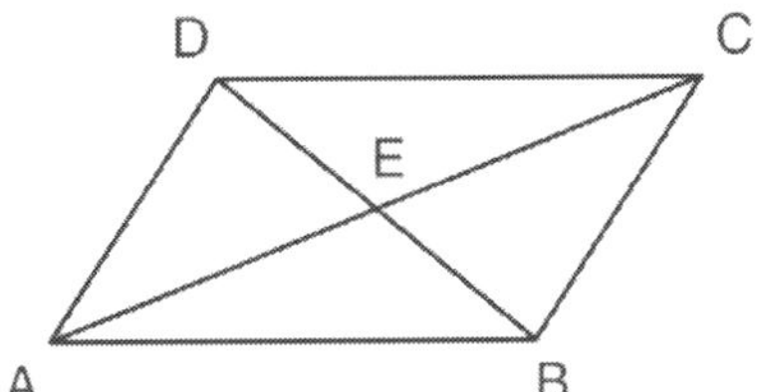

Which expression is *not* always true?

(1) $\angle DAE \cong \angle BCE$
(2) $\angle DEC \cong \angle BEA$
(3) $\overline{AC} \cong \overline{DB}$
(4) $\overline{DE} \cong \overline{EB}$

9. In rhombus $ABCD$, the diagonals $\overline{AC}$ and $\overline{BD}$ intersect at E. If $AE = 5$ and $BE = 12$, what is the length of $\overline{AB}$?

(1) 7
(2) 10
(3) 13
(4) 17

10. Which quadrilateral has diagonals that always bisect its angles and also bisect each other?

(1) rhombus
(2) rectangle
(3) parallelogram
(4) isosceles trapezoid

11. The diagonals of a quadrilateral are congruent but do not bisect each other. This quadrilateral is

(1) an isosceles trapezoid
(2) a parallelogram
(3) a rectangle
(4) a rhombus

12. Given three distinct quadrilaterals, a square, a rectangle, and a rhombus, which quadrilaterals must have perpendicular diagonals?
    (1) the rhombus, only
    (2) the rectangle and the square
    (3) the rhombus and the square
    (4) the rectangle, the rhombus, and the square

13. In the diagram below, $MATH$ is a rhombus with diagonals $\overline{AH}$ and $\overline{MT}$.

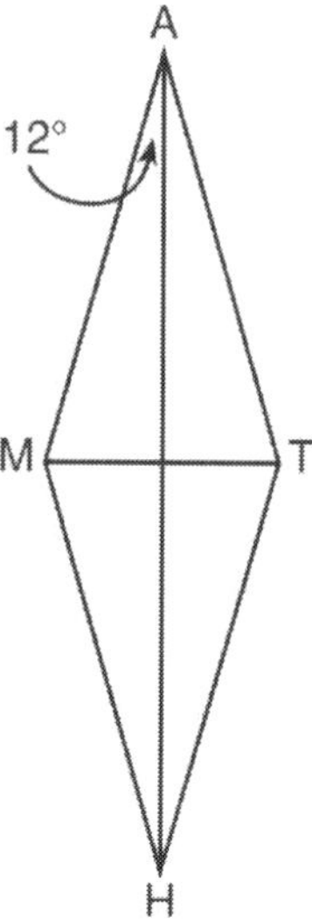

If $m\angle HAM = 12$, what is $m\angle AMT$?
    (1) 12 (3) 84
    (2) 78 (4) 156

14. What is the perimeter of a rhombus whose diagonals are 16 and 30?
    (1) 92 (3) 60
    (2) 68 (4) 17

15. Which quadrilateral does *not* always have congruent diagonals?
    (1) isosceles trapezoid (3) rhombus
    (2) rectangle (4) square

16. In rhombus $ABCD$, with diagonals $\overline{AC}$ and $\overline{DB}$, $AD = 10$.

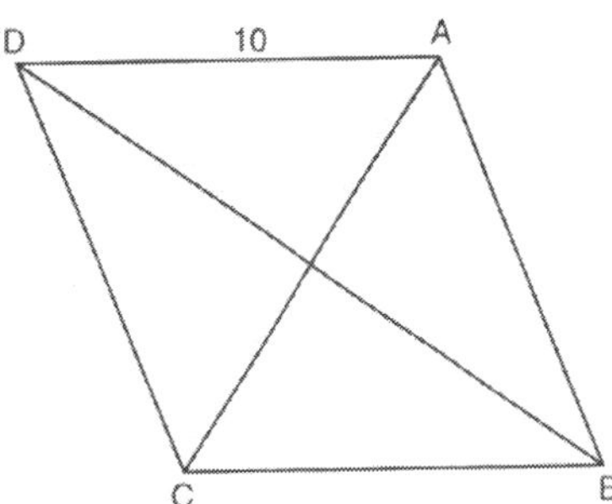

If the length of diagonal $\overline{AC}$ is 12, what is the length of $\overline{DB}$?
    (1) 8 (3) $\sqrt{44}$
    (2) 16 (4) $\sqrt{136}$

17. In parallelogram $QRST$, diagonal $\overline{QS}$ is drawn. Which statement must always be true?

(1) $\triangle QRS$ is an isosceles triangle.
(2) $\triangle STQ$ is an acute triangle.
(3) $\triangle STQ \cong \triangle QRS$
(4) $\overline{QS} \cong \overline{QT}$

18. In quadrilateral $ABCD$, the diagonals bisect its angles. If the diagonals are *not* congruent, quadrilateral $ABCD$ must be a

(1) square
(2) rectangle
(3) rhombus
(4) trapezoid

19. In the diagram below of rhombus $ABCD$, the diagonals $\overline{AC}$ and $\overline{BD}$ intersect at E.

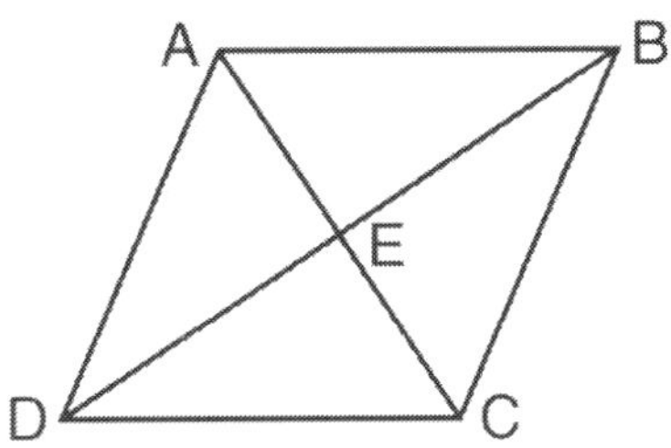

If $AC = 18$ and $BD = 24$, what is the length of one side of rhombus $ABCD$?

(1) 15
(2) 18
(3) 24
(4) 30

20. Parallelogram $ABCD$ with diagonals $\overline{AC}$ and $\overline{BD}$ intersecting at E is shown below.

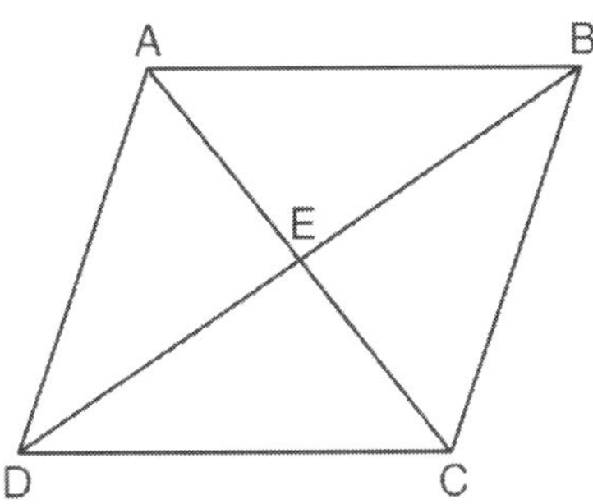

Which statement must be true?

(1) $\overline{BE} \cong \overline{CE}$
(2) $\angle BAE \cong \angle DCE$
(3) $\overline{AB} \cong \overline{BC}$
(4) $\angle DAE \cong \angle CBE$

21. In quadrilateral $ABCD$, each diagonal bisects opposite angles. If $m\angle DAB = 70$, then $ABCD$ must be a

(1) rectangle
(2) trapezoid
(3) rhombus
(4) square

22. Which quadrilateral has diagonals that are always perpendicular bisectors of each other?

(1) square
(2) rectangle
(3) trapezoid
(4) parallelogram

23. CC A parallelogram must be a rectangle when its
    (1) diagonals are perpendicular
    (2) diagonals are congruent
    (3) opposite sides are parallel
    (4) opposite sides are congruent

24. CC In the diagram of parallelogram *FRED* shown below, $\overline{ED}$ is extended to *A*, and $\overline{AF}$ is drawn such that $\overline{AF} \cong \overline{DF}$.

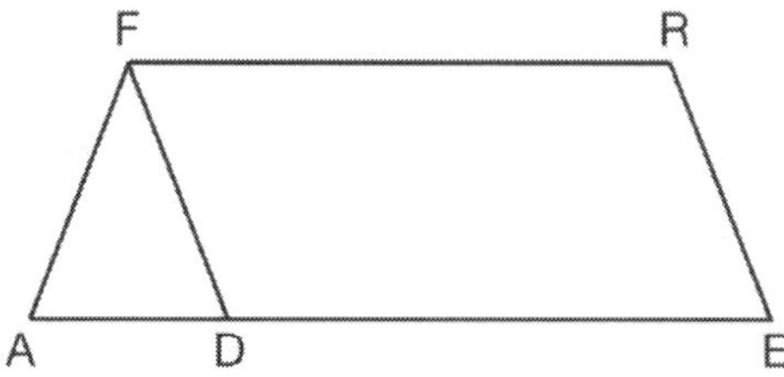

If $m\angle R = 124°$, what is $m\angle AFD$?
    (1) 124°
    (2) 112°
    (3) 68°
    (4) 56°

25. CC In parallelogram *QRST* shown below, diagonal $\overline{TR}$ is drawn, *U* and *V* are points on $\overline{TS}$ and $\overline{QR}$, respectively, and $\overline{UV}$ intersects $\overline{TR}$ at *W*.

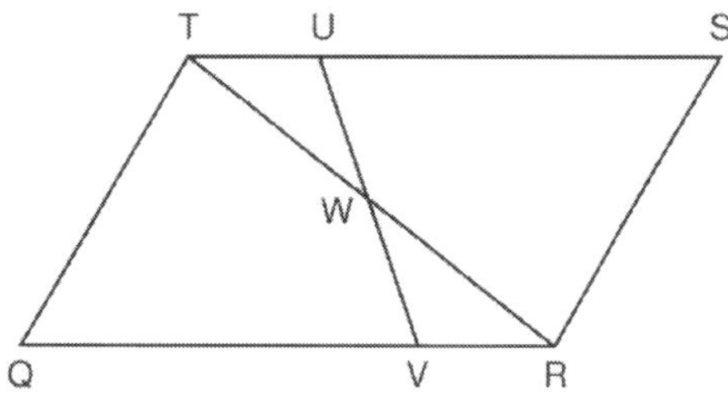

If $m\angle S = 60°$, $m\angle SRT = 83°$, and $m\angle TWU = 35°$, what is $m\angle WVQ$?
    (1) 37°
    (2) 60°
    (3) 72°
    (4) 83°

26. CC In the diagram below, *ABCD* is a parallelogram, $\overline{AB}$ is extended through *B* to *E*, and $\overline{CE}$ is drawn.

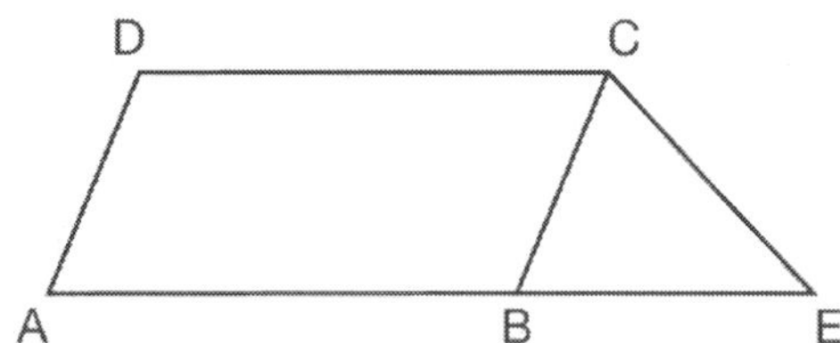

If $\overline{CE} \cong \overline{BE}$ and $m\angle D = 112°$, what is $m\angle E$?
    (1) 44°
    (2) 56°
    (3) 68°
    (4) 112°

27. (CC) A parallelogram is always a rectangle if
    (1) the diagonals are congruent
    (2) the diagonals bisect each other
    (3) the diagonals intersect at right angles
    (4) the opposite angles are congruent

Constructed Response

28. As shown in the accompanying diagram, a rectangular gate has two diagonal supports. If $m\angle 1 = 42$, what is $m\angle 2$?

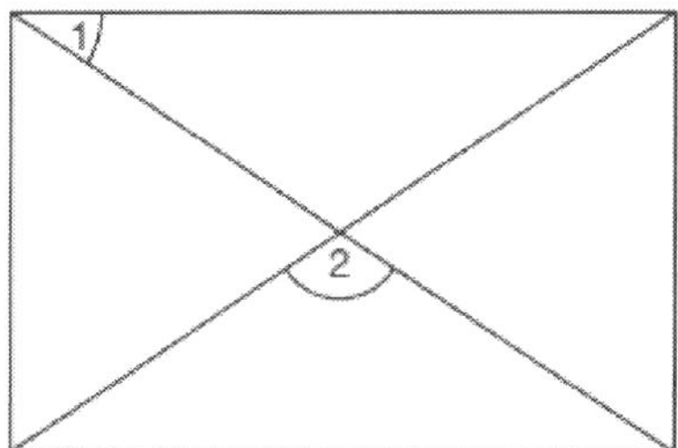

29. (CC) The diagram below shows parallelogram $LMNO$ with diagonal $\overline{LN}$, $m\angle M = 118°$, and $m\angle LNO = 22°$.

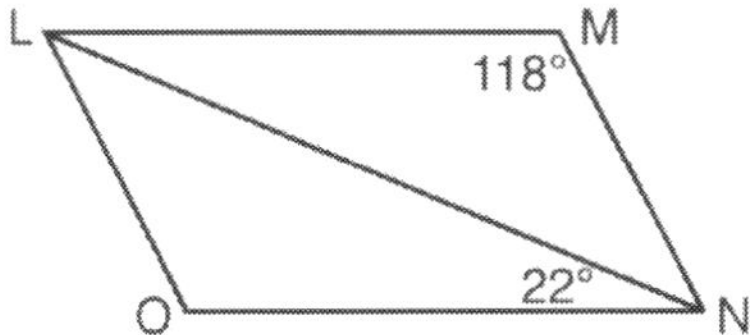

Explain why $m\angle NLO$ is 40 degrees.

7.2 Trapezoids

Key Terms and Concepts

As we have seen, a **trapezoid** is a quadrilateral with at least one pair of parallel sides. Opposite sides that are parallel are called **bases** and opposite sides that are not parallel are called **legs**.

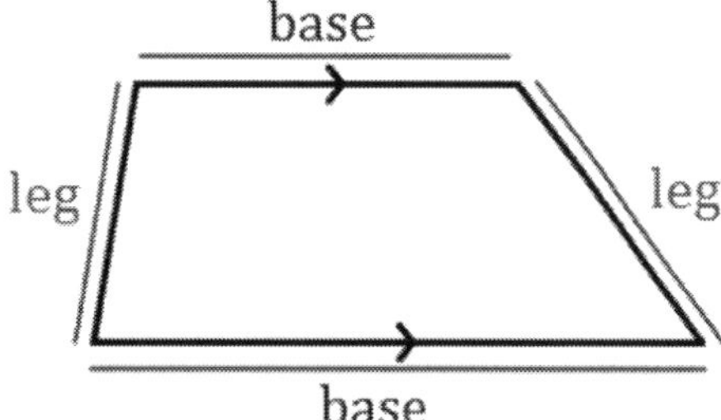

An **altitude** (or *height*) of a trapezoid is a segment drawn from one base to the opposite base and perpendicular to both bases. Since the bases of a trapezoid are parallel, all altitudes between a pair of bases are congruent. The formula for the area of a trapezoid is $A=\left(\frac{b_1+b_2}{2}\right)h$.

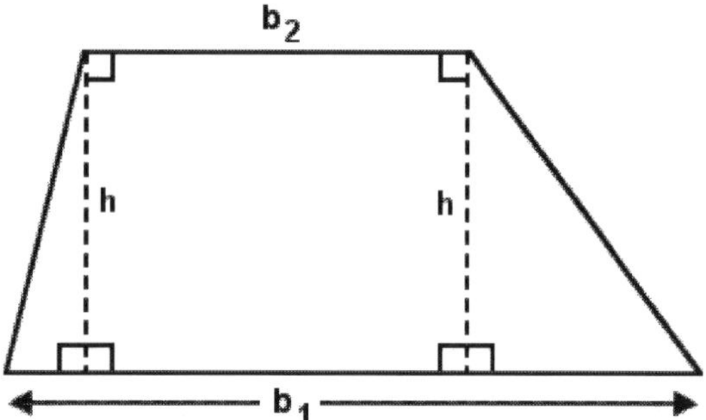

A trapezoid with at least one pair of opposite sides that are congruent is an *isosceles trapezoid.* Both pairs of base angles of an isosceles trapezoid are congruent. In an isosceles trapezoid, opposite angles are supplementary. Also, the diagonals of an isosceles trapezoid are congruent.

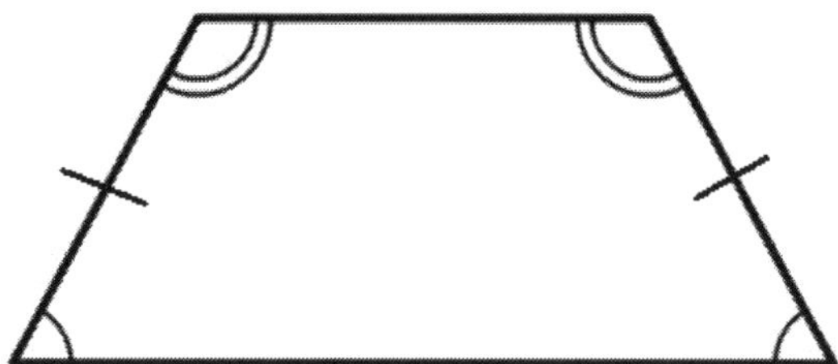

Model Problem:

The cross section of an attic is in the shape of an isosceles trapezoid, as shown in the accompanying figure. If the height of the attic is 9 feet, $BC = 12$ feet, and $AD = 28$ feet, find the length of $\overline{AB}$ to the *nearest foot*.

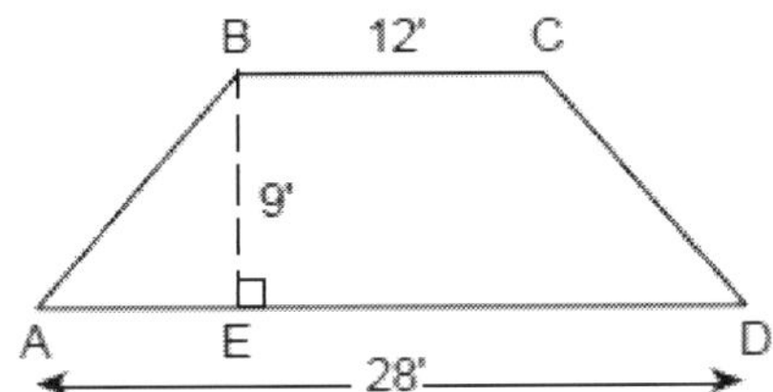

Solution:

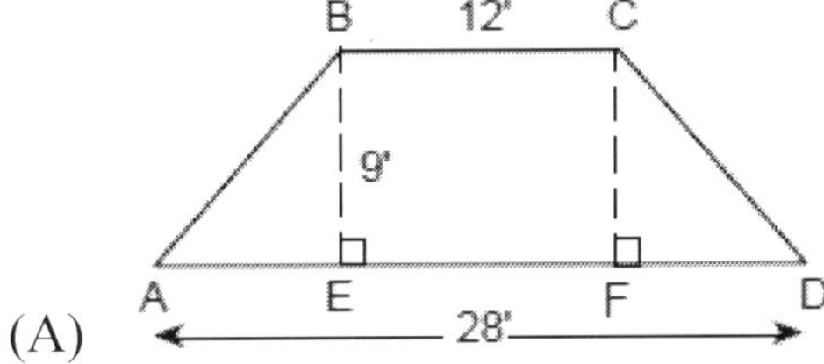

(A)

(B) $AE = \dfrac{28-12}{2} = 8$

(C) $AB^2 = 8^2 + 9^2 = 145$
$AB = \sqrt{145} \approx 12$

Explanation of steps:

(A) By drawing an altitude $\overline{CF}$, we know $\overline{CF} \cong \overline{BE}$. Since it is an isosceles triangle, we also know $\overline{AB} \cong \overline{CD}$. So, $\triangle ABE \cong \triangle DCF$ by HL. Therefore, $\overline{AE} \cong \overline{DF}$ by CPCTC.

(B) $BCFE$ is a rectangle, so $EF = 12$. Since $AE = DF$, we can find AE by subtracting $AD - EF$ and dividing by 2.

(C) The legs of $\triangle ABE$ are 8 and 9, so we can find the hypotenuse AB by using the Pythagorean Theorem.

Practice Problems

1. Given isosceles trapezoid $ABCD$ with legs $\overline{AD}$ and $\overline{BC}$. 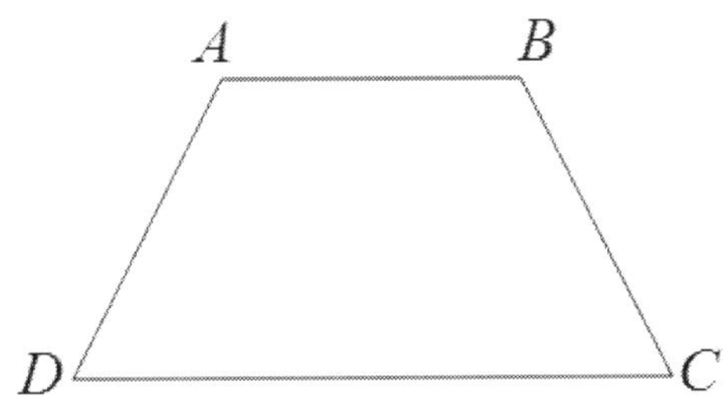 (a) If $AC = 25$, find BD. (b) If $m\angle A = 75$, find $m\angle D$.	2. In isosceles trapezoid $ABCD$, $\overline{AB} \cong \overline{CD}$. If $BC = 20$, $AD = 36$, and $AB = 17$, what is the length of the altitude of the trapezoid?

REGENTS QUESTIONS

Multiple Choice

1. In the diagram below of isosceles trapezoid *ABCD*, $AB = CD = 25$, $AD = 26$, and $BC = 12$.

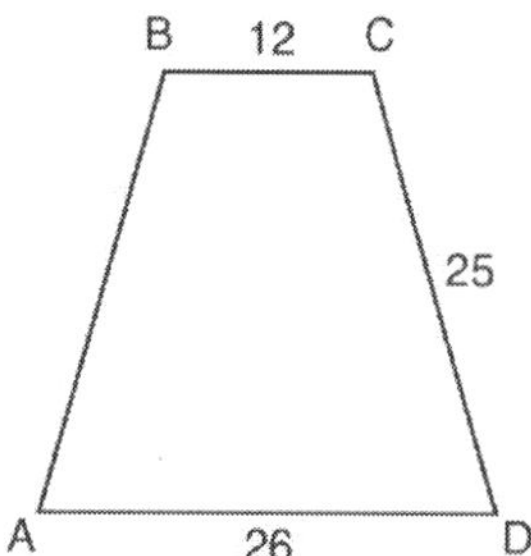

What is the length of an altitude of the trapezoid?

(1) 7 (3) 19
(2) 14 (4) 24

2. In the diagram below, *LATE* is an isosceles trapezoid with $\overline{LE} \cong \overline{AT}$, $LA = 24$, $ET = 40$, and $AT = 10$. Altitudes $\overline{LF}$ and $\overline{AG}$ are drawn.

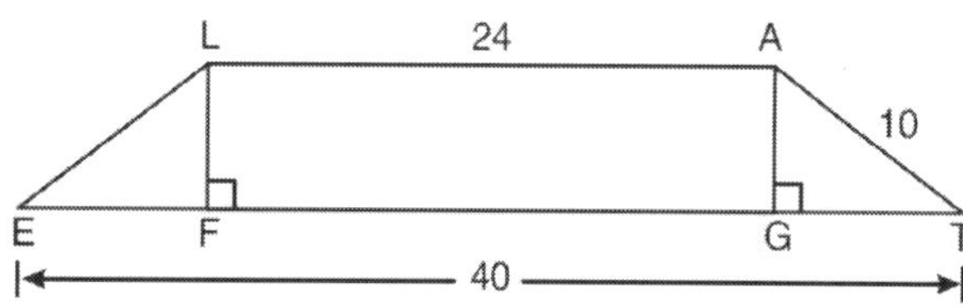

What is the length of $\overline{LF}$?

(1) 6 (3) 3
(2) 8 (4) 4

3. In trapezoid *RSTV* with bases $\overline{RS}$ and $\overline{VT}$, diagonals $\overline{RT}$ and $\overline{SV}$ intersect at *Q*.

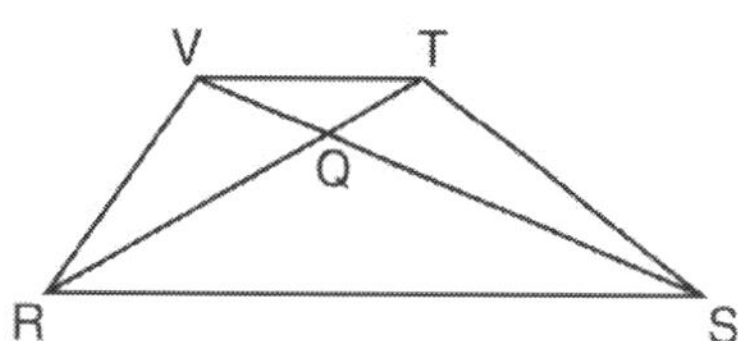

If trapezoid *RSTV* is *not* isosceles, which triangle is equal in area to $\triangle RSV$?

(1) $\triangle RQV$ (3) $\triangle RVT$
(2) $\triangle RST$ (4) $\triangle SVT$

Constructed Response

4. The accompanying diagram shows ramp $\overline{RA}$ leading to level platform $\overline{AM}$, forming an angle of 45° with level ground. If platform $\overline{AM}$ measures 2 feet and is 6 feet above the ground, explain why the exact length of ramp $\overline{RA}$ is $6\sqrt{2}$ feet.

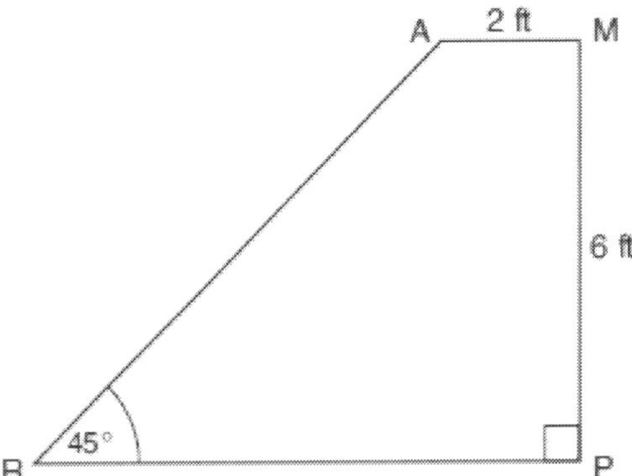

7.3 Parallelogram Proofs

Key Terms and Concepts

To prove that a *quadrilateral is a parallelogram*, it is sufficient to show any one of these:

- both pairs of opposite sides are parallel
- both pairs of opposite sides are congruent
- both pairs of opposite angles are congruent
- one pair of opposite sides are both parallel and congruent
- diagonals bisect each other

To prove that a *parallelogram is a rectangle*, it is sufficient to show any one of these:

- any one of its angles is a right angle
- one pair of consecutive angles are congruent
- diagonals are congruent

To prove that a *parallelogram is a rhombus*, it is sufficient to show any one of these:

- one pair of consecutive sides are congruent
- diagonals are perpendicular
- either diagonal is an angle bisector

Model Problem:
Given: $\angle UQV \cong \angle RVQ$ and $\angle TUQ \cong \angle SRV$
Prove: $QRVU$ is a parallelogram

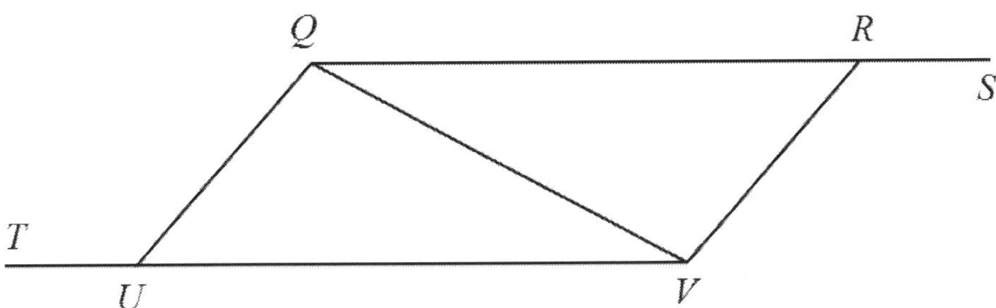

Solution:

Statements		*Reasons*
$\angle UQV \cong \angle RVQ$, $\angle TUQ \cong \angle SRV$	(A)	Given
$\overline{UQ} \parallel \overline{VR}$		If two lines cut by a transversal form congruent alternate interior angles, then they are parallel
$\angle TUQ$ and $\angle QUV$ are a linear pair, $\angle SRV$ and $\angle VRQ$ are a linear pair		Definition of linear pair
$\angle TUQ$ and $\angle QUV$ are supplementary, $\angle SRV$ and $\angle VRQ$ are supplementary		Linear pairs are supplementary
$\angle QUV \cong \angle VRQ$	(A)	Supplements of congruent angles are congruent
$\overline{QV} \cong \overline{QV}$	(S)	Reflexive Property
$\triangle QUV \cong \triangle VRQ$		AAS
$\overline{UQ} \cong \overline{VR}$		CPCTC
$QRVU$ is a parallelogram		If a pair of opposite sides of a quadrilateral are parallel and congruent, then the quadrilateral is a parallelogram

Important Proofs

Proof that opposite sides and angles of a parallelogram are congruent

In this proof, we will first show that a diagonal divides the figure into two congruent triangles.

Given: In $\square ABCD$, diagonal $\overline{BD}$ is drawn.

Prove: $\overline{AD} \cong \overline{BC}$ and $\overline{AB} \cong \overline{DC}$

$\angle A \cong \angle C$ and $\angle ADC \cong \angle ABC$

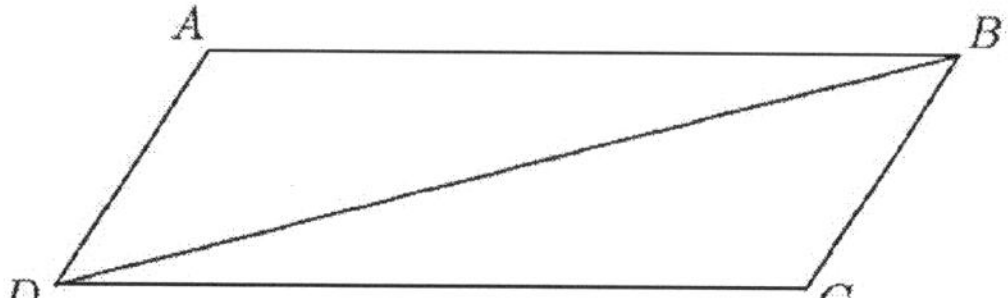

Statements	*Reasons*
$\square ABCD$	Given
$\overline{AB} \parallel \overline{DC}$ and $\overline{AD} \parallel \overline{BC}$	Definition of parallelogram
$\angle ADB \cong \angle CBD$ and (A) $\angle ABD \cong \angle CDB$ (A)	Parallel lines cut by a transversal form congruent alternate interior angles
$m\angle ADB = m\angle CBD$ and $m\angle ABD = m\angle CDB$	Definition of congruence
$m\angle ADB + m\angle CDB = m\angle ABD + m\angle CBD$	Addition Property
$m\angle ADC = m\angle ADB + m\angle CDB$ and $m\angle ABC = m\angle ABD + m\angle CBD$	Partition Postulate
$m\angle ADC = m\angle ABC$	Substitution
$\angle ADC \cong \angle ABC$	Definition of congruence
$\overline{BD} \cong \overline{BD}$ (S)	Reflexive property
$\triangle ABD \cong \triangle CDB$	ASA
$\overline{AD} \cong \overline{BC}$ and $\overline{AB} \cong \overline{DC}$	CPCTC
$\angle A \cong \angle C$	CPCTC

Proof that the diagonals of a parallelogram bisect each other

Given: In $\square ABCD$, diagonals $\overline{BD}$ and $\overline{AC}$ are drawn, intersecting at E

Prove: $\overline{BD}$ bisects $\overline{AC}$ and $\overline{AC}$ bisects $\overline{BD}$

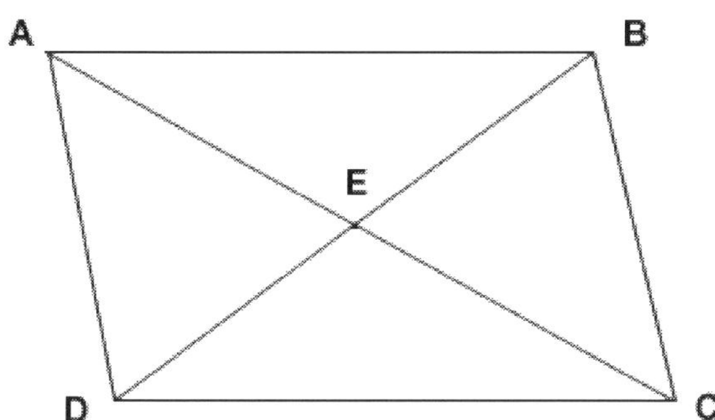

Statements		*Reasons*
$\square ABCD$		Given
$\angle BAE \cong \angle DCE$ and $\angle ABE \cong \angle CDE$	(A) (A)	Parallel lines cut by a transversal form congruent alternate interior angles
$\overline{AB} \cong \overline{CD}$	(S)	Opposite sides of a parallelogram are congruent
$\triangle ABE \cong \triangle CDE$		ASA
$\overline{AE} \cong \overline{CE}$ and $\overline{BE} \cong \overline{DE}$		CPCTC
$\overline{BD}$ bisects $\overline{AC}$ and $\overline{AC}$ bisects $\overline{BD}$		Definition of bisector

Proof that a parallelogram with congruent diagonals is a rectangle

Given: In $\square ABCD$, diagonals $\overline{AC}$ and $\overline{BD}$ are congruent

Prove: $\square ABCD$ is a rectangle

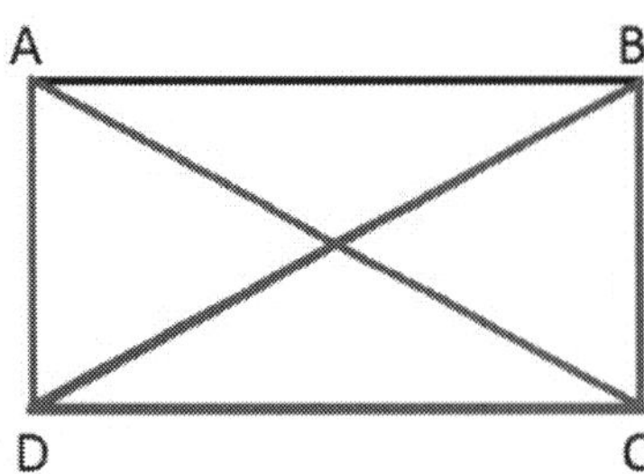

Statements	*Reasons*
$\square ABCD$	Given
$\overline{AC} \cong \overline{BD}$ (S)	Given
$\overline{AD} \cong \overline{BC}$ (S)	Opposite sides of a parallelogram are congruent
$\overline{DC} \cong \overline{DC}$ (S)	Reflexive Property
$\triangle ADC \cong \triangle BCD$	SSS
$\angle ADC$ and $\angle BCD$ are supplementary	Consecutive angles of a parallelogram are supplementary
$\angle ADC \cong \angle BCD$	CPCTC
$\angle ADC$ and $\angle BCD$ are right angles	If supplementary angles are congruent, then they are right angles
$\angle ADC \cong \angle ABC$ and $\angle BCD \cong \angle BAD$	Opposite angles of a parallelogram are congruent
$\angle ABC$ and $\angle BAD$ are right angles	If an angle is congruent to a right angle, then it is a right angle
$\square ABCD$ is a rectangle	A rectangle is a parallelogram with four right angles (definition of rectangle)

Practice Problems

1. Based on the markings, determine if the figure is a parallelogram. If so, justify your answer. 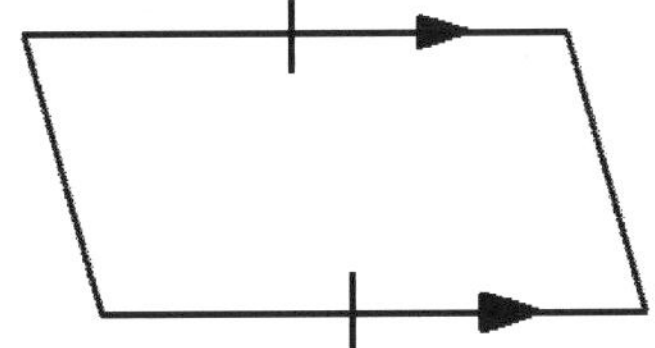	2. Based on the markings, determine if the figure is a parallelogram. If so, justify your answer. 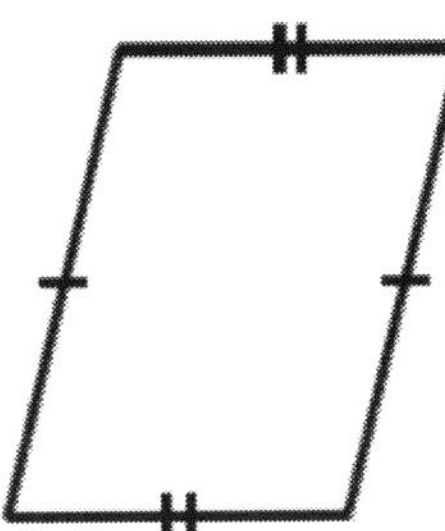
3. Given: $ABCD$ is a parallelogram Prove: $\triangle ABD \cong \triangle CDB$	4. Given: $\triangle AOB \cong \triangle COD$ Prove: $ABCD$ is a parallelogram 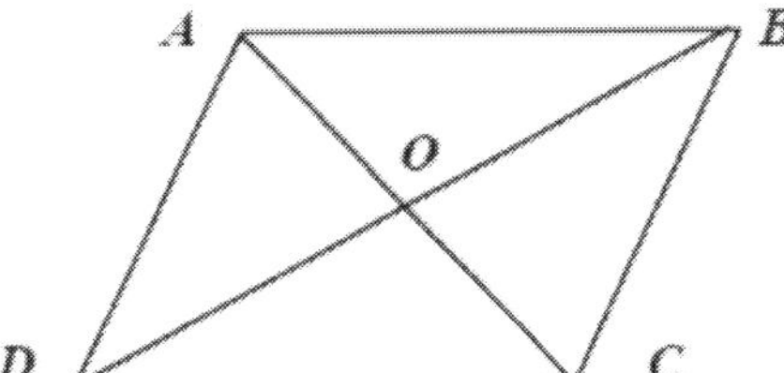

5. Given: *ABCD* is a parallelogram, $DF = EB$

   Prove: *AECF* is a parallelogram

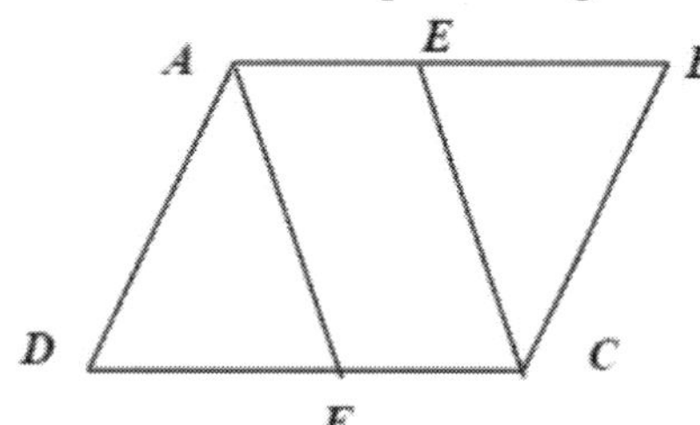

6. Given: *ABCD* is a parallelogram, and *CEBF* is a rhombus

   Prove: *ABCD* is a rectangle

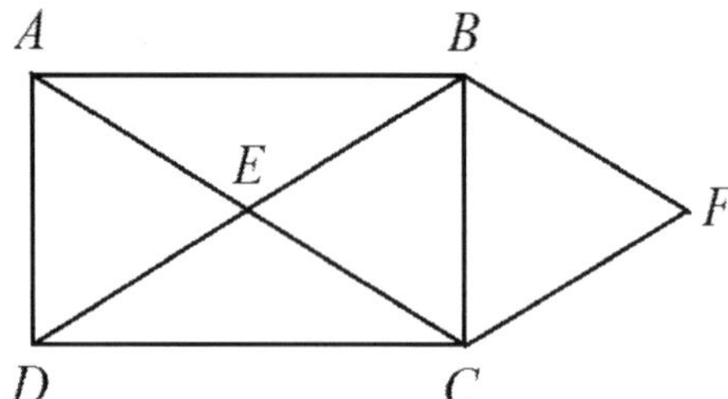

REGENTS QUESTIONS

Multiple Choice

1. Which reason could be used to prove that a parallelogram is a rhombus?
   (1) Diagonals are congruent.
   (2) Opposite sides are parallel.
   (3) Diagonals are perpendicular.
   (4) Opposite angles are congruent.

2. (CC) Quadrilateral $ABCD$ has diagonals $\overline{AC}$ and $\overline{BD}$. Which information is *not* sufficient to prove $ABCD$ is a parallelogram?
   (1) $\overline{AC}$ and $\overline{BD}$ bisect each other.
   (2) $\overline{AB} \cong \overline{CD}$ and $\overline{BC} \cong \overline{AD}$
   (3) $\overline{AB} \cong \overline{CD}$ and $\overline{AB} \| \overline{CD}$
   (4) $\overline{AB} \cong \overline{CD}$ and $\overline{BC} \| \overline{AD}$

3. (CC) In parallelogram $ABCD$, diagonals $\overline{AC}$ and $\overline{BD}$ intersect at E. Which statement does *not* prove parallelogram $ABCD$ is a rhombus?
   (1) $\overline{AC} \cong \overline{DB}$
   (2) $\overline{AB} \cong \overline{BC}$
   (3) $\overline{AC} \perp \overline{DB}$
   (4) $\overline{AC}$ bisects $\angle DCB$

4. (CC) Quadrilateral $ABCD$ with diagonals $\overline{AC}$ and $\overline{BD}$ is shown in the diagram below.

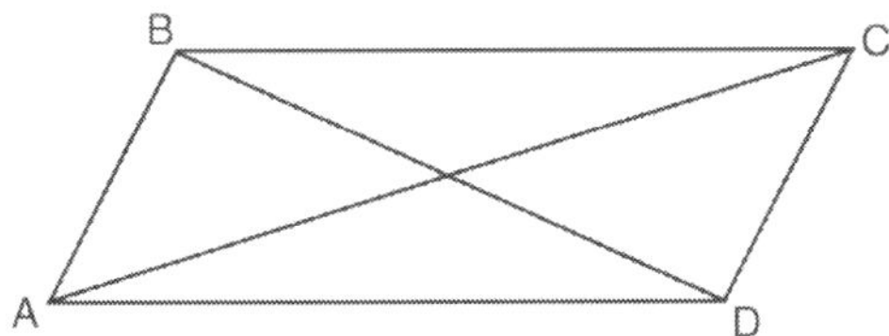

   Which information is *not* enough to prove $ABCD$ is a parallelogram?
   (1) $\overline{AB} \cong \overline{CD}$ and $\overline{AB} \| \overline{DC}$
   (2) $\overline{AB} \cong \overline{CD}$ and $\overline{BC} \cong \overline{DA}$
   (3) $\overline{AB} \cong \overline{CD}$ and $\overline{BC} \| \overline{AD}$
   (4) $\overline{AB} \| \overline{DC}$ and $\overline{BC} \| \overline{AD}$

5. (CC) In the diagram below, if $\triangle ABE \cong \triangle CDF$ and $\overline{AEFC}$ is drawn, then it could be proven that quadrilateral $ABCD$ is a

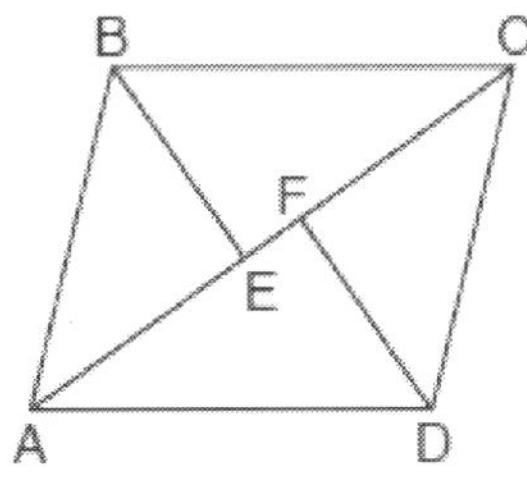

   (1) square
   (2) rhombus
   (3) rectangle
   (4) parallelogram

Constructed Response

6. (CC) In parallelogram $ABCD$ shown below, diagonals $\overline{AC}$ and $\overline{BD}$ intersect at E.

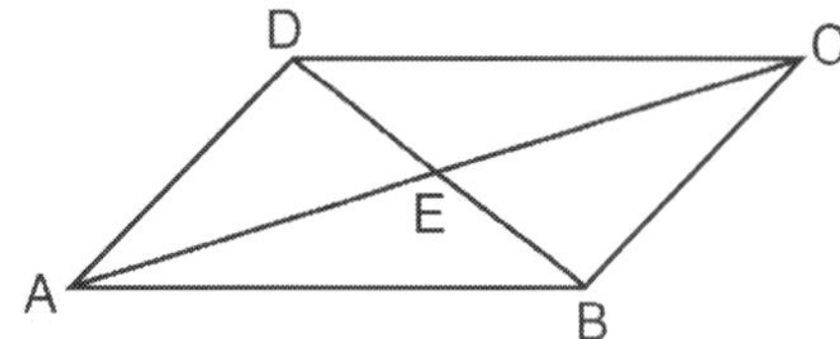

Prove: $\angle ACD \cong \angle CAB$

7. (CC) In the diagram of parallelogram $ABCD$ below, $\overline{BE} \perp \overline{CED}$, $\overline{DF} \perp \overline{BFC}$, $\overline{CE} \cong \overline{CF}$.

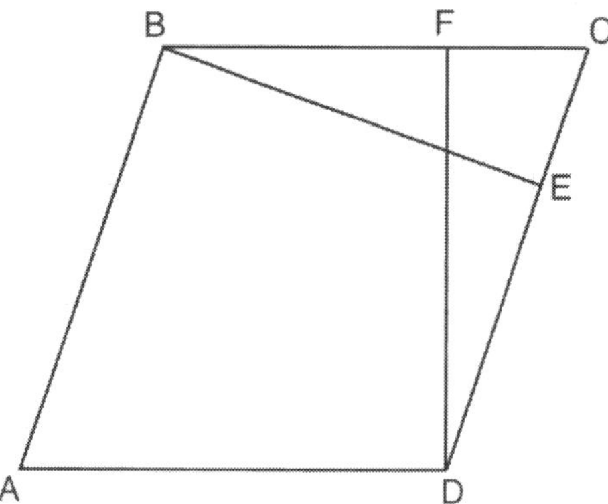

Prove $ABCD$ is a rhombus.

8. (CC) Given: Parallelogram $ANDR$ with $\overline{AW}$ and $\overline{DE}$ bisecting $\overline{NWD}$ and $\overline{REA}$ at points W and E, respectively

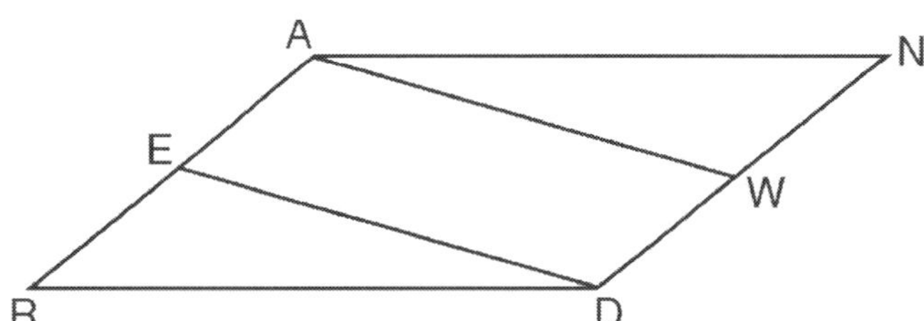

Prove that $\triangle ANW \cong \triangle DRE$. Prove that quadrilateral $AWDE$ is a parallelogram.

9. (CC) Given: Parallelogram $ABCD$, $\overline{EFG}$, and diagonal $\overline{DFB}$

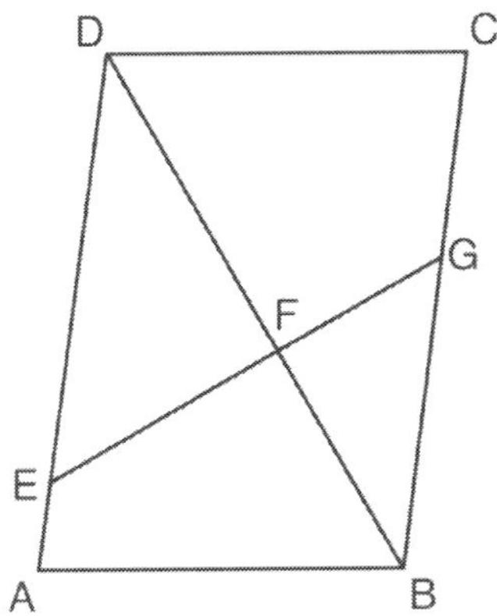

Prove: $\triangle DEF \sim \triangle BGF$

10. (CC) Given: Quadrilateral $ABCD$ with diagonals $\overline{AC}$ and $\overline{BD}$ that bisect each other, and $\angle 1 \cong \angle 2$.

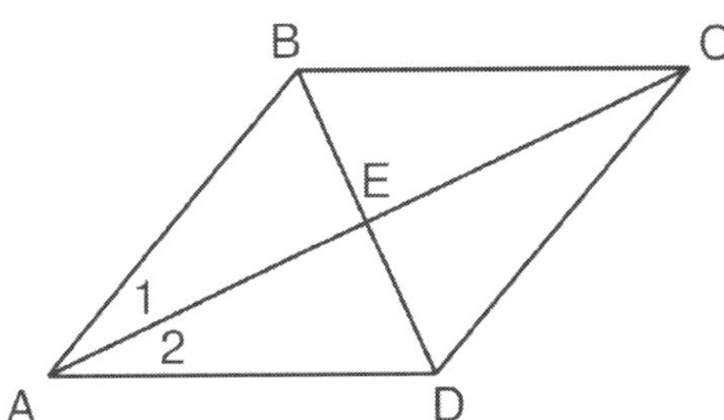

Prove: $\triangle ACD$ is an isosceles triangle and $\triangle AEB$ is a right triangle

11. (CC) In quadrilateral $ABCD$, $\overline{AB} \cong \overline{CD}$, $\overline{AB} \parallel \overline{CD}$, and $\overline{BF}$ and $\overline{DE}$ are perpendicular to diagonal $\overline{AC}$ at points F and E.

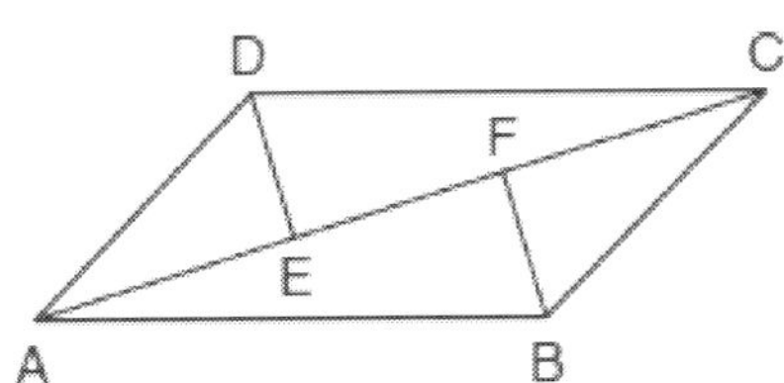

Prove: $\overline{AE} \cong \overline{CF}$

Chapter 8. Coordinate Geometry

8.1 Forms of Linear Equations

Key Terms and Concepts

A **coordinate plane** (also known as a *Cartesian plane*) is defined by two perpendicular number lines: the x-axis, which is horizontal, and the y-axis, which is vertical. Each point on the coordinate plane can be represented by an **ordered pair** stating its x-value and y-value as (x,y).

A **line** consists of a set of points representing the **solution set** of a **linear equation** involving the two variables, x and y.

You can **determine whether a point is on a line** by substituting the x-value and y-value for the variables x and y in the equation and then checking if these values make the equation true.
For example: (4,13) is on the line $y = 3x + 1$ because substituting 4 for x and 13 for y, we get $13 = 3(4) + 1$, which is true.

The **slope** of the line passing through the points (x_1, y_1) and (x_2, y_2) can be calculated as:

$$m = \frac{y_2 - y_1}{x_2 - x_1}$$

The equation of a line is most commonly written in **slope-intercept form**: $y = mx + b$.
In this form, m represents the *slope* of the line and b is the *y-intercept*. The y-intercept is the value of y at the point where the line intersects the y-axis.

However, there are other common ways of writing linear equations.

In **point-slope form**, $y - y_1 = m(x - x_1)$, the m represents the slope and x_1 and y_1 represent the coordinates of a point, (x_1, y_1), on the line.

Example: In point-slope form, the equation of the line that passes through (–6,3) and has a slope of 2 is written as $y - 3 = 2(x + 6)$.

We can convert an equation from point-slope form to slope-intercept form by solving for y.

Example: To convert $y - 2 = 3(x - 1)$ to slope-intercept form, apply the distributive property to the right side to get $y - 2 = 3x - 3$, then isolate y to get $y = 3x - 1$.

In **standard form**, $Ax + By = C$, we express the equation is such a way that A, B, and C are integers and A is non-negative.

To convert from slope-intercept to standard form:

1. If m or b are fractions, multiply both sides by their least common denominator (LCD).
2. Subtract the x term from both sides.
3. If this results in a negative coefficient of x, multiply both sides by –1.

Model Problem:

Convert the equation $y=\frac{5}{6}x+\frac{7}{4}$ to standard form.

Solution:

(A) $12y=12\left(\frac{5}{6}x+\frac{7}{4}\right)$

$12y=10x+21$

(B) $-10x+12y=21$

(C) $10x-12y=-21$

Explanation of steps:

(A) If m or b are fractions, multiply both sides by their least common denominator *[LCD of 6 and 4 is 12]*.

(B) Subtract the x term from both sides.

(C) If this results in a negative coefficient of x, multiply both sides by –1. *[Negate all terms.]*

Practice Problems

1. Determine whether the point (–2,3) lies on the line whose equation is $y=3x+15$.	2. Write an equation of the line whose slope is –4 and y-intercept is 5.
3. Write an equation of the line through the point (1,–2) with a slope of –3.	4. Write an equation of the line through the points (–2,–3) and (5,–5).
5. Rewrite the equation $y=2x-5$ in standard form.	6. Rewrite the equation $y=\frac{3}{4}x+\frac{1}{2}$ in standard form.

REGENTS QUESTIONS

Multiple Choice

1. What is an equation of the line that passes through the points (1,3) and (8,5)?

(1) $y+1=\frac{2}{7}(x+3)$ (3) $y-1=\frac{2}{7}(x+3)$

(2) $y-5=\frac{2}{7}(x-8)$ (4) $y+5=\frac{2}{7}(x-8)$

Constructed Response

2. Write an equation that represents the line that passes through the points (5,4) and (–5,0).

8.2 Parallel and Perpendicular Lines

Key Terms and Concepts

If two distinct lines have the **same slope**, they are **parallel**. If two lines have **slopes that are opposite reciprocals**, they are **perpendicular**.

So, to determine whether two lines are parallel or perpendicular, we could write each equation in *slope-intercept form* to determine their slopes (the coefficients of x). If their slopes are the same, but their y-intercepts are different, then they are parallel. If their slopes are opposite reciprocals, then they are perpendicular.

Examples: (a) $y = 3x + 5$ and $y = 3x - 2$ are equations of parallel lines.

(b) $y = 3x + 5$ and $y = -\frac{1}{3}x - 2$ are equations of perpendicular lines.

Note the exceptions: A vertical line has no slope and an equation of the form $x = k$. A horizontal line has a zero slope and an equation of the form $y = k$. All vertical lines are parallel to each other, and all horizontal lines are parallel to each other. Also, a vertical line is perpendicular to a horizontal line, and vice versa.

Even when we are not given the slope, if we are told that a line is *parallel* or *perpendicular* to another line, then we know the lines have the *same* or *opposite reciprocal* slopes, respectively.

Example: Line ℓ has an equation of $y = \frac{1}{2}x + 5$. Therefore, its slope is $\frac{1}{2}$.

(a) The equation of a line that passes through (4,1) and is parallel to ℓ is $y - 1 = \frac{1}{2}(x - 4)$, or in slope-intercept form, $y = \frac{1}{2}x - 1$.

(b) The equation of a line that passes through (4,1) and is perpendicular to ℓ is $y - 1 = -2(x - 4)$, or in slope-intercept form, $y = -2x + 9$.

If an equation is written in *standard form*, the slope is $-\frac{A}{B}$. We can see this by transforming the general equation, $Ax + By = C$, into slope-intercept form:

$$Ax + By = C$$
$$By = -Ax + C$$
$$y = -\frac{A}{B}x + \frac{C}{B}$$

If two equations can be written in standard form with the same coefficients of x and y on the left sides, but different constants on the right sides, they are parallel.

Model Problem 1: *slope-intercept form*

The equations of two distinct lines are $y = 3x - 6$ and $2y = 3x + 6$. Are the lines parallel?

Solution:

(A) For $y = 3x - 6$, the slope $m = 3$.

Solving $2y = 3x + 6$ for *y*:

$$\frac{2y}{2} = \frac{3x+6}{2}$$

$y = \frac{3}{2}x + 3$, so the slope $m = \frac{3}{2}$.

(B) The lines are *not* parallel because the slopes are not equal.

Explanation of steps:

(A) Write each equation in slope-intercept form to determine the slope of each line.
[The first equation is already in slope-intercept form, $y = mx + b$*, so the slope* $m = 3$*.*
The second equation needed to be transformed, showing a slope of $\frac{3}{2}$*.]*

(B) If the slopes are equal, the lines are parallel.
[These slopes are 3 and $\frac{3}{2}$*, so they are not parallel.]*

Model Problem 2: *standard form*

The equations of two distinct lines are $2x - 3y = 10$ and $4x - 6y = 10$. Are the lines parallel?

Solution:

Divide both sides of the second equation by 2:

$$\frac{4x-6y}{2} = \frac{10}{2} \qquad 2x - 3y = 5$$

Yes, they are parallel. The slopes of both lines are $-\frac{A}{B} = \frac{2}{3}$.

Explanation of steps:

The two lines are parallel because they can be written in standard form with the same coefficients of *x* and *y* but different constants.

Important Proofs

Informal Proof that two parallel lines have the same slope

In the diagram below, we are given $\ell \parallel m$. A vertical transversal, *t*, is drawn, and horizontal segments *b* and *d* are drawn forming right angles with *t*, as shown.

$\angle 1 \cong \angle 2$ since they are corresponding angles formed by parallel lines cut by a transversal. Therefore, the two triangles are similar by AA. Corresponding sides of similar triangles are in proportion, so $\frac{a}{b} = \frac{c}{d}$.

The slope of a line may be expressed as $\frac{rise}{run}$. So, the slope of $\ell = \frac{a}{b}$ and the slope of $m = \frac{c}{d}$. Therefore, the slopes are the two lines are equal.

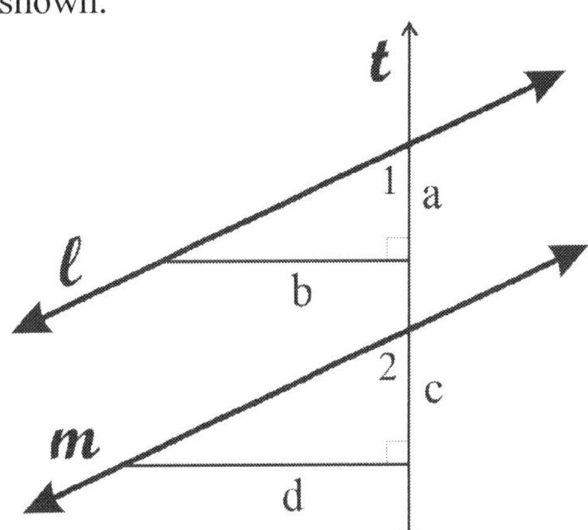

Informal Proof that two perpendicular lines have slopes that are opposite reciprocals

In the diagram below, we are given $L \perp M$. Through the point of intersection of *L* and *M*, lines *T* and *V* are drawn such that *T* is parallel to the *y*-axis and *V* is parallel to the *x*-axis (ie, $T \perp V$).

Right triangles are drawn as shown, with $b = d$. Since $m\angle 1 = 90 - m\angle 3$ and $m\angle 2 = 90 - m\angle 3$, we know that $m\angle 1 = m\angle 2$ (and therefore, $\angle 1 \cong \angle 2$). So, the two triangles are congruent by ASA. Therefore, $a = c$.

The slope of *L*, represented as $\frac{rise}{run}$, is $-\frac{b}{a}$.

The slope of *M*, represented as $\frac{rise}{run}$, is $\frac{c}{d}$.

By substituting *a* for *c* and *b* for *d*, the slope of *M* may be rewritten as $\frac{a}{b}$. Therefore, the slopes of *L* and *M* are opposite reciprocals.

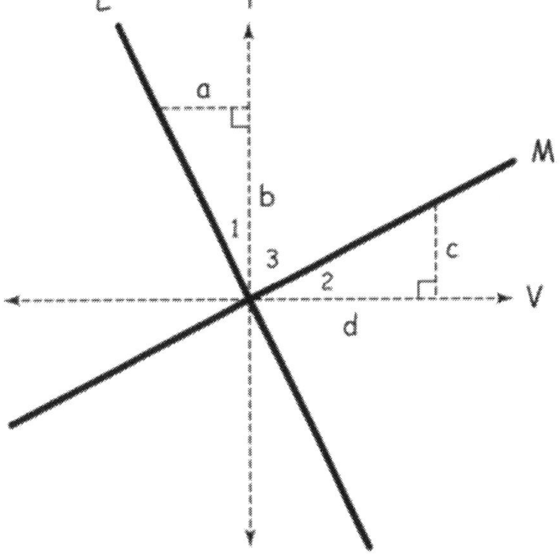

Practice Problems

1. Which equation below represents a line that is parallel to the line, $y = -x + 4$?
   (1) $2y + 2x = 6$
   (2) $2y - x = 6$

2. Which equation below represents a line that is parallel to the line, $4x + 6y = 5$?
   (1) $-3y = 2x + 5$
   (2) $-6y + 4x = 5$

3. Which equation below represents a line that is perpendicular to the line, $y = 2x - 7$?
   (1) $y = 2x + \frac{1}{7}$
   (2) $y = -\frac{1}{2}x + 1$

4. Which equation below represents a line that is perpendicular to the line, $y = -5x + 2$?
   (1) $x - 5y = 25$
   (2) $5x - y = 5$

5. Line ℓ has an equation of $y = -2x - 5$. Write the equation of a line that is parallel to line ℓ but has a y-intercept of 2.

6. Line ℓ has an equation of $y = \frac{1}{2}x + 2$. Write an equation of a line perpendicular to line ℓ and passing through the origin.

REGENTS QUESTIONS

Multiple Choice

1. Line ℓ passes through the point (5,3) and is parallel to line k whose equation is $5x + y = 6$. An equation of line ℓ is

   (1) $y = \frac{1}{5}x + 2$ (3) $y = \frac{1}{5}x - 2$

   (2) $y = -5x + 28$ (4) $y = -5x - 28$

2. The equation of a line is $3y + 2x = 12$. What is the slope of the line perpendicular to the given line?

   (1) $\frac{2}{3}$ (3) $-\frac{2}{3}$

   (2) $\frac{3}{2}$ (4) $-\frac{3}{2}$

3. What is the equation of a line passing through the point (4,–1) and parallel to the line whose equation is $2y - x = 8$?

   (1) $y = \frac{1}{2}x - 3$ (3) $y = -2x + 7$

   (2) $y = \frac{1}{2}x - 1$ (4) $y = -2x + 2$

4. Which equation represents a line that is parallel to the line whose equation is $y = -3x$?

   (1) $\frac{1}{3}x + y = 4$ (3) $6x + 2y = 4$

   (2) $-\frac{1}{3}x + y = 4$ (4) $-6x + 2y = 4$

5. What is an equation of the line that passes through (–9,12) and is perpendicular to the line whose equation is $y = \frac{1}{3}x + 6$?

   (1) $y = \frac{1}{3}x + 15$ (3) $y = \frac{1}{3}x - 13$

   (2) $y = -3x - 15$ (4) $y = -3x + 27$

6. What is the slope of a line perpendicular to the line whose equation is $3x - 7y + 14 = 0$?

   (1) $\frac{3}{7}$ (3) 3

   (2) $-\frac{7}{3}$ (4) $-\frac{1}{3}$

7. The equations of lines *k*, *p*, and *m* are given below:

$k: \ x+2y=6$

$p: \ 6x+3y=12$

$m: \ -x+2y=10$

Which statement is true?

(1) $p \perp m$
(2) $m \perp k$
(3) $k \parallel p$
(4) $m \parallel k$

8. The lines represented by the equations $4x+6y=6$ and $y=\frac{2}{3}x-1$ are

(1) parallel
(2) the same line
(3) perpendicular
(4) intersecting, but *not* perpendicular

9. What is an equation of the line that passes through the point (2,4) and is perpendicular to the line whose equation is $3y=6x+3$?

(1) $y=-\frac{1}{2}x+5$
(2) $y=-\frac{1}{2}x+4$
(3) $y=2x-6$
(4) $y=2x$

10. The equations of lines *k*, *m*, and *n* are given below:

$k: \ 3y+6=2x$

$m: \ 3y+2x+6=0$

$n: \ 2y=3x+6$

Which statement is true?

(1) $k \parallel m$
(2) $n \parallel m$
(3) $m \perp k$
(4) $m \perp n$

11. What is an equation of the line that passes through the point (4,5) and is parallel to the line whose equation is $y=\frac{2}{3}x-4$?

(1) $2y+3x=11$
(2) $2y+3x=22$
(3) $3y-2x=2$
(4) $3y-2x=7$

12. The graphs of the lines represented by the equations $y=\frac{1}{3}x+7$ and $y=-\frac{1}{3}x-2$ are

(1) parallel
(2) horizontal
(3) perpendicular
(4) intersecting, but *not* perpendicular

13. (CC) Which equation represents a line that is perpendicular to the line represented by $2x-y=7$?

(1) $y=-\frac{1}{2}x+6$
(2) $y=\frac{1}{2}x+6$
(3) $y=-2x+6$
(4) $y=2x+6$

14. **CC** Given $\overline{MN}$ shown below, with $M(-6,1)$ and $N(3,-5)$, what is an equation of the line that passes through point $P(6,1)$ and is parallel to $\overline{MN}$?

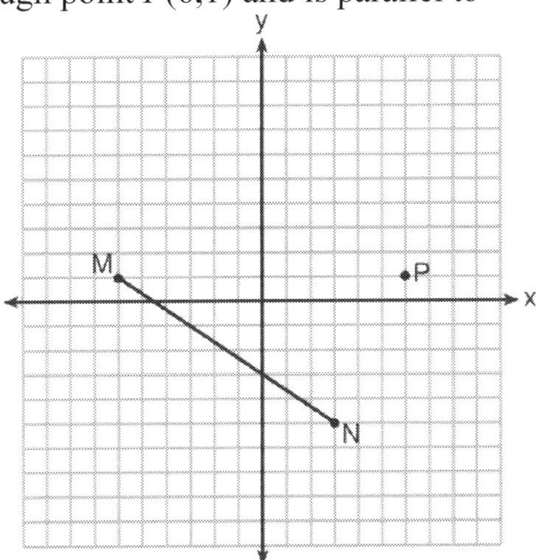

(1) $y=-\frac{2}{3}x+5$ (3) $y=\frac{3}{2}x+7$

(2) $y=-\frac{2}{3}x-3$ (4) $y=\frac{3}{2}x-8$

15. **CC** An equation of a line perpendicular to the line represented by the equation $y=-\frac{1}{2}x-5$ and passing through $(6,-4)$ is

(1) $y=-\frac{1}{2}x+4$ (3) $y=2x+14$

(2) $y=-\frac{1}{2}x-1$ (4) $y=2x-16$

16. **CC** In the diagram below, $\triangle ABC$ has vertices $A(4,5)$, $B(2,1)$, and $C(7,3)$.

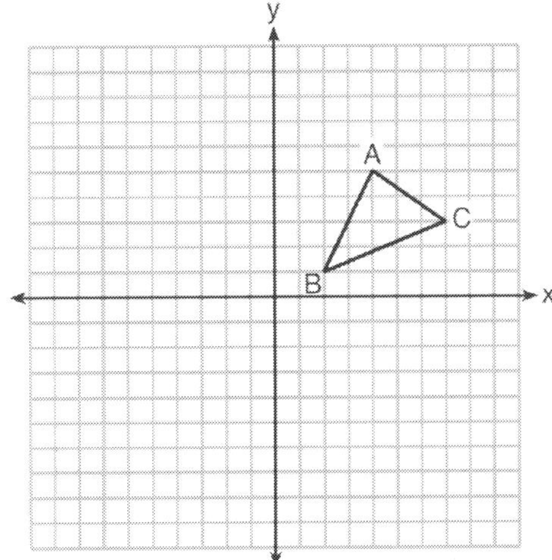

What is the slope of the altitude drawn from A to $\overline{BC}$?

(1) $\frac{2}{5}$ (3) $-\frac{1}{2}$

(2) $\frac{3}{2}$ (4) $-\frac{5}{2}$

17. (CC) Which equation represents the line that passes through the point (–2,2) and is parallel to $y = \frac{1}{2}x + 8$?

(1) $y = \frac{1}{2}x$

(2) $y = -2x - 3$

(3) $y = \frac{1}{2}x + 3$

(4) $y = -2x + 3$

Constructed Response

18. Determine whether the two lines represented by the equations $y = 2x + 3$ and $2y + x = 6$ are parallel, perpendicular, or neither. Justify your response.

19. Two lines are represented by the equations $x + 2y = 4$ and $4y - 2x = 12$. Determine whether these lines are parallel, perpendicular, or neither. Justify your answer.

20. Write an equation of a line that is parallel to the line whose equation is $3y = x + 6$ and that passes through the point (–3,4).

21. The slope of $\overline{QR}$ is $\dfrac{x-1}{4}$ and the slope of $\overline{ST}$ is $\frac{8}{3}$. If $\overline{QR} \perp \overline{ST}$, determine and state the value of x.

8.3 Distance Formula

Key Terms and Concepts

The formula for the distance between two points on a plane is derived from the Pythagorean Theorem. Suppose we want to find the distance, *d*, between points *A* and *B*, below. Point *A* has coordinates (x_1, y_1) and point *B* has coordinates (x_2, y_2).

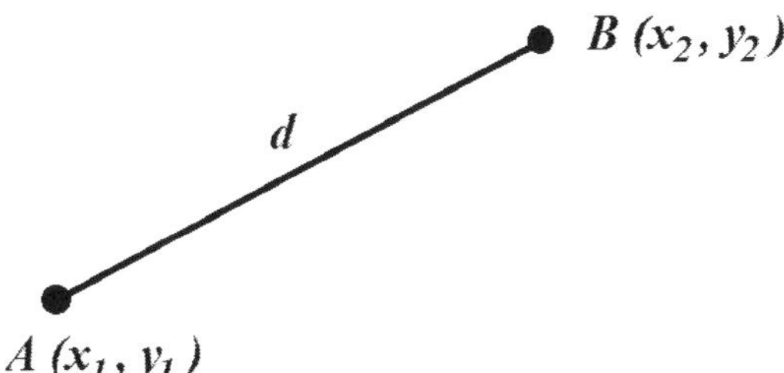

To find the distance, we can create a right triangle with $\overline{AB}$ as its hypotenuse. Draw a vertical altitude from *B* and a horizontal altitude from *A* that meets at a right angle at point *C*, shown below. *C* has the same *x*-coordinate as *B* and the same *y*-coordinate as *A*, or (x_2, y_1).

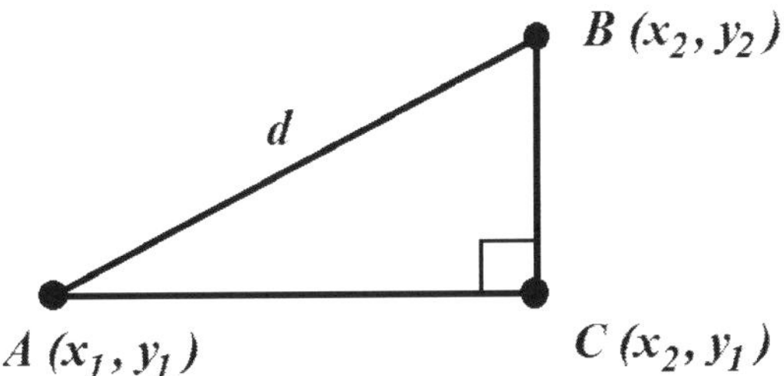

The length of $\overline{AC}$ is the difference between the *x*-values, or $x_2 - x_1$, and the length of $\overline{BC}$ is the difference between the *y*-values, or $y_2 - y_1$.

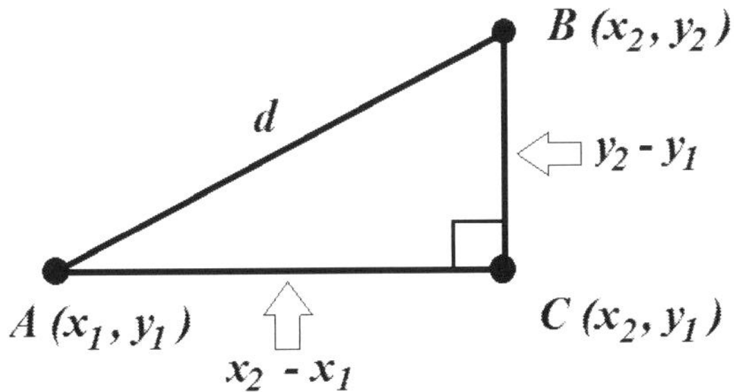

So, by the Pythagorean Theorem, $d^2 = (x_2 - x_1)^2 + (y_2 - y_1)^2$. Taking the positive square root of both sides, we get the **distance formula**:

$$d = \sqrt{(x_2 - x_1)^2 + (y_2 - y_1)^2}$$

The distance formula is used to find the distance between points (x_1, y_1) and (x_2, y_2).

Example: The distance between (–2,1) and (5,–3) is calculated as

$$d=\sqrt{(5-(-2))^2+(-3-1)^2}=\sqrt{7^2+(-4)^2}=\sqrt{49+16}=\sqrt{65}$$

Model Problem:

Find the distance between the point $A(-2,-4)$ and the line $y=-\frac{1}{3}x+2$.

Solution:

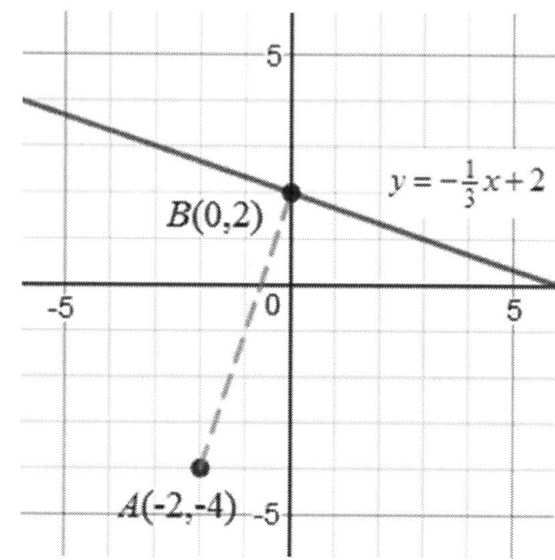

(A) For $y=-\frac{1}{3}x+2$, $m=-\frac{1}{3}$, so $m_{\perp}=3$

(B) Equation of the perpendicular line through $A(-2,-4)$ is $y+4=3(x+2)$, or in slope-intercept form, $y=3x+2$

(C) $-\frac{1}{3}x+2=3x+2$, or $x=0$

$y=3(0)+2$, or $y=2$

So, intersection of lines is at $B(0,2)$

(D) 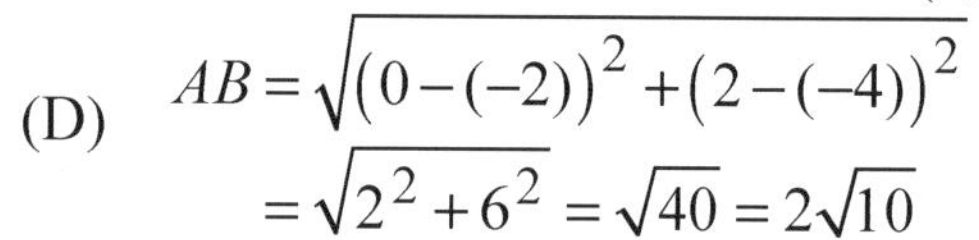

$$AB=\sqrt{(0-(-2))^2+(2-(-4))^2}=\sqrt{2^2+6^2}=\sqrt{40}=2\sqrt{10}$$

Explanation of steps:

(A) The shortest distance between a point and a line is the length of the line segment from the point to the line, perpendicular to that line. This segment will have the opposite reciprocal slope. *[The opposite reciprocal of* $-\frac{1}{3}$ *is 3.]*

(B) Write the equation of the line segment in point-slope form, using the given point and the slope from part (A).

(C) To find the point of intersection, solve the system of equation by substitution. *[Given* $y=-\frac{1}{3}x+2$ *and* $y=3x+2$*, substitute for y in the second equation, and solve to get x = 0. Then substitute for x in the second equation and solve to get y = 2. These coordinates give us the point of intersection, (0,2).]*

(D) Use the distance formula to find the distance between the two points *[A(–2,–4) and B(0,2)].*

Practice Problems

1. Find the distance between the points (–2,3) and (6,–3).

2. Find the distance between the points (3,5) and (8,10), in simplest radical form.

3. The endpoints of $\overline{PQ}$ are $P(-3,1)$ and $Q(4,25)$. Find the length of $\overline{PQ}$.

4. What is the length, in simplest radical form, of the line segment joining the points (–4,2) and (146,52)?

5. Find the distance between the point (6,–2) and the line $y=\frac{1}{5}x+2$.

REGENTS QUESTIONS

Multiple Choice

1. What is the length of the line segment whose endpoints are $A(-1,9)$ and $B(7,4)$?
   (1) $\sqrt{61}$ (3) $\sqrt{205}$
   (2) $\sqrt{89}$ (4) $\sqrt{233}$

2. What is the length of the line segment whose endpoints are (1,–4) and (9,2)?
   (1) 5 (3) 10
   (2) $2\sqrt{17}$ (4) $2\sqrt{26}$

3. A line segment has endpoints (4,7) and (1,11). What is the length of the segment?
   (1) 5 (3) 16
   (2) 7 (4) 25

4. What is the length of $\overline{AB}$ with endpoints $A(-1,0)$ and $B(4,-3)$?
   (1) $\sqrt{6}$ (3) $\sqrt{34}$
   (2) $\sqrt{18}$ (4) $\sqrt{50}$

5. Square $ABCD$ has vertices $A(-2,-3)$, $B(4,-1)$, $C(2,5)$, and $D(-4,3)$. What is the length of a side of the square?
   (1) $2\sqrt{5}$ (3) $4\sqrt{5}$
   (2) $2\sqrt{10}$ (4) $10\sqrt{2}$

6. What is the length of $\overline{RS}$ with $R(-2,3)$ and $S(4,5)$?
   (1) $2\sqrt{2}$ (3) $2\sqrt{10}$
   (2) 40 (4) $2\sqrt{17}$

7. Line segment AB has endpoint A located at the origin. Line segment AB is longest when the coordinates of B are
   (1) (3,7) (3) (–6,4)
   (2) (2,–8) (4) (–5,–5)

8. What is the length of a line segment whose endpoints have coordinates (5,3) and (1,6)?
   (1) 5 (3) $\sqrt{17}$
   (2) 25 (4) $\sqrt{29}$

9. (CC) The center of circle Q has coordinates (3,–2). If circle Q passes through $R(7,1)$, what is the length of its diameter?
   (1) 50 (3) 10
   (2) 25 (4) 5

10. (CC) The endpoints of one side of a regular pentagon are (–1,4) and (2,3). What is the perimeter of the pentagon?

(1) $\sqrt{10}$ (3) $5\sqrt{2}$

(2) $5\sqrt{10}$ (4) $25\sqrt{2}$

Constructed Response

11. Katrina hikes 5 miles north, 7 miles east, and then 3 miles north again. To the *nearest tenth of a mile,* how far, in a straight line, is Katrina from her starting point?

12. To get from his high school to his home, Jamal travels 5.0 miles east and then 4.0 miles north. When Sheila goes to her home from the same high school, she travels 8.0 miles east and 2.0 miles south. What is the measure of the shortest distance, to the *nearest tenth of a mile,* between Jamal's home and Sheila's home? [The use of the accompanying grid is optional.]

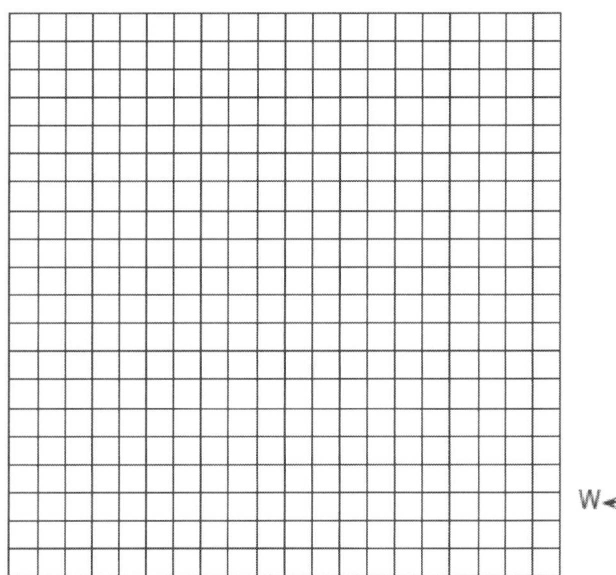

N
W E
S

13. Two hikers started at the same location. One traveled 2 miles east and then 1 mile north. The other traveled 1 mile west and then 3 miles south. At the end of their hikes, how many miles apart are the two hikers? [The use of the accompanying grid is optional.]

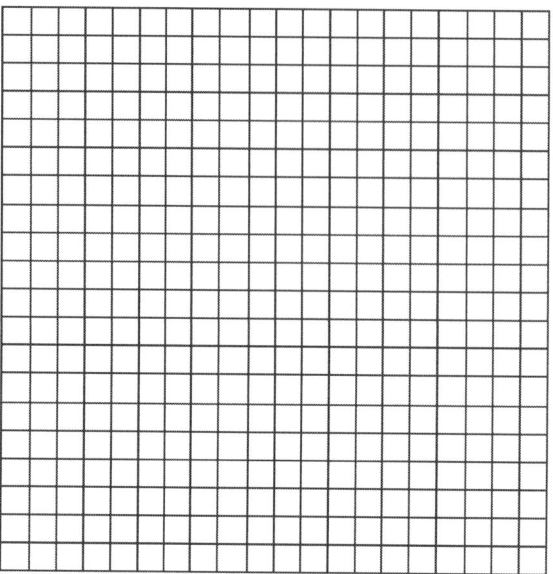

14. The coordinates of the endpoints of $\overline{FG}$ are (–4,3) and (2,5). Find the length of $\overline{FG}$ in simplest radical form.

15. Find, in simplest radical form, the length of the line segment with endpoints whose coordinates are (–1,4) and (3,–2).

16. The endpoints of $\overline{AB}$ are $A(3,-4)$ and $B(7,2)$. Determine and state the length of $\overline{AB}$ in simplest radical form.

17. The coordinates of the endpoints of $\overline{CD}$ are C(3,8) and D(6,–1). Find the length of $\overline{CD}$ in simplest radical form.

8.4 Midpoint Formula

Key Terms and Concepts

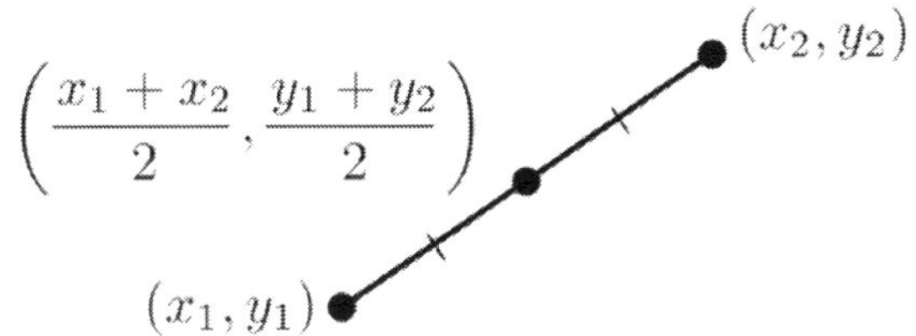

Given a line segment with endpoints of (x_1, y_1) and (x_2, y_2), the coordinates of the midpoint of the segment are (*the average of* x_1 *and* x_2, *the average of* y_1 *and* y_2).

The **midpoint formula** can be written as $\left(\frac{x_1+x_2}{2}, \frac{y_1+y_2}{2}\right)$.

Example: To find the midpoint of the line segment whose endpoints are (–2,1) and (5,–3), calculate $\left(\frac{x_1+x_2}{2}, \frac{y_1+y_2}{2}\right) = \left(\frac{(-2)+5}{2}, \frac{1+(-3)}{2}\right) = \left(\frac{3}{2}, -1\right)$.

Given an endpoint and midpoint of a line segment, you can find the other endpoint by adding an equal distance beyond the midpoint. If one endpoint is (x_1, y_1) and the midpoint is (x_M, y_M), then we can find the *x*-coordinate of the other endpoint by finding the next term in the arithmetic sequence $x_1, x_M, \ldots$. The *y*-coordinate is determined in the same way, by extending the arithmetic sequence $y_1, y_M, \ldots$ to a third term.

Example: The coordinates of *A* are (8,–4). The midpoint of $\overline{AB}$ is (6,–8).
We can find the *x*-coordinate of *B* by extending 8, 6, … to a third term, 4.
We can find the *y*-coordinate of *B* by extending –4, –8, … to a third term, –12.
Therefore, the coordinates of *B* are (4,–12).

An alternative method for finding the other endpoint, given one endpoint and the midpoint, is to substitute into the midpoint formula and solve. In other words, substitute for x_1 and x_M in the equation $\frac{x_1+x_2}{2} = x_M$ and substitute for y_1 and y_M in the equation $\frac{y_1+y_2}{2} = y_M$.

Example: The coordinates of *A* are (8,–4). The midpoint of $\overline{AB}$ is (6,–8).
To find the coordinates of *B*, solve both $\frac{8+x_2}{2} = 6$ and $\frac{-4+y_2}{2} = -8$.
The equations give us $x_2 = 4$ and $y_2 = -12$, so *B* is (4,–12).

Model Problem:

The endpoints of $\overline{AB}$ are $A(-2,5)$ and $B(4,11)$. Find the coordinates of the midpoint of $\overline{AB}$.

Solution:

$$\left(\frac{x_1+x_2}{2},\frac{y_1+y_2}{2}\right)=\left(\frac{-2+4}{2},\frac{5+11}{2}\right)=(1,8)$$

Explanation:

Substitute the coordinates (x_1, y_1) *[(–2,5)]* and (x_2, y_2) *[(4,11)]* into the midpoint formula, and simplify.

Practice Problems

1. Find the midpoint of the segment whose endpoints are (–2,3) and (6,–3).	2. The diameter of a circle has endpoints of (3,5) and (8,10). Find the coordinates of the center of the circle.
3. The midpoint of $\overline{AB}$ is $M(3,5)$. Given $A(1,-5)$, find the coordinates of B.	4. In triangle ABC, median AM is drawn. Given $A(3,4)$, $B(4,-3)$, and $M(0,-2)$, find the coordinates of C.

REGENTS QUESTIONS

Multiple Choice

1. Square $LMNO$ is shown in the diagram below.

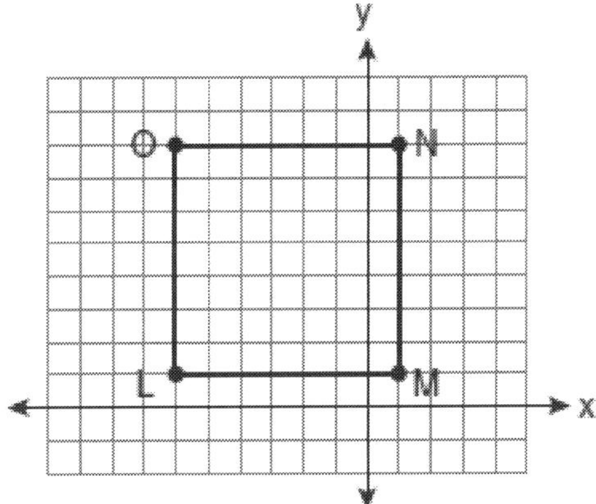

What are the coordinates of the midpoint of diagonal $\overline{LN}$?

(1) $\left(4\frac{1}{2},-2\frac{1}{2}\right)$ (3) $\left(-2\frac{1}{2},3\frac{1}{2}\right)$
(2) $\left(-3\frac{1}{2},3\frac{1}{2}\right)$ (4) $\left(-2\frac{1}{2},4\frac{1}{2}\right)$

2. The endpoints of $\overline{CD}$ are $C(-2,-4)$ and $D(6,2)$. What are the coordinates of the midpoint of $\overline{CD}$?

(1) (2,3) (3) (4,–2)
(2) (2,–1) (4) (4,3)

3. A line segment has endpoints $A(7,-1)$ and $B(-3,3)$. What are the coordinates of the midpoint of $\overline{AB}$?

(1) (1,2) (3) (–5,2)
(2) (2,1) (4) (5,–2)

4. Segment AB is the diameter of circle M. The coordinates of A are (–4,3). The coordinates of M are (1,5). What are the coordinates of B?

(1) (6,7) (3) (–3,8)
(2) (5,8) (4) (–5,2)

5. Point M is the midpoint of $\overline{AB}$. If the coordinates of A are (–3,6) and the coordinates of M are (–5,2), what are the coordinates of B?

(1) (1,2) (3) (–4,4)
(2) (7,10) (4) (–7,–2)

6. Line segment AB is a diameter of circle O whose center has coordinates (6,8). What are the coordinates of point B if the coordinates of point A are (4,2)?

(1) (1,3) (3) (8,14)
(2) (5,5) (4) (10,10)

7. What are the coordinates of the center of a circle if the endpoints of its diameter are $A(8,-4)$ and $B(-3,2)$?

(1) (2.5,1)	(3) (5.5,–3)
(2) (2.5,–1)	(4) (5.5,3)

8. The midpoint of $\overline{AB}$ is $M(4,2)$. If the coordinates of A are (6,–4), what are the coordinates of B?

(1) (1,–3)	(3) (5,–1)
(2) (2,8)	(4) (14,0)

9. In the diagram below, quadrilateral $ABCD$ has vertices $A(-5,1)$, $B(6,-1)$, $C(3,5)$, and $D(-2,7)$.

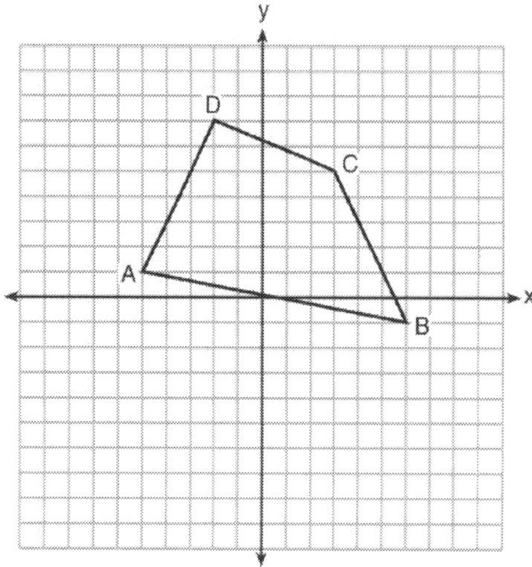

What are the coordinates of the midpoint of diagonal $\overline{AC}$?

(1) (–1,3)	(3) (1,4)
(2) (1,3)	(4) (2,3)

10. In the diagram below, parallelogram $ABCD$ has vertices $A(1,3)$, $B(5,7)$, $C(10,7)$, and $D(6,3)$. Diagonals $\overline{AC}$ and $\overline{BD}$ intersect at E.

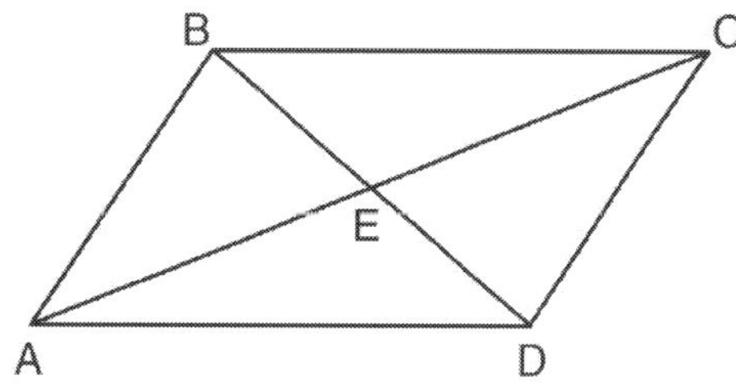

(Not drawn to scale)

What are the coordinates of point E?

(1) (0.5,2)	(3) (5.5,5)
(2) (4.5,2)	(4) (7.5,7)

11. What are the coordinates of the midpoint of the line segment with endpoints (2,–5) and (8,3)?

(1) (3,–4)	(3) (5,–4)
(2) (3,–1)	(4) (5,–1)

12. Point M is the midpoint of $\overline{AB}$. If the coordinates of M are (2,8) and the coordinates of A are (10,12), what are the coordinates of B?
    (1) (6,10)
    (2) (–6,4)
    (3) (–8,–4)
    (4) (18,16)

Constructed Response

13. In a circle whose center is (2,3), one endpoint of a diameter is (–1,5). Find the coordinates of the other endpoint of that diameter.

14. One endpoint of a line segment is (6,2). The midpoint of the segment is (2,0). Find the coordinates of the other endpoint.

15. In the diagram below of circle C, $\overline{QR}$ is a diameter, and $Q(1,8)$ and $C(3.5,2)$ are points on a coordinate plane. Find and state the coordinates of point R.

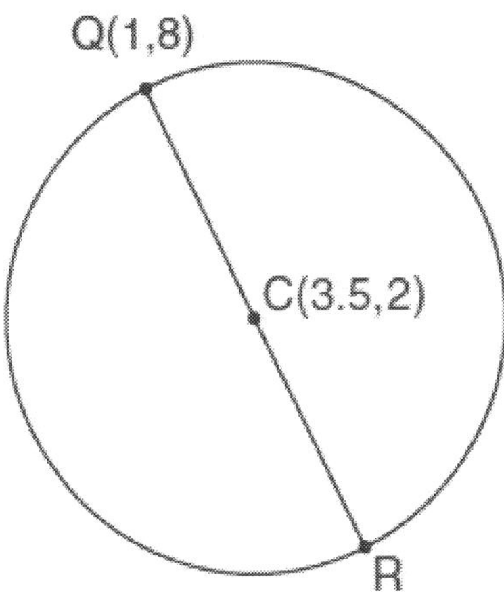

8.5 Perpendicular Bisectors

Key Terms and Concepts

The **perpendicular bisector** of a line segment is a line that is perpendicular to the line segment and passes through its midpoint. If we want to determine the equation of the perpendicular bisector of a line segment, we will need to complete these steps:

1. Find the coordinates of the midpoint.
2. Find the slope of the line segment.
3. Write an equation of the line that passes through the midpoint and has an opposite reciprocal slope. We can use the point-slope form, $y - y_1 = m(x - x_1)$.

Remember the exceptions: A vertical line has no slope and an equation of the form $x = k$, while a horizontal line has a zero slope and an equation of the form $y = k$. A vertical line is perpendicular to a horizontal line, and vice versa.

Model Problem:

Write an equation for the perpendicular bisector of $\overline{PQ}$, shown in the graph to the right.

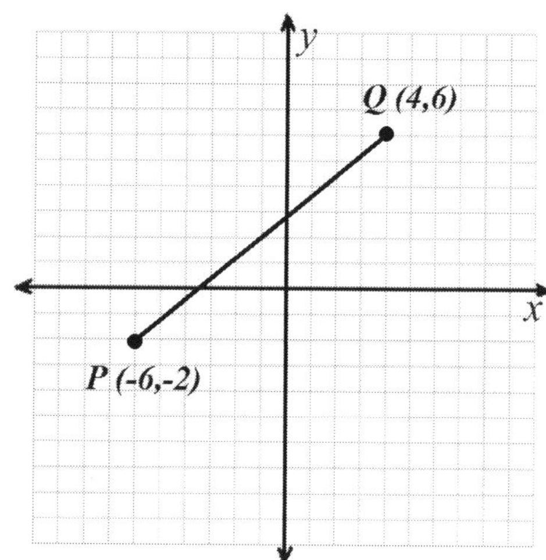

Solution:

(A) Midpoint is $\left(\frac{-6+4}{2}, \frac{-2+6}{2}\right) = (-1, 2)$.

(B) Slope is $\frac{6-(-2)}{4-(-6)} = \frac{8}{10} = \frac{4}{5}$.

(C) Slope of perpendicular line is $-\frac{5}{4}$.

Equation of perpendicular bisector is $y - 2 = -\frac{5}{4}(x+1)$.

Explanation of steps:

(A) Find the coordinates of the midpoint *[using the midpoint formula].*

(B) Find the slope of the line segment *[using the slope formula].*

(C) Write an equation of the line that passes through the midpoint and has an opposite reciprocal slope. We can use the point-slope form, $y - y_1 = m(x - x_1)$. *[Substitute the midpoint coordinates (–1,2) for x_1 and y_1 and the slope for m.]*

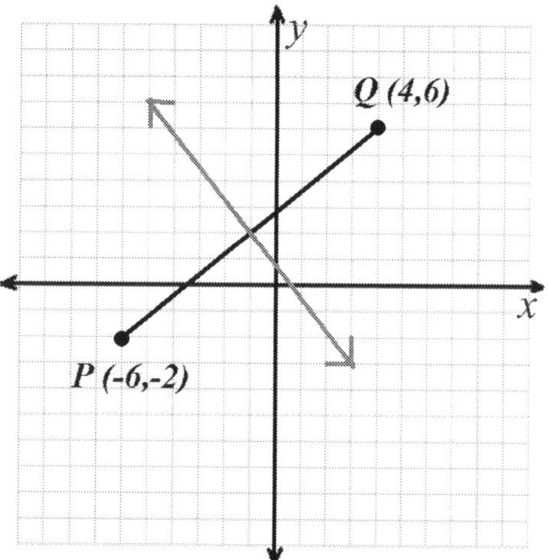

Practice Problems

1. Write an equation of the perpendicular bisector of the segment whose endpoints are (3,5) and (9,17).	2. Write an equation of the perpendicular bisector of the segment whose endpoints are (–2,3) and (6,–3).
3. Given $A(2,6)$ and $B(8,12)$, write an equation of $\overleftrightarrow{CD}$, the perpendicular bisector of $\overline{AB}$. What is the y-intercept of $\overleftrightarrow{CD}$?	4. Write an equation of the perpendicular bisector of the segment whose endpoints are (–4,5) and (2,5).

REGENTS QUESTIONS

Multiple Choice

1. Which equation represents the perpendicular bisector of $\overline{AB}$ whose endpoints are $A(8,2)$ and $B(0,6)$?

(1) $y=2x-4$ (3) $y=-\frac{1}{2}x+6$

(2) $y=-\frac{1}{2}x+2$ (4) $y=2x-12$

2. The coordinates of the endpoints of $\overline{AB}$ are $A(0,0)$ and $B(0,6)$. The equation of the perpendicular bisector of $\overline{AB}$ is

(1) $x=0$ (3) $y=0$

(2) $x=3$ (4) $y=3$

3. Triangle ABC has vertices $A(0,0)$, $B(6,8)$, and $C(8,4)$. Which equation represents the perpendicular bisector of $\overline{BC}$?

(1) $y=2x-6$ (3) $y=\frac{1}{2}x+\frac{5}{2}$

(2) $y=-2x+4$ (4) $y=-\frac{1}{2}x+\frac{19}{2}$

4. (CC) Line segment NY has endpoints $N(-11,5)$ and $Y(5,-7)$. What is the equation of the perpendicular bisector of $\overline{NY}$?

(1) $y+1=\frac{4}{3}(x+3)$ (3) $y-6=\frac{4}{3}(x-8)$

(2) $y+1=-\frac{3}{4}(x+3)$ (4) $y-6=-\frac{3}{4}(x-8)$

Constructed Response

5. Write an equation of the perpendicular bisector of the line segment whose endpoints are $(-1,1)$ and $(7,-5)$.

6. Write an equation of the line that is the perpendicular bisector of the line segment having endpoints $(3,-1)$ and $(3,5)$.

7. If $\overline{AB}$ is defined by the endpoints $A(4,2)$ and $B(8,6)$, write an equation of the line that is the perpendicular bisector of $\overline{AB}$.

8.6 Directed Line Segments

Key Terms and Concepts

A **directed line segment** is a portion of a line that has both a length and *direction*. It is a line segment that extends *from* one point *to* another point on the coordinate plane.
Example: Directed line segment AB extends from point A to point B.
A is called the *tail* and B is called the *head*.

In the diagram below, directed line segment AB is divided into five equal parts by the points shown. Point P is the point that divides (or *partitions*) AB into a 2:3 **ratio**.

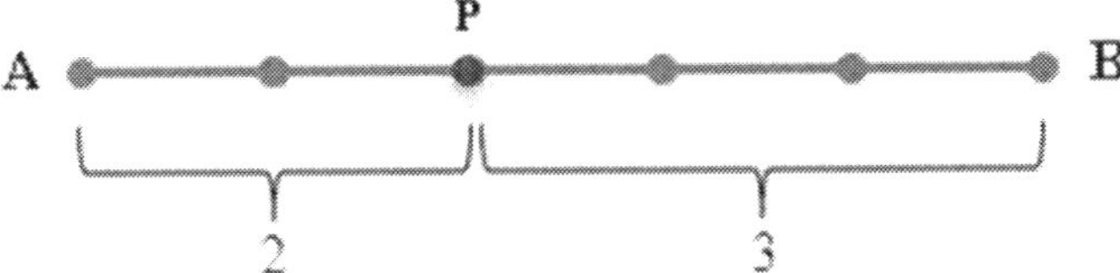

Note that in the diagram above, we could also refer to directed line segment BA, which has the same length but opposite direction from AB. Point P divides BA into a 3:2 ratio.

To find the coordinates of point P that partitions a directed line segment AB into a certain ratio:

1. Convert the ratio into a fraction, k, that represents a part of a whole. In the above example, since AP:PB is in the ratio 2:3, we can say that P lies $\frac{2}{5}$ of the distance along the segment. (The denominator is the sum of the two parts.)
2. The x-coordinate of P will be the x-coordinate of A plus the fraction k of the **run** (the difference between the x-coordinates) from A to B. That is, $P_x = A_x + k(B_x - A_x)$
3. The y-coordinate of P will be the y-coordinate of A plus the fraction k of the **rise** (the difference between the y-coordinates) from A to B. That is, $P_y = A_y + k(B_y - A_y)$

Note that the sign of the rise or the run may be *positive or negative*, since we are working with a directed line segment AB. The rise or run will be *positive* if the y or x coordinate of B, respectively, is greater than that of A, or *negative* otherwise.

Model Problem:

Given points $A(3,4)$ and $B(6,10)$, find the coordinates of point P along the directed line segment AB so that the ratio of AP to PB is 3:2.

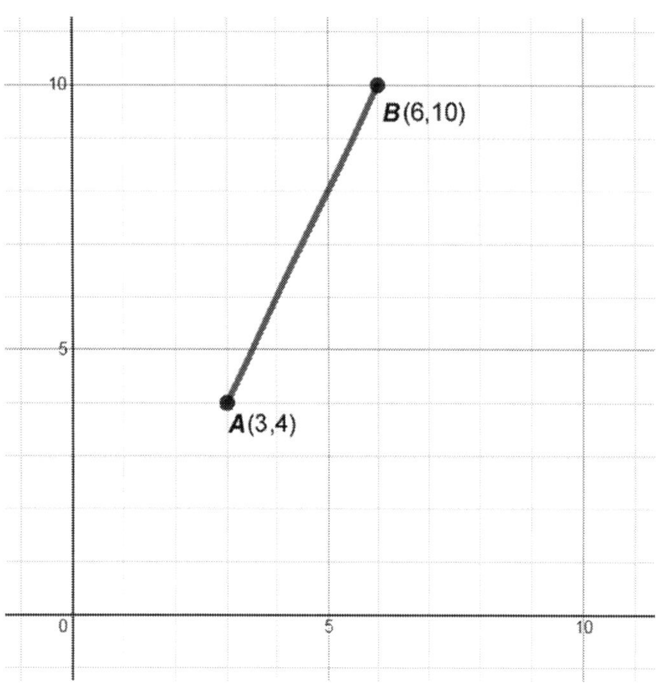

Solution:

(A) $AP = \frac{3}{5}AB$, so $k = \frac{3}{5}$.

(B) $P_x = A_x + \frac{3}{5}(B_x - A_x) = 3 + \frac{3}{5}(6-3) = 4.8$

(C) $P_y = A_y + \frac{3}{5}(B_y - A_y) = 4 + \frac{3}{5}(10-4) = 7.6$

(D) Point P is (4.8, 7.6)

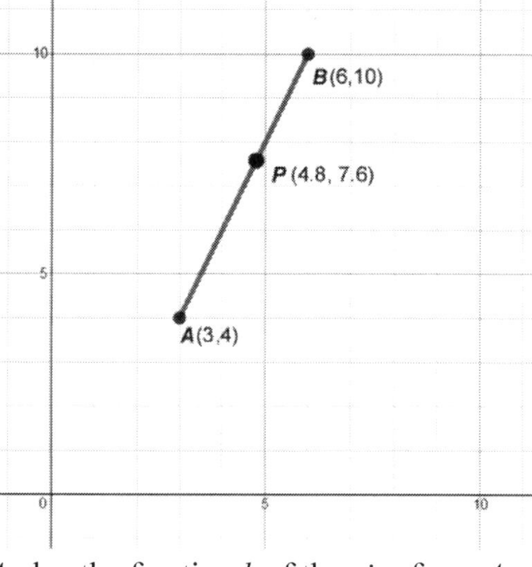

Explanation of steps:

(A) Convert the ratio *[3:2]* to a fraction, k, representing part over whole *[*$\frac{3}{5}$*]*.

(B) The x-coordinate of P will be the x-coordinate of A plus the fraction k of the *run* from A to B.

(C) The y-coordinate of P will be the y-coordinate of A plus the fraction k of the *rise* from A to B.

(D) State the coordinates of point P.

Practice Problems

1. What are the coordinates of the point P on the directed line segment from $A(3,5)$ to $B(9,17)$ that partitions the segment into the ratio of 5:1.	2. What are the coordinates of the point S on the directed line segment from $R(-2,2)$ to $T(3,-8)$ that partitions the segment into the ratio of 3:2.
3. What are the coordinates of the point P on the directed line segment from $L(-2,3)$ to $M(6,-3)$ such that $LP{:}PM$ is 3:5.	4. What are the coordinates of the point G on the directed line segment from $F(1,-3)$ to $H(6,5)$ such that $FG{:}GH$ is 2:3.

REGENTS QUESTIONS

Multiple Choice

1. CC What are the coordinates of the point on the directed line segment from $K(-5,-4)$ to $L(5,1)$ that partitions the segment into a ratio of 3 to 2?

   (1) $(-3,-3)$ (3) $\left(0,-\frac{3}{2}\right)$

   (2) $(-1,-2)$ (4) $(1,-1)$

2. CC Point P is on the directed line segment from point $X(-6,-2)$ to point $Y(6,7)$ and divides the segment in the ratio 1:5. What are the coordinates of point P?

   (1) $\left(4,5\frac{1}{2}\right)$ (3) $\left(-4\frac{1}{2},0\right)$

   (2) $\left(-\frac{1}{2},-4\right)$ (4) $\left(-4,-\frac{1}{2}\right)$

3. CC Point Q is on $\overline{MN}$ such that $MQ:QN=2:3$. If M has coordinates (3,5) and N has coordinates (8,–5), the coordinates of Q are

   (1) (5,1) (3) (6,–1)

   (2) (5,0) (4) (6,0)

Constructed Response

4. CC The coordinates of the endpoints of $\overline{AB}$ are $A(-6,-5)$ and $B(4,0)$. Point P is on $\overline{AB}$. Determine and state the coordinates of point P, such that $AP:PB$ is 2:3.
   [The use of the set of axes below is optional.]

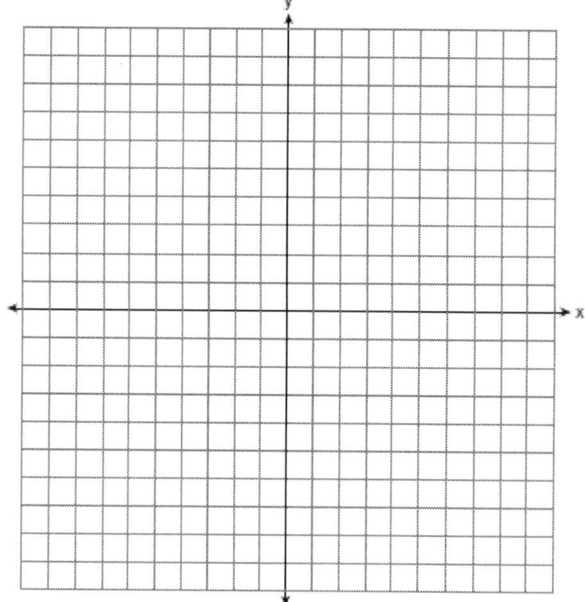

5. CC The endpoints of $\overline{DEF}$ are $D(1,4)$ and $F(16,14)$. Determine and state the coordinates of point E, if $DE:EF = 2:3$

6. (CC) Directed line segment PT has endpoints whose coordinates are $P(-2,1)$ and $T(4,7)$. Determine the coordinates of point J that divides the segment in the ratio 2 to 1. [The use of the set of axes below is optional.]

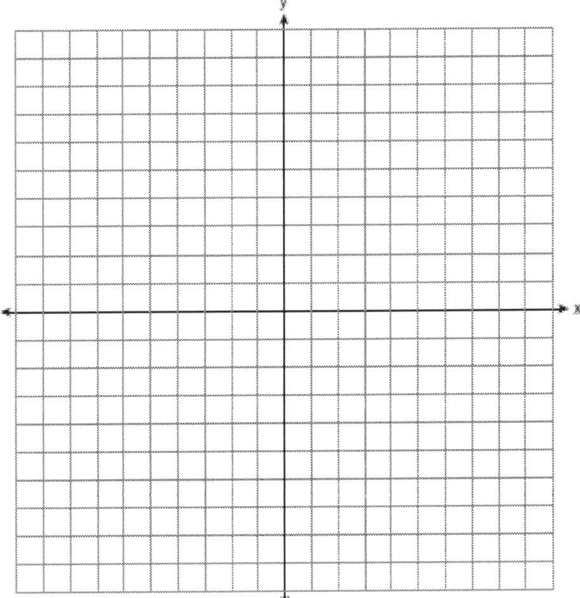

7. (CC) Point P is on segment AB such that $AP:PB$ is 4:5. If A has coordinates (4,2), and B has coordinates (22,2), determine and state the coordinates of P.

Chapter 9. Polygons in the Coordinate Plane

9.1 Triangles in the Coordinate Plane

Key Terms and Concepts

We can use our knowledge of coordinate geometry to examine or prove properties of triangles given the coordinates of their vertices in the coordinate plane.

We can determine whether a triangle is a **right triangle** by examining the slopes of its sides. If two sides have opposite reciprocal slopes, then they are perpendicular. This means the angle at their shared vertex is a right angle; therefore, the triangle is a right triangle.

Example: Given triangle ABC with $A(1,2)$, $B(3,3)$, and $C(4,1)$, we can determine if the triangle is a right triangle by finding the slopes of the sides:

$$m_{\overline{AB}} = \frac{3-2}{3-1} = \tfrac{1}{2} \qquad m_{\overline{BC}} = \frac{1-3}{4-3} = -2 \qquad m_{\overline{AC}} = \frac{1-2}{4-1} = -\tfrac{1}{3}$$

Since the slopes of $\overline{AB}$ and $\overline{BC}$ are opposite reciprocals, they are perpendicular. Therefore, $\angle B$ is a right angle and $\triangle ABC$ is a right triangle.

To determine whether a triangle is **scalene**, **isosceles**, or **equilateral**, we can use the distance formula to find the lengths of its sides.

- A *scalene* triangle has no congruent sides.
- An *isosceles* triangle has two congruent sides.
- An *equilateral* triangle has three congruent sides.

Example: Given the same triangle ABC with $A(1,2)$, $B(3,3)$, and $C(4,1)$, we can use the distance formula to find the lengths of its sides:

$$AB = \sqrt{(3-1)^2 + (3-2)^2} = \sqrt{5}$$

$$BC = \sqrt{(4-3)^2 + (1-3)^2} = \sqrt{5}$$

$$AC = \sqrt{(4-1)^2 + (1-2)^2} = \sqrt{10}$$

Two sides, AB and BC, are equal in length. Therefore, $\triangle ABC$ is an isosceles triangle with AB and BC as its legs and $\angle B$ as its vertex angle. This also means that $\angle A \cong \angle C$, since base angles of isosceles triangles are congruent.

Model Problem:

Given $\triangle KLM$ with $K(1,2)$, $L(3,3)$, $M(2,y)$, and $\angle L$ is a right angle, find the value of y. Prove that $\angle K \cong \angle M$.

Solution:

(A) $m_{\overline{KL}} = \dfrac{3-2}{3-1} = \dfrac{1}{2}$

(B) $m_{\overline{LM}} = -2$

(C) $\dfrac{y-3}{2-3} = -2$

(D) $y-3=2$; $y=5$ $M(2,5)$

(E) $LM = \sqrt{(2-3)^2+(5-3)^2} = \sqrt{5}$

$KL = \sqrt{(3-1)^2+(3-2)^2} = \sqrt{5}$

(F) Since $\overline{LM} \cong \overline{KL}$ and the angles opposite congruent sides in a triangle are congruent, then $\angle K \cong \angle M$.

Explanation of steps:

(A) The legs of the right triangle *[$\overline{KL}$ and $\overline{LM}$]* are perpendicular. Find the slope of one leg *[slope of $\overline{KL}$ is $\frac{1}{2}$].*

(B) The slope of the legs are opposite reciprocals *[so, the slope of $\overline{LM}$ is –2].*

(C) Substitute the slope formula using the coordinates of the endpoints *[L and M].*

(D) Solve for *y*. This gives us the coordinates of the third vertex *[M].*

(E) To prove that two angles of a triangle are congruent, we can show that their opposite sides are congruent. Use the distance formula to find the lengths of the opposite sides. Congruent sides are equal in length.

(F) State the conclusion and reason.

Practice Problems

1. Determine whether $\triangle ABC$ is a right or oblique triangle. Justify your answer.

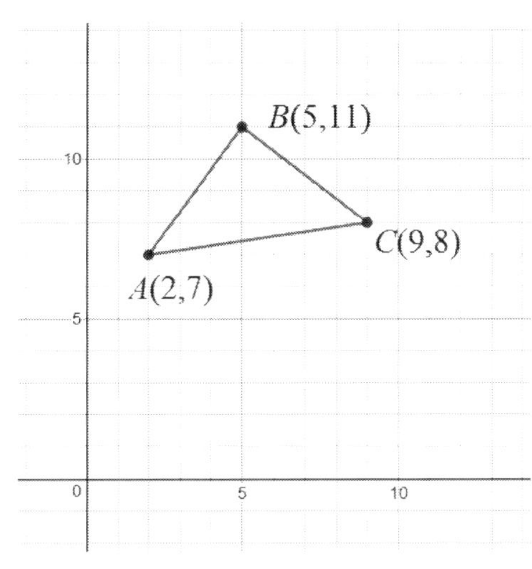

2. For $\triangle ABC$ shown in question 1, determine whether the triangle is scalene, isosceles, or equilateral.

3. Given $\triangle DEF$ with $D(4,-2)$, $E(5,5)$, and $F(-1,3)$. Determine whether $\triangle DEF$ is scalene, isosceles, or equilateral.

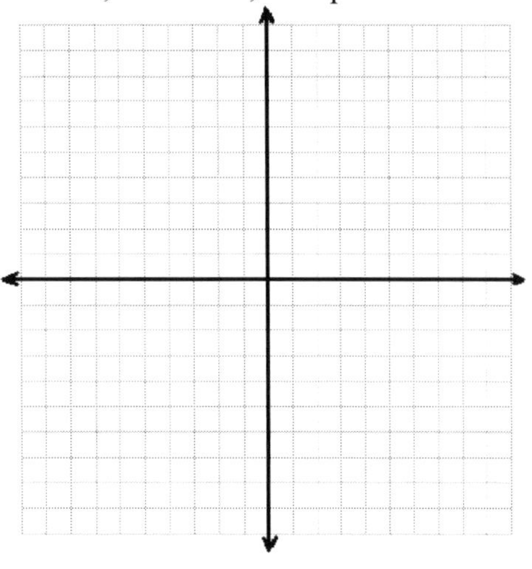

4. Right triangle JKL has a right angle at $\angle K$ and vertices $J(-2,4)$, $K(6,6)$, and $L(x,-2)$. Find x.

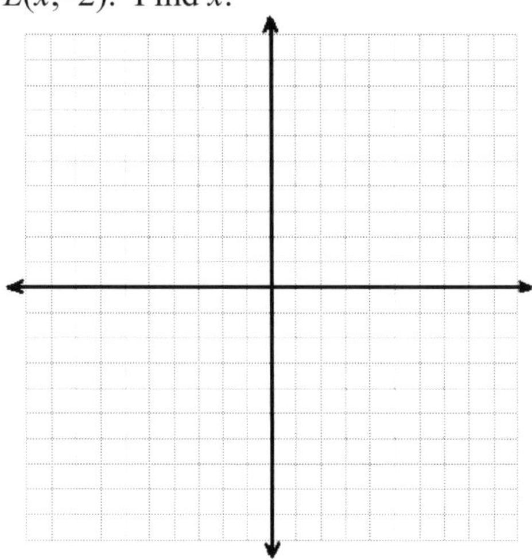

5. Find the area of triangle PQR with vertices $P(4,-2)$, $Q(-6,4)$, and $R(8,-2)$.

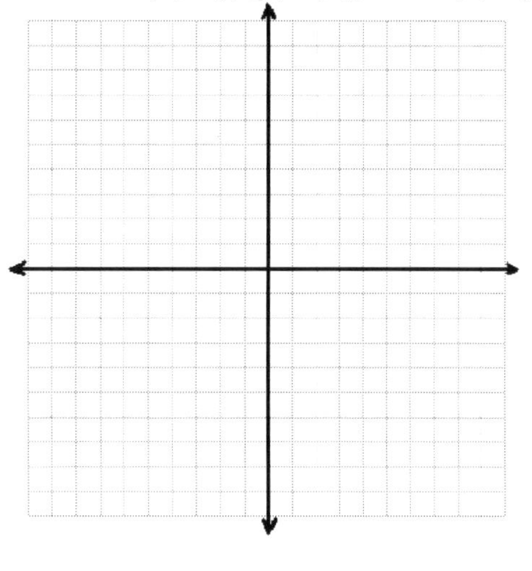

6. Given: $A(2,2)$, $B(5,1)$, $C(4,5)$, $D(1,-4)$, $E(4,-5)$, $F(3,-1)$
   Prove: $\triangle ABC \cong \triangle DEF$

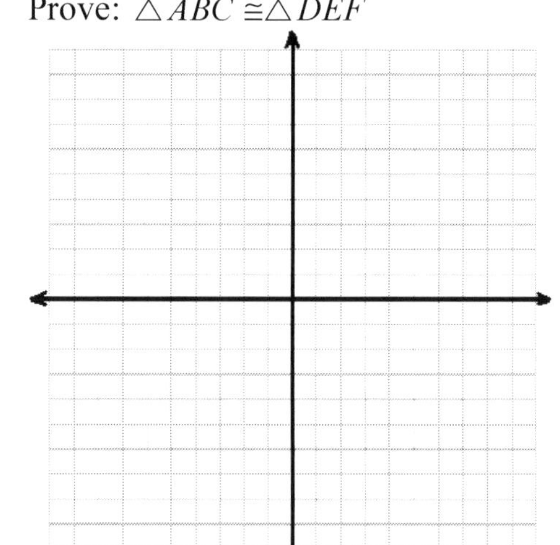

REGENTS QUESTIONS

Multiple Choice

1. Triangle ABC has vertices $A(0,0)$, $B(3,2)$, and $C(0,4)$. The triangle may be classified as
   (1) equilateral (3) right
   (2) isosceles (4) scalene

2. Which type of triangle can be drawn using the points $(-2,3)$, $(-2,-7)$, and $(4,-5)$?
   (1) scalene (3) equilateral
   (2) isosceles (4) no triangle can be drawn

3. If the vertices of $\triangle ABC$ are $A(-2,4)$, $B(-2,8)$, and $C(-5,6)$, then $\triangle ABC$ is classified as
   (1) right (3) isosceles
   (2) scalene (4) equilateral

4. (CC) The coordinates of the vertices of $\triangle RST$ are $R(-2,-3)$, $S(8,2)$, and $T(4,5)$. Which type of triangle is $\triangle RST$?
   (1) right (3) obtuse
   (2) acute (4) equiangular

5. (CC) The coordinates of vertices A and B of $\triangle ABC$ are $A(3,4)$ and $B(3,12)$. If the area of $\triangle ABC$ is 24 square units, what could be the coordinates of point C?
   (1) (3,6) (3) (–3,8)
   (2) (8,–3) (4) (6,3)

Constructed Response

6. Given: $J(-4,1)$, $E(-2,-3)$, $N(2,-1)$
   Prove: $\triangle JEN$ is an isosceles right triangle.
   [The use of the grid is optional.]

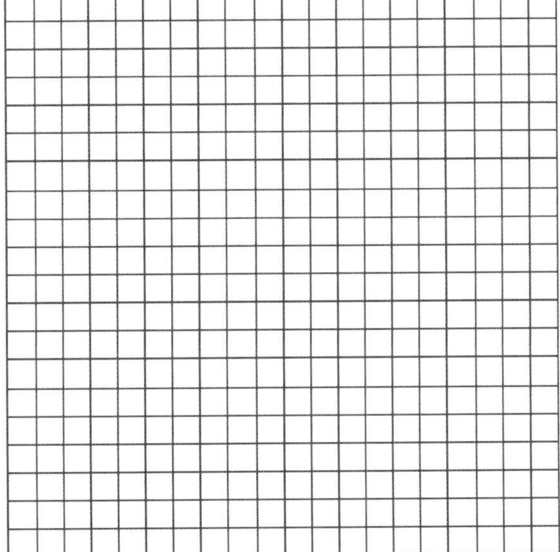

7. Triangle ABC has vertices at $A(3,0)$, $B(9,-5)$, and $C(7,-8)$. Find the length of $\overline{AC}$ in simplest radical form.

8. Given: Triangle RST has coordinates $R(-1,7)$, $S(3,-1)$, and $T(9,2)$
   Prove: $\triangle RST$ is a right triangle
   [The use of the set of axes below is optional.]

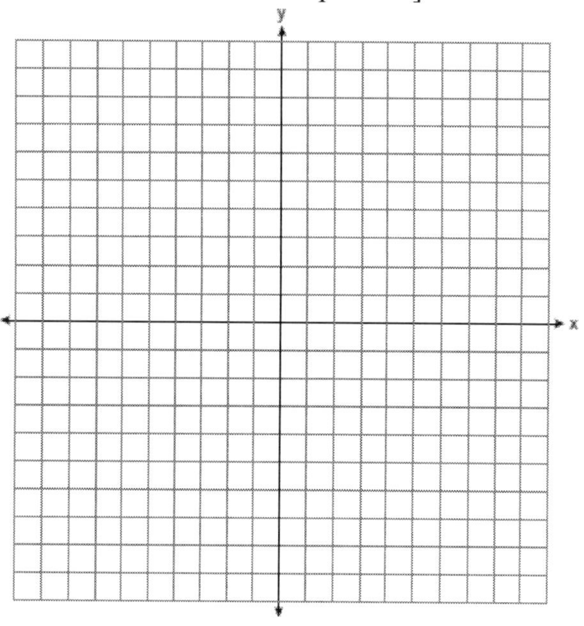

9. (CC) Triangle ABC has vertices with $A(x,3)$, $B(-3,-1)$, and $C(-1,-4)$. Determine and state a value of x that would make triangle ABC a right triangle. Justify why $\triangle ABC$ is a right triangle. [The use of the set of axes below is optional.]

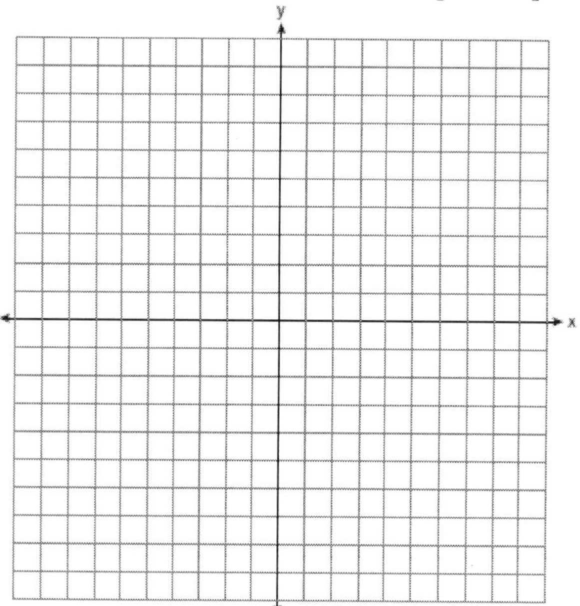

9.2 Quadrilaterals in the Coordinate Plane

Key Terms and Concepts

We can use our knowledge of coordinate geometry and of the properties of quadrilaterals to examine quadrilaterals in the coordinate plane.

As a reminder, here are some important properties of quadrilaterals:

A **trapezoid** is a quadrilateral with
- at least one pair of parallel sides

In a **parallelogram**,
- both pairs of opposite sides are parallel
- both pairs of opposite sides are congruent
- diagonals bisect each other

(To prove a quadrilateral is a parallelogram, it's sufficient to show one pair of sides that are both parallel and congruent.)

A **rectangle** is a parallelogram in which
- all four angles are right angles
- diagonals are congruent

(To prove a parallelogram is a rectangle, it's sufficient to show any one right angle.)

A **rhombus** is a parallelogram in which
- all four sides are congruent
- diagonals are perpendicular

(To prove a parallelogram is a rhombus, it's sufficient to show one pair of consecutive sides congruent.)

A **square** has the properties of a rectangle and a rhombus.

Once you've established what needs to be shown in order to prove a quadrilateral is of a certain type, use the table below, which explains how to show it.

To Show	**Use**
sides or diagonals are congruent	the distance formula to show their lengths are equal
sides are parallel	the slope formula to show their slopes are equal
diagonals bisect each other	the midpoint formula to show that their point of intersection is the midpoint of each diagonal
an angle is a right angle	the slope formula to show that the sides forming that vertex are perpendicular (i.e., their slopes are opposite reciprocals)

Model Problem:

Given: Parallelogram *ABCD* with A(–2,6), B(4,3), C(2,–1), D(–4,2).
Prove: *ABCD* is a rectangle.

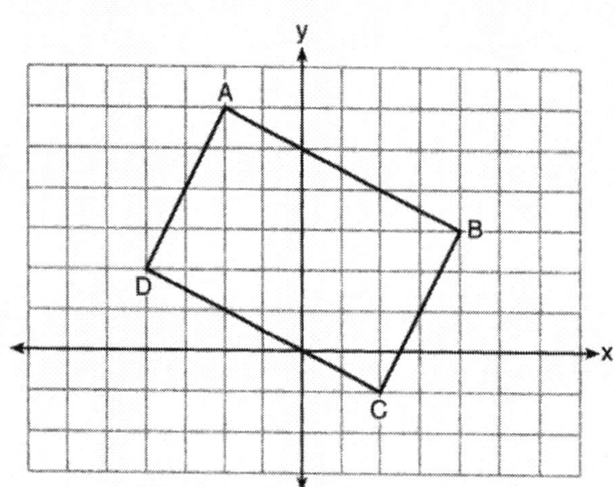

Solution:

(A)
Method 1: Show any angle is a right angle.

$m_{\overline{AB}} = \dfrac{3-6}{4+2} = -\frac{1}{2} \qquad m_{\overline{BC}} = \dfrac{-1-3}{2-4} = 2$

The slopes of $\overline{AB}$ and $\overline{BC}$ are opposite reciprocals, so $\overline{AB} \perp \overline{BC}$ and $\angle B$ is a right angle. A parallelogram with a right angle is a rectangle, so *ABCD* is a rectangle.

(B)
Method 2: Show diagonals are congruent.

$AC = \sqrt{(2+2)^2 + (-1-6)^2} = \sqrt{65}$

$BD = \sqrt{(-4-4)^2 + (2-3)^2} = \sqrt{65}$

Diagonals $\overline{AC}$ and $\overline{BD}$ have the same length, so $\overline{AC} \cong \overline{BD}$. A parallelogram with congruent diagonals is a rectangle, so *ABCD* is a rectangle.

Explanation of steps:

We can use either one of the following methods to prove *ABCD* is a rectangle:
(A) show that any angle is a right angle, or
(B) show that the diagonals are congruent.

(A) We can show an angle *[such as $\angle B$]* is a right angle by showing the sides that form the vertex *[$\overline{AB}$ and $\overline{BC}$]* are perpendicular. Use the slope formula to show their slopes are opposite reciprocals.

(B) We can show the diagonals *[$\overline{AC}$ and $\overline{BD}$]* are congruent by using the distance formula to find their lengths. Diagonals with equal lengths are congruent.

Practice Problems

1. Given: Quadrilateral $ABCD$ with $A(-5,0)$, $B(-1,-8)$, $C(7,-4)$, $D(3,4)$.
   Prove: $ABCD$ is a rectangle.

2. Quadrilateral $ABCD$ has vertices $A(-6,-3)$, $B(1,0)$, $C(4,7)$, and $D(-3,4)$. Classify $ABCD$ using the most precise name.

3. Quadrilateral $ABCD$ has vertices $A(-5,-6)$, $B(2,0)$, $C(11,9)$, and $D(4,3)$. Classify $ABCD$ using the most precise name.

4. Quadrilateral $ABCD$ has vertices $A(1,1)$, $B(5,2)$, $C(6,-2)$, and $D(2,-3)$. Classify $ABCD$ using the most precise name.

5. The vertices of square $PQRS$ are $P(-4, 0)$, $Q(4, 3)$, $R(7, -5)$, and $S(-1, -8)$. Show that the diagonals of square $PQRS$ are congruent perpendicular bisectors of each other.

REGENTS QUESTIONS

Multiple Choice

1. The coordinates of the vertices of parallelogram $ABCD$ are $A(-3,2)$, $B(-2,-1)$, $C(4,1)$, and $D(3,4)$. The slopes of which line segments could be calculated to show that $ABCD$ is a rectangle?
   (1) $\overline{AB}$ and $\overline{DC}$
   (2) $\overline{AB}$ and $\overline{BC}$
   (3) $\overline{AD}$ and $\overline{BC}$
   (4) $\overline{AC}$ and $\overline{BD}$

2. Parallelogram $ABCD$ has coordinates $A(1,5)$, $B(6,3)$, $C(3,-1)$, and $D(-2,1)$. What are the coordinates of E, the intersection of diagonals $\overline{AC}$ and $\overline{BD}$?
   (1) (2,2)
   (2) (4.5,1)
   (3) (3.5,2)
   (4) (−1,3)

3. (CC) A quadrilateral has vertices with coordinates (−3,1), (0,3), (5,2), and (−1,−2). Which type of quadrilateral is this?
   (1) rhombus
   (2) rectangle
   (3) square
   (4) trapezoid

4. (CC) The diagonals of rhombus $TEAM$ intersect at $P(2,1)$. If the equation of the line that contains diagonal $\overline{TA}$ is $y=-x+3$, what is the equation of a line that contains diagonal EM?
   (1) $y=x-1$
   (2) $y=x-3$
   (3) $y=-x-1$
   (4) $y=-x-3$

5. (CC) Parallelogram $ABCD$ has coordinates $A(0,7)$ and $C(2,1)$. Which statement would prove that $ABCD$ is a rhombus?
   (1) The midpoint of $\overline{AC}$ is (1,4).
   (2) The length of $\overline{BD}$ is $\sqrt{40}$.
   (3) The slope of $\overline{BD}$ is $\frac{1}{3}$.
   (4) The slope of $\overline{AB}$ is $\frac{1}{3}$.

Constructed Response

6. Ashanti is surveying for a new parking lot shaped like a parallelogram. She knows that three of the vertices of parallelogram $ABCD$ are $A(0,0)$, $B(5,2)$, and $C(6,5)$. Find the coordinates of point D and sketch parallelogram $ABCD$ on the accompanying set of axes. Justify mathematically that the figure you have drawn is a parallelogram.

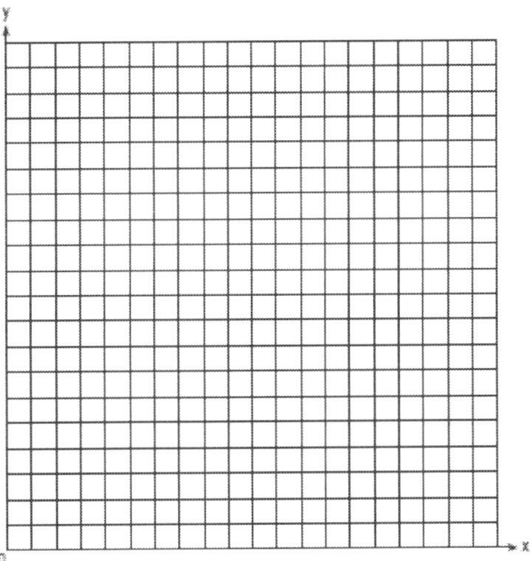

7. The coordinates of quadrilateral $ABCD$ are $A(-1,-5)$, $B(8,2)$, $C(11,13)$, and $D(2,6)$. Using coordinate geometry, prove that quadrilateral $ABCD$ is a rhombus.
[The use of the grid is optional.]

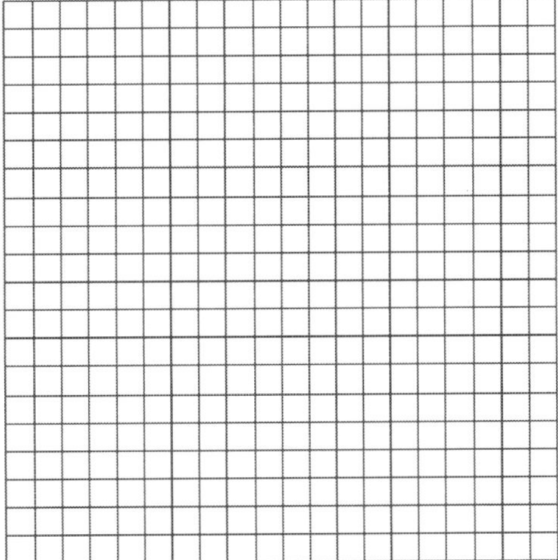

8. Given: $A(-2,2)$, $B(6,5)$, $C(4,0)$, $D(-4,-3)$
   Prove: $ABCD$ is a parallelogram but not a rectangle.
   [The use of the grid is optional.]

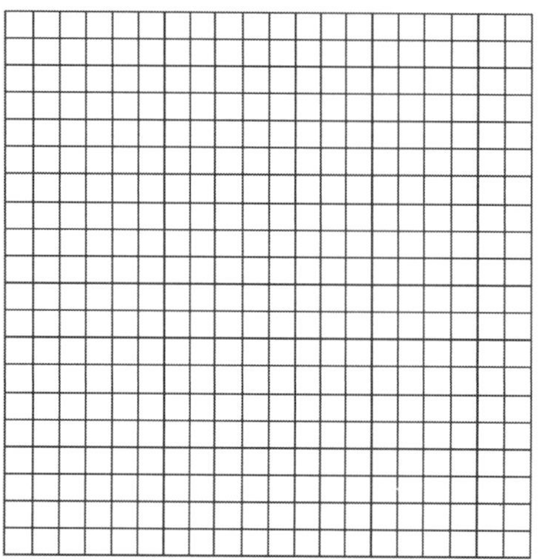

9. Given: Quadrilateral $ABCD$ has vertices $A(-5,6)$, $B(6,6)$, $C(8,-3)$, and $D(-3,-3)$.
   Prove: Quadrilateral $ABCD$ is a parallelogram but is neither a rhombus nor a rectangle.
   [The use of the grid below is optional.]

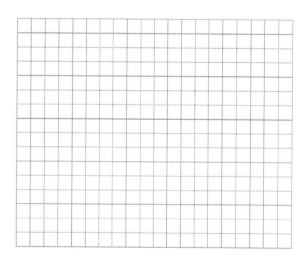

10. Quadrilateral *MATH* has coordinates $M(1,1)$, $A(-2,5)$, $T(3,5)$, and $H(6,1)$. Prove that quadrilateral *MATH* is a rhombus and prove that it is *not* a square.
[The use of the grid is optional.]

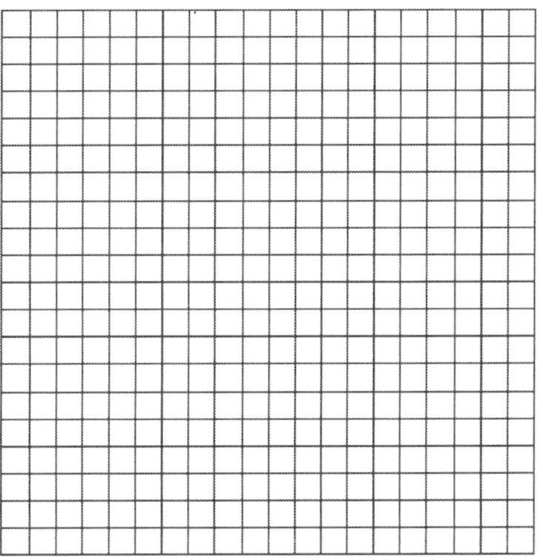

11. Given: $\triangle ABC$ with vertices $A(-6,-2)$, $B(2,8)$, and $C(6,-2)$.
$\overline{AB}$ has midpoint D, $\overline{BC}$ has midpoint E, and $\overline{AC}$ has midpoint F.
Prove: *ADEF* is a parallelogram
ADEF is *not* a rhombus
[The use of the grid is optional.]

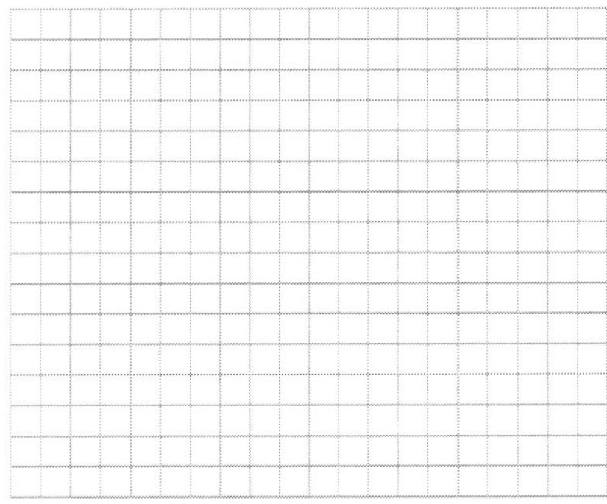

12. The coordinates of two vertices of square *ABCD* are $A(2,1)$ and $B(4,4)$. Determine the slope of side $\overline{BC}$.

13. Quadrilateral $ABCD$ with vertices $A(-7,4)$, $B(-3,6)$, $C(3,0)$, and $D(1,8)$ is graphed on the set of axes below. Quadrilateral $MNPQ$ is formed by joining M, N, P, and Q, the midpoints of $\overline{AB}$, $\overline{BC}$, $\overline{CD}$, and $\overline{AD}$, respectively.
Prove that quadrilateral $MNPQ$ is a parallelogram.
Prove that quadrilateral $MNPQ$ is *not* a rhombus.

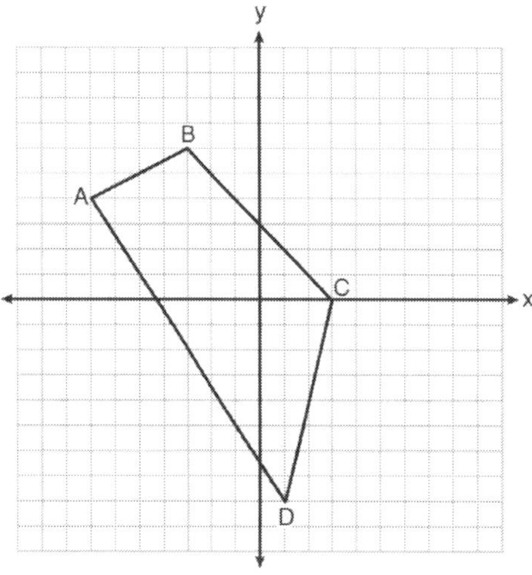

14. The vertices of quadrilateral $JKLM$ have coordinates $J(-3,1)$, $K(1,-5)$, $L(7,-2)$, and $M(3,4)$. Prove that $JKLM$ is a parallelogram. Prove that $JKLM$ is *not* a rhombus.
[The use of the set of axes below is optional.]

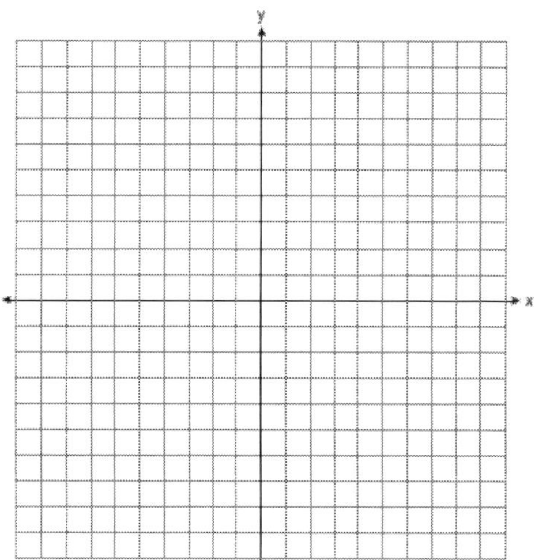

15. Rectangle $KLMN$ has vertices $K(0,4)$, $L(4,2)$, $M(1,-4)$, and $N(-3,-2)$. Determine and state the coordinates of the point of intersection of the diagonals.

16. (CC) In rhombus $MATH$, the coordinates of the endpoints of the diagonal $\overline{MT}$ are $M(0,-1)$ and $T(4,6)$. Write an equation of the line that contains diagonal $\overline{AH}$.
[Use of the set of axes below is optional.]

Using the given information, explain how you know that your line contains diagonal $\overline{AH}$.

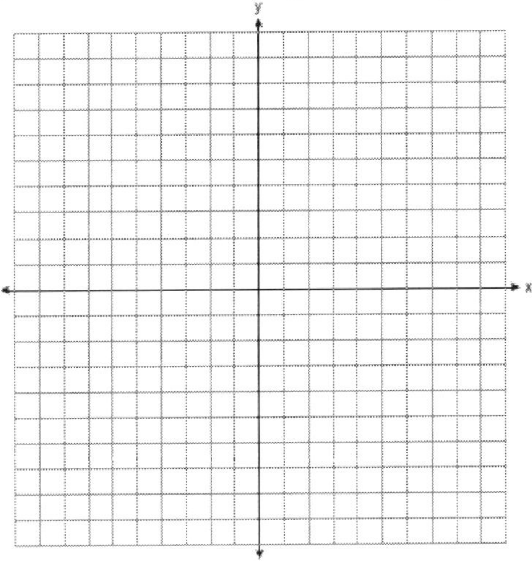

17. (CC) In the coordinate plane, the vertices of $\triangle RST$ are $R(6,-1)$, $S(1,-4)$, and $T(-5,6)$. Prove that $\triangle RST$ is a right triangle. [The use of the set of axes below is optional.]

State the coordinates of point P such that quadrilateral $RSTP$ is a rectangle.

Prove that your quadrilateral $RSTP$ is a rectangle.

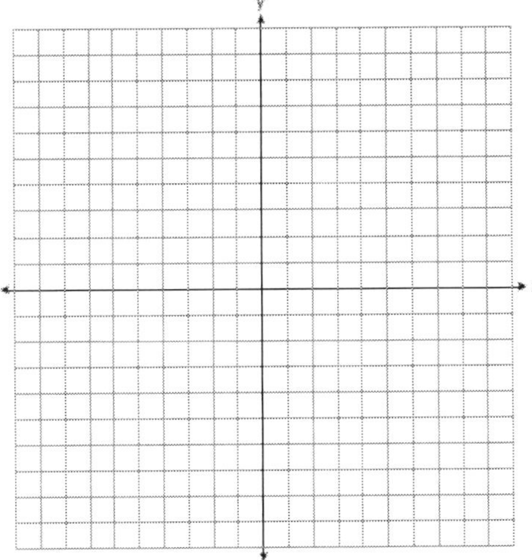

18. (CC) In square $GEOM$, the coordinates of G are (2,–2) and the coordinates of O are (–4,2). Determine and state the coordinates of vertices E and M.
[The use of the set of axes below is optional.]

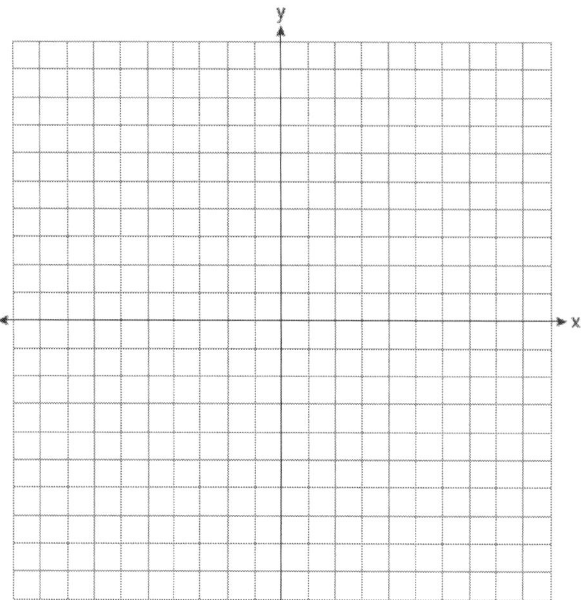

9.3 Perimeter and Area using Coordinates

Key Terms and Concepts

We can use the distance formula to find the length of any line segment, given its endpoints. This can help us to find the **perimeter** or **area** of a polygon on a coordinate graph.

Example: The perimeter of triangle *ABC* is found by adding the lengths of the three sides, which are calculated using the distance formula, $d=\sqrt{(x_2-x_1)^2+(y_2-y_1)^2}$.

$AB=\sqrt{11^2+3^2}=\sqrt{130}\approx 11.40$

$BC=\sqrt{2^2+9^2}=\sqrt{85}\approx 9.22$

$AC=\sqrt{9^2+(-6)^2}=\sqrt{117}\approx 10.82$

Perimeter ≈ 11.40 + 9.22 + 10.82 ≈ 31.4

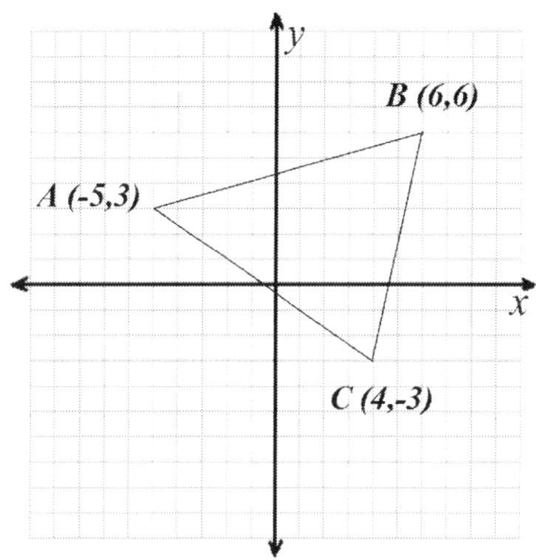

To calculate the area of a polygon given the coordinates of its vertices, we can sometimes determine the lengths of sides by the distance formula and then use them in the area formula.

Examples: The formula for the area of a *rectangle* is $A = bh$, and the formula for the area of a *right triangle* is $A=\frac{1}{2}bh$, where b and h are the lengths of sides (since one of the legs of a right triangle is also a height).

However, for some polygons, it may be easier to enclose the polygon in a rectangle, and then subtract the areas of the triangles outside the polygon from the area of the rectangle.

Example: For this same triangle, we don't know the length of any of its altitudes, so we cannot calculate the area directly. Instead, we can draw a rectangle around it, as shown. This forms three triangles, whose areas are:

$Area\,I=\frac{1}{2}(11)(3)=16.5$

$Area\,II=\frac{1}{2}(2)(9)=9$

$Area\,III=\frac{1}{2}(9)(6)=27$

The area of the rectangle is $(11)(9)=99$.

Therefore, the area of triangle *ABC* is $99-16.5-9-27=46.5$ square units.

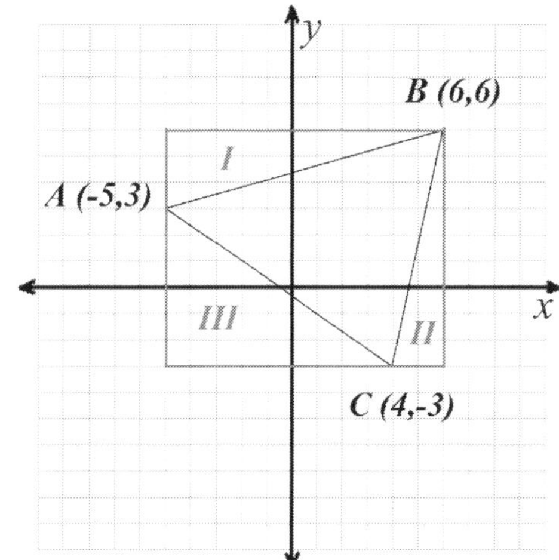

Using the shoelace method to find the area of a polygon
Not all polygons can be enclosed within a rectangle. Fortunately, there is a procedure for calculating the area of *any* polygon, known as the **shoelace formula**:

$$A=\frac{|(x_1\cdot y_2-y_1\cdot x_2)+(x_2\cdot y_3-y_2\cdot x_3)+...+(x_n\cdot y_1-y_n\cdot x_1)|}{2}$$, for a polygon with n vertices

Although the formula seems complex, it's not difficult to calculate. First, create a table listing the coordinates of the vertices of the polygon in clockwise order, ending with the starting vertex. Then, calculate the "upper lace" products minus the "lower lace" products. Finally, add the differences, take the absolute value, and divide by 2.

Example: Let's use the same triangle ABC above, with coordinates $A(-5,3)$, $B(6,6)$, and $C(4,-3)$. Start by listing the coordinates in a table, duplicating point A at the end.

vertex	*x*	*y*	*upper*	*lower*	*difference*
A	–5	3			
B	6	6			
C	4	–3			
A	–5	3			

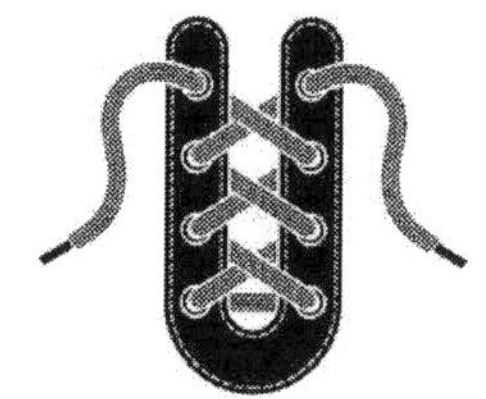

Now, calculate the "upper lace" products, as shown by the arrows drawn.

vertex	*x*	*y*	*upper*	*lower*	*difference*
A	–5	3	–30		
B	6	6	–18		
C	4	–3	12		
A	–5	3			

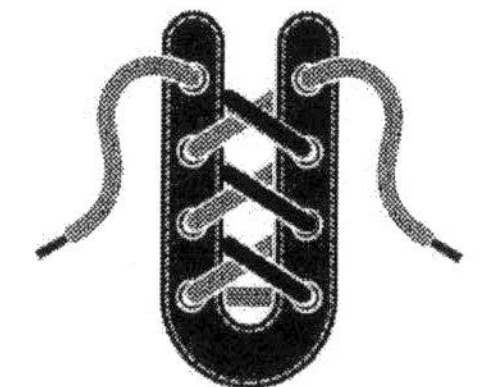

Then, calculate the "lower lace" products, as shown by the arrows drawn.

vertex	*x*	*y*	*upper*	*lower*	*difference*
A	–5	3	–30	18	
B	6	6	–18	24	
C	4	–3	12	15	
A	–5	3			

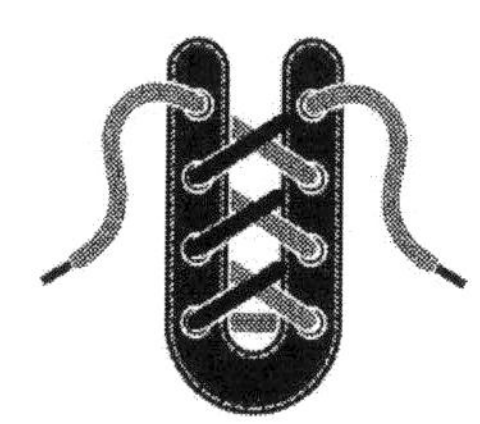

Find each difference, *upper – lower*.

vertex	*x*	*y*	*upper*	*lower*	*difference*
A	–5	3	–30	18	–48
B	6	6	–18	24	–42
C	4	–3	12	15	–3
A	–5	3			

Finally, add the differences, take the absolute value, and divide by 2.

$$A=\frac{|(-48)+(-42)+(-3)|}{2}=\frac{|-93|}{2}=46.5 \text{ square units}$$

Model Problem 1: *using the distance formula*

Find the perimeter and area of rectangle *ABCD,* whose vertices are points $A(-3,0)$, $B(3,2)$, $C(4,-1)$, and $D(-2,-3)$.

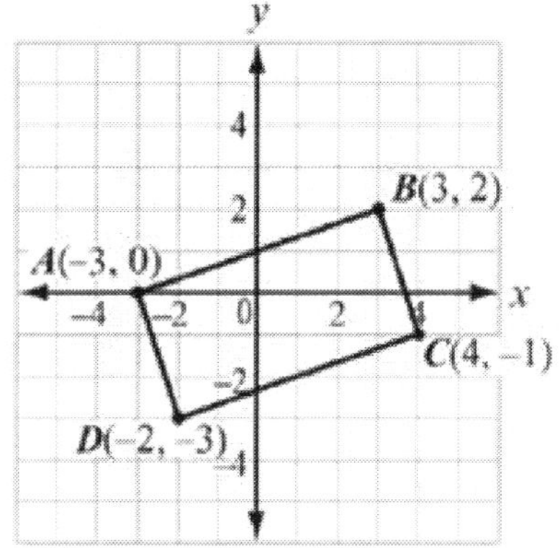

Solution:

(A) $AB = \sqrt{6^2 + 2^2} = \sqrt{40} = 2\sqrt{10}$ $\qquad CD = 2\sqrt{10}$

(B) $BC = \sqrt{1^2 + (-3)^2} = \sqrt{10}$ $\qquad AD = \sqrt{10}$

(C) Perimeter $= 2\sqrt{10} + 2\sqrt{10} + \sqrt{10} + \sqrt{10} = 6\sqrt{10} \approx 19.0$

(D) Area $= 2\sqrt{10} \cdot \sqrt{10} = 20$ square units

Explanation of steps:

(A) Use the distance formula to find the base *[AB]*. Since it is a rectangle, the opposite side *[CD]* is the same length.

(B) Use the distance formula to find the height *[BC]*. In a rectangle, its opposite side *[AD]* is also equal in length.

(C) Add the two bases and two heights to find the perimeter.

(D) Multiply the base and height to find the area. *[We are able to calculate $A = bh$ directly.]*

Model Problem 2: *using the shoelace method*

Find the area of the quadrilateral *ABCD*, whose vertices are points $A(-3,0)$, $B(2,4)$, $C(3,1)$, and $D(-4,-3)$, using the shoelace method.

Solution:

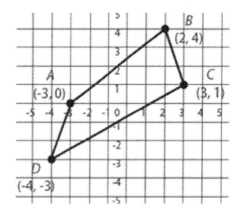

(A)			(B)		(C)
vertex	*x*	*y*	*upper*	*lower*	*difference*
A	–3	0	–12	0	–12
B	2	4	2	12	–10
C	3	1	–9	–4	–5
D	–4	–3	0	9	–9
A	–3	0			

(D) $A = \dfrac{|(-12)+(-10)+(-5)+(-9)|}{2} = \dfrac{|-36|}{2} = 18$ square units

Explanation of steps:

(A) Create a table listing the coordinates of the vertices of the polygon in clockwise order, ending with the starting vertex.

(B) Calculate the "upper lace" products *[*$-3 \times 4 = -12$; $2 \times 1 = 2$; $3 \times -3 = -9$; $-4 \times 0 = 0$*]* and the "lower lace" products *[*$0 \times 2 = 0$; $4 \times 3 = 12$; $1 \times -4 = -4$; $-3 \times -3 = 9$*]*.

(C) Calculate each difference, *upper – lower [*$(-12) - 0 = -12$; $2 - 12 = -10$; *etc.]*.

(D) Add the differences, take the absolute value of the sum, and divide by 2.

Practice Problems

1. Find the perimeter of $\triangle ABC$ below.

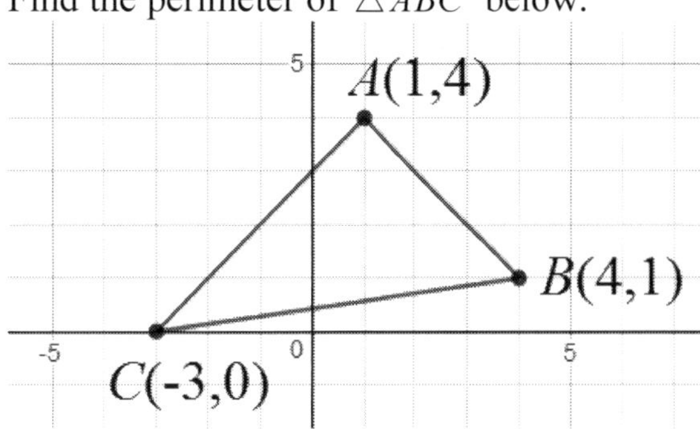

2. Find the area of $\triangle ABC$ below.

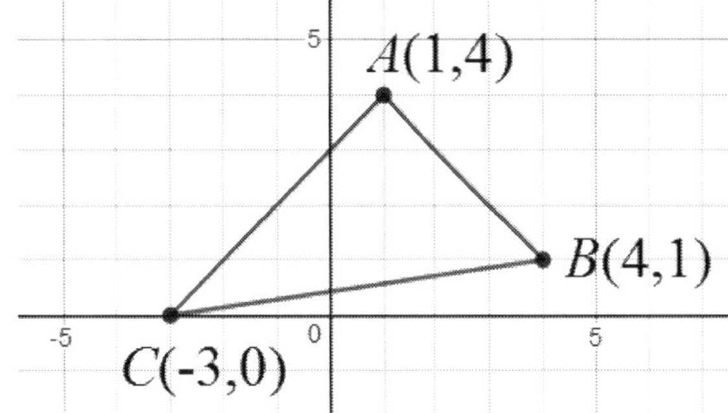

3. Parallelogram $EFGH$ has vertices $E(3,6)$, $F(6,10)$, $G(18,5)$, and $H(15,1)$. Find its perimeter and area.

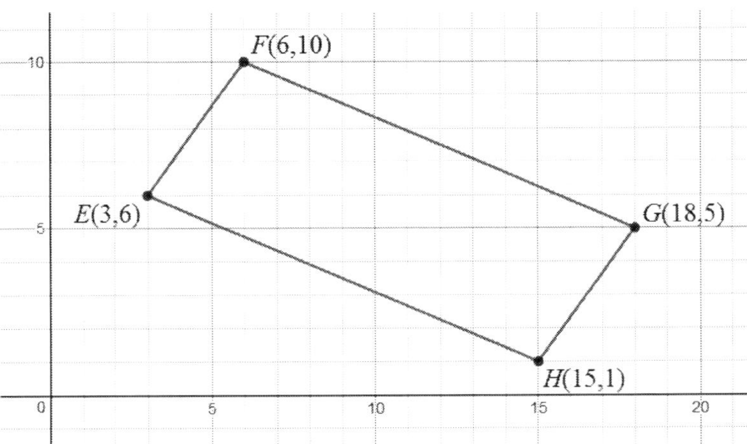

4. Parallelogram $KLMN$ has vertices $K(-7,-7)$, $L(-5,2)$, $M(3,6)$, and $N(1,-3)$. Find its area using the shoelace method.

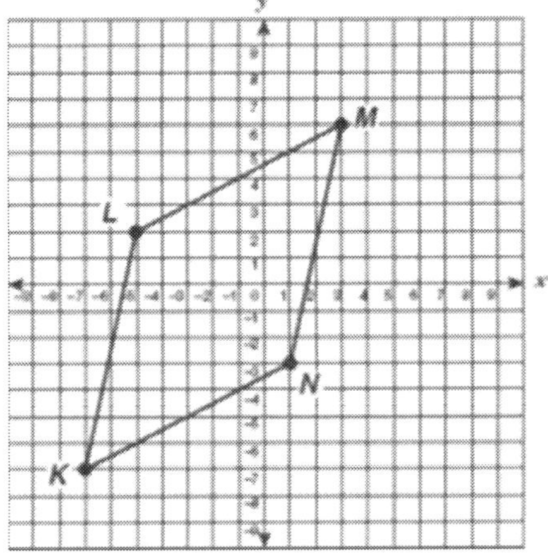

REGENTS QUESTIONS

Multiple Choice

1. (CC) Triangle RST is graphed on the set of axes below.

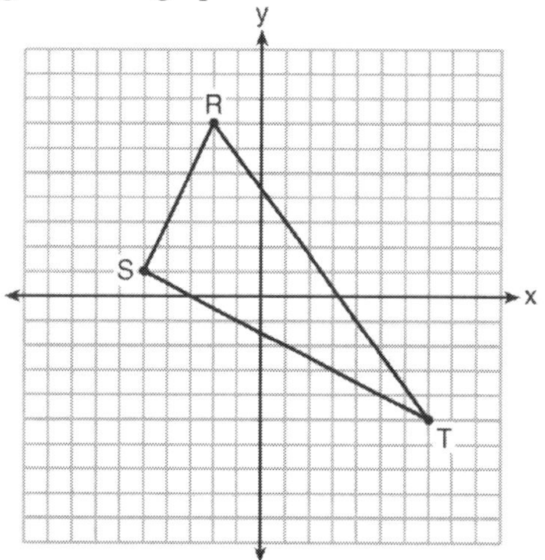

How many square units are in the area of $\triangle RST$?

(1) $9\sqrt{3}+15$ (3) 45

(2) $9\sqrt{5}+15$ (4) 90

Constructed Response

2. Triangle ABC has coordinates $A(-6,2)$, $B(-3,6)$, and $C(5,0)$. Find the perimeter of the triangle. Express your answer in simplest radical form.
[The use of the grid below is optional.]

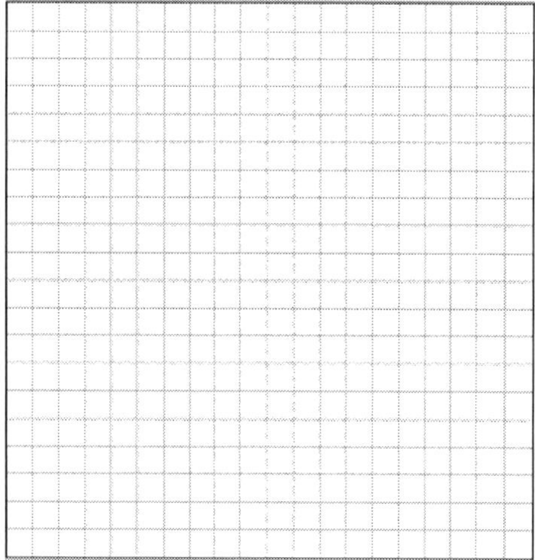

Chapter 10. Rigid Motions

10.1 Translations

Key Terms and Concepts

A **transformation** is a general term for four specific ways to manipulate a figure in a plane: *translation*, *reflection*, *rotation*, and *dilation*. The original figure is called the **pre-image** and the figure after the transformation is called the **image**.

A **rigid motion** (also called an **isometry**) is a transformation that preserves both *distance* and *angles*. A rigid motion is said to preserve distance because the distance between any two points of the image is equal to the distance between their corresponding points in the pre-image. For this reason, each line segment of the image is the same length as its pre-image. A rigid motion also preserves angles in that each angle of the image has the same measure as its pre-image.

Since distance and angles are preserved, we can therefore say that rigid motion preserves *congruency*; that is, the image of a rigid motion is congruent to the original pre-image.

Example: If a rigid motion is performed on a polygon, the length of each side and the measure of each angle is preserved, so the image is congruent to the pre-image.

In this unit, we look at three types of rigid motions – translation, reflection, and rotation. As we'll see in later units, a dilation is *not* a rigid motion because it doesn't preserve distance.

In this section, we look at translations. A **translation** "slides" every point in the pre-image by the same distance in the same direction.

Example: This graph shows a translation of $\triangle ABC$ by 6 units to the right and 1 unit up. We can write this using the notation $T_{6,1}$. The values are the changes in x and y.

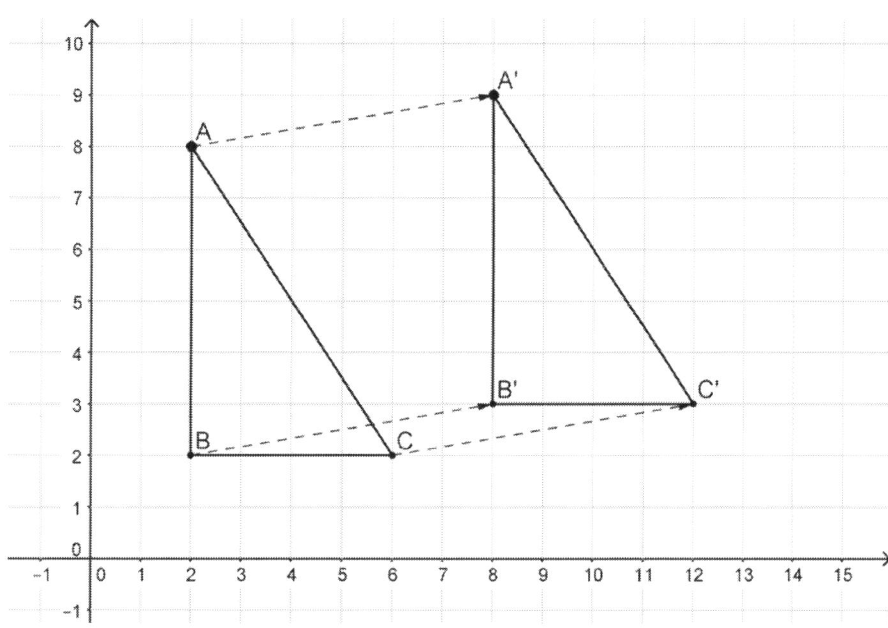

We will often use a **prime symbol** (typed as a single quote or apostrophe) to represent the image of a point. For example, the image of vertex A above is written as A'.

In general, $T_{a,b}$ represents a translation where a is the change in x-values and b is the change in y-values (a and b may be negative). In other words, each point (x, y) maps to an image at $(x+a, y+b)$. This can be written as $T_{a,b} : (x, y) \rightarrow (x+a, y+b)$.

Example: In the graph above for $T_{6,1}$, notice that A is (2,8) and its image A' is (8,9). We could have determined A' by adding 6 and 1 to the coordinates of A.

In the translation rule $(x, y) \rightarrow (x+a, y+b)$, the variable a represents how far the image moves to the **right** $(a > 0)$ or to the **left** $(a < 0)$, and the variable b represents how far the image moves **up** $(b > 0)$ or **down** $(b < 0)$.

Model Problem:

A translation maps $P(3,-2)$ to $P'(1,2)$. Under the same translation, find the coordinates of Q', the image of $Q(-3,2)$.

Solution:

(A) Translation is $T_{-2,4}$.

(B) $Q'(-5,6)$.

Explanation of steps:

(A) Find the change in x and y by subtracting the coordinates of a point in the pre-image from its image. *[Subtracting the coordinates of P from P' gives us* $1-3=-2$ *and* $2-(-2)=4$*]*

(B) Determine the image of the other point. *[Add –2 and 4 to the coordinates of Q to find Q']*

Practice Problems

1. What is the image of point (2,4) under the translation $T_{-6,1}$?	2. A translation maps $P(3,-2)$ onto $P'(5,0)$. Find the coordinates of the image of $Q(4,-6)$ under the same translation.

3. The vertices of a rectangle are $R(-5,-5)$, $S(-1,-5)$, $T(-1,1)$ and $U(-5,1)$. After a translation, R' is the point (3, 0). Describe the translation and state the coordinates of U'.

4. On the grid below, graph and label $\triangle K'L'M'$, the image of $\triangle KLM$ after a translation of $T_{6,-4}$.

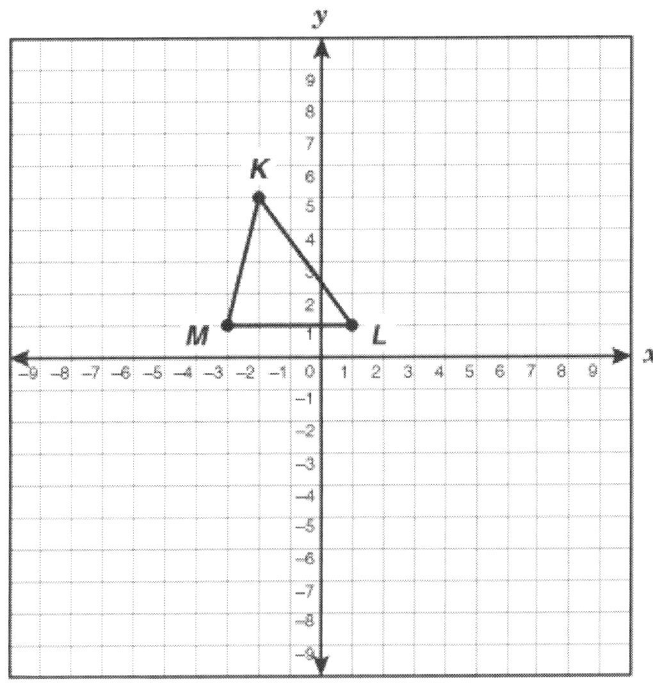

REGENTS QUESTIONS

Multiple Choice

1. A translation moves $P(3,5)$ to $P'(6,1)$. What are the coordinates of the image of point (–3,–5) under the same translation?
   (1) (0,–9) (3) (–6,–1)
   (2) (–5,–3) (4) (–6,–9)

2. What is the image of (x, y) after a translation of 3 units right and 7 units down?
   (1) $(x+3, y-7)$ (3) $(x-3, y-7)$
   (2) $(x+3, y+7)$ (4) $(x-3, y+7)$

3. The image of point (3,–5) under the translation that shifts (x, y) to $(x-1, y-3)$ is
   (1) (–4,8) (3) (2,8)
   (2) (–3,15) (4) (2,–8)

4. The image of the origin under a certain translation is the point (2,–6). The image of point (–3,–2) under the same translation is the point
   (1) (–6,12) (3) $\left(-\frac{3}{2}, \frac{1}{3}\right)$
   (2) (–5,4) (4) (–1,–8)

5. The image of point (–2,3) under translation T is (3,–1). What is the image of point (4,2) under the same translation?
   (1) (–1,6) (3) (5,4)
   (2) (0,7) (4) (9,–2)

6. What is the image of point (–3,4) under the translation that shifts (x, y) to $(x-3, y+2)$?
   (1) (0,6) (3) (–6,8)
   (2) (6,6) (4) (–6,6)

7. Triangle ABC has vertices $A(1,3)$, $B(0,1)$, and $C(4,0)$. Under a translation, A', the image point of A, is located at (4,4). Under this same translation, point C' is located at
   (1) (7,1) (3) (3,2)
   (2) (5,3) (4) (1,–1)

8. A polygon is transformed according to the rule: $(x, y) \rightarrow (x+2, y)$. Every point of the polygon moves two units in which direction?
   (1) up (3) left
   (2) down (4) right

9. What is the image of the point (–5,2) under the translation $T_{3,-4}$?
   (1) (–9,5) (3) (–2,–2)
   (2) (–8,6) (4) (–15,–8)

10. What are the coordinates of the image of point A(2,–7) under the translation $(x, y) \to (x-3, y+5)$?

(1) (–1,–2) (3) (5,–12)
(2) (–1,2) (4) (5,12)

11. What are the coordinates of P', the image of point P after translation $T_{4,4}$?

(1) $(x-4, y-4)$ (3) $(4x, 4y)$
(2) $(x+4, y+4)$ (4) $(4, 4)$

12. When the transformation $T_{2,-1}$ is performed on point A, its image is point $A'(-3,4)$. What are the coordinates of A?

(1) (5,–5) (3) (–1,3)
(2) (–5,5) (4) (–6,–4)

Constructed Response

13. A design was constructed by using two rectangles *ABDC* and *A'B'C'D'*. Rectangle *A'B'C'D'* is the result of a translation of rectangle *ABDC*. The table of translations is shown below. Find the coordinates of points *B* and *D'*.

Rectangle *ABDC*	Rectangle *A'B'D'C'*
A (2,4)	A' (3,1)
B	B' (-5,1)
C (2,-1)	C' (3,-4)
D (-6,-1)	D'

14. Triangle *TAP* has coordinates T(–1,4), A(2,4), and P(2,0). On the set of axes below, graph and label $\triangle T'A'P'$, the image of $\triangle TAP$ after the translation $(x, y) \rightarrow (x-5, y-1)$.

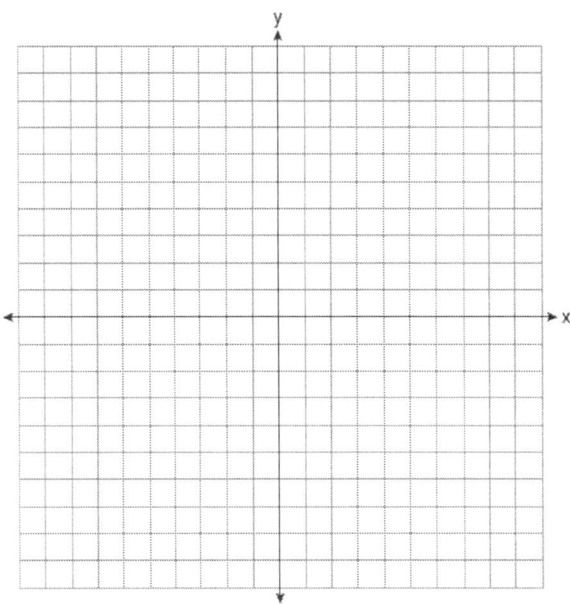

15. The image of $\triangle ABC$ under a translation is $\triangle A'B'C'$. Under this translation, *B*(3,–2) maps onto *B'*(1,–1). Using this translation, the coordinates of image *A'* are (–2,2). Determine and state the coordinates of point *A*.

10.2 Reflections

Key Terms and Concepts

A reflection is another type of rigid motion. In a **reflection** (*also known as line reflection*), points in the pre-image are reflected over a line, called the **axis of reflection** (or *line of reflection*). The reflection of an object appears as the mirror image of the object. If you imagine the graph as paper, the image after a reflection is where the pre-image would land if the paper were folded at the axis of reflection.

Example: $\triangle A'B'C'$ is the image of $\triangle ABC$ after a reflection in the x-axis.

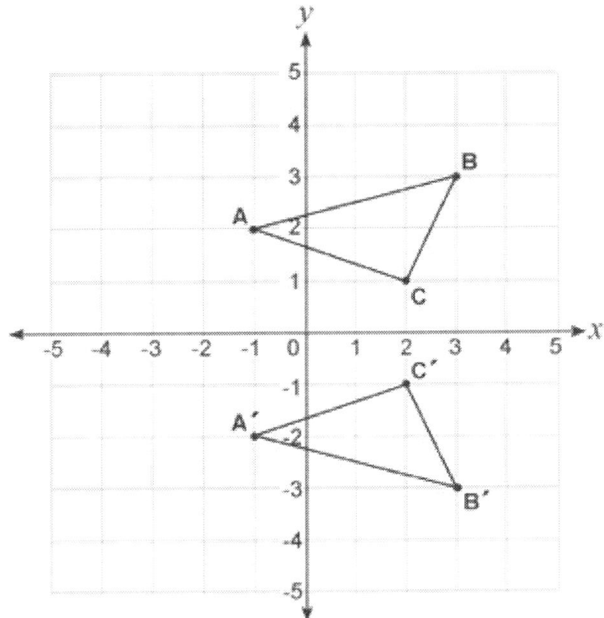

A reflection is a *rigid motion*, which means it preserves both distance and angles, thereby preserving congruency. However, they have *opposite orientations*.

Example: The vertices of $\triangle ABC$ above can be traced in a clockwise direction from A to B to C, but the corresponding vertices of its image $\triangle A'B'C'$ can be traced in a counterclockwise direction; therefore, the objects have opposite orientations.

The axis of reflection is the *perpendicular bisector* of any line segment drawn between a point and its reflection over that line.

Example: In the grid to the right, P' is the image of P after a reflection in line m. Note that m is the perpendicular bisector of $\overline{PP'}$.

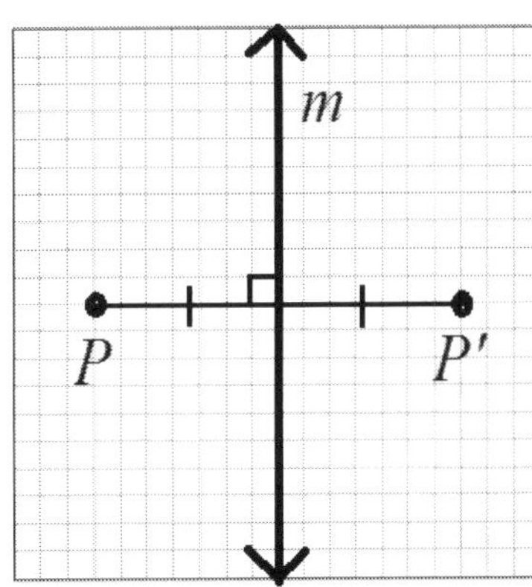

A lower-case r is often used to represent a reflection, with the line of reflection in subscript.

Examples: r_{x-axis} represents a reflection in (or over) the x-axis, r_{y-axis} means a reflection in the y-axis, and $r_{y=x}$ represents a reflection over the line, $y = x$.

Coordinates of a point after a reflection over a line:

$r_{x-axis}(x,y) = (x,-y)$ When a point (x,y) is reflected over the x-axis, the image has the same x-coordinate but the opposite y-coordinate.

$r_{y-axis}(x,y) = (-x,y)$ When a point (x,y) is reflected over the y-axis, the image has the same y-coordinate but the opposite x-coordinate.

$r_{y=x}(x,y) = (y,x)$ When a point (x,y) is reflected over the line $y = x$, swap the coordinates of the point to find the coordinates of its image.

$r_{y=-x}(x,y) = (-y,-x)$ When a point (x,y) is reflected over the line $y = -x$, swap and negate the coordinates of the point to find its image.

Model Problem:

Graph and label $\triangle G'L'Q'$, the image of $\triangle GLQ$ under the transformation r_{y-axis}.

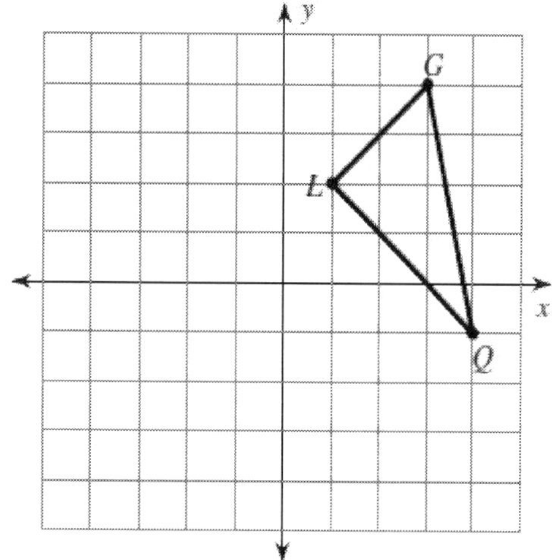

Solution:

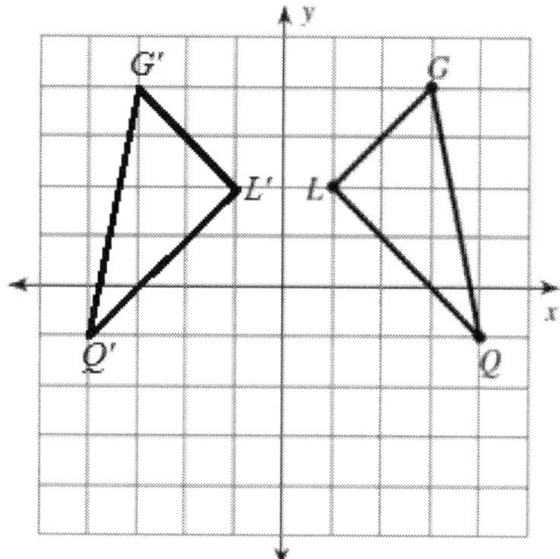

Explanation of steps:

When reflecting over the y-axis, negate the x-coordinate of each point but keep the same y-coordinates.

[$L(1,2) \to L'(-1,2)$, $G(3,4) \to G'(-3,4)$, and $Q(4,-1) \to Q'(-4,-1)$]

Practice Problems

1. What is the image of point (2,–3) after it is reflected over the x-axis?	2. If $P(-4,-1)$ is reflected in the x-axis, what are the coordinates of P', the image of P?
3. If $M(-2,8)$ is reflected in the y-axis, what are the coordinates of M', the image of M?	4. What is the image of point (5,–2) under the transformation $r_{y=x}$?
5. Graph and label $\triangle A'B'C'$, the image of $\triangle ABC$ after a reflection over the line $y=-1$. 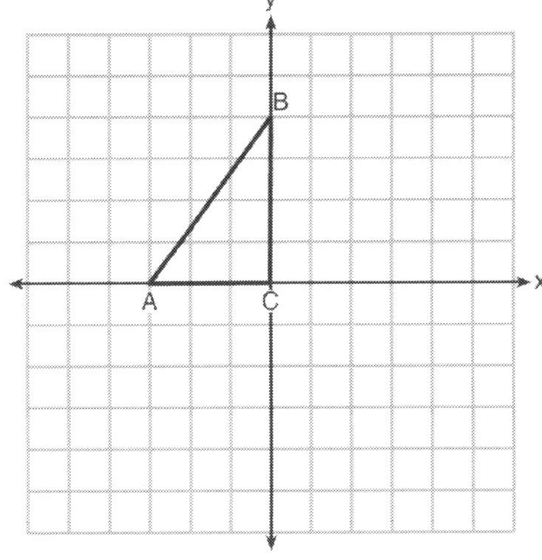	6. Graph and label $\triangle A'B'C'$, the image of $\triangle ABC$ after a reflection over the line $y=x$. 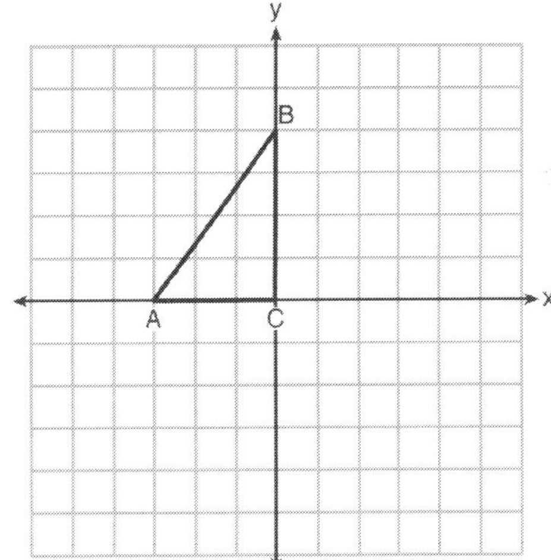

REGENTS QUESTIONS

Multiple Choice

1. The image of point (3,4) when reflected in the y-axis is
   (1) (–3,–4) (3) (3,–4)
   (2) (–3,4) (4) (4,3)

2. When the point (2,–5) is reflected in the x-axis, what are the coordinates of its image?
   (1) (–5,2) (3) (2,5)
   (2) (–2,5) (4) (5,2)

3. What are the coordinates of point P, the image of point (3,–4), after a reflection in the line $y = x$?
   (1) (3,4) (3) (4,–3)
   (2) (–3,4) (4) (–4,3)

4. What are the coordinates of point (2,–3) after it is reflected over the x-axis?
   (1) (2,3) (3) (–2,–3)
   (2) (–2,3) (4) (–3,2)

5. What is the image of point (–3,7) after a reflection in the x-axis?
   (1) (3,7) (3) (3,–7)
   (2) (–3,–7) (4) (7,–3)

6. Point A is located at (4,–7). The point is reflected in the x-axis. Its image is located at
   (1) (–4,7) (3) (4,7)
   (2) (–4,–7) (4) (7,–4)

7. What is the image of the point (2,–3) after the transformation r_{y-axis}?
   (1) (2,3) (3) (–2,3)
   (2) (–2,–3) (4) (–3,2)

8. The coordinates of point A are $(-3a,4b)$. If point A' is the image of point A reflected over the line $y = x$, the coordinates of A' are
   (1) $(4b,-3a)$ (3) $(-3a,-4b)$
   (2) $(3a,4b)$ (4) $(-4b,-3a)$

Constructed Response

9. The coordinates of the endpoints of $\overline{AB}$ are $A(0,2)$ and $B(4,6)$. Graph and state the coordinates of A' and B', the images of A and B after $\overline{AB}$ is reflected in the x-axis.

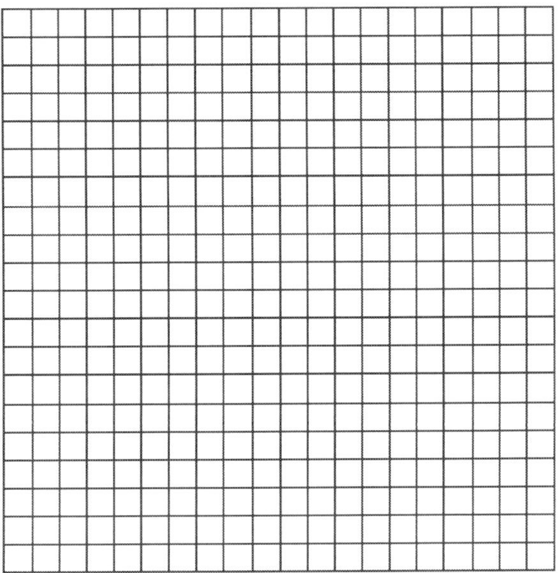

10. Triangle SUN has coordinates $S(0,6)$, $U(3,5)$, and $N(3,0)$. On the accompanying grid, draw and label $\triangle SUN$. Then, graph and state the coordinates of $\triangle S'U'N'$, the image of $\triangle SUN$ after a reflection in the y-axis.

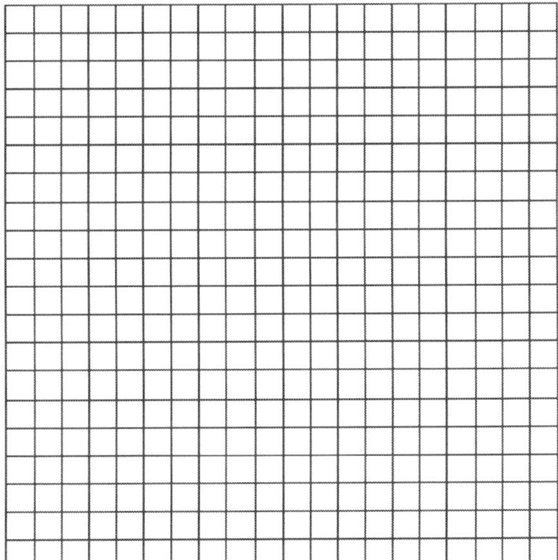

11. On the accompanying set of axes, draw the reflection of $ABCD$ in the y-axis. Label and state the coordinates of the reflected figure.

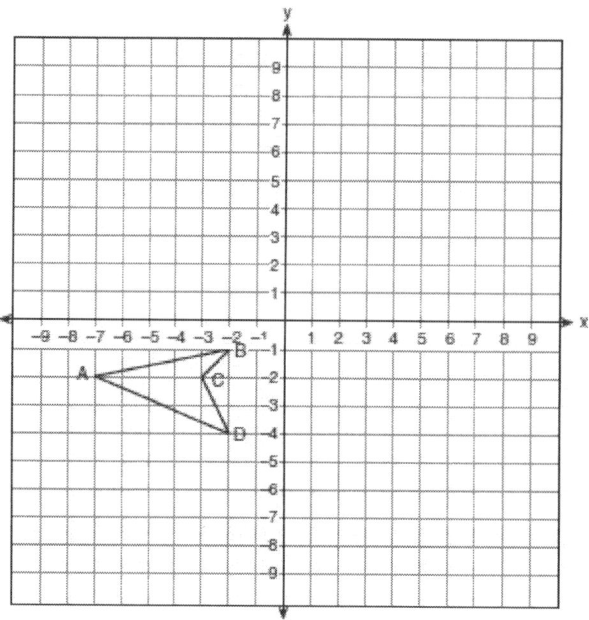

12. Triangle ABC has coordinates A(2,0), B(1,7), and C(5,1). On the accompanying set of axes, graph, label, and state the coordinates of $\triangle A'B'C'$, the reflection of $\triangle ABC$ in the y-axis.

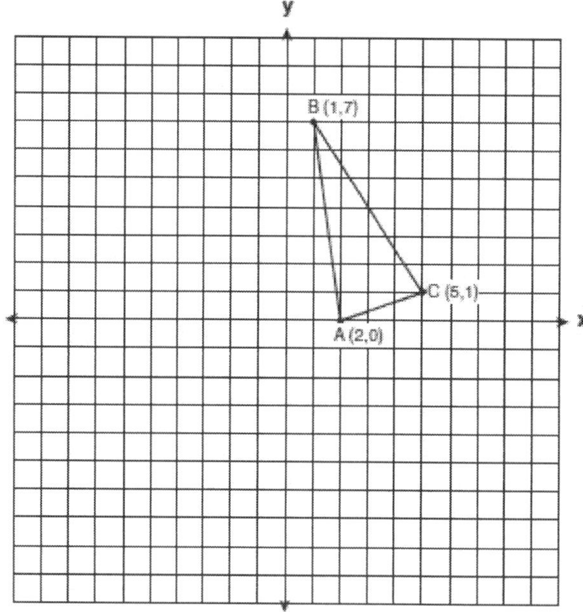

13. Carson is a decorator. He often sketches his room designs on the coordinate plane. He has graphed a square table on his grid so that its corners are at the coordinates $A(2,6)$, $B(7,8)$, $C(9,3)$, and $D(4,1)$. To graph a second identical table, he reflects $ABCD$ over the y-axis. On the accompanying set of coordinate axes, sketch and label $ABCD$ and its image $A'B'C'D'$, which show the locations of the two tables.

    Then find the number of square units in the area of $ABCD$.

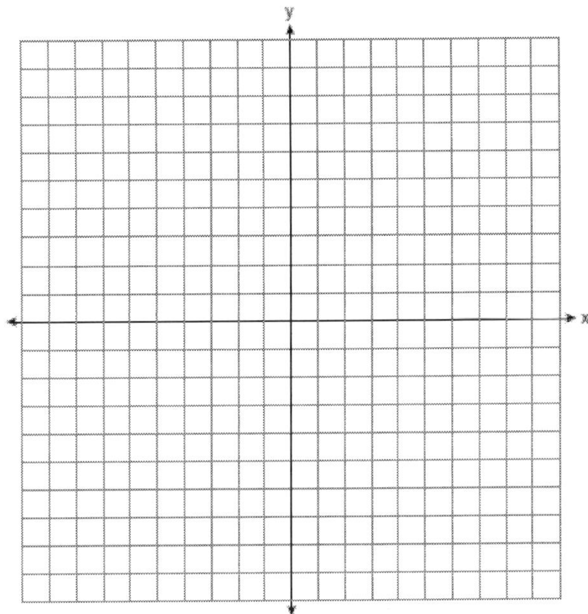

14. Triangle XYZ, shown in the diagram below, is reflected over the line $x = 2$. State the coordinates of $\triangle X'Y'Z'$, the image of $\triangle XYZ$.

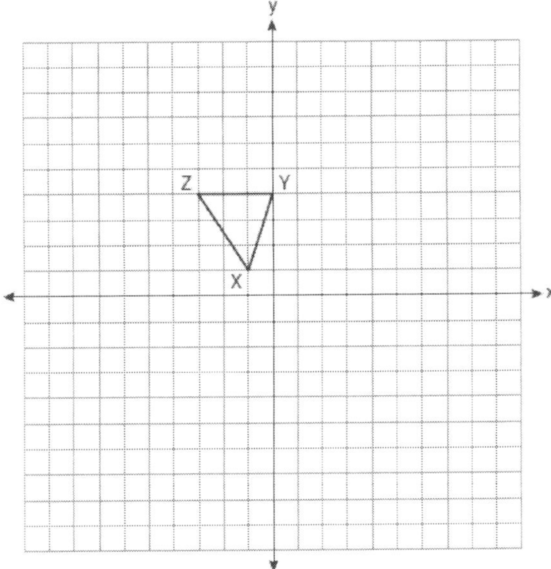

15. Triangle ABC has vertices $A(-2,2)$, $B(-1,-3)$, and $C(4,0)$. Find the coordinates of the vertices of $\triangle A'B'C'$, the image of $\triangle ABC$ after the transformation r_{x-axis}.
[The use of the grid is optional.]

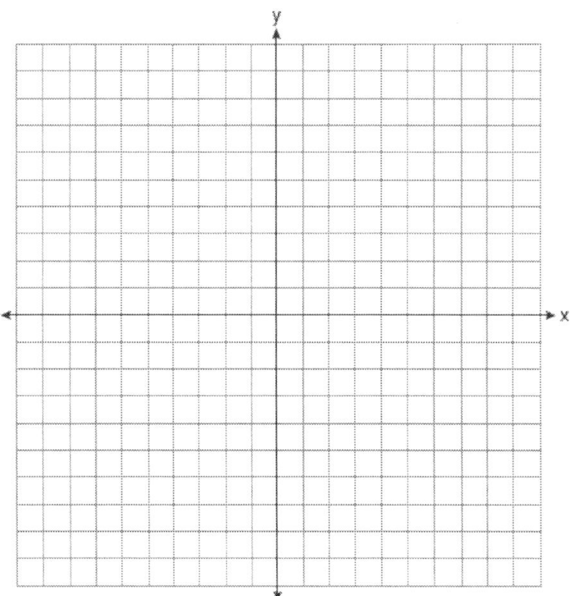

16. Triangle ABC has vertices $A(-1,1)$, $B(1,3)$, and $C(4,1)$. The image of $\triangle ABC$ after the transformation $r_{y=x}$ is $\triangle A'B'C'$. State and label the coordinates of $\triangle A'B'C'$.
[The use of the set of axes below is optional.]

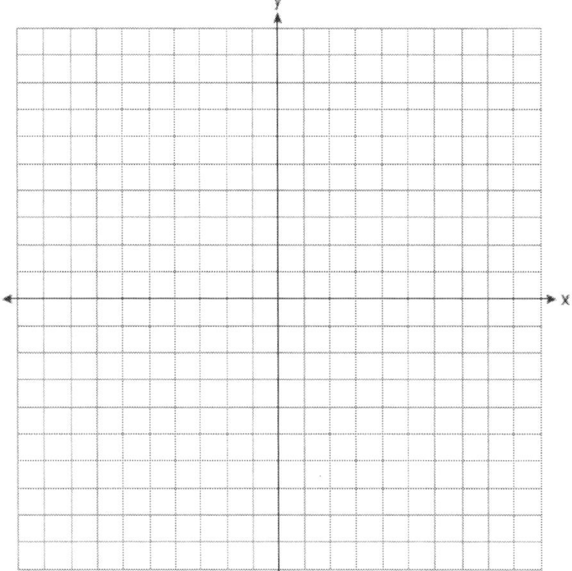

17. The image of $\overline{RS}$ after a reflection through the origin is $\overline{R'S'}$. If the coordinates of the endpoints of $\overline{RS}$ are $R(2,-3)$ and $S(5,1)$, state and label the coordinates of R' and S'. [The use of the set of axes below is optional.]

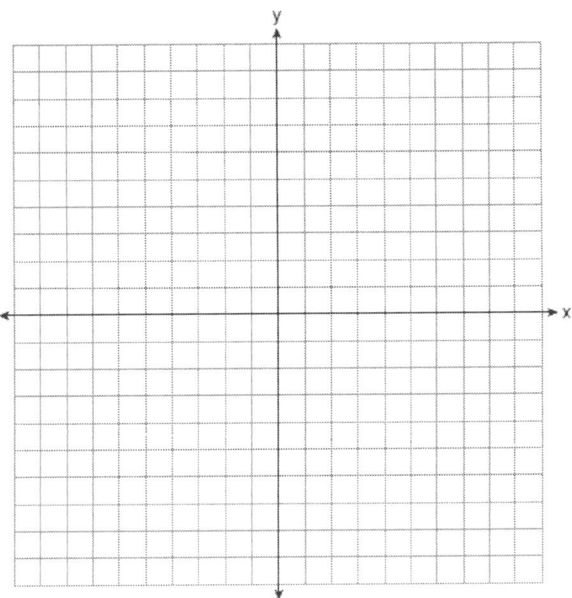

18. Ⓒⓒ Triangle ABC is graphed on the set of axes below. Graph and label $\triangle A'B'C'$, the image of $\triangle ABC$ after a reflection over the line $x = 1$.

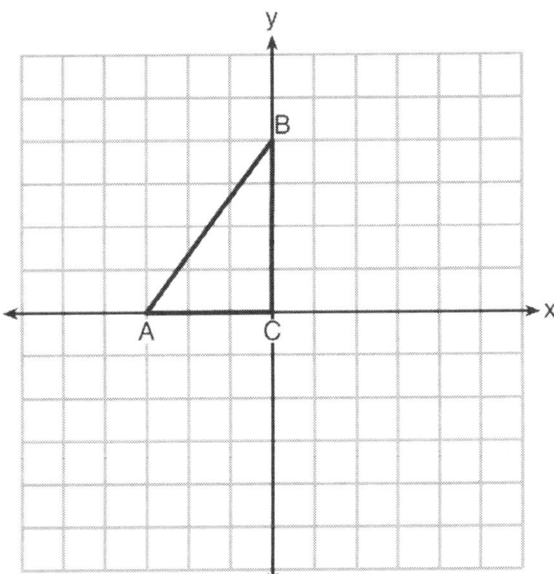

10.3 Rotations

Key Terms and Concepts

A third type of rigid motion is a rotation. In a **rotation**, each point is rotated a given number of degrees, called the **angle of rotation**, along a circle whose **center point** is given.

Example: The graph below shows the point $A(2,1)$ and its image $A'(-1,2)$ after a counterclockwise rotation of 90° around the origin, (0,0).

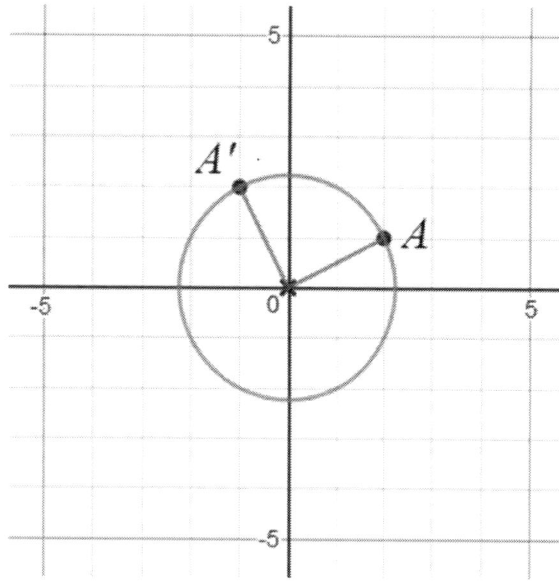

Note that the distances from the center to the point and to its image are equal, since they are radii of the same circle.

If a center point is not specified, it is assumed to be the origin. If a direction is not specified, a counterclockwise rotation is assumed, as shown in the circle to the right.

It is helpful to remember that 90° is a quarter of a circle, 180° is half of a circle, and 270° is three-quarters of a circle.

A clockwise rotation of n degrees is equivalent to a counterclockwise rotation of $(360 - n)$ degrees.

Example: A clockwise rotation of 90° will have the same result as a counterclockwise rotation of $(360 - 90)°$, or 270°.

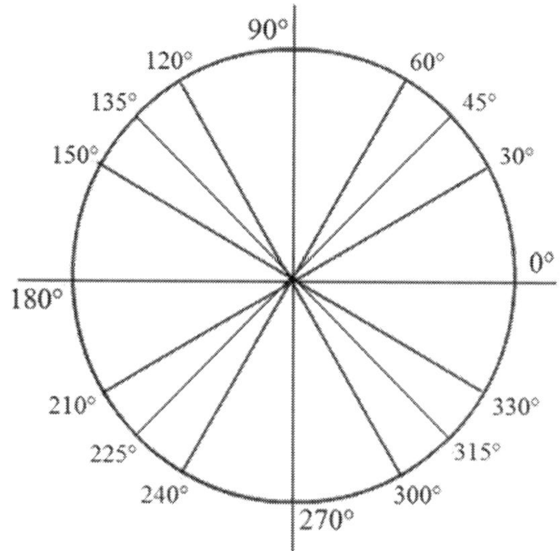

We often use a capital *R*, with the angle in subscript, to represent rotation, as in R_{90°. Note that a capital *R* represents rotation but a lowercase *r* represents reflection.

When a point is rotated *around the origin* by 90°, 180°, or 270°, we can use the following rules to find its image:

$R_{90^\circ} : (x, y) \rightarrow (-y, x)$

$R_{180^\circ} : (x, y) \rightarrow (-x, -y)$

$R_{270^\circ} : (x, y) \rightarrow (y, -x)$

Examples: $R_{90^\circ} : (2,1) \rightarrow (-1,2)$ and $R_{180^\circ} : (8,3) \rightarrow (-8,-3)$

To rotate a polygon, rotate each of its vertices by the same angle around the same center point.

Example: This graph shows a R_{270° of $\triangle DEF$ around the origin. We can use the rule $R_{270^\circ} : (x, y) \rightarrow (y, -x)$ to determine that $D(-1,6) \rightarrow D'(6,1)$, $E(1,3) \rightarrow E'(3,-1)$, and $F(6,3) \rightarrow F'(3,-6)$.

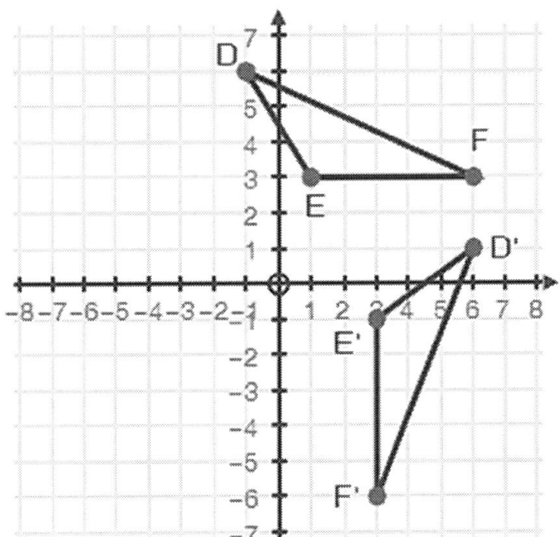

Rotation is a *rigid motion*, which means it preserves distance and angles, and thereby preserves congruency. Whenever a rigid motion (a translation, reflection, or rotation) is performed on an object, the result is a congruent image. Therefore, in the graph above, $\triangle DEF \cong \triangle D'E'F'$. Unlike reflections, rotations will also preserve orientation.

When the origin is not the center point, these rules cannot be applied. However, we may still determine the image of each vertex.

Example: $\triangle ABC$ below is rotated 90° around point $X(-1,1)$.

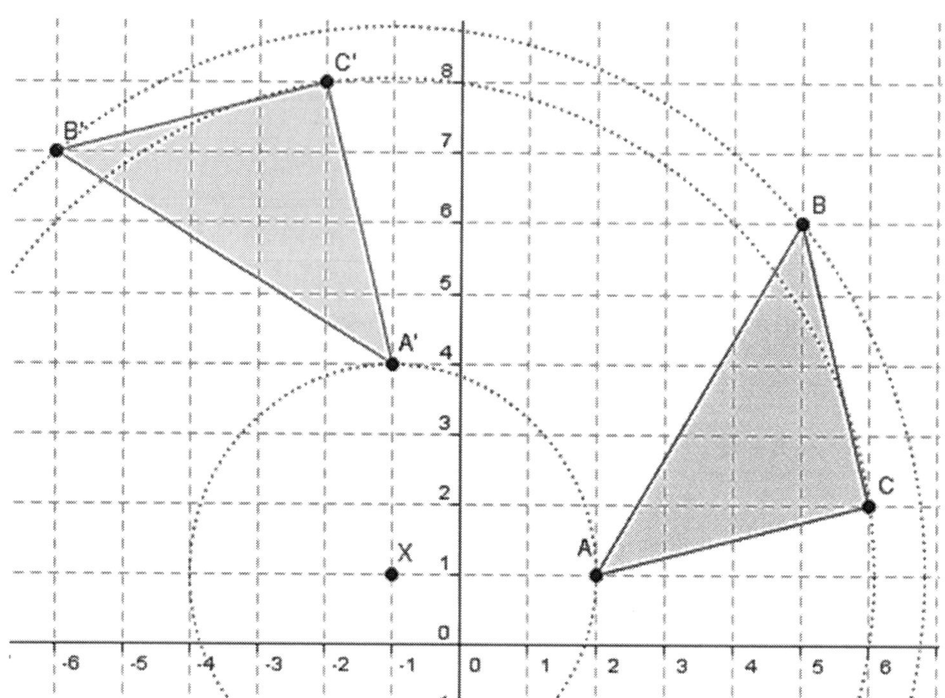

a) $\angle AXA'$ must be a right angle, since we are rotating 90°. Therefore, since $\overline{AX}$ is horizontal, $\overline{A'X}$ must be vertical. Also, $AX = A'X$, since the distances from the center to a point and its image are equal (they are both radii of the same circle). Since A is 3 units from X, A' must also be 3 units from X. Therefore, A' is $(-1,4)$.

b) Similarly, $\angle BXB'$ must be a right angle. $\overline{BX}$ has a slope of $\frac{5}{6}$, so $\overline{B'X}$ must have a slope of $-\frac{6}{5}$. Since we can travel up 5 and to the right 6 (*rise over run*) to get from X to B, we can travel the same distance but up 6 and to the left 5 to get from X to B'. This puts B' at $(-6,7)$.

c) Finally, we can travel up 1 and to the right 7 (a slope of $\frac{1}{7}$) to get from X to C. Therefore, we can travel the same distance but along a perpendicular line by going up 7 and to the left 1 (a slope of -7) to get from X to C'. Therefore, C' is $(-2,8)$.

It is also possible to rotate a polygon around one of its vertices.
Example: $\triangle ABC$ is rotated 90° around vertex A to produce the image, $\triangle AB'C'$.

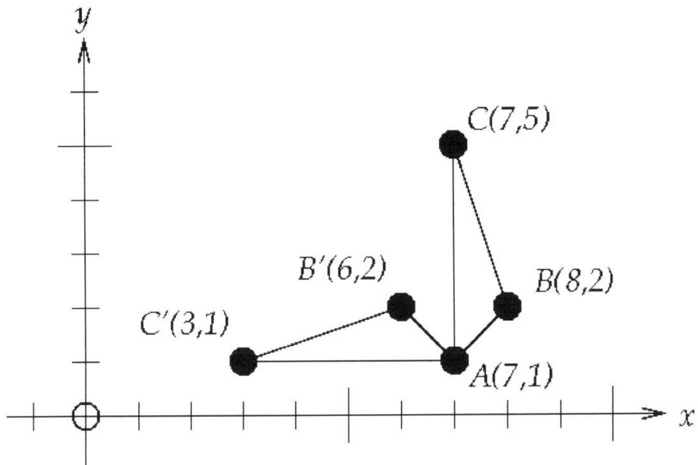

An object may also be rotated around a point inside the object.
Example: The heart shape below is rotated 90° around a central point.

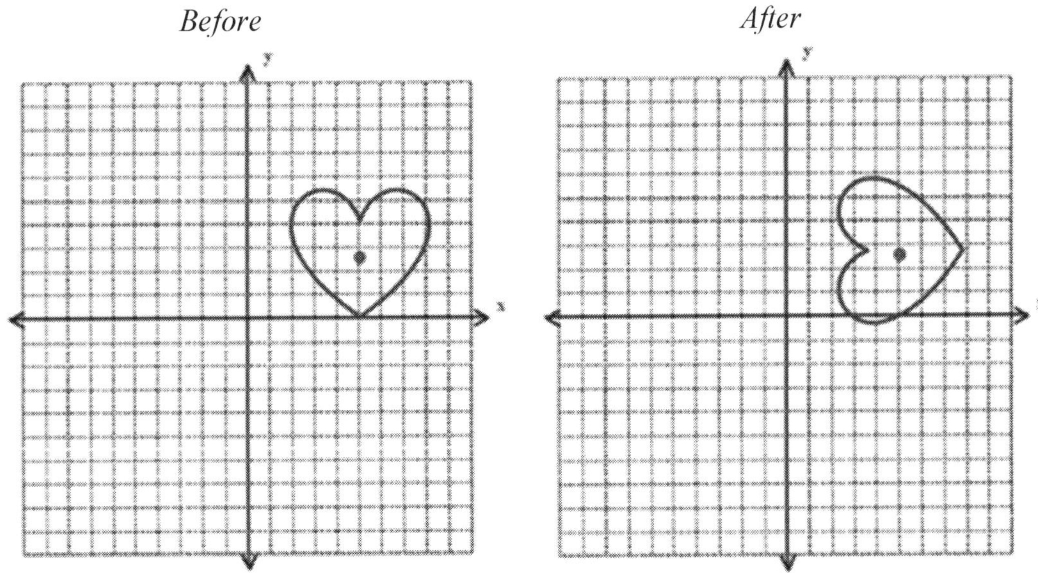

Model Problem:

Which figure represents a rotation of triangle *A*?

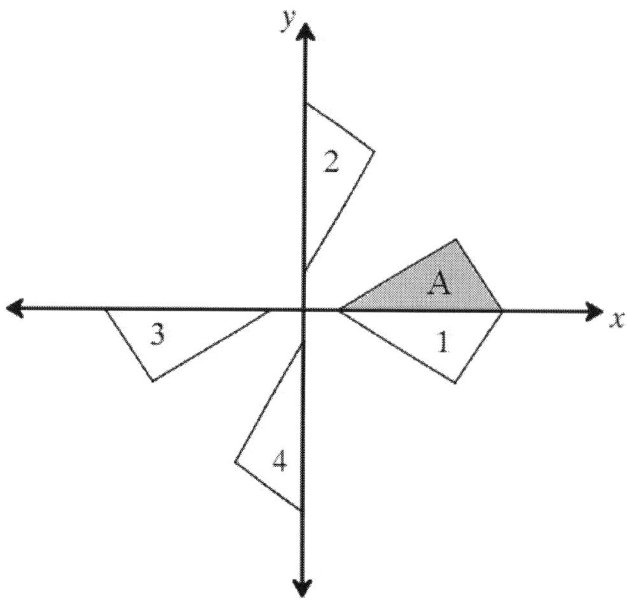

Solution:

Figure 3

Explanation:

A rotation will always preserve orientation. *[Figure 3 is the image after* R_{180°*. It is the only figure among the four that has the same orientation as triangle A.]*

Practice Problems

1. What are the coordinates of M', the image of $M(2,4)$, after a counterclockwise rotation of 90° about the origin?

2. What are the coordinates of the image of $P(-2,5)$ after a *clockwise* rotation of 90° about the origin?

3. The point $(-2,1)$ is rotated 180° about the origin in a *clockwise* direction. What are the coordinates of its image?

4. Rectangle $ABCD$ has vertices $A(0,-4)$, $B(4,-2)$, $C(5,-4)$, and $D(1,-6)$. Find the coordinates of the vertices of $A'B'C'D'$, the image of $ABCD$ after a rotation of 180° about the origin.

5.

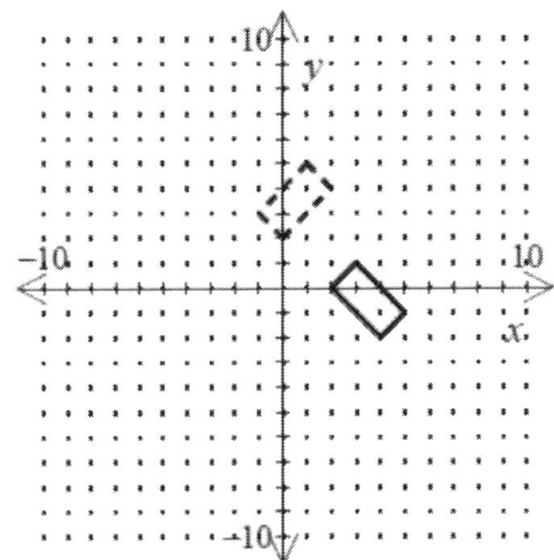

In the graph above, the dashed rectangle is the image of the solid rectangle under a rotation
(1) 180° about the origin.
(2) 90° counterclockwise about the origin.
(3) 90° clockwise about the origin.
(4) 270° counterclockwise about the origin.

6. Which graph represents the image of trapezoid $ABCD$ after a rotation of 180° around vertex D?

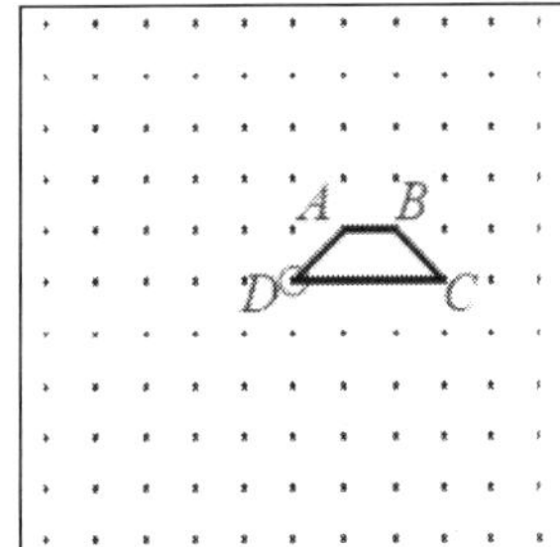

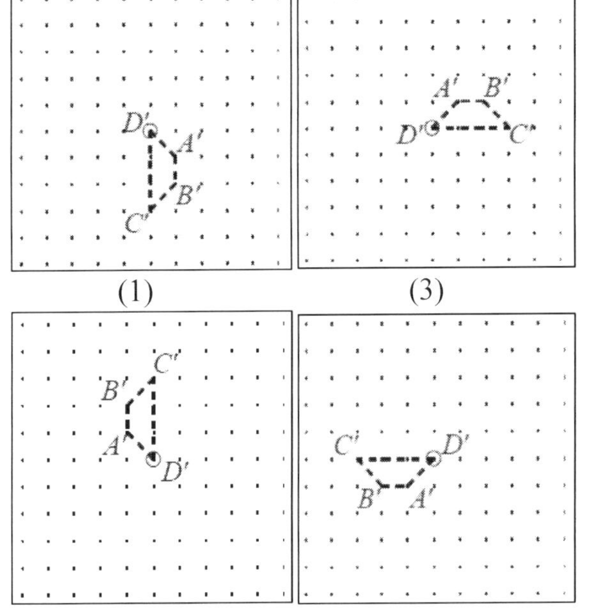

(2) (4)

REGENTS QUESTIONS

Multiple Choice

1. The accompanying diagram shows the starting position of the spinner on a board game.

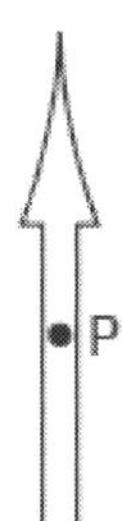

How does this spinner appear after a 270° counterclockwise rotation about point P?

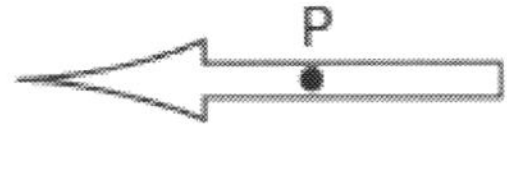

(1)

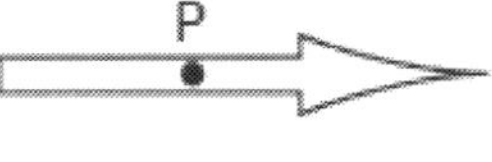

(3)

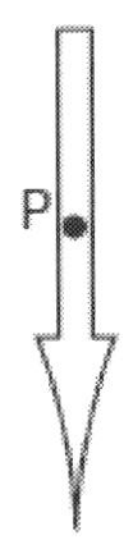

(2)

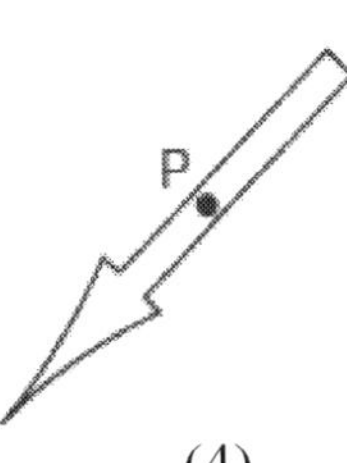

(4)

2. If point (5,2) is rotated counterclockwise 90° about the origin, its image will be point
   (1) (2,5) (3) (–2,5)
   (2) (2,–5) (4) (–5,–2)

3. What are the coordinates of A', the image of $A(-3,4)$, after a rotation of 180° about the origin?
   (1) (4,–3) (3) (3,4)
   (2) (–4,–3) (4) (3,–4)

4. The coordinates of point P are (7,1). What are the coordinates of the image of P after $R_{90°}$ about the origin?
   (1) (1,7) (3) (1,–7)
   (2) (–7,–1) (4) (–1,7)

5. (CC) The image of $\triangle ABC$ after a rotation of 90° clockwise about the origin is $\triangle DEF$, as shown below.

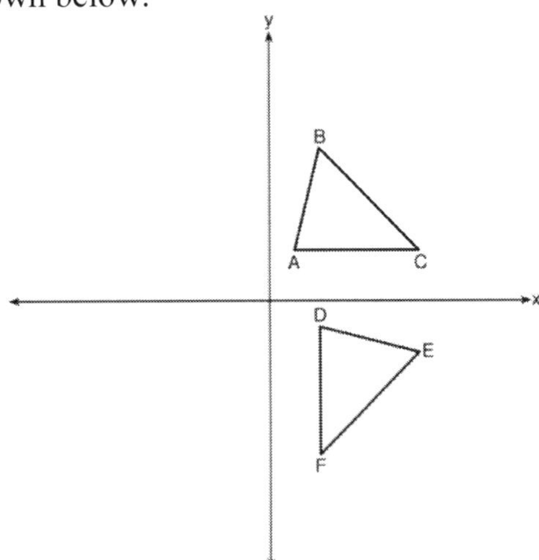

Which statement is true?

(1) $\overline{BC} \cong \overline{DE}$
(2) $\overline{AB} \cong \overline{DF}$
(3) $\angle C \cong \angle E$
(4) $\angle A \cong \angle D$

6. (CC) Quadrilateral $ABCD$ is graphed on the set of axes below.

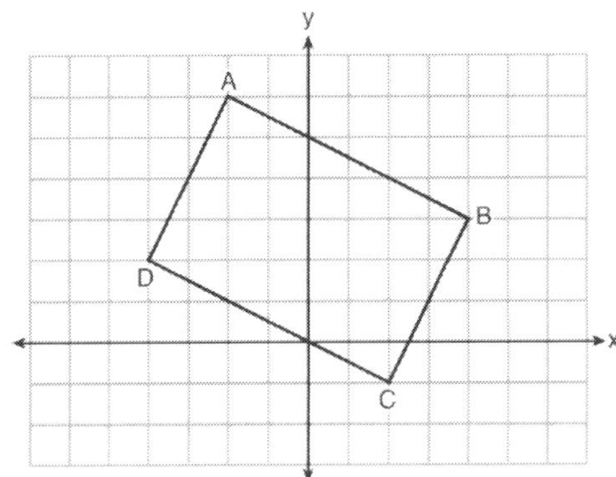

When $ABCD$ is rotated 90° in a counterclockwise direction about the origin, its image is quadrilateral $A'B'C'D'$. Is distance preserved under this rotation, and which coordinates are correct for the given vertex?

(1) no and $C'(1,2)$
(2) no and $D'(2,4)$
(3) yes and $A'(6,2)$
(4) yes and $B'(-3,4)$

7. (CC) Which point shown in the graph below is the image of point P after a counterclockwise rotation of 90° about the origin?

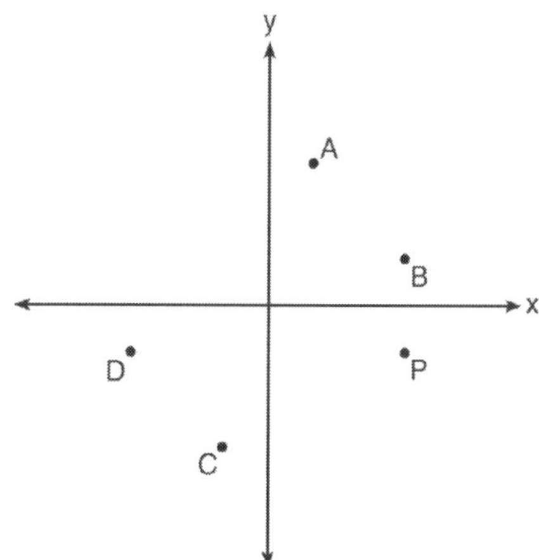

(1) A
(2) B
(3) C
(4) D

Constructed Response

8. The coordinates of the vertices of $\triangle RST$ are $R(-2,3)$, $S(4,4)$, and $T(2,-2)$. Triangle $R'S'T'$ is the image of $\triangle RST$ after a rotation of 90° about the origin. State the coordinates of the vertices of $\triangle R'S'T'$. [The use of the set of axes below is optional.]

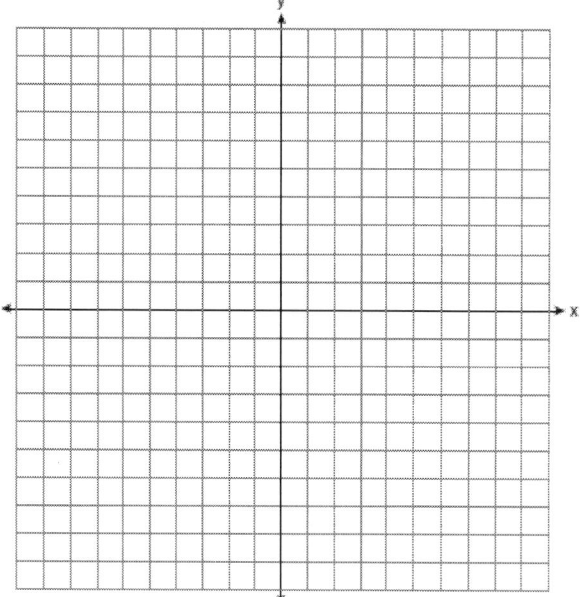

9. The coordinates of the vertices of $\triangle ABC$ are $A(1,2)$, $B(-4,3)$, and $C(-3,-5)$. State the coordinates of $\triangle A'B'C'$, the image of $\triangle ABC$ after a rotation of 90° about the origin. [The use of the set of axes below is optional.]

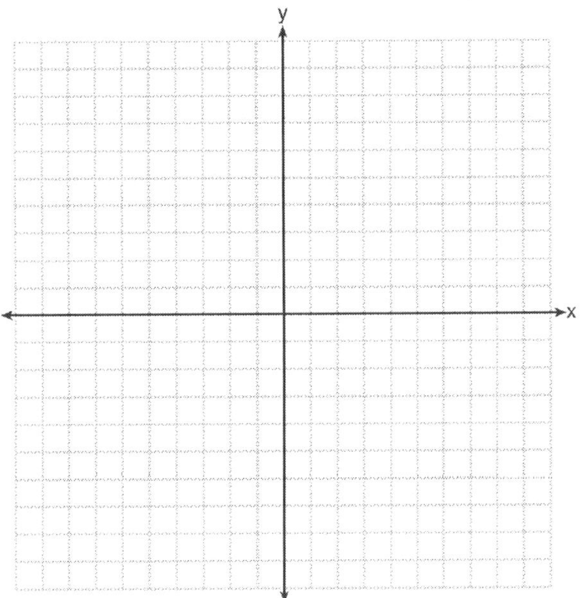

10. The coordinates of the endpoints of $\overline{BC}$ are $B(5,1)$ and $C(-3,-2)$. Under the transformation R_{90°, the image of $\overline{BC}$ is $\overline{B'C'}$. State the coordinates of points B' and C'.

11. (CC) Triangle MNP is the image of triangle JKL after a 120° counterclockwise rotation about point Q. If the measure of angle L is 47° and the measure of angle N is 57°, determine the measure of angle M. Explain how you arrived at your answer.

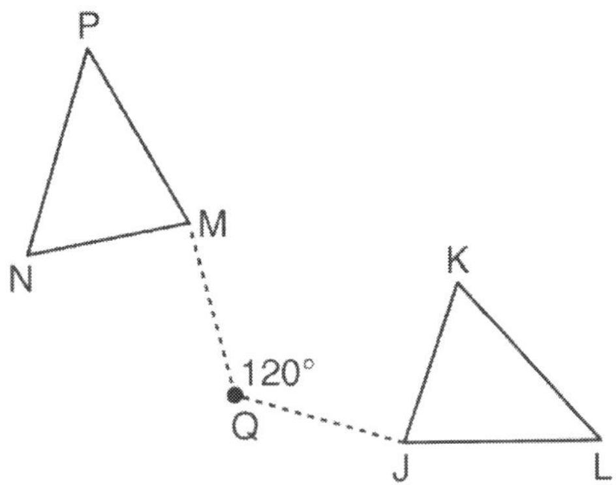

10.4 Map a Polygon onto Itself

Key Terms and Concepts

A polygon **maps onto itself** when a transformation maps each vertex of the original polygon onto a vertex of the image. In other words, the image appears identical to the original shape.

A polygon will map onto itself when it is *reflected over a line of symmetry.*

Examples: (a) The triangle in *Graph 1* maps onto itself when reflected over the x-axis.
(b) The triangle in *Graph 2* maps onto itself when reflected over the y-axis.
(c) A reflection over the line $x = 1$ will map the trapezoid in *Graph 3* onto itself.
(d) The rectangle in *Graph 4* maps onto itself by either a reflection over the line $x = -1$ or a reflection over the line $y = -2$. It has two lines of symmetry.

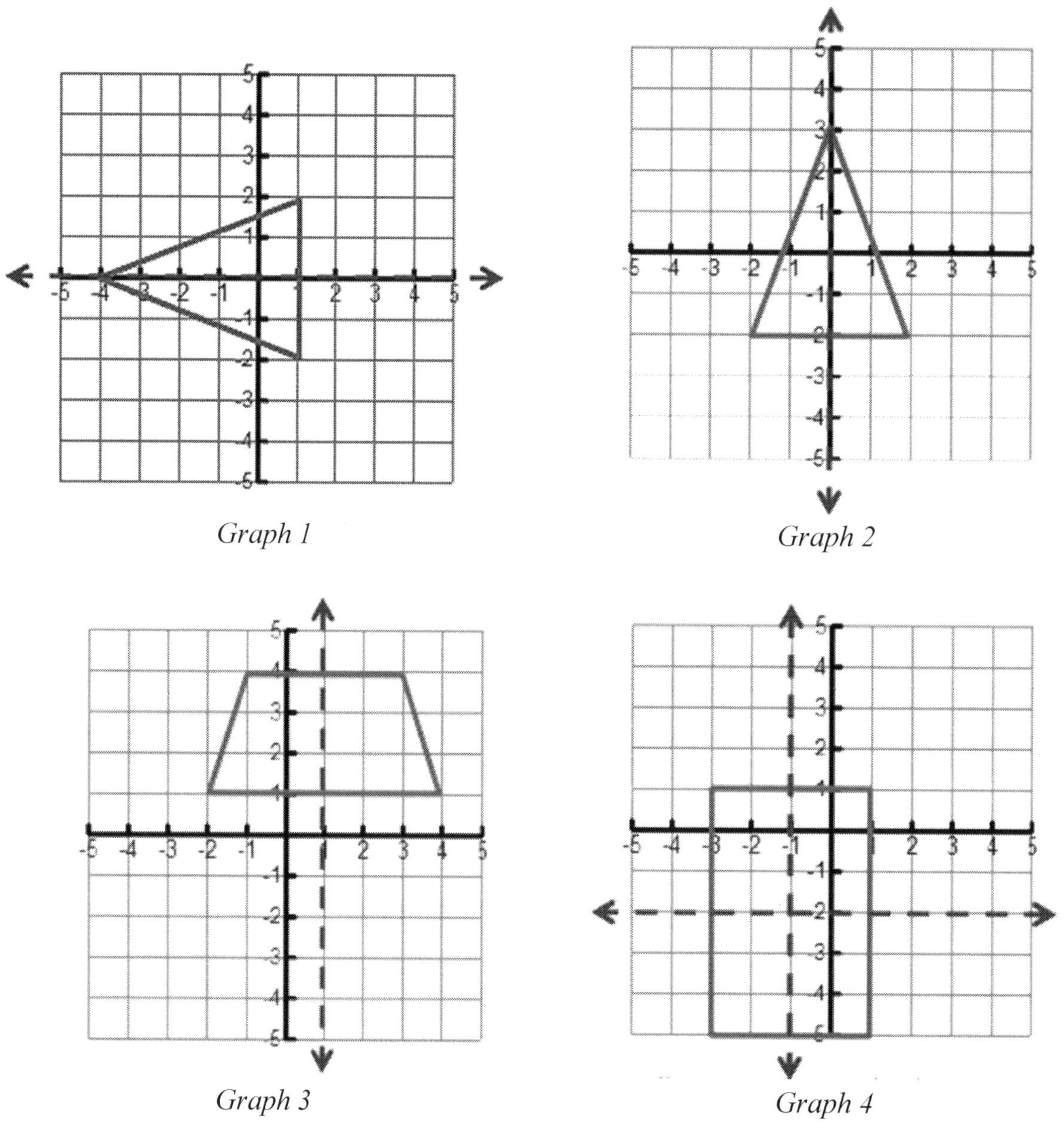

Graph 1 *Graph 2*

Graph 3 *Graph 4*

A polygon can also map onto itself by a rotation.
Example: The rectangle below, when rotated 180° around its center (2,1), maps onto itself.

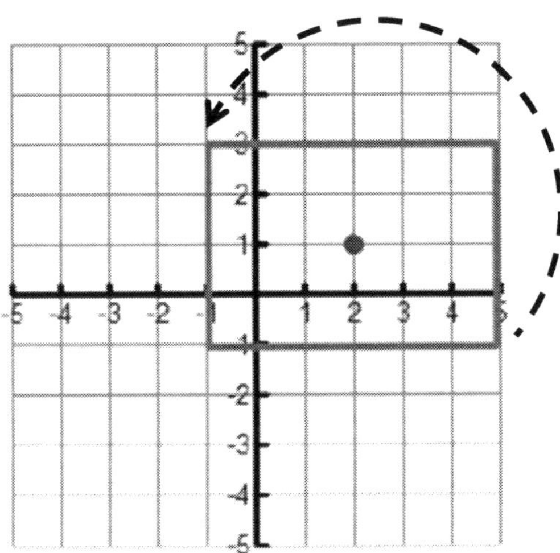

Regular polygons map onto themselves by rotating the polygon around its center by a number of degrees equal to 360° divided by the number of sides (or by any multiple of this).
Example: The regular hexagon below, centered at the origin, maps onto itself when it is reflected $\frac{360°}{6} = 60°$ around the origin, either clockwise or counterclockwise. It also maps onto itself on a rotation of any multiple of 60° (ie, 120°, 180°, etc.).

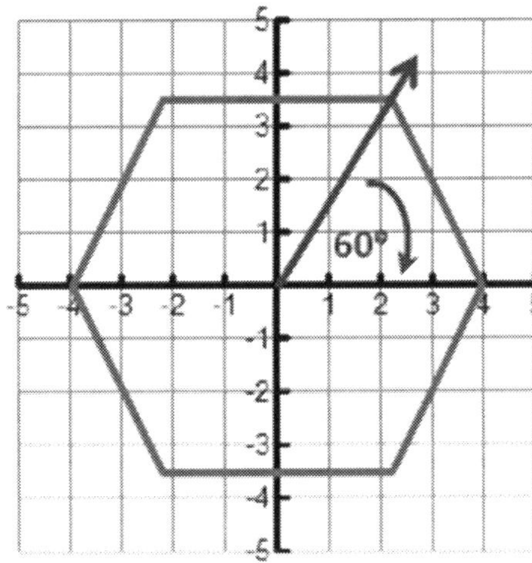

Model Problem:

To the right is an image of a nautical steering wheel. If the wheel is turned, what is the minimum number of degrees that it would need to rotate in order to map onto itself?

Solution:

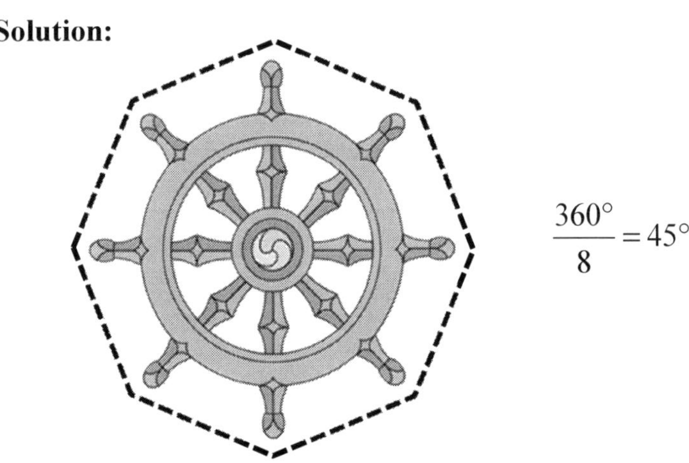

$$\frac{360°}{8} = 45°$$

Explanation of steps:

For a regular polygon, divide 360° by the number of sides. *[By connecting the tips of the spokes of the wheel, we form a regular octagon, so there are 8 sides.]*

Practice Problems

1. Which reflection would map this regular pentagon onto itself?

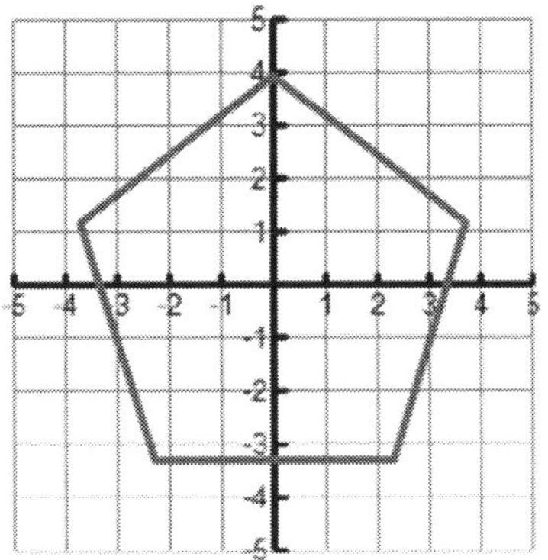

(1) reflection over $y = 0$
(2) reflection over $y = 1$
(3) reflection over $x = 0$
(4) reflection over $y = x$

2. What angle of rotation around (–1,2) would map this rectangle onto itself?

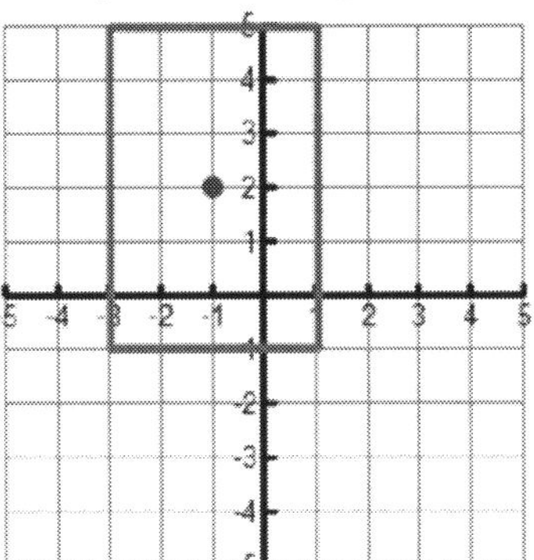

3. What is the minimum number of degrees that the airplane propeller shown below would need to rotate in order to map onto itself?

4. The bottom car of the Ferris wheel below is positioned for riders to board. How many degrees must the wheel turn for riders to board an adjacent car?

REGENTS QUESTIONS

Multiple Choice

1. (CC) A regular pentagon is shown in the diagram below.

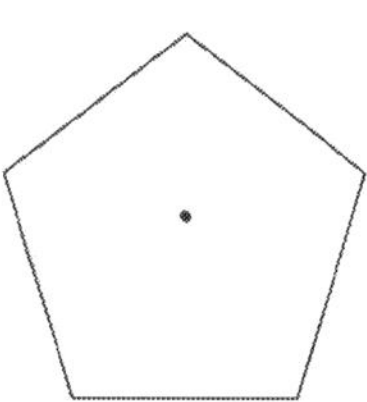

If the pentagon is rotated clockwise around its center, the minimum number of degrees it must be rotated to carry the pentagon onto itself is

(1) 54° (3) 108°
(2) 72° (4) 360°

2. (CC) Which regular polygon has a minimum rotation of 45° to carry the polygon onto itself?

(1) octagon (3) hexagon
(2) decagon (4) pentagon

3. (CC) In the diagram below, a square is graphed in the coordinate plane.

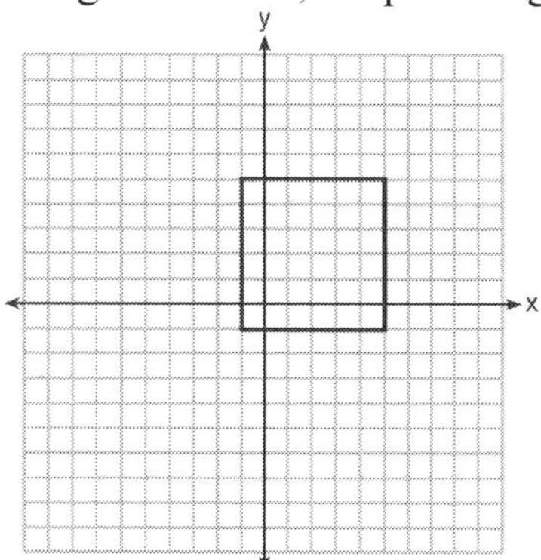

A reflection over which line does *not* carry the square onto itself?

(1) $x = 5$ (3) $y = x$
(2) $y = 2$ (4) $x + y = 4$

4. (CC) Which rotation about its center will carry a regular decagon onto itself?

(1) 54° (3) 198°
(2) 162° (4) 252°

Constructed Response

5. (CC) A regular hexagon is rotated in a counterclockwise direction about its center. Determine and state the minimum number of degrees in the rotation such that the hexagon will coincide with itself.

Chapter 11. Dilations

11.1 Dilations of Line Segments

Key Terms and Concepts

A dilation is another type of transformation. A **dilation** will transform an object by enlarging it or reducing it, depending on a given **scale factor** (which is restricted to *positive* real numbers for the purposes of this course). A capital D is usually used to represent dilation, with the scale factor in subscript, as in D_2.

The effect of a dilation depends on the positive scale factor (also known as the *ratio*):

- If the scale factor is **greater than 1**, the image is *enlarged*, as each point will move away from the given center point.
- If the scale factor is **less than 1**, the image is *reduced*, with each point moving closer to the given center point.
- For a scale factor of 1, the image will remain the *same size* as the pre-image.

The two diagrams to the right show dilations of point P to its image P' from a center point, O. In the top diagram, the scale factor is greater than 1, and in the bottom diagram the scale factor is less than 1.

In both cases, if the scale factor is r, then the ratio $\dfrac{OP'}{OP} = r$.

If the **center point** of a dilation is not specified, the origin (0,0) is assumed. When the center of dilation is the origin, we can calculate the coordinates of the points in the image by multiplying both the x and y coordinates by the scale factor, r. In other words, $D_r : (x, y) \to (rx, ry)$.

Examples: Shown below, A' is the image of A under the dilation D_2 and B' is the image of B under the dilation $D_{\frac{1}{3}}$. Note that $A(4,2) \to A'(8,4)$ by multiplying both coordinates by 2, and $B(-3,9) \to B'(-1,3)$ by multiplying both coordinates by $\frac{1}{3}$.

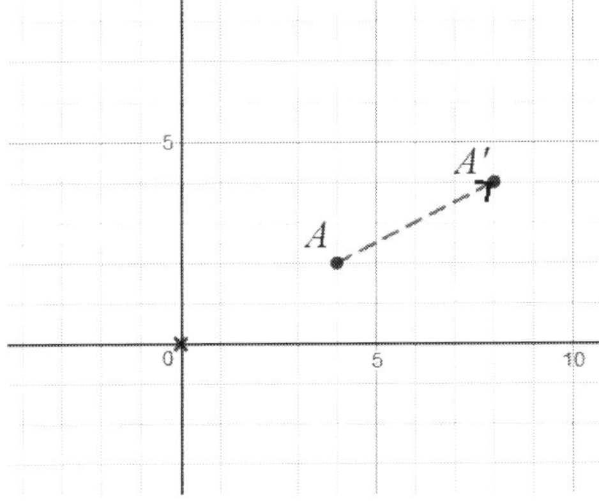

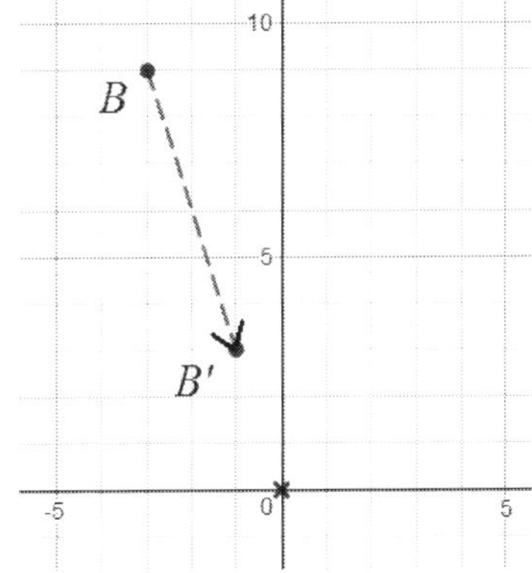

When a line segment is dilated from an external (*non-collinear*) center point, the image is a parallel line segment. To determine the image of the dilation, simply dilate each endpoint separately. If the center point is the *origin*, we can use $(x, y) \rightarrow (rx, ry)$ to dilate each point, where r is the scale factor.

The length of the image is longer or shorter in the ratio given by the scale factor, r.

Example: A dilation of $\overline{AB}$, with the center as the origin, O, and the scale factor (*ratio*) of r, results in: $OA' = r(OA)$, $OB' = r(OB)$, and $A'B' = r(AB)$

In other words, $r = \dfrac{OA'}{OA} = \dfrac{OB'}{OB} = \dfrac{A'B'}{AB}$.

The graph to the right shows a dilation of $\overline{AB}$ with the origin, O, as the center and a scale factor of 2.

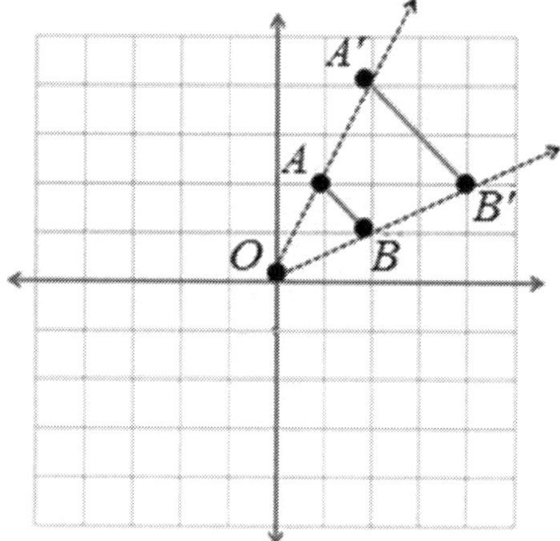

The scale factor determines the endpoints of the image: $A(1,2) \rightarrow A'(2,4)$ and $B(2,1) \rightarrow B'(4,2)$.

Note that $OA' = 2(OA)$ and $OB' = 2(OB)$.
Also, $A'B' = 2\sqrt{2}$ and $AB = \sqrt{2}$, so $A'B' = 2(AB)$.

Both $\overline{AB}$ and $\overline{A'B'}$ have the same slope, and are therefore parallel.

Model Problem:

$\overline{M'N'}$ is the image of $\overline{MN}$ under a dilation with the origin as the center and a scale factor of 2.5. Given $M(4,2)$ and $N(-2,0)$, find the coordinates of M' and N'.

Solution:
$M(4,2) \rightarrow M'(10,5)$ and $N(-2,0) \rightarrow N'(-5,0)$

Explanation of steps:
Use $(x, y) \rightarrow (rx, ry)$ to multiply the coordinates of each endpoint by the scale factor *[* $r = 2.5$ *]*.

Practice Problems

1. What is the image of (3,–2) under the dilation D_5?	2. Find the image of (3, –2) for a dilation centered at the origin with scale factor $\frac{1}{2}$.

3. Which graph shows a dilation that has a scale factor of $\frac{1}{2}$ and center at the origin?

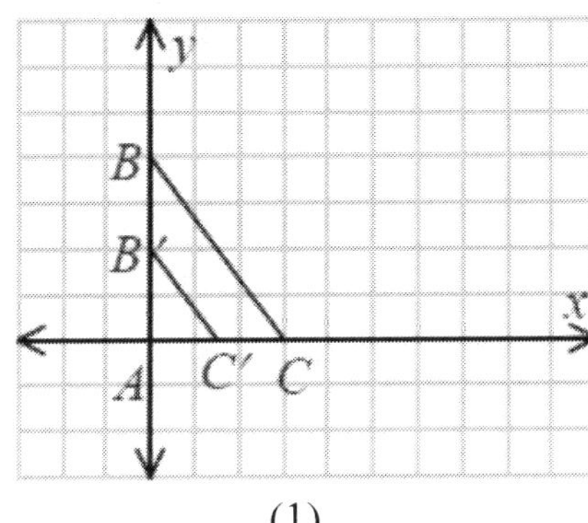

(1)

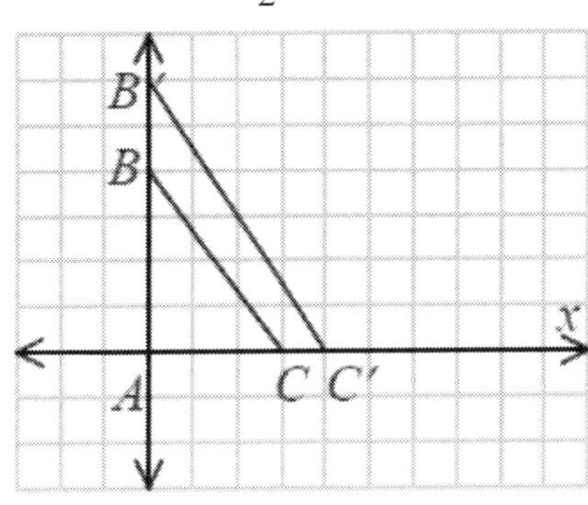

(3)

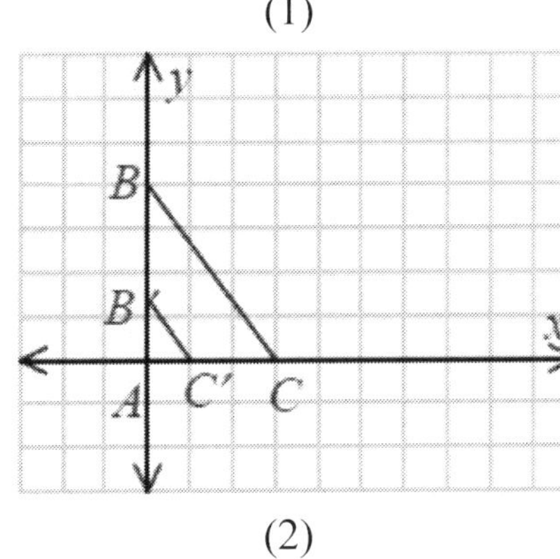

(2)

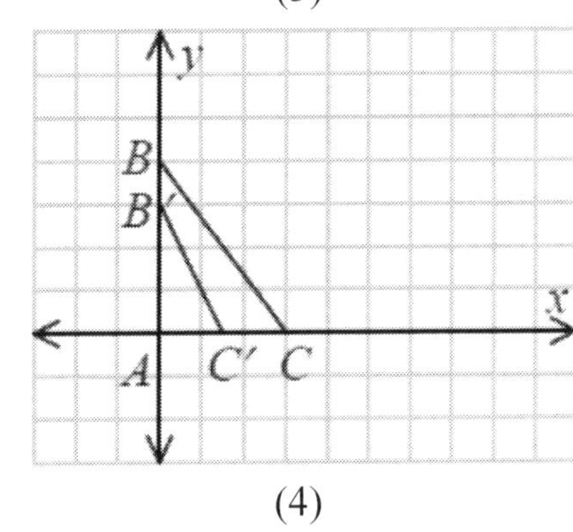

(4)

4. Graph $\overline{JK}$ with $J(2, 3)$ and $K(-4, 4)$. Then graph its dilation with a scale factor of 2.

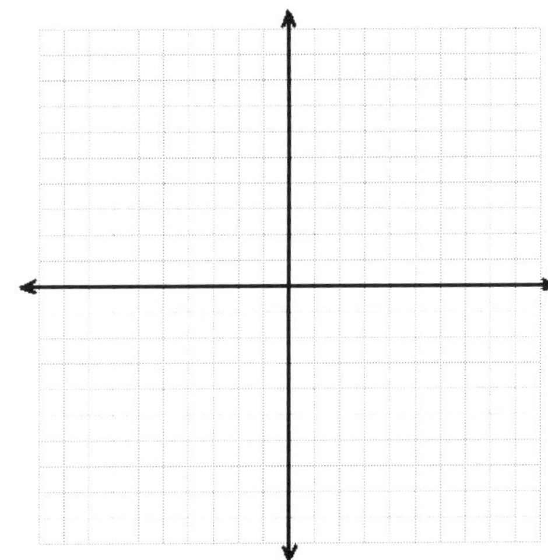

REGENTS QUESTIONS

Multiple Choice

1. The image of point A after a dilation of 3 is (6,15). What was the original location of point A?
   (1) (2,5)
   (2) (3,12)
   (3) (9,18)
   (4) (18,45)

2. Under a dilation where the center of dilation is the origin, the image of A(–2,–3) is $A'(-6,-9)$. What are the coordinates of B', the image of B(4,0) under the same dilation?
   (1) (–12,0)
   (2) (12,0)
   (3) (–4,0)
   (4) (4,0)

3. (CC) In the diagram below, $\overline{CD}$ is the image of $\overline{AB}$ after a dilation of scale factor k with center E.

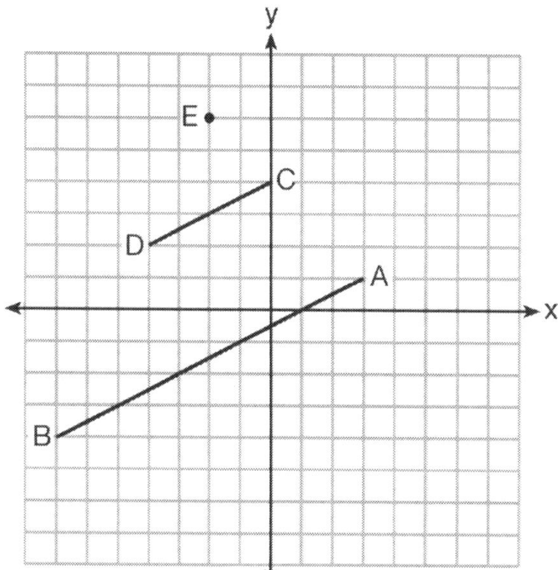

   Which ratio is equal to the scale factor k of the dilation?
   (1) $\frac{EC}{EA}$
   (2) $\frac{BA}{EA}$
   (3) $\frac{EA}{BA}$
   (4) $\frac{EA}{EC}$

4. (CC) A line that passes through the points whose coordinates are (1,1) and (5,7) is dilated by a scale factor of 3 and centered at the origin. The image of the line
   (1) is perpendicular to the original line
   (2) is parallel to the original line
   (3) passes through the origin
   (4) is the original line

5. (CC) A three-inch line segment is dilated by a scale factor of 6 and centered at its midpoint. What is the length of its image?
   (1) 9 inches
   (2) 2 inches
   (3) 15 inches
   (4) 18 inches

6. (CC) Line segment $A'B'$, whose endpoints are (4,–2) and (16,14), is the image of $\overline{AB}$ after a dilation of $\frac{1}{2}$ centered at the origin. What is the length of $\overline{AB}$?
   (1) 5
   (2) 10
   (3) 20
   (4) 40

7. (CC) On the graph below, point $A(3,4)$ and $\overline{BC}$ with coordinates $B(4,3)$ and $C(2,1)$ are graphed.

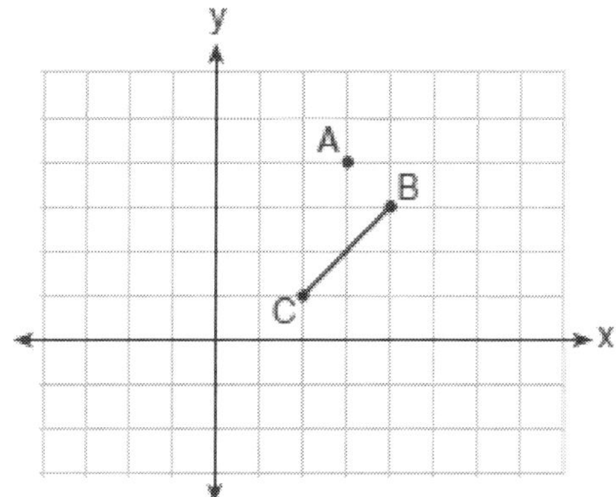

What are the coordinates of B' and C' after $\overline{BC}$ undergoes a dilation centered at point A with a scale factor of 2?
   (1) $B'(5,2)$ and $C'(1,-2)$
   (2) $B'(6,1)$ and $C'(0,-1)$
   (3) $B'(5,0)$ and $C'(1,-2)$
   (4) $B'(5,2)$ and $C'(3,0)$

Constructed Response

8. What is the image of point $A(1,3)$ after a dilation with the center at the origin and a scale factor of 4?

11.2 Dilations of Polygons

Key Terms and Concepts

A **dilation** will enlarge or reduce an object by a given scale factor. A dilation it is *not* a rigid motion, since the image may be a different size. Dilations do *preserve angles*, however, so the image after a dilation of an object is always **similar** to the original object.

Just as we did with the endpoints of dilated line segments, we can determine the dilated image of a polygon by dilating each vertex separately.

Example: The graph below shows the image of $\triangle ABC$ under the dilation D_3.
Note that $A(4,5) \to A'(12,15)$, $B(3,2) \to B'(9,6)$, and $C(6,3) \to C'(18,9)$.
Dilations preserve angles, so $\triangle ABC \sim \triangle A'B'C'$.

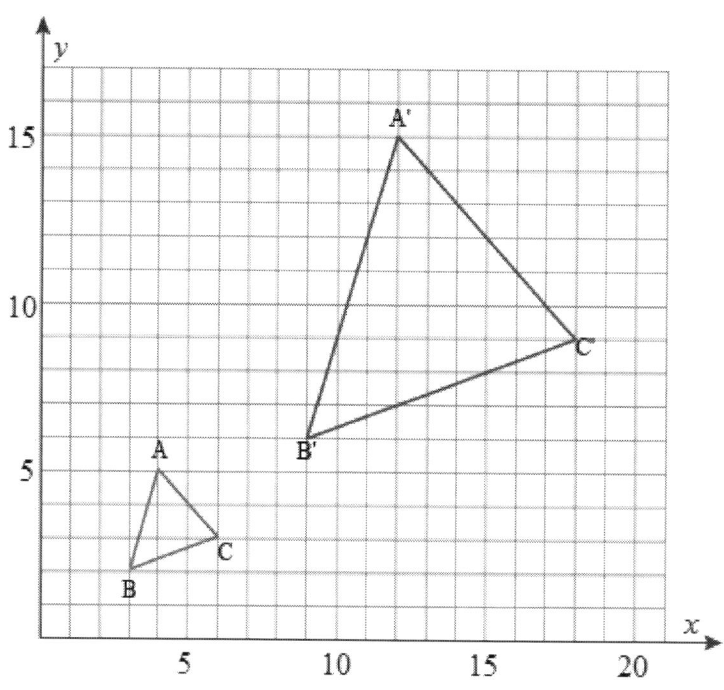

If given the graph of a dilation, we can determine the center point and scale factor.

Example: In the graph below, $\triangle A'B'C'$ is the image of $\triangle ABC$ after a dilation.

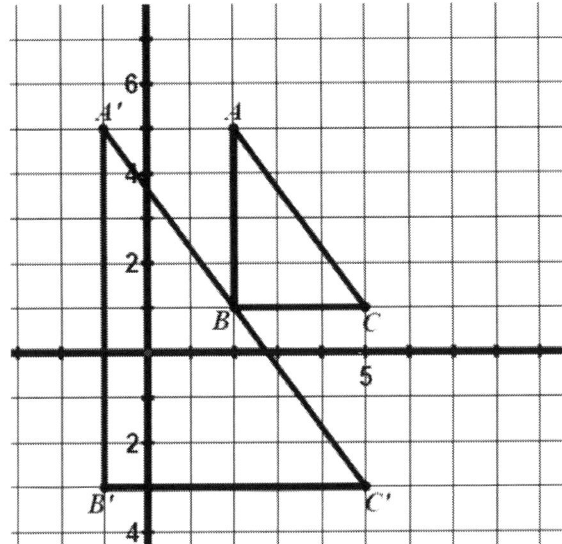

We can determine the center of dilation by extending the segments formed by any two pairs of corresponding vertices to find their point of intersection. In this case, we can easily extend $\overline{A'A}$ and $\overline{C'C}$ to see that they intersect at (5,5), as shown below, which we'll label point D. Point D is the center of dilation.

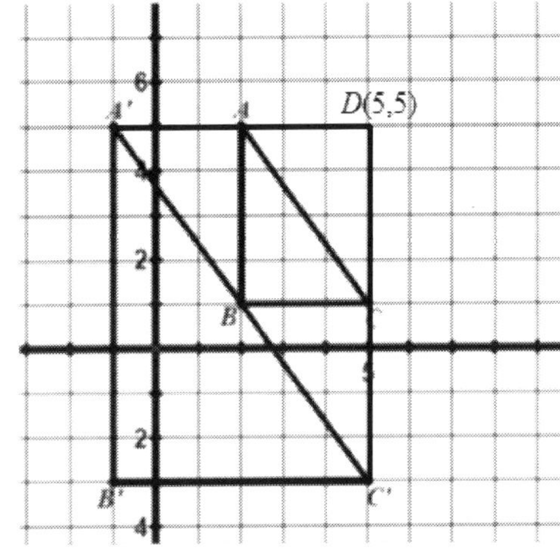

We can now easily see the scale factor by recognizing that $A'D = 2(AD)$ and $C'D = 2(CD)$. Therefore, the scale factor must be 2.

Model Problem:

Graph the dilation of quadrilateral *MNOP* using a scale factor of 3 and the origin as the center.

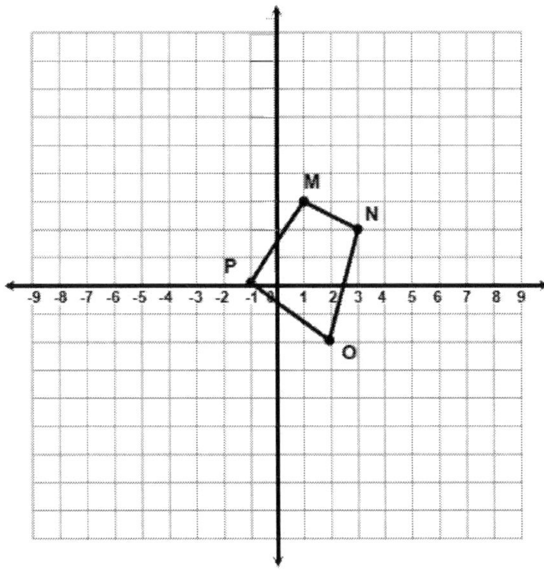

Solution:

(A)

$M(1,3) \rightarrow M'(3,9)$
$N(3,2) \rightarrow N'(9,6)$
$O(2,-2) \rightarrow O'(6,-6)$
$P(-1,0) \rightarrow P'(-3,0)$

(B)

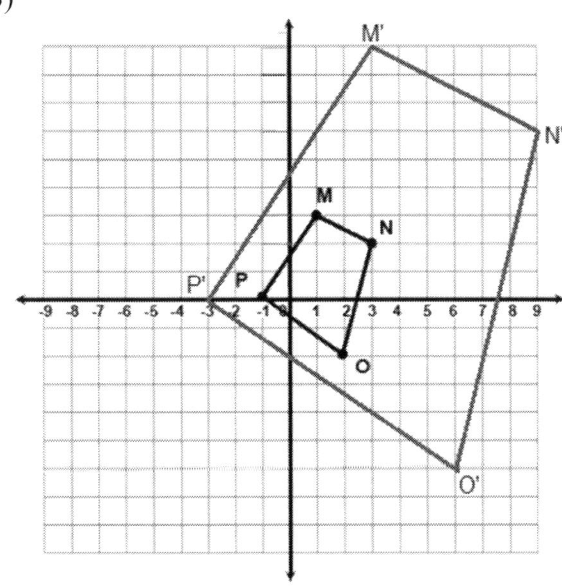

Explanation of steps:

(A) Find the image of each vertex by multiplying the coordinates of the vertex by the scale factor. *[Since the scale factor is 3,* $(x, y) \rightarrow (3x, 3y)$ *]*

(B) Plot and label the vertices of the image and draw the sides.

Practice Problems

1. Graph the dilation of $\triangle ABC$ under D_2.

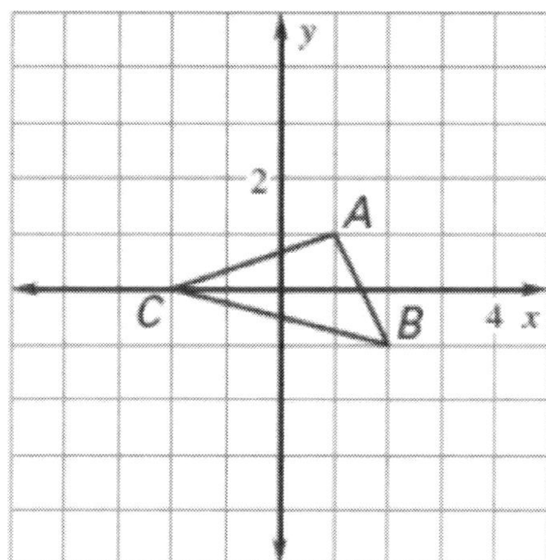

2. Graph the dilation of the quadrilateral using a scale factor of $\frac{1}{2}$.

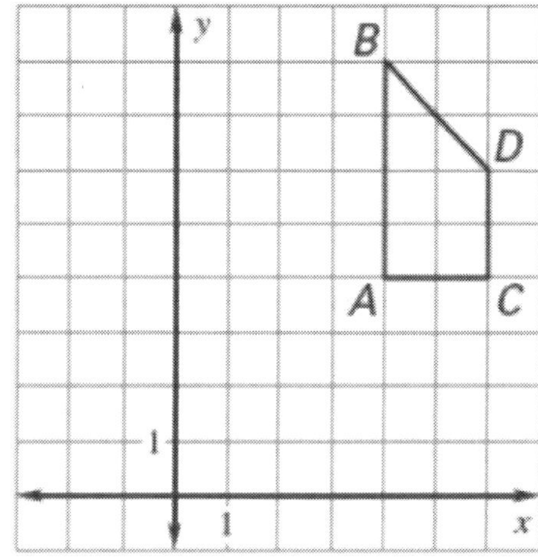

3. Graph $\triangle ABC$ with $A(-5,5)$, $B(-5,10)$, and $C(10,0)$. Then graph its dilation with a scale factor of $\frac{3}{5}$.

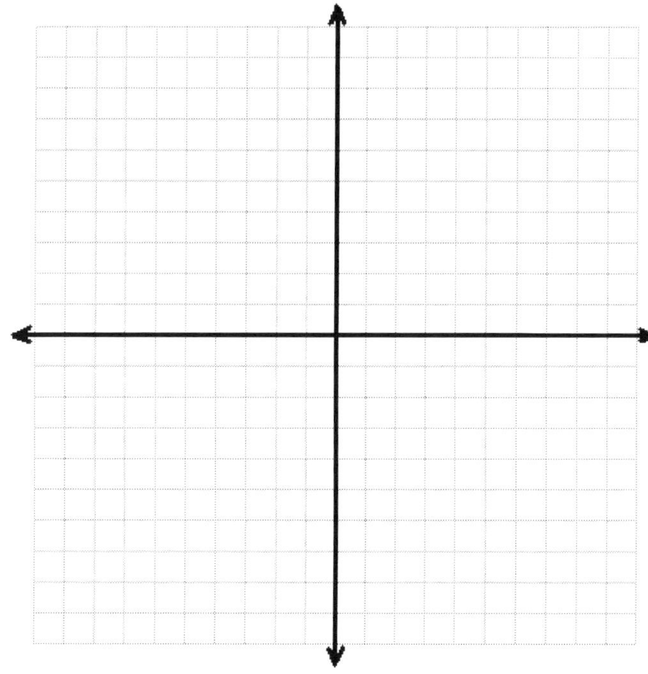

REGENTS QUESTIONS

Multiple Choice

1. Triangle $A'B'C'$ is the image of $\triangle ABC$ after a dilation of 2. Which statement is true?
   (1) $AB = A'B'$
   (2) $BC = 2(B'C')$
   (3) $m\angle B = m\angle B'$
   (4) $m\angle A = \frac{1}{2}(m\angle A')$

2. CC If $\triangle ABC$ is dilated by a scale factor of 3, which statement is true of the image $\triangle A'B'C'$?
   (1) $3A'B' = AB$
   (2) $B'C' = 3BC$
   (3) $m\angle A' = 3(m\angle A)$
   (4) $3(m\angle C') = m\angle C$

3. CC The image of $\triangle ABC$ after a dilation of scale factor k centered at point A is $\triangle ADE$, as shown in the diagram below.

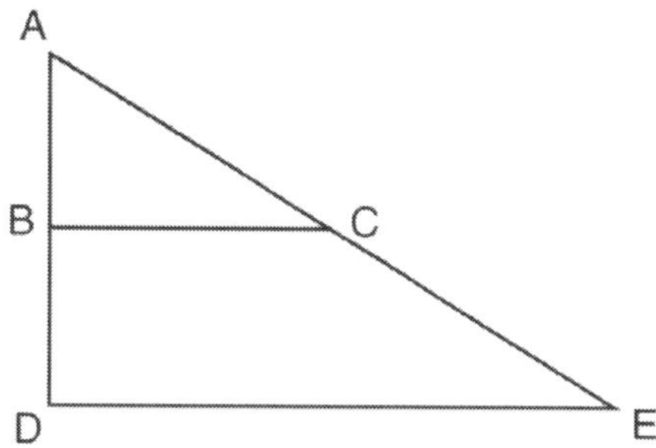

   Which statement is always true?
   (1) $2AB = AD$
   (2) $\overline{AD} \perp \overline{DE}$
   (3) $AC = CE$
   (4) $\overline{BC} \parallel \overline{DE}$

4. CC A triangle is dilated by a scale factor of 3 with the center of dilation at the origin. Which statement is true?
   (1) The area of the image is nine times the area of the original triangle.
   (2) The perimeter of the image is nine times the perimeter of the original triangle.
   (3) The slope of any side of the image is three times the slope of the corresponding side of the original triangle.
   (4) The measure of each angle in the image is three times the measure of the corresponding angle of the original triangle.

5. (CC) In the diagram below, $\triangle ABE$ is the image of $\triangle ACD$ after a dilation centered at the origin. The coordinates of the vertices are $A(0,0)$, $B(3,0)$, $C(4.5,0)$, $D(0,6)$, and $E(0,4)$.

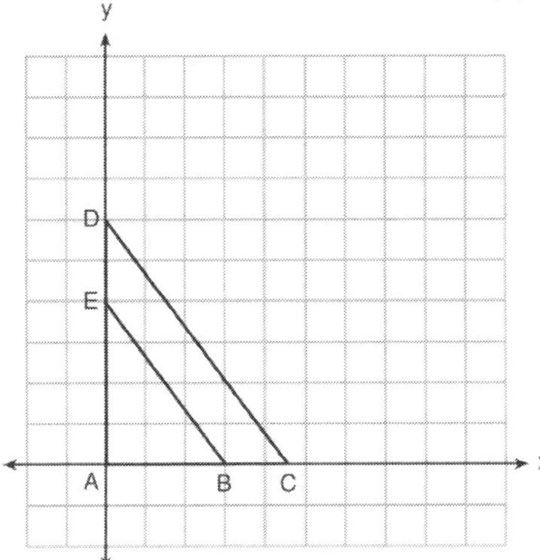

The ratio of the lengths of $\overline{BE}$ to $\overline{CD}$ is

(1) $\frac{2}{3}$

(2) $\frac{3}{2}$

(3) $\frac{3}{4}$

(4) $\frac{4}{3}$

Constructed Response

6. On the accompanying set of axes, graph $\triangle ABC$ with coordinates $A(-1,2)$, $B(0,6)$, and $C(5,4)$. Then graph $\triangle A'B'C'$, the image of $\triangle ABC$ after a dilation of 2.

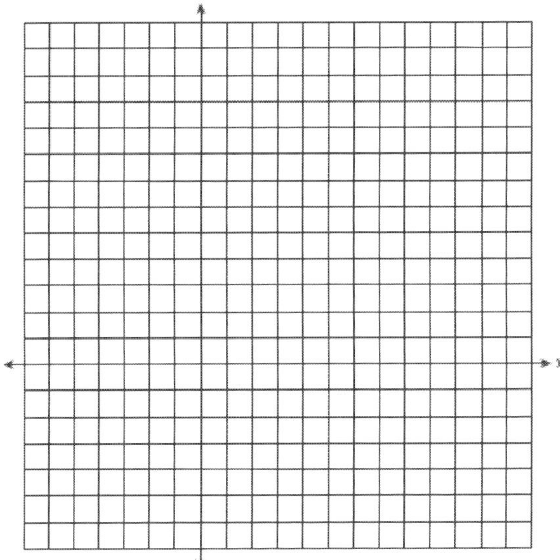

7. Given: quadrilateral $ABCD$ with vertices A(–2,2), B(8,–4), C(6,–10), and D(–4,–4). State the coordinates of $A'B'C'D'$, the image of quadrilateral $ABCD$ under a dilation of factor $\frac{1}{2}$. Prove that $A'B'C'D'$ is a parallelogram. [The use of the grid is optional.]

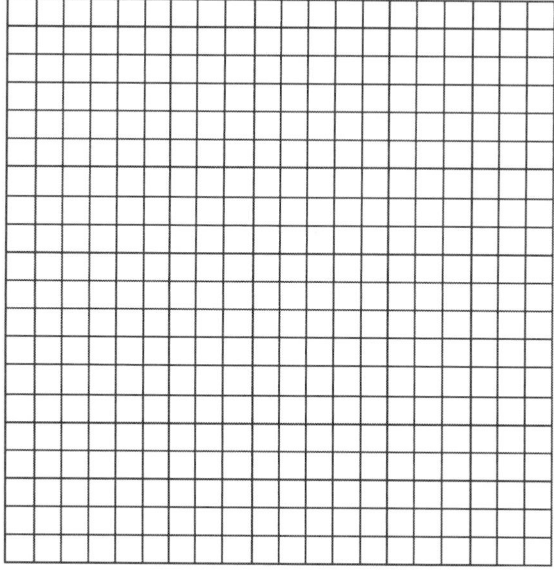

8. On the accompanying grid, graph and label quadrilateral $ABCD$, whose coordinates are $A(-1,3)$, $B(2,0)$, $C(2,-1)$, and $D(-3,-1)$. Graph, label, and state the coordinates of $A'B'C'D'$, the image of $ABCD$ under a dilation of 2, where the center of dilation is the origin.

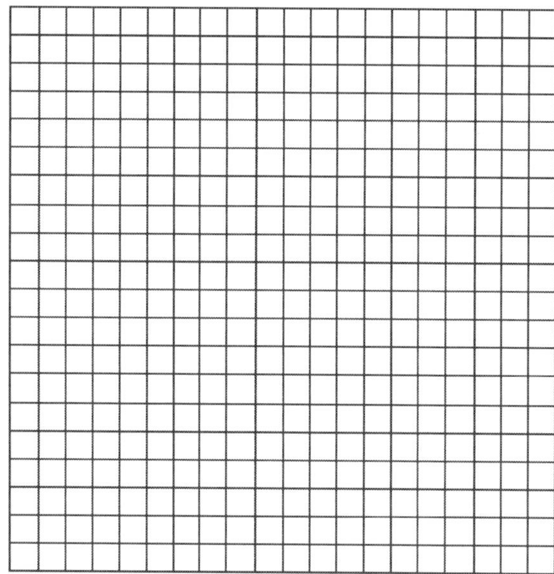

9. Triangle ABC has vertices A(6,6), B(9,0), and C(3,–3). State and label the coordinates of $\triangle A'B'C'$, the image of $\triangle ABC$ after a dilation of $D_{\frac{1}{3}}$.

10. Triangle ABC has coordinates $A(-2,1)$, $B(3,1)$, and $C(0,-3)$. On the set of axes below, graph and label $\triangle A'B'C'$, the image of $\triangle ABC$ after a dilation of 2.

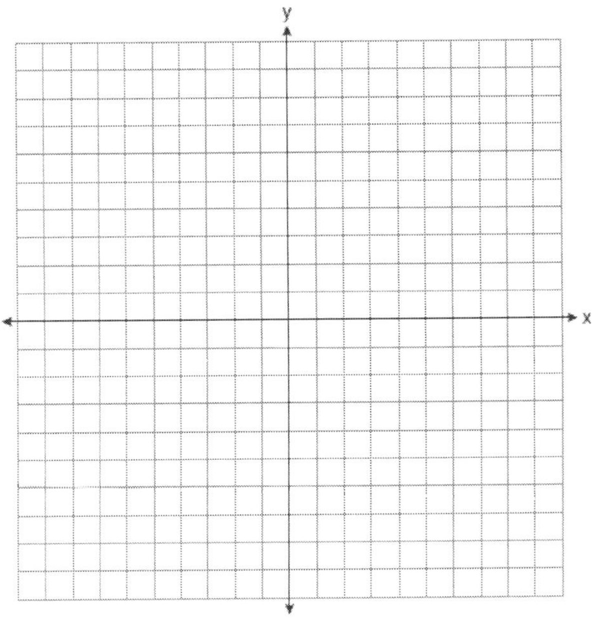

11. (CC) Triangle QRS is graphed on the set of axes below.

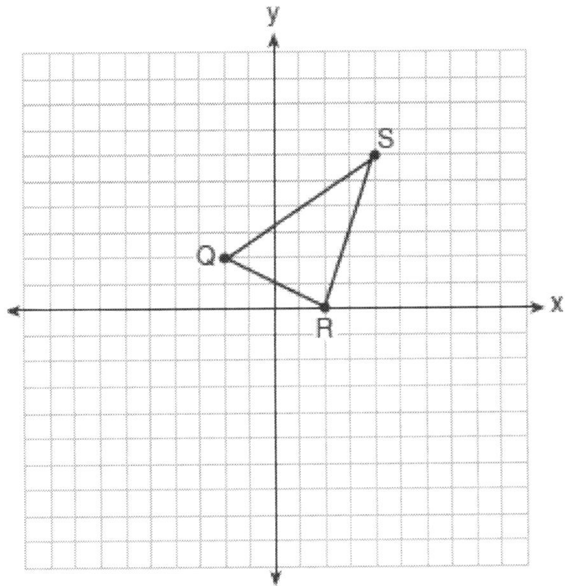

On the same set of axes, graph and label $\triangle Q'R'S'$, the image of $\triangle QRS$ after a dilation with a scale factor of $\frac{3}{2}$ centered at the origin. Use slopes to explain why $\overline{Q'R'} \parallel \overline{QR}$.

11.3 Dilations of Lines

Key Terms and Concepts

The **dilation of a line** will result in an image that is also a line. If the center of dilation is a point that is *not on the line*, the image will be a *parallel line*. However, if the center of dilation is a point *on the line*, then the image is the *same line*.

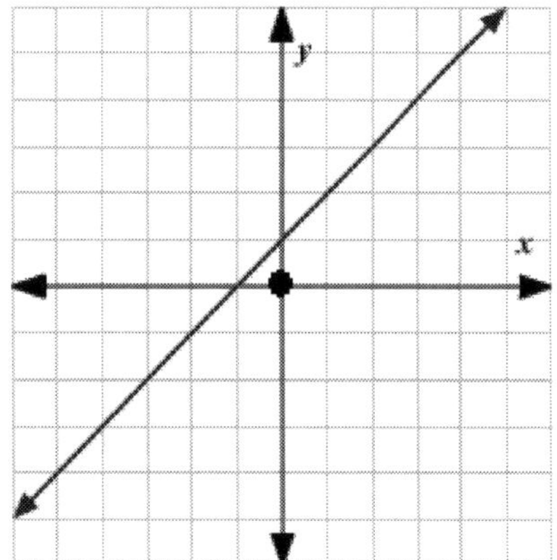

Take the graph of the line $y = x + 1$, shown to the left. Any dilation of this line with the origin as the center point will result in an image that is a parallel line. Therefore, the image will have the same slope.

Also note that the *y*-intercept of this line is 1. Since the origin is the center, we can find the *y*-intercept of the image by multiplying this by the scale factor.

For example, a dilation of $y = x + 1$ with the origin as the center and a scale factor of 3 will have a *y*-intercept of $1 \times 3 = 3$. The equation of the image would be $y = x + 3$.

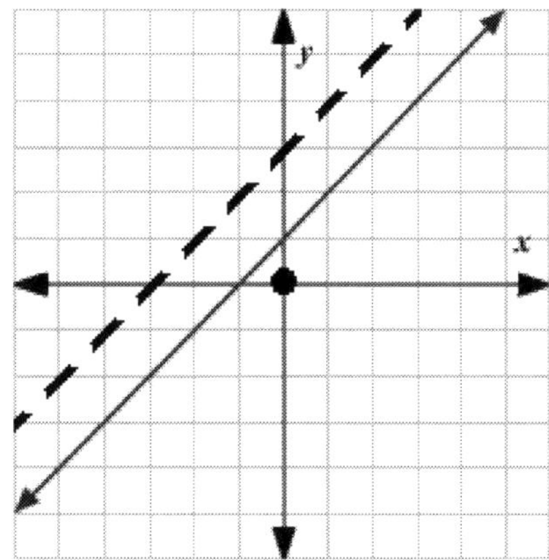

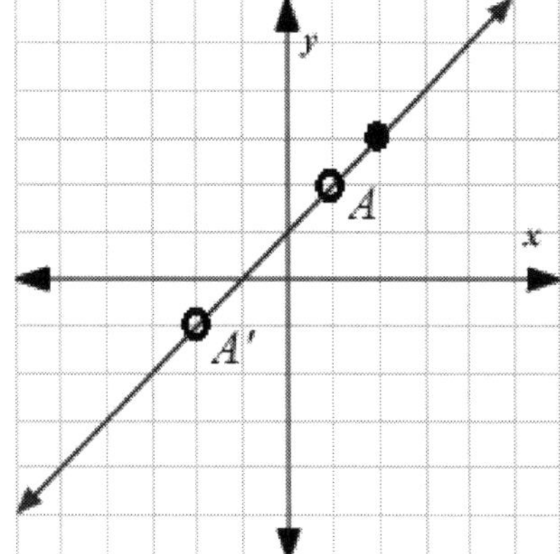

However, if the center point is on the line, then any dilation of the line would be the same line, $y = x + 1$.

For example, suppose the center is (2,3) and a scale factor of 4 is used. As shown to the left, since (2,3) is on the line, any other point, such as A(1,2), would be dilated to another point that lies on the same line, $A'(-2,-1)$.

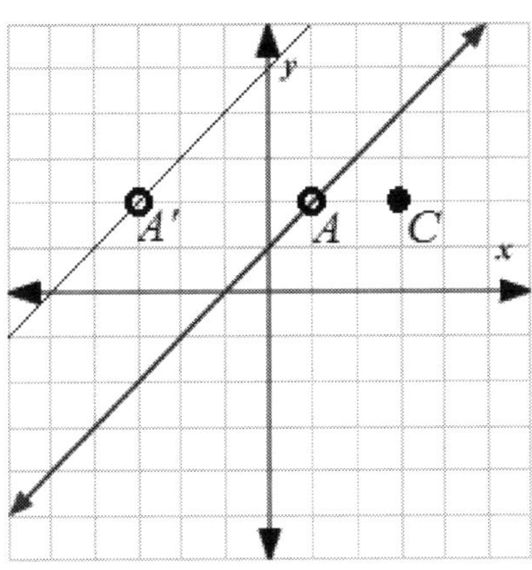

When the center point is an external point, but not the origin, we can determine the image of the dilation by selecting and dilating any point on the line.

For example, with $C(3,2)$ as the center of dilation, we can select point $A(1,2)$ on the line. (It helps to select a point with the same x or y value as the center point.)

Dilating A by a scale factor of 3 results in $A'(-3,2)$, since we know $CA' = 3(CA)$. The line parallel to $y = x + 1$ but passing through $(-3,2)$ is $y = x + 5$.

Model Problem:

The line $y - 2x = 6$ is dilated by a scale factor of $\frac{1}{3}$ and centered at the origin. Write the equation of the image of the line after the dilation.

Solution:

(A) $y - 2x = 6 \rightarrow y = 2x + 6$

(B) Image is $y = 2x + 2$

Explanation of steps:

(A) Transform the original linear equation into slope-intercept form.

(B) The dilation of a line, with an external center point, results in a parallel line, so the slope of the image *[2]* remains the same. We can find the y-intercept of the image by multiplying the original y-intercept by the scale factor *[* $6 \times \frac{1}{3} = 2$ *]*.

Practice Problems

1. The line $y = 3x - 4$ is dilated by a scale factor of 5 and centered at the origin. What is the equation of its image?	2. The line $2x + 3y = 4$ is dilated by a scale factor of 3 and centered at the origin. What is the equation of its image?
3. The line $y = 3x - 4$ is dilated by a scale factor of 2 with a center at $C(1,-1)$. What is the equation of its image?	4. The line $y = 2x + 2$ is dilated by a scale factor of 2 with a center at $C(1,0)$. What is the equation of its image?

REGENTS QUESTIONS

Multiple Choice

1. (CC) The line $y=2x-4$ is dilated by a scale factor of $\frac{3}{2}$ and centered at the origin. Which equation represents the image of the line after the dilation?
   (1) $y=2x-4$ (3) $y=3x-4$
   (2) $y=2x-6$ (4) $y=3x-6$

2. (CC) The equation of line *h* is $2x+y=1$. Line *m* is the image of line *h* after a dilation of scale factor 4 with respect to the origin. What is the equation of the line *m*?
   (1) $y=-2x+1$ (3) $y=2x+4$
   (2) $y=-2x+4$ (4) $y=2x+1$

3. (CC) The line $3y=-2x+8$ is transformed by a dilation centered at the origin. Which linear equation could be its image?
   (1) $2x+3y=5$ (3) $3x+2y=5$
   (2) $2x-3y=5$ (4) $3x-2y=5$

4. (CC) Line $y=3x-1$ is transformed by a dilation with a scale factor of 2 and centered at (3,8). The line's image is
   (1) $y=3x-8$ (3) $y=3x-2$
   (2) $y=3x-4$ (4) $y=3x-1$

Constructed Response

5. (CC) Line ℓ is mapped onto line *m* by a dilation centered at the origin with a scale factor of 2. The equation of line ℓ is $3x-y=4$. Determine and state an equation for line *m*.

Chapter 12. Transformation Proofs

12.1 Properties of Transformations

Key Terms and Concepts

The following table summarizes important properties of the four types of transformations.

Transformation	Rigid Motion (Isometry)	Preserves Distance	Preserves Angles	Preserves Orientation	Image is Congruent	Image is Similar
Translation	Yes	Yes	Yes	Yes	Yes	Yes
Reflection	Yes	Yes	Yes	**NO**	Yes	Yes
Rotation	Yes	Yes	Yes	Yes	Yes	Yes
Dilation	**NO**	**NO**	Yes	Yes	**NO**	Yes

This table of properties can help us to identify transformations.

Examples: A reflection is the only transformation that would result in an image having an opposite orientation. Also, a dilation is the only transformation that could result in a larger or smaller image that is not congruent to the pre-image.

Any of the **rigid motions** – **translation**, **reflection**, and **rotation** – will *always* preserve **distance** (the *size* of a figure) and **angles** (the *shape* of a figure). The image after any rigid motion is always congruent to the pre-image. Another word for a rigid motion is an *isometry*.

Generally, **dilations** do *not* preserve distance, unless we consider the special case of D_1. With the exception of D_1, the image after a dilation is similar, but not congruent, to the pre-image. In the case of D_1, the scale factor is 1, so the image will be unchanged and therefore congruent.

Remember that **orientation** describes the order in which the vertices of a polygon are read. Only after a reflection will a polygon and its image have opposite orientations.

Example: The vertices of $\triangle ABC$ below are read $A \rightarrow B \rightarrow C$ in a counterclockwise order. However, after a reflection, the vertices of $\triangle A'B'C'$ are read $A' \rightarrow B' \rightarrow C'$ in a clockwise order.

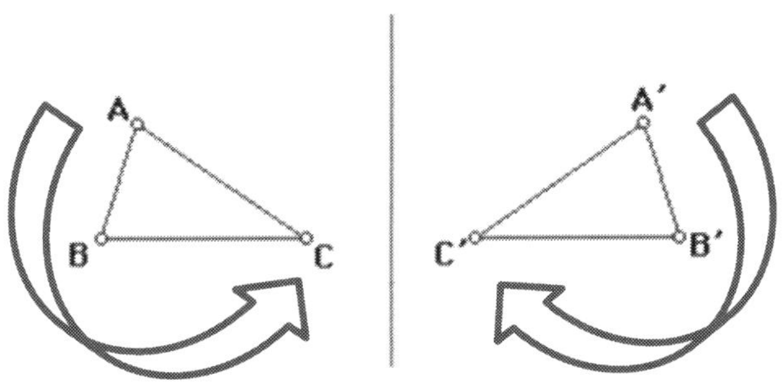

You may not be able to distinguish between a translation and a rotation from the table. However, one important difference is that **translations preserve direction** (or *slope*), while rotations do not (except, of course, for a 360° rotation).

Example: In the translation of the right triangle (*below left*), the shorter leg remains horizontal, the longer leg remains vertical, and the hypotenuse keeps its negative slope. However, in the 270° rotation of the same right triangle (*below right*), the slopes of all three sides of the triangle have changed.

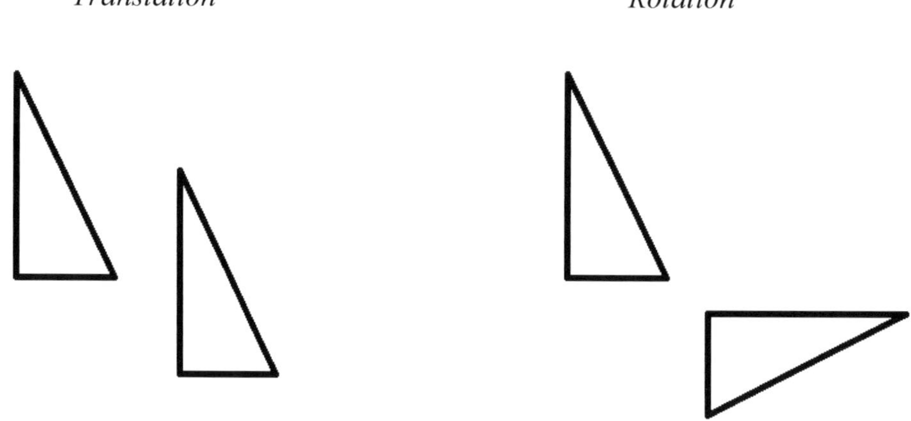

Model Problem:

In $\triangle KLM$, $m\angle K = 36$ and $KM = 5$. The transformation D_2 is performed on $\triangle KLM$ to form $\triangle K'L'M'$. Find $m\angle K'$ and the length of $\overline{K'M'}$.

Solution:

(A) (B)

$m\angle K' = 36$ and $K'M' = 10$

Explanation:

(A) Dilations preserve angle measures *[$\angle K'$ has the same measure as $\angle K$].*

(B) Dilations do not preserve distance. In fact, each side or a polygon will be enlarged or reduced based on the scale factor *[a scale factor of 2 doubles the length of $\overline{KM}$].*

Practice Problems

1. Identify each transformation as a translation, reflection, rotation, or dilation.

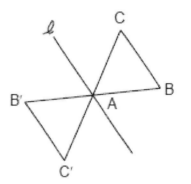

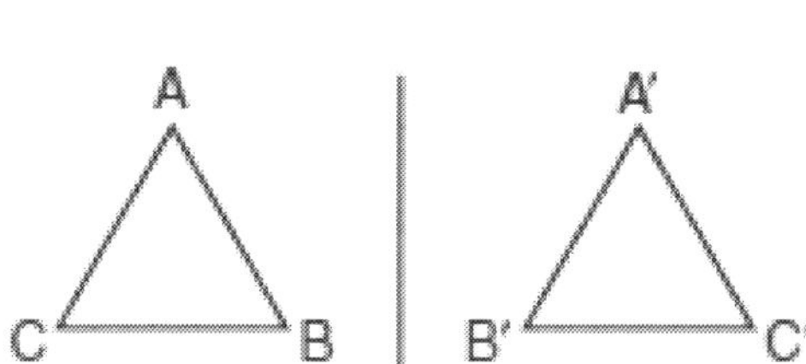

(a) ______________________________ (b) ______________________________

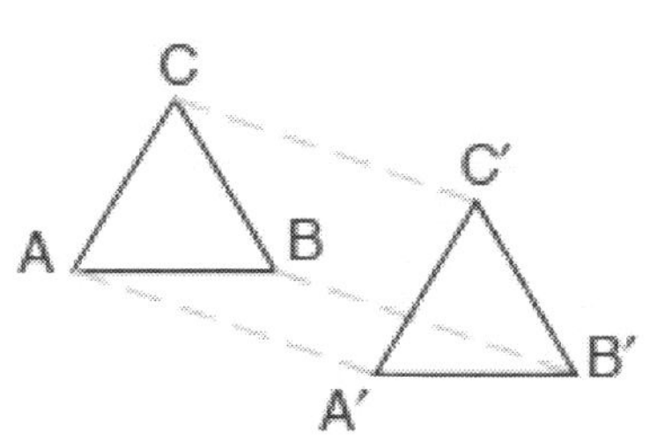

(c) ______________________________ (d) ______________________________

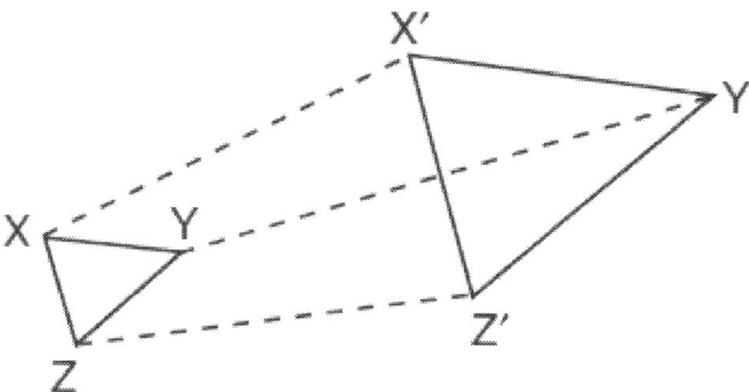

(e) ______________________________

REGENTS QUESTIONS

Multiple Choice

1. The perimeter of $\triangle A'B'C'$, the image of $\triangle ABC$, is twice as large as the perimeter of $\triangle ABC$. What type of transformation has taken place?
   (1) dilation
   (2) translation
   (3) rotation
   (4) reflection

2. Which expression best describes the transformation shown in the diagram below?

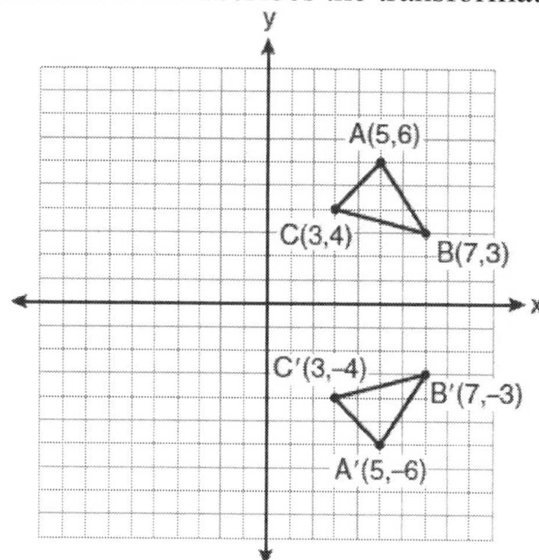

   (1) same orientation; reflection
   (2) opposite orientation; reflection
   (3) same orientation; translation
   (4) opposite orientation; translation

3. The diagram below shows $\overline{AB}$ and $\overline{DE}$.

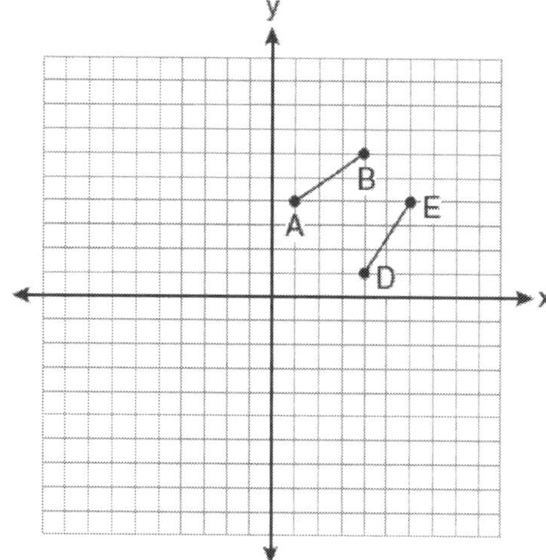

   Which transformation will move $\overline{AB}$ onto $\overline{DE}$ such that point D is the image of point A and point E is the image of point B?
   (1) $T_{3,-3}$
   (2) $D_{\frac{1}{2}}$
   (3) R_{90°
   (4) $r_{y=x}$

4. Triangle ABC is graphed on the set of axes below.

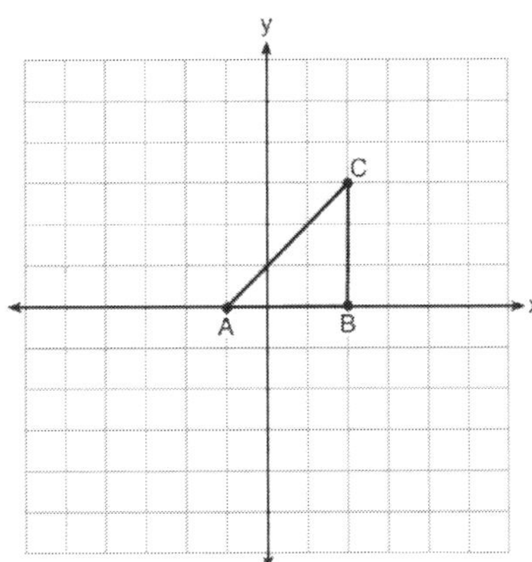

Which transformation produces an image that is similar to, but *not* congruent to, $\triangle ABC$?

(1) $T_{2,3}$ (3) $r_{y=x}$

(2) D_2 (4) R_{90°

5. In the diagram below, under which transformation is $\triangle A'B'C'$ the image of $\triangle ABC$?

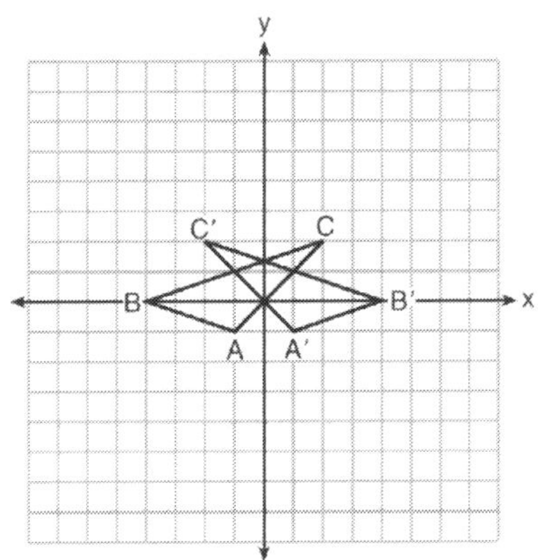

(1) D_2 (3) r_{y-axis}

(2) r_{x-axis} (4) $(x, y) \rightarrow (x-2, y)$

6. In the diagram below, under which transformation is $\triangle X'Y'Z'$ the image of $\triangle XYZ$?

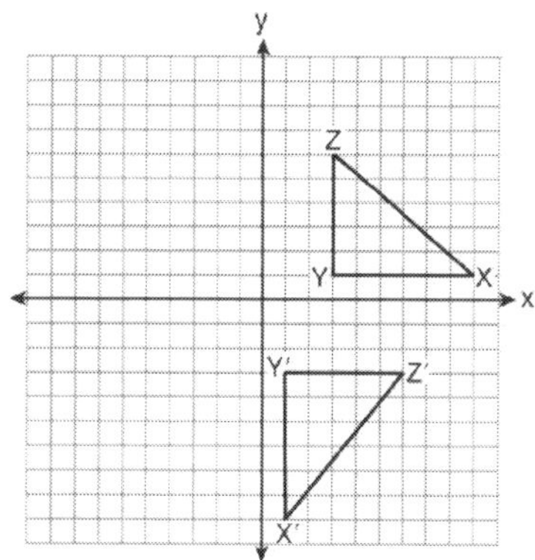

(1) dilation
(2) reflection
(3) rotation
(4) translation

7. As shown in the diagram below, when hexagon $ABCDEF$ is reflected over line m, the image is hexagon $A'B'C'D'E'F'$.

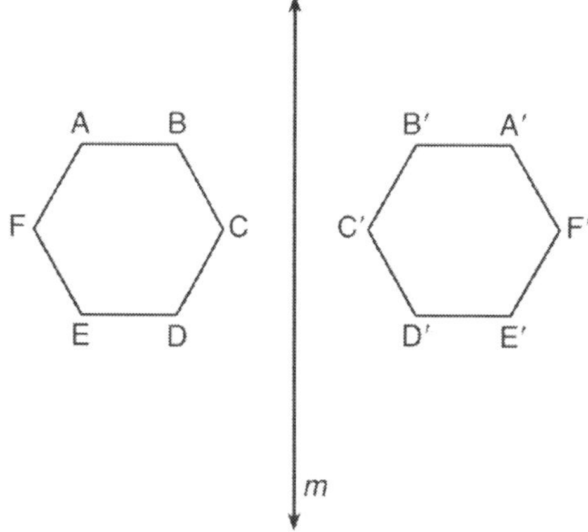

Under this transformation, which property is *not* preserved?
(1) area
(2) distance
(3) orientation
(4) angle measure

8. Quadrilateral $ABCD$ undergoes a transformation, producing quadrilateral $A'B'C'D'$. For which transformation would the area of $A'B'C'D'$ *not* be equal to the area of $ABCD$?
(1) a rotation of 90° about the origin
(2) a reflection over the y-axis
(3) a dilation by a scale factor of 2
(4) a translation defined by $(x, y) \rightarrow (x+4, y-1)$

9. Triangle *JTM* is shown on the graph below.

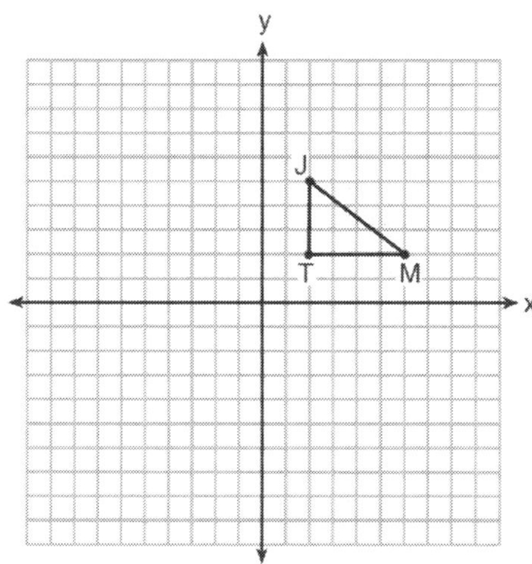

Which transformation would result in an image that is *not* congruent to $\triangle JTM$?

(1) $r_{y=x}$ (3) $T_{0,-3}$

(2) R_{90° (4) D_2

10. CC In the diagram below, which single transformation was used to map triangle *A* onto triangle *B*?

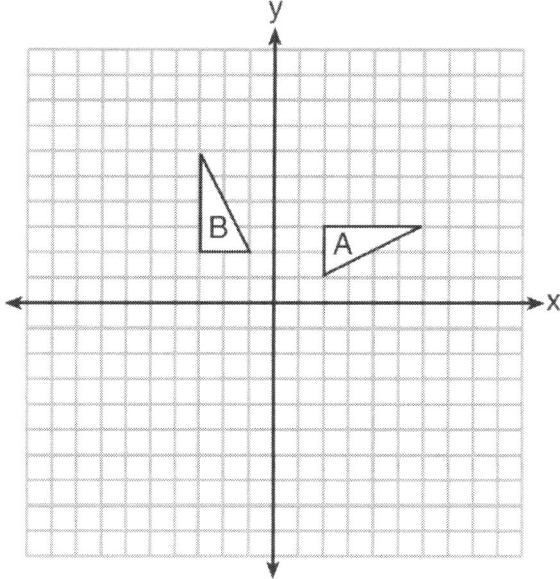

(1) line reflection (3) dilation

(2) rotation (4) translation

11. CC Which transformation would result in the perimeter of a triangle being different from the perimeter of its image?

(1) $(x,y) \rightarrow (y,x)$ (3) $(x,y) \rightarrow (4x,4y)$

(2) $(x,y) \rightarrow (x,-y)$ (4) $(x,y) \rightarrow (x+2,y-5)$

12. (CC) Which transformation of $\overline{OA}$ would result in an image parallel to $\overline{OA}$?

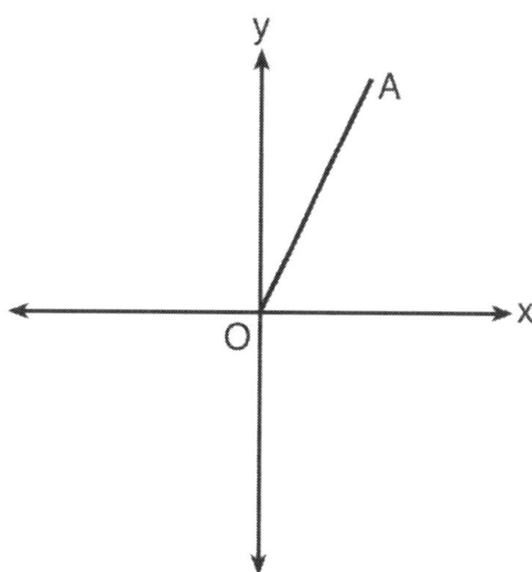

(1) a translation of two units down
(2) a reflection over the x-axis
(3) a reflection over the y-axis
(4) a clockwise rotation of 90° about the origin

13. (CC) Which transformation would *not* always produce an image that would be congruent to the original figure?
(1) translation
(2) dilation
(3) rotation
(4) reflection

12.2 Sequences of Transformations

Key Terms and Concepts

Up to now, we've considered the result of a single transformation on an object. In this section, we'll look at how a **sequence of transformations** can be applied.

We've used a single prime symbol, as in P', to represent the image of a point after a single transformation. When we have a sequence of transformations, we can use a double or triple prime symbol, and so on, for the image after each additional step.
Example: $P \to P' \to P'' \to P'''$

The composition of a translation and a reflection is sometimes called a *glide reflection.*
Example: In the graph below, a translation of $T_{-5,0}$ maps $\triangle QRS \to \triangle Q'R'S'$ and then a reflection of r_{x-axis} maps $\triangle Q'R'S' \to \triangle Q''R''S''$.

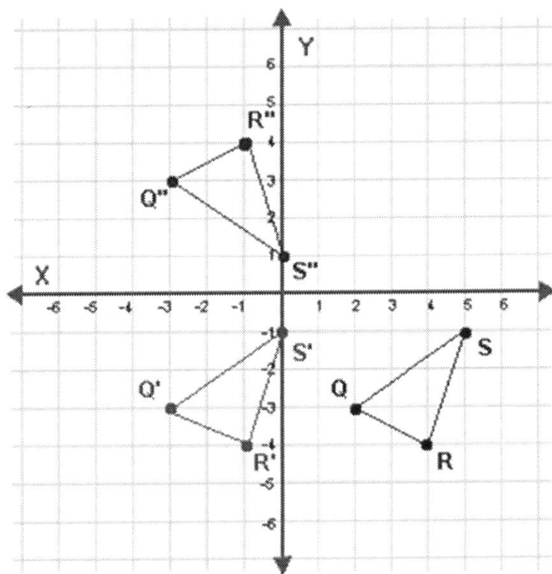

The order of the sequence of transformations is important.
Example: Consider the graphic of the potted plant below, which starts upright at the origin. The diagram on the left shows the result of a 45° rotation around the origin followed by a translation to the right. The diagram on the right shows a translation to the right followed by a 45° rotation around the origin.

Model Problem:

Reflect the triangle over the line $x=-1$ and then rotate it 90° with the center at the origin.

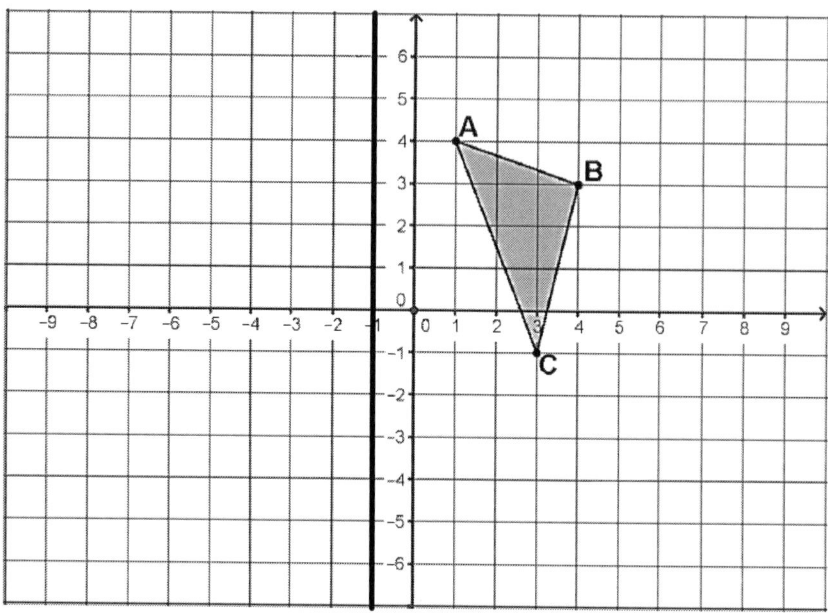

Solution:

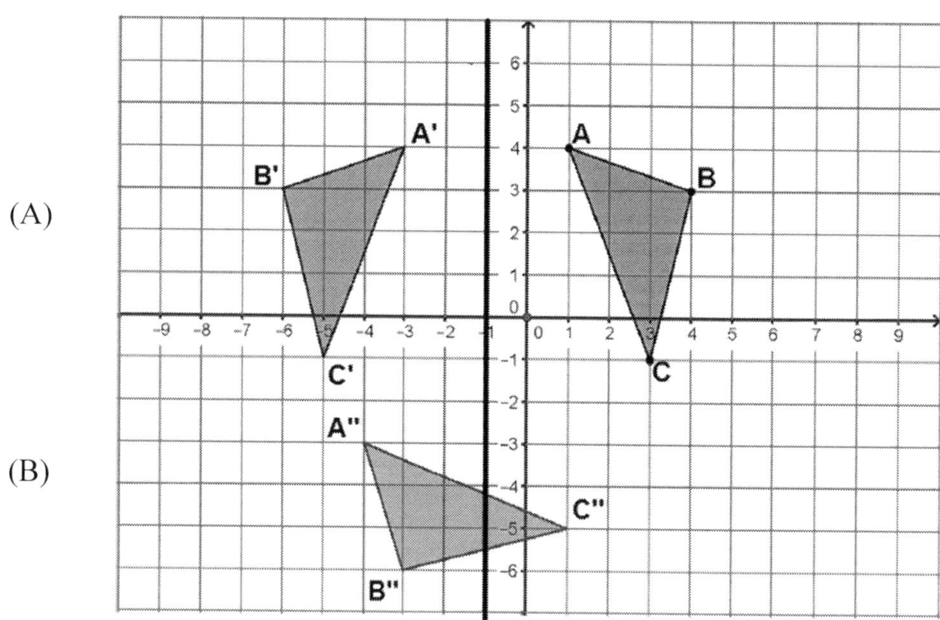

Explanation of steps:

(A) Determine the vertices after the reflection by using the line of reflection as the perpendicular bisector of $\overline{AA'}$, $\overline{BB'}$, and $\overline{CC'}$.

(B) For the rotation, use the rule $R_{90^\circ}:(x,y)\rightarrow(-y,x)$.
[For example, $A'(-3,4)\rightarrow A''(-4,-3)$.]

Practice Problems

1. Describe the sequence of transformations applied in the graph below.

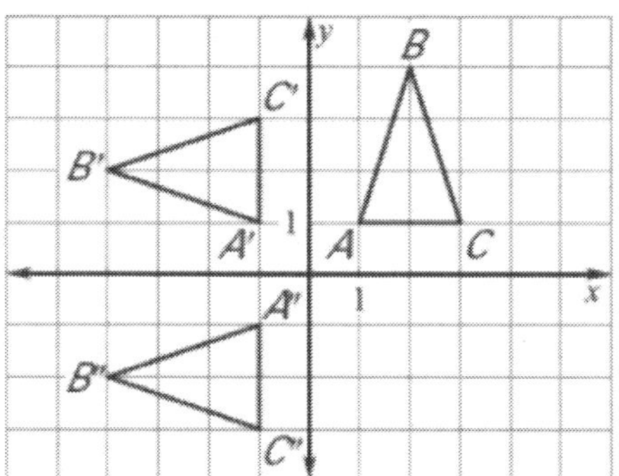

2. Describe the sequence of transformations applied in the graph below.

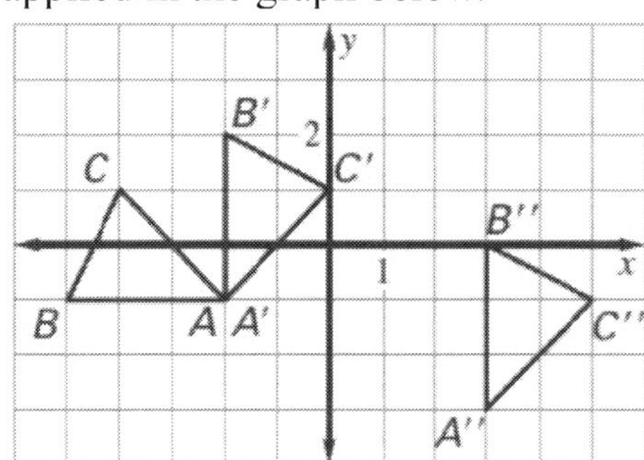

3. Describe the sequence of transformations applied in the graph below.

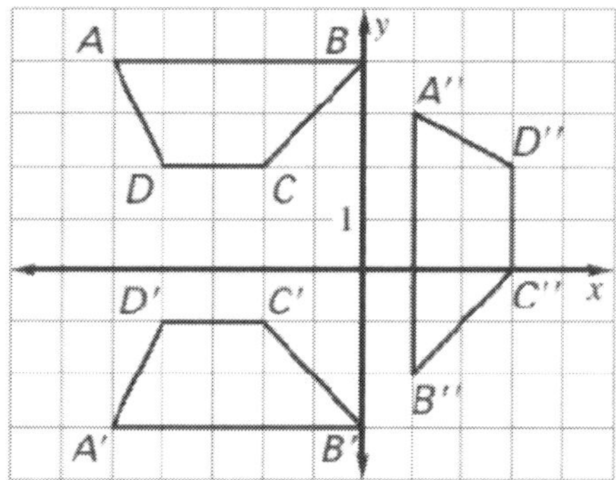

4. Graph the image of $\triangle GLQ$ after r_{y-axis} followed by $T_{4,-4}$.

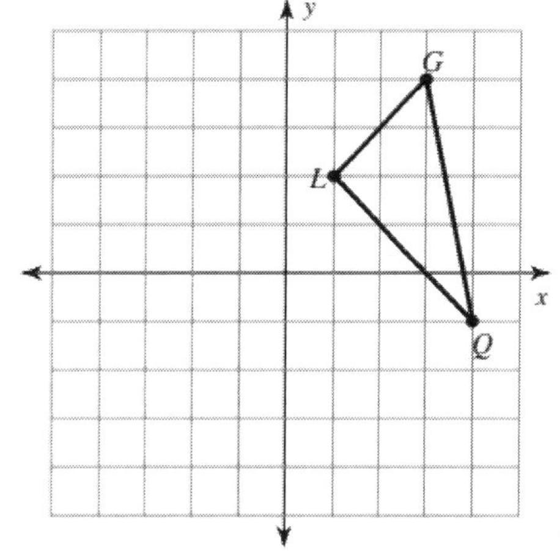

5. Graph the image of $\triangle LUX$ after $r_{x=1}$ followed by $D_{\frac{2}{5}}$.

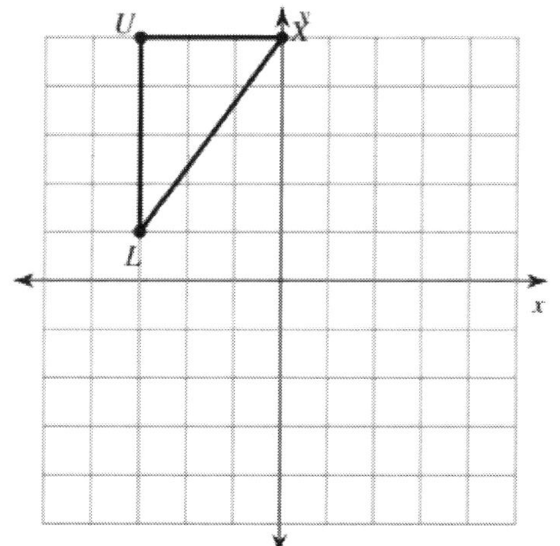

6. Graph the image of MPDZ after $T_{-3,0}$ followed by R_{180°.

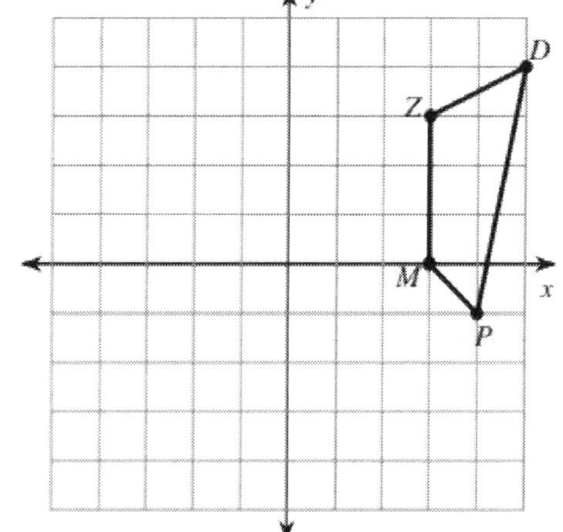

REGENTS QUESTIONS

Multiple Choice

1. The point (3,–2) is rotated 90° about the origin and then dilated by a scale factor of 4. What are the coordinates of the resulting image?
   (1) (–12,8) (3) (8,12)
   (2) (12,–8) (4) (–8,–12)

2. In the diagram below, $\triangle A'B'C'$ is a transformation of $\triangle ABC$, and $\triangle A''B''C''$ is a transformation of $\triangle A'B'C'$.

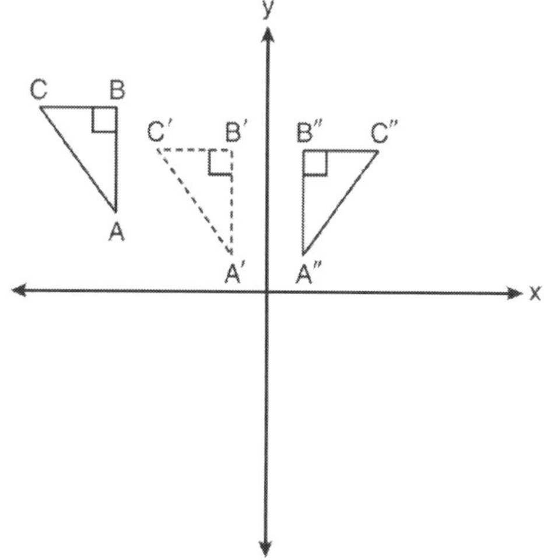

The composite transformation of $\triangle ABC$ to $\triangle A''B''C''$ is an example of a
   (1) reflection followed by a rotation
   (2) reflection followed by a translation
   (3) translation followed by a rotation
   (4) translation followed by a reflection

3. (CC) In the diagram below, congruent figures 1, 2, and 3 are drawn.

   Which sequence of transformations maps figure 1 onto figure 2 and then figure 2 onto figure 3?
   (1) a reflection followed by a translation
   (2) a rotation followed by a translation
   (3) a translation followed by a reflection
   (4) a translation followed by a rotation

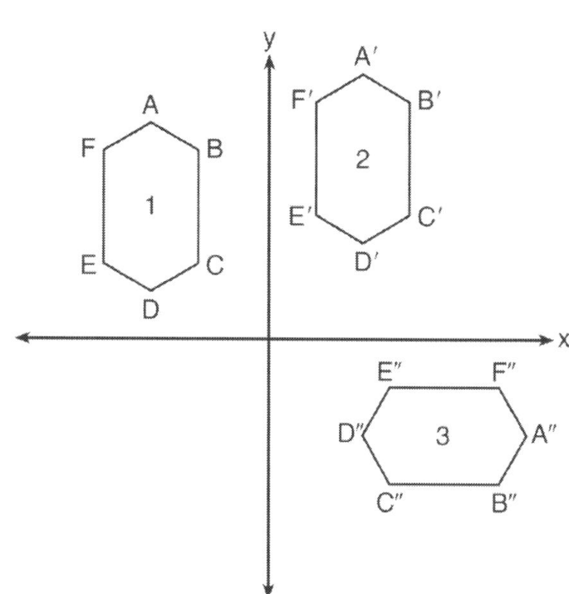

4. (CC) A sequence of transformations maps rectangle $ABCD$ onto rectangle $A''B''C''D''$, as shown in the diagram below.

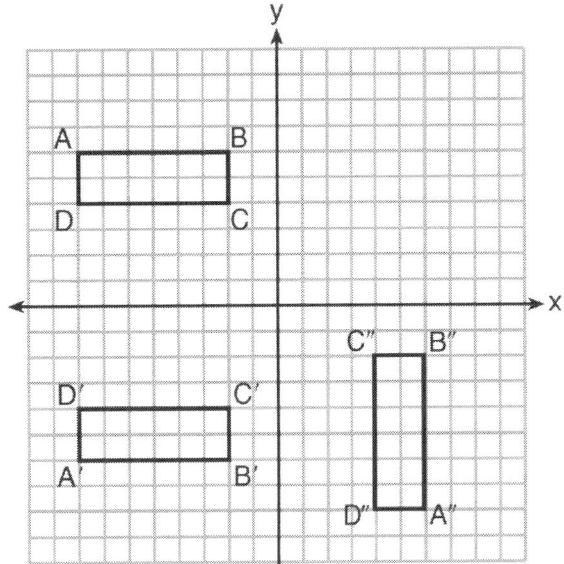

Which sequence of transformations maps $ABCD$ onto $A'B'C'D'$ and then maps $A'B'C'D'$ onto $A''B''C''D''$?

(1) a reflection followed by a rotation
(2) a reflection followed by a translation
(3) a translation followed by a rotation
(4) a translation followed by a reflection

5. (CC) Triangle ABC and triangle DEF are graphed on the set of axes below.

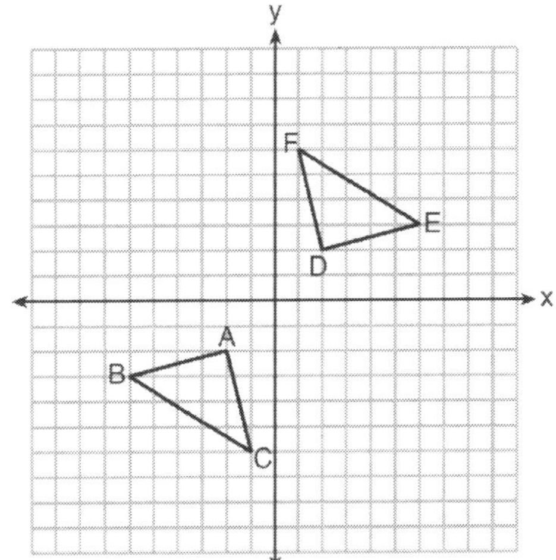

Which sequence of transformations maps triangle ABC onto triangle DEF?

(1) a reflection over the x-axis followed by a reflection over the y-axis
(2) a 180° rotation about the origin followed by a reflection over the line $y = x$
(3) a 90° clockwise rotation about the origin followed by a reflection over the y-axis
(4) a translation 8 units to the right and 1 unit up followed by a 90° counterclockwise rotation about the origin

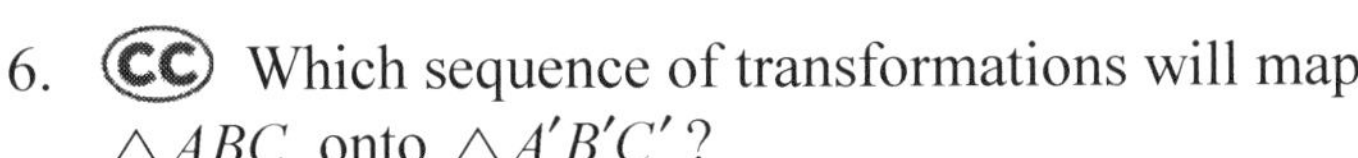

6. (CC) Which sequence of transformations will map $\triangle ABC$ onto $\triangle A'B'C'$?

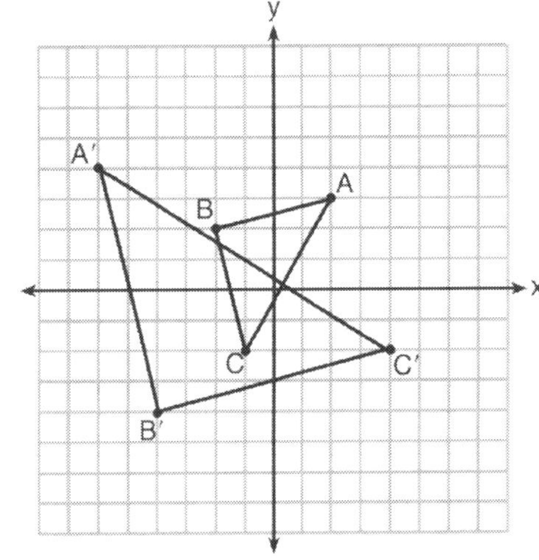

(1) reflection and translation
(2) rotation and reflection
(3) translation and dilation
(4) dilation and rotation

7. (CC) Identify which sequence of transformations could map pentagon $ABCDE$ onto pentagon $A''B''C''D''E''$, as shown below.

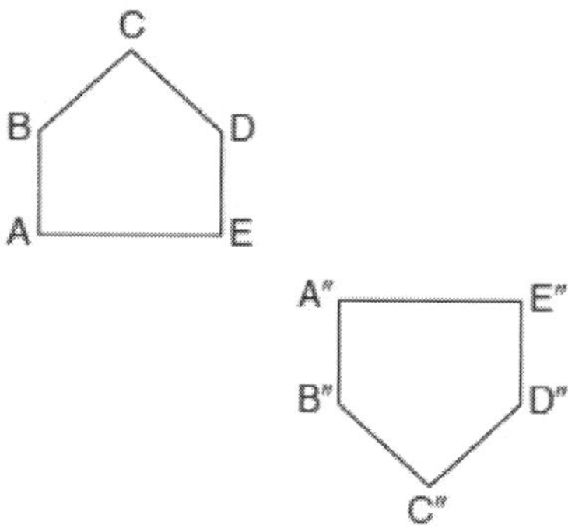

(1) dilation followed by a rotation
(2) translation followed by a rotation
(3) line reflection followed by a translation
(4) line reflection followed by a line reflection

Constructed Response

8. The coordinates of the endpoints of $\overline{AB}$ are $A(2,6)$ and $B(4,2)$. Is the image $\overline{A''B''}$ the same if it is reflected in the x-axis, then dilated by $\frac{1}{2}$ as the image is if it is dilated by $\frac{1}{2}$, then reflected in the x-axis? Justify your answer. (The use of the accompanying grid is optional.)

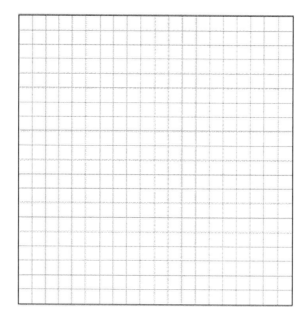

9. Given $\triangle ABC$ with points A(4,3), B(4,–2), and C(2,3). On the grid below, sketch $\triangle ABC$. On the same set of axes, graph and state the coordinates of $\triangle A'B'C'$, the image of $\triangle ABC$ after a reflection in the line $y = x$. On the same set of axes, graph and state the coordinates of $\triangle A''B''C''$, the image of $\triangle A'B'C'$ after the translation $T_{-4,3}$.

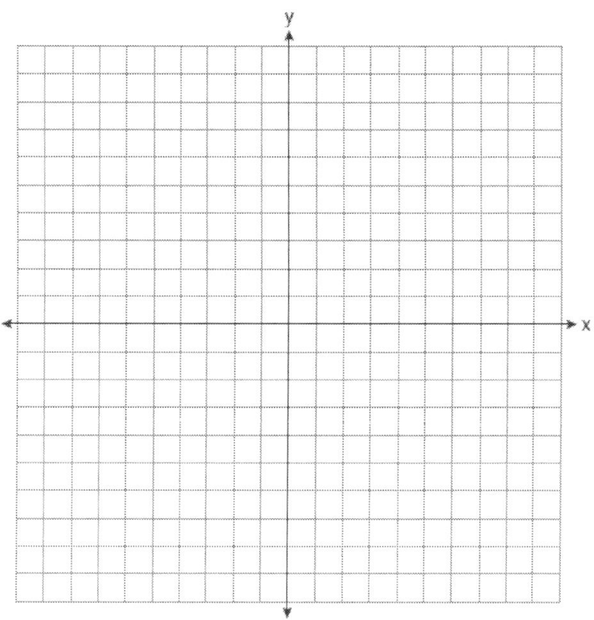

10. The coordinates of $\triangle ABC$, shown on the graph below, are A(2,5), B(5,7), and C(4,1). Graph and label $\triangle A'B'C'$, the image of $\triangle ABC$ after it is reflected over the y-axis.

Graph and label $\triangle A''B''C''$, the image of $\triangle A'B'C'$ after it is reflected over the x-axis.

State a single transformation that will map $\triangle ABC$ onto $\triangle A''B''C''$.

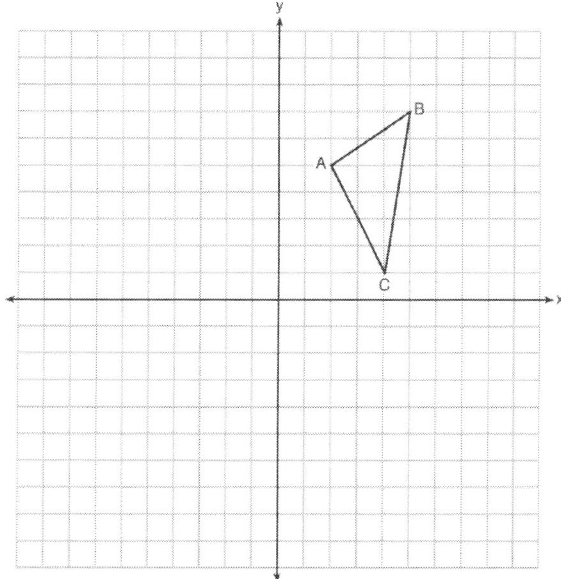

11. CC Describe a sequence of transformations that will map $\triangle ABC$ onto $\triangle DEF$ as shown below.

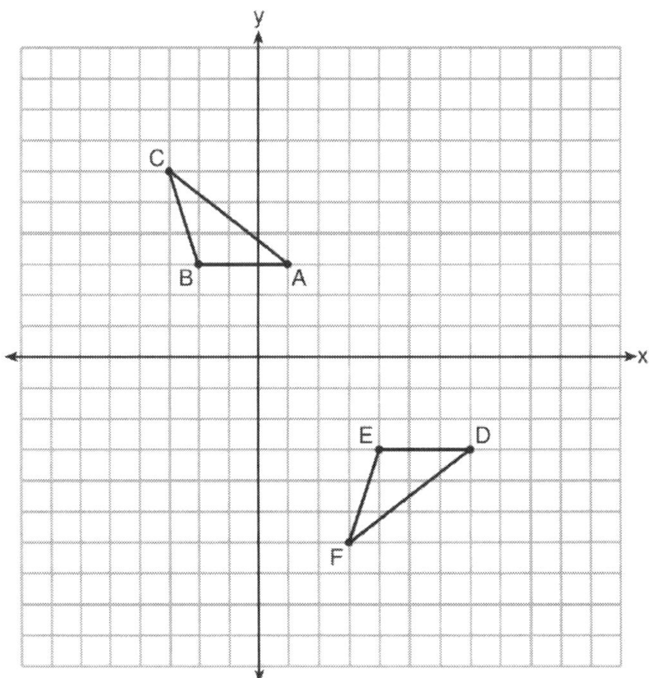

12. CC In the diagram below, $\triangle ABC$ has coordinates A(1,1), B(4,1), and C(4,5). Graph and label $\triangle A''B''C''$, the image of $\triangle ABC$ after the translation five units to the right and two units up followed by the reflection over the line $y = 0$.

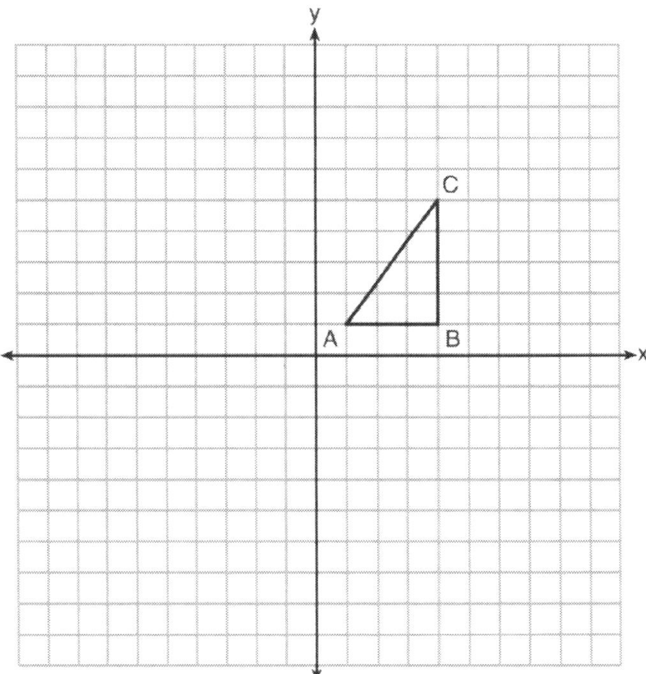

13. CC The graph below shows $\triangle ABC$ and its image, $\triangle A''B''C''$.

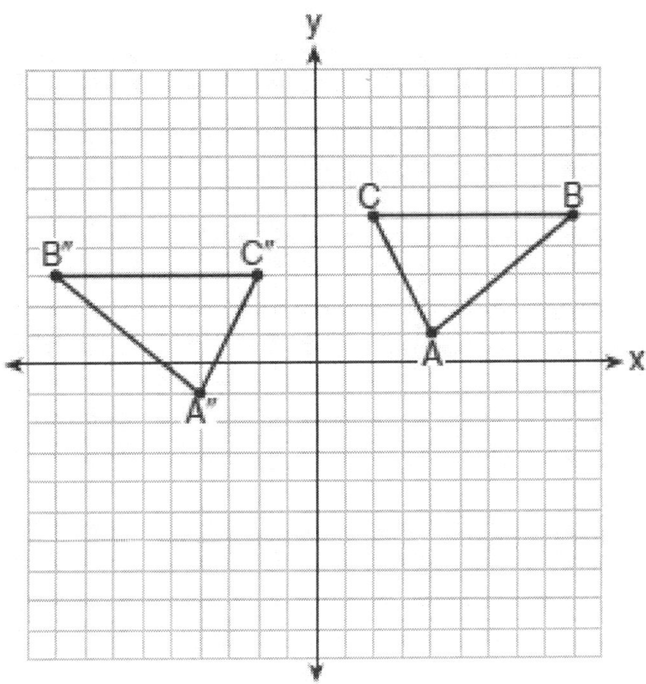

Describe a sequence of rigid motions which would map $\triangle ABC$ onto $\triangle A''B''C''$.

12.3 Transformations and Congruence

Key Terms and Concepts

Since **rigid motions preserve congruence**, we can prove two polygons on a coordinate plane are congruent by showing that one of the polygons is the image of the other after any **sequence of rigid motions**. It is important to remember that dilation is *not* a rigid motion.

Example: We know that triangle 2 is congruent to triangle 1 because triangle 2 is the image of triangle 1 after $T_{0,3}$ followed by r_{y-axis}.

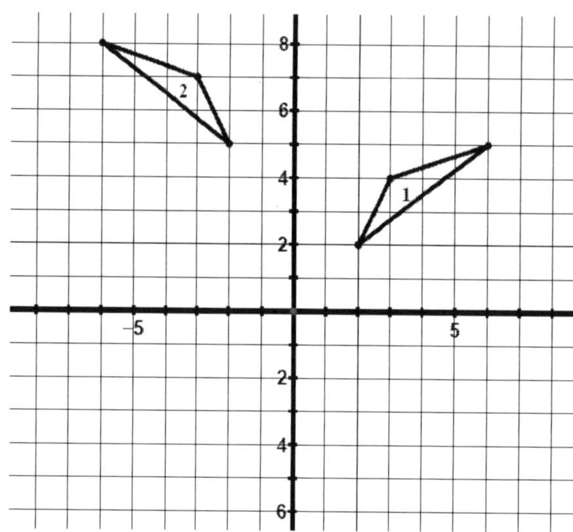

Model Problem:

In the diagram below, $\triangle ABC \cong \triangle DEF$. Describe a sequence of rigid motions that will map $\triangle ABC$ onto $\triangle DEF$.

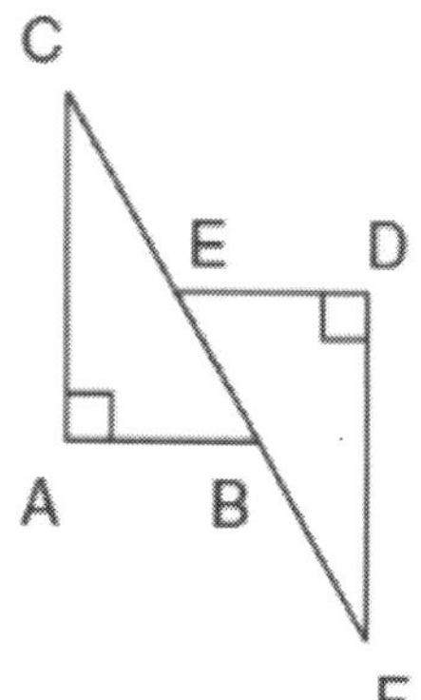

Solution:
Translate $\triangle ABC$ so that B coincides with E.
Then rotate $\triangle ABC$ 180° around B.

Explanation:
See diagram at right.

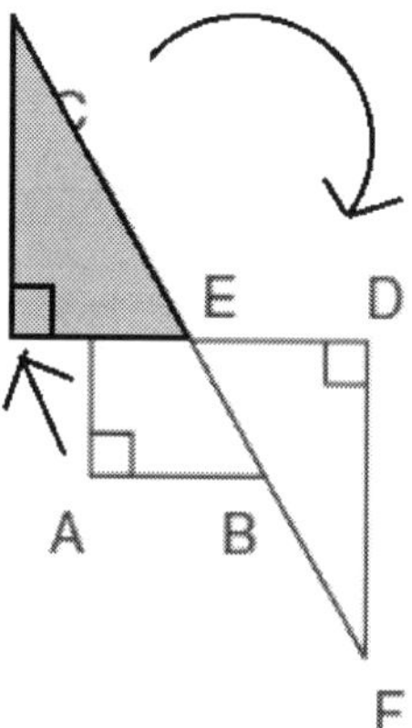

Practice Problems

1. Given $\triangle ABC \cong \triangle DEF$ and $\overline{ADBE}$. Describe a rigid motion that maps $\triangle ABC$ onto $\triangle DEF$.

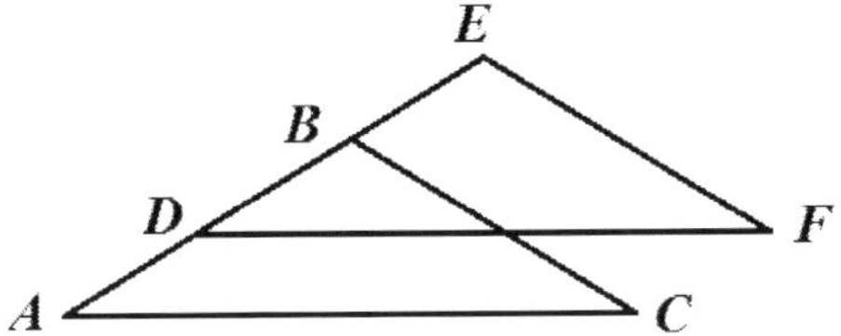

2. Given $\triangle ABC \cong \triangle ADC$. Describe a rigid motion that maps $\triangle ABC$ onto $\triangle ADC$.

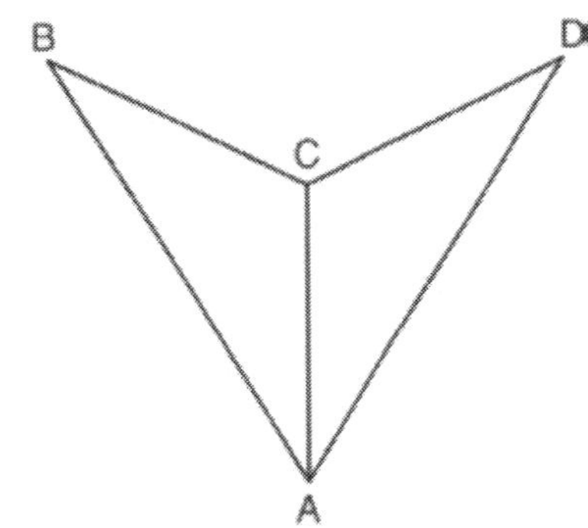

3. Given $\overline{HK}$ and $\overline{IL}$, $\triangle HIJ \cong \triangle KLJ$. Describe a rigid motion that maps $\triangle HIJ$ onto $\triangle KLJ$.

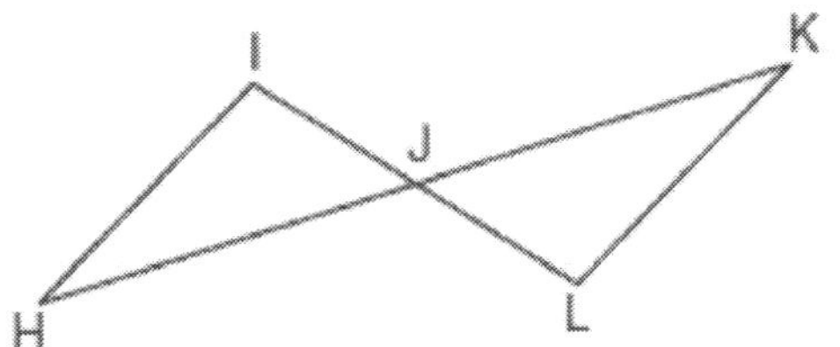

4. Given $\triangle HIK \cong \triangle KJH$. Describe a rigid motion that maps $\triangle HIK$ onto $\triangle KJH$.

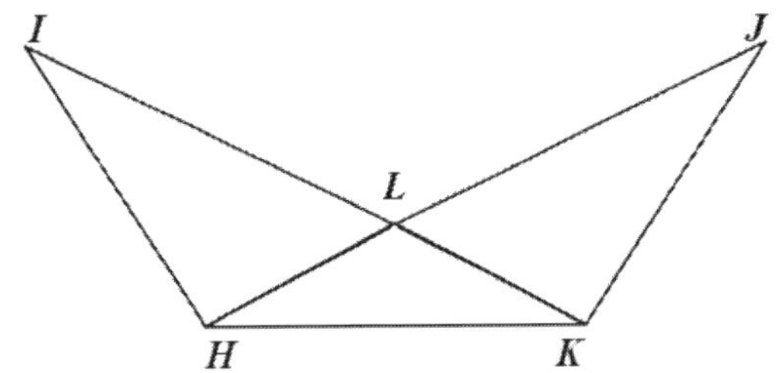

5. Given $\triangle PRQ \cong \triangle ABC$. Describe a sequence of rigid motions that maps $\triangle PRQ$ to $\triangle ABC$. Is it possible to show the congruence using only translations and rotations?

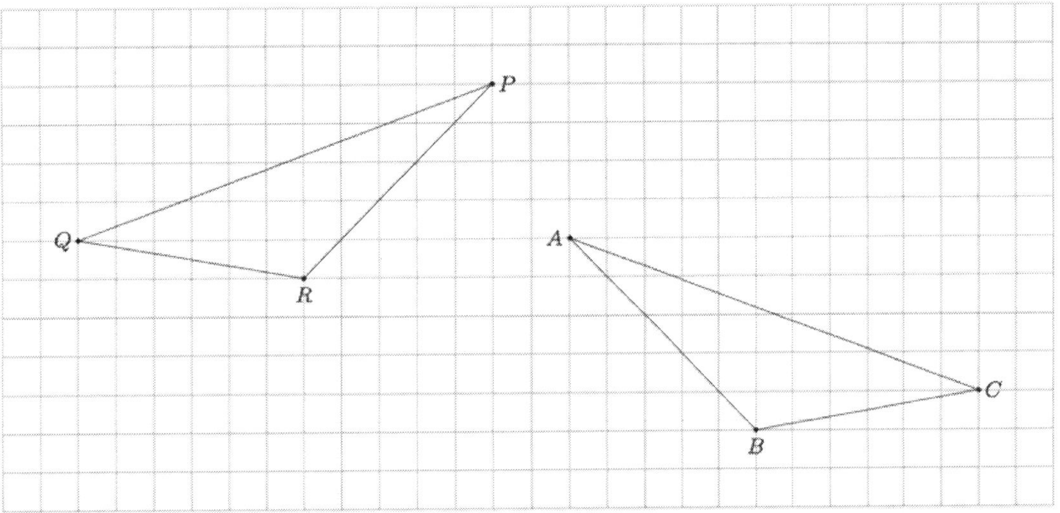

REGENTS QUESTIONS

Multiple Choice

1. (CC) Which statement is sufficient evidence that $\triangle DEF$ is congruent to $\triangle ABC$?

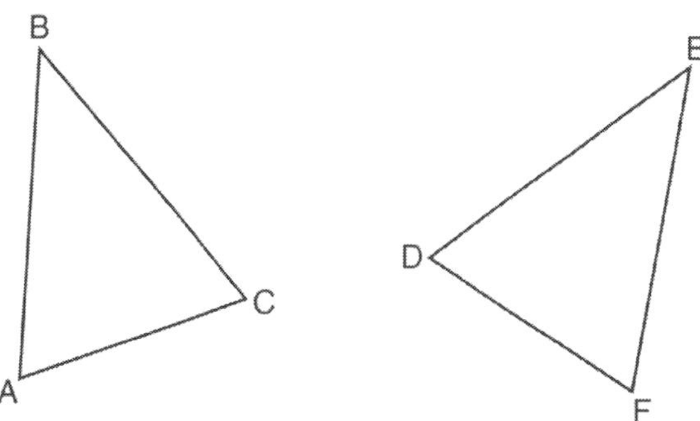

(1) $AB = DE$ and $BC = EF$
(2) $\angle D \cong \angle A$, $\angle B \cong \angle E$, $\angle C \cong \angle F$
(3) There is a sequence of rigid motions that maps $\overline{AB}$ onto $\overline{DE}$, $\overline{BC}$ onto $\overline{EF}$, and $\overline{AC}$ onto $\overline{DF}$.
(4) There is a sequence of rigid motions that maps point A onto point D, $\overline{AB}$ onto $\overline{DE}$, and $\angle B$ onto $\angle E$.

2. (CC) On the set of axes below, rectangle $ABCD$ can be proven congruent to rectangle $KLMN$ using which transformation?

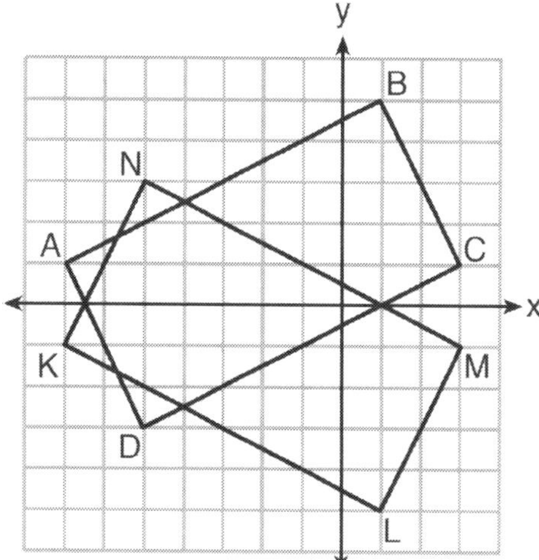

(1) rotation
(2) translation
(3) reflection over the x-axis
(4) reflection over the y-axis

3. (CC) Under which transformation would $\triangle A'B'C'$, the image of $\triangle ABC$, *not* be congruent to $\triangle ABC$?
   (1) reflection over the y-axis
   (2) rotation of 90° clockwise about the origin
   (3) translation of 3 units right and 2 units down
   (4) dilation with a scale factor of 2 centered at the origin

Constructed Response

4. (CC) Given right triangles ABC and DEF where $\angle C$ and $\angle F$ are right angles, $\overline{AC} \cong \overline{DF}$ and $\overline{CB} \cong \overline{FE}$. Describe a precise sequence of rigid motions which would show $\triangle ABC \cong \triangle DEF$.

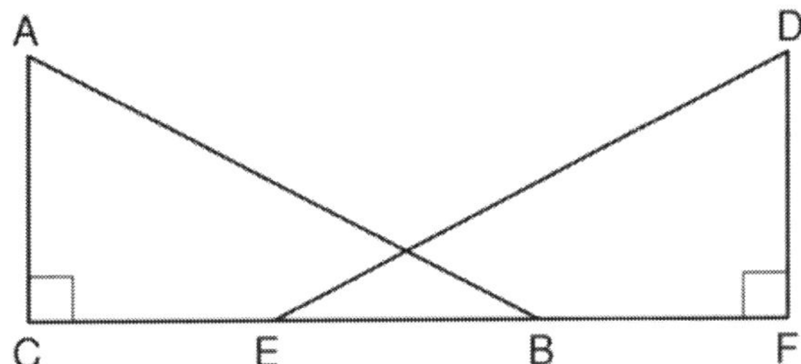

5. (CC) In the diagram of $\triangle LAC$ and $\triangle DNC$ below, $\overline{LA} \cong \overline{DN}$, $\overline{CA} \cong \overline{CN}$, and $\overline{DAC} \perp \overline{LCN}$.

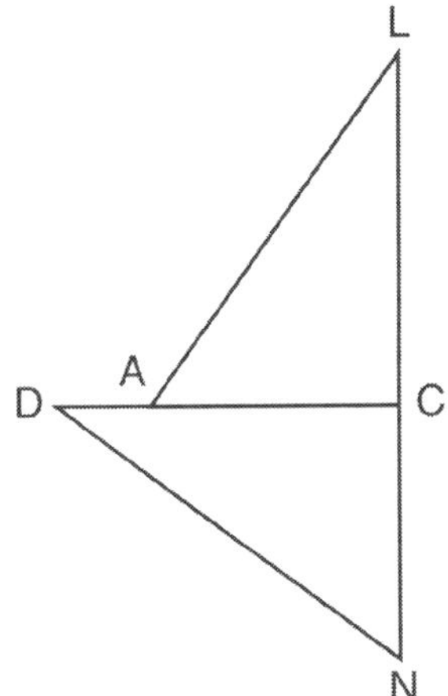

   a) Prove that $\triangle LAC \cong \triangle DNC$.
   b) Describe a sequence of rigid motions that will map $\triangle LAC$ onto $\triangle DNC$.

6. (CC) After a reflection over a line, $\triangle A'B'C'$ is the image of $\triangle ABC$. Explain why triangle ABC is congruent to triangle $A'B'C'$.

7. (CC) Given: Quadrilateral $ABCD$ is a parallelogram with diagonals $\overline{AC}$ and $\overline{BD}$ intersecting at E

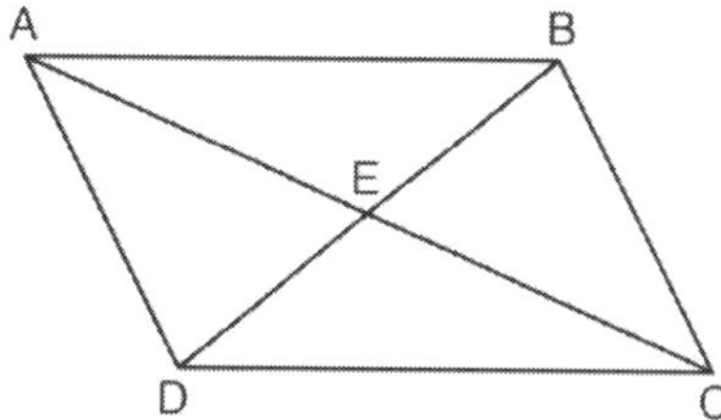

Prove: $\triangle AED \cong \triangle CEB$

Describe a single rigid motion that maps $\triangle AED$ onto $\triangle CEB$.

8. (CC) In the diagram below, $\triangle ABC$ and $\triangle XYZ$ are graphed.

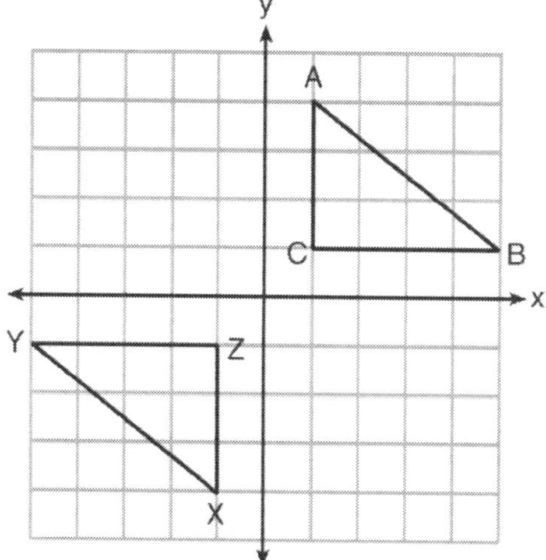

Use the properties of rigid motions to explain why $\triangle ABC \cong \triangle XYZ$.

9. (CC) In the diagram below, $\overline{AC} \cong \overline{DF}$ and points A, C, D, and F are collinear on line ℓ.

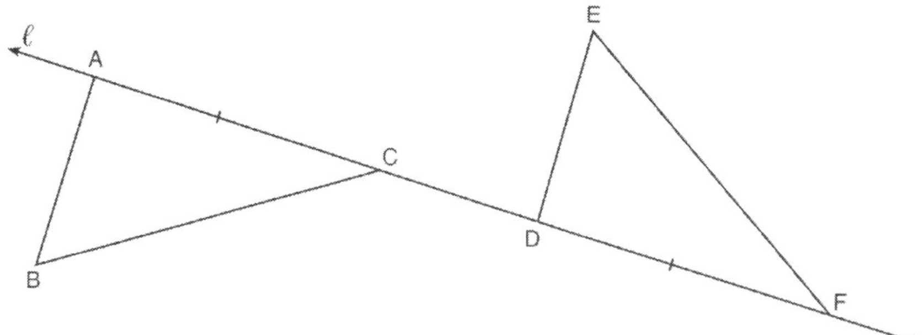

Let $\triangle D'E'F'$ be the image of $\triangle DEF$ after a translation along ℓ, such that point D is mapped onto point A. Determine and state the location of F'. Explain your answer.

Let $\triangle D''E''F''$ be the image of $\triangle D'E'F'$ after a reflection across line ℓ. Suppose that E'' is located at B. Is $\triangle DEF$ congruent to $\triangle ABC$? Explain your answer.

10. (CC) As graphed on the set of axes below, $\triangle A'B'C'$ is the image of $\triangle ABC$ after a sequence of transformations.

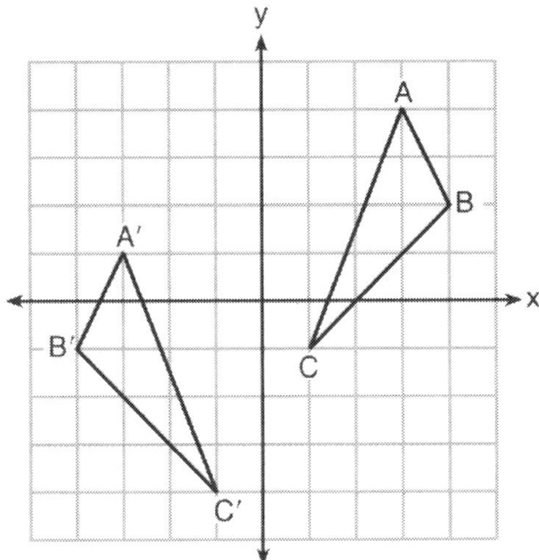

Is $\triangle A'B'C'$ congruent to $\triangle ABC$? Use the properties of rigid motion to explain your answer.

11. (CC) The grid below shows $\triangle ABC$ and $\triangle DEF$.

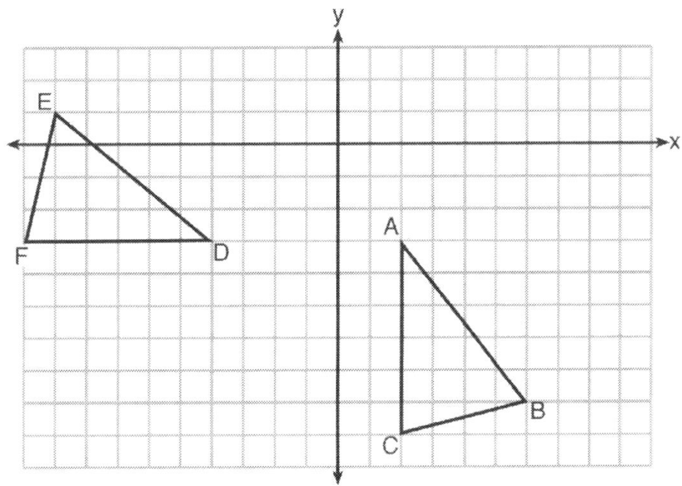

Let $\triangle A'B'C'$ be the image of $\triangle ABC$ after a rotation about point A. Determine and state the location of B' if the location of point C' is $(8,-3)$. Explain your answer.

Is $\triangle DEF$ congruent to $\triangle A'B'C'$? Explain your answer.

12.4 Transformations and Similarity

Key Terms and Concepts

The three rigid motions all preserve congruency, so they also preserve similarity. **Dilations do not preserve congruency, but they do preserve similarity**. Remember, two figures are similar if they have the same shape but not necessarily the same size.

Example: Below is a pair of similar shapes.

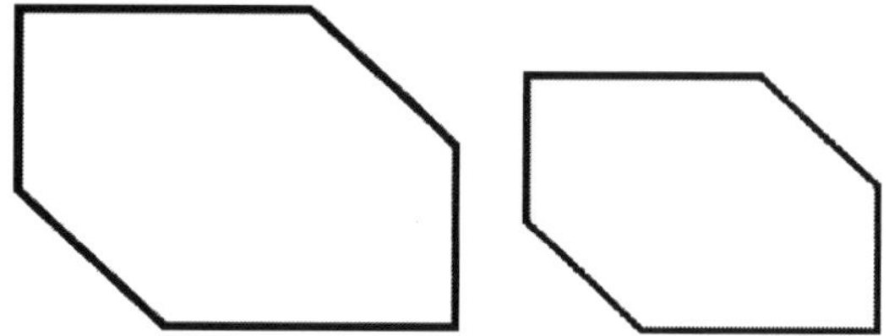

Any sequence of transformations that includes a dilation (other than D_1) results in an image that is *similar, but not congruent*, to the original figure.

Important Proofs

Informal proof that all circles are similar
All circles have the same shape, and we can prove they are similar by transformations.

Consider any two circles, A and B, shown to the right, where circle A has a radius of m and circle B has a radius of n and $n \geq m$.

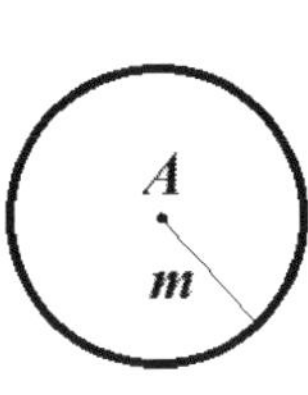

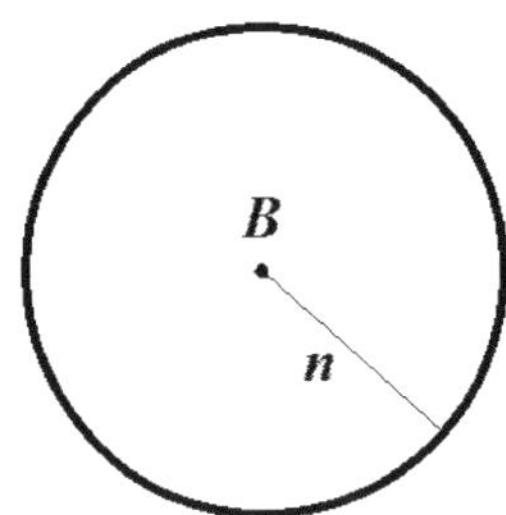

We can prove their similarity by showing a sequence of transformations that maps circle A to circle B.

First, translate circle A so that point A maps onto point B. The circles now have the same center.

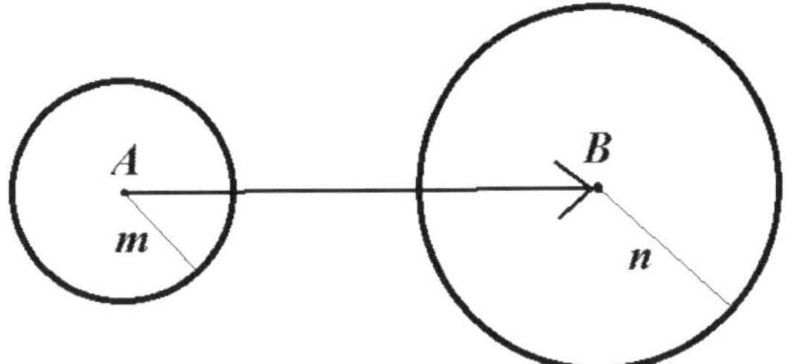

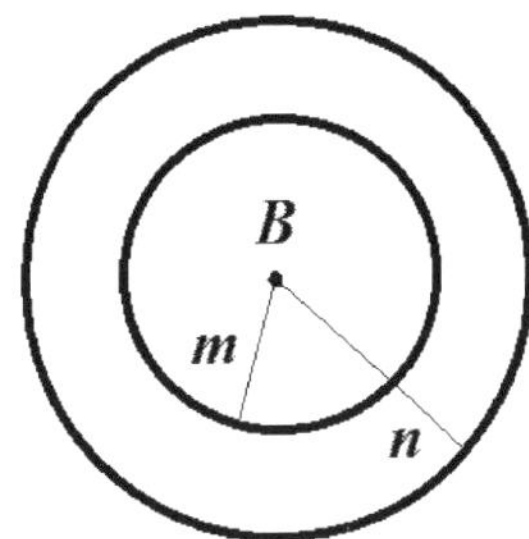

We can then dilate circle A by a scale factor of $\frac{n}{m}$ from its center. A radius of m, when scaled by a factor of $\frac{n}{m}$, will result in a radius of $m \cdot \frac{n}{m} = n$. Therefore, circle A maps onto circle B by a sequence of transformations – a translation and dilation – to prove that the circles are similar.

Model Problem:

Prove $\triangle LMN \sim \triangle STR$ by giving a sequence of transformations that maps $\triangle LMN$ onto $\triangle STR$.

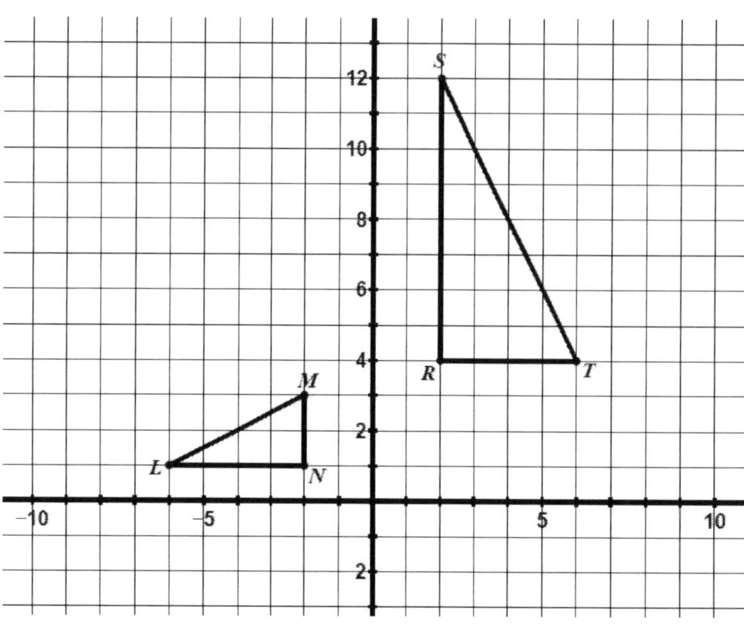

Solution:

(A) (B)

Rotation R_{270° followed by dilation D_2.

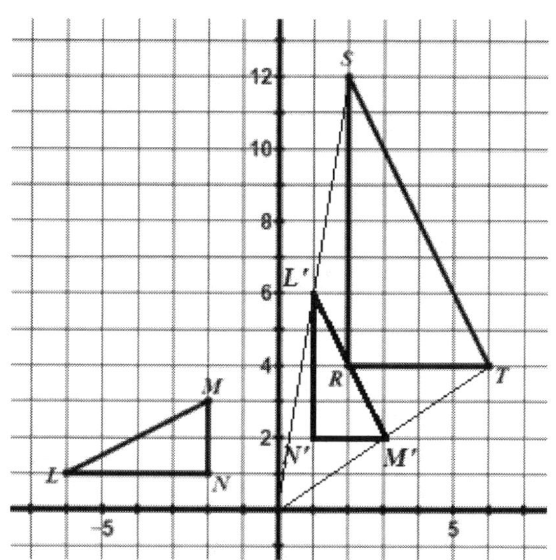

Explanation of steps:

(A) To align $\triangle LMN$ to the same direction as $\triangle STR$, we need to rotate it. *[We can either rotate it 90° clockwise or 270° counterclockwise to achieve the same result. Since $R_{270^\circ}:(x,y)\rightarrow(y,-x)$, this will map $N(-2,1)\rightarrow N'(1,2)$.]*

(B) Since $\triangle STR$ is larger than $\triangle LMN$, a dilation is needed. *[A dilation of scale factor 2 would map $N'(1,2)$ to R(2,4). This will also map L' onto S and M' onto T.]*

Practice Problems

1. Prove $\triangle ABC \sim \triangle A'B'C'$ by giving a sequence of transformations that maps $\triangle ABC$ onto $\triangle A'B'C'$. 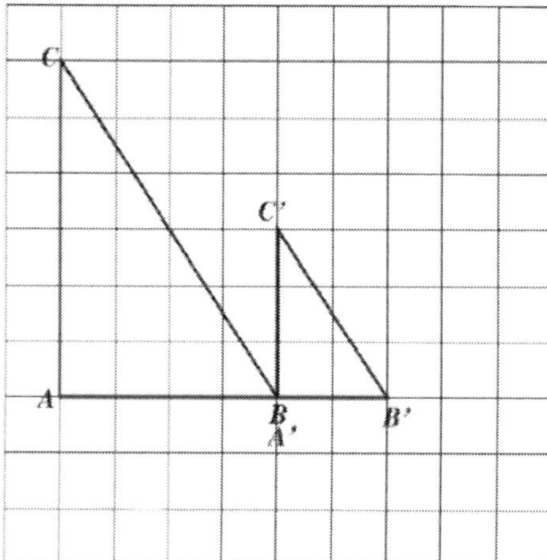	2. Prove $\triangle ABC \sim \triangle A'B'C'$ by giving a sequence of transformations that maps $\triangle ABC$ onto $\triangle A'B'C'$. 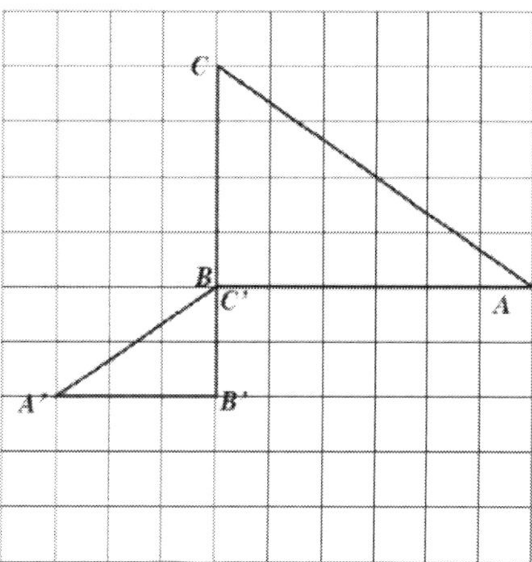

REGENTS QUESTIONS

Multiple Choice

1. (CC) The vertices of $\triangle JKL$ have coordinates J(5,1), K(–2,–3), and L(–4,1). Under which transformation is the image $\triangle J'K'L'$ *not* congruent to $\triangle JKL$?
   (1) a translation of two units to the right and two units down
   (2) a counterclockwise rotation of 180 degrees around the origin
   (3) a reflection over the x-axis
   (4) a dilation with a scale factor of 2 and centered at the origin

2. (CC) If $\triangle A'B'C'$ is the image of $\triangle ABC$, under which transformation will the triangles *not* be congruent?
   (1) reflection over the x-axis
   (2) translation to the left 5 and down 4
   (3) dilation centered at the origin with scale factor 2
   (4) rotation of 270° counterclockwise about the origin

3. (CC) In the diagram below, $\triangle DEF$ is the image of $\triangle ABC$ after a clockwise rotation of 180° and a dilation where $AB = 3$, $BC = 5.5$, $AC = 4.5$, $DE = 6$, $FD = 9$, and $EF = 11$.

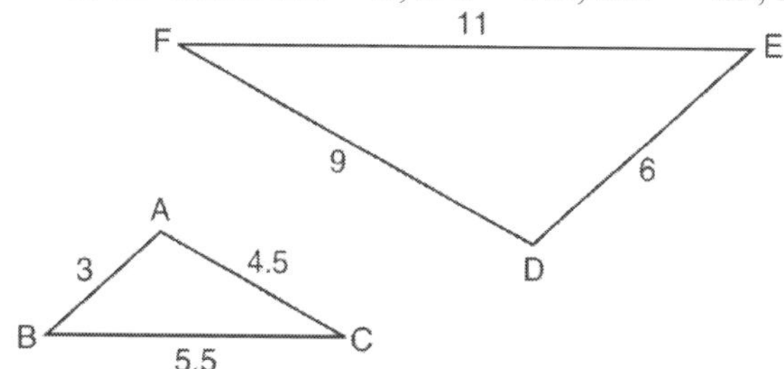

Which relationship must always be true?

(1) $\dfrac{m\angle A}{m\angle D} = \dfrac{1}{2}$

(2) $\dfrac{m\angle C}{m\angle F} = \dfrac{2}{1}$

(3) $\dfrac{m\angle A}{m\angle C} = \dfrac{m\angle F}{m\angle D}$

(4) $\dfrac{m\angle B}{m\angle E} = \dfrac{m\angle C}{m\angle F}$

4. (CC) Given: $\triangle AEC$, $\triangle DEF$, and $\overline{FE} \perp \overline{CE}$.

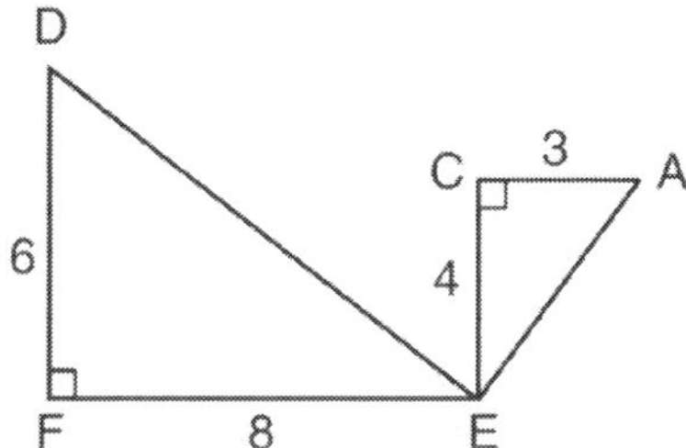

What is a correct sequence of similarity transformations that shows $\triangle AEC \sim \triangle DEF$?

(1) a rotation of 180 degrees about point *E* followed by a horizontal translation
(2) a counterclockwise rotation of 90 degrees about point *E* followed by a horizontal translation
(3) a rotation of 180 degrees about point *E* followed by a dilation with a scale factor of 2 centered at point *E*
(4) a counterclockwise rotation of 90 degrees about point *E* followed by a dilation with a scale factor of 2 centered at point *E*

5. (CC) In the diagram below, $\triangle ADE$ is the image of $\triangle ABC$ after a reflection over the line *AC* followed by a dilation of scale factor $\frac{AE}{AC}$ centered at point *A*.

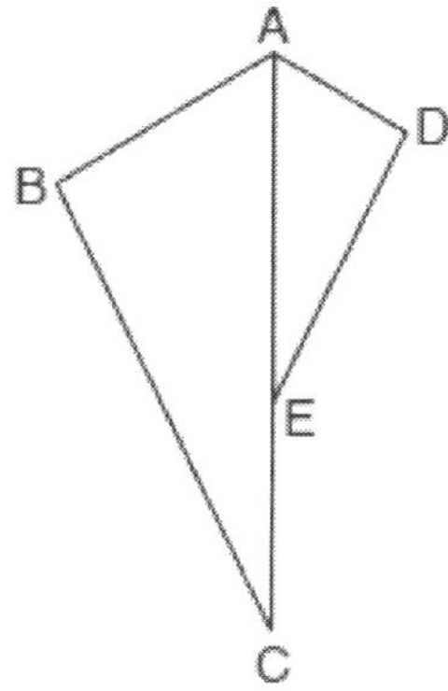

Which statement must be true?

(1) $m\angle BAC = m\angle AED$
(2) $m\angle ABC = m\angle ADE$
(3) $m\angle DAE = \frac{1}{2}m\angle BAC$
(4) $m\angle ACB = \frac{1}{2}m\angle DAB$

Constructed Response

6. (CC) As shown in the diagram below, circle A has a radius of 3 and circle B has a radius of 5.

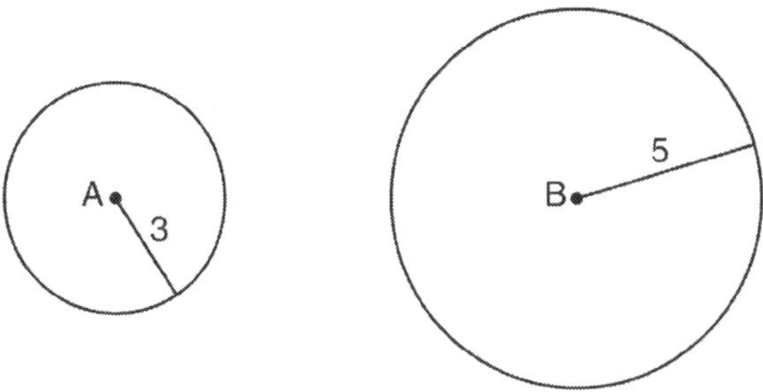

Use transformations to explain why circles A and B are similar.

7. (CC) In the diagram below, triangles XYZ and UVZ are drawn such that $\angle X \cong \angle U$ and $\angle XZY \cong \angle UZV$.

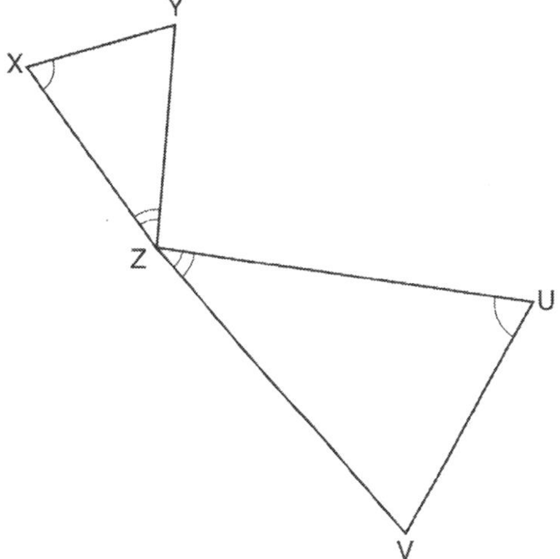

Describe a sequence of similarity transformations that shows $\triangle XYZ$ is similar to $\triangle UVZ$.

8. (CC) In the diagram below, $\triangle A'B'C'$ is the image of $\triangle ABC$ after a transformation.

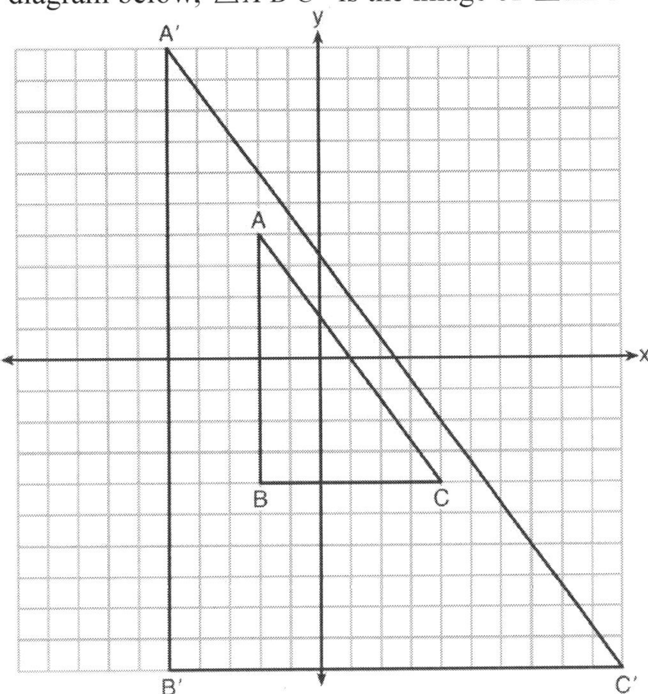

Describe the transformation that was performed.

Explain why $\triangle A'B'C' \sim \triangle ABC$.

Chapter 13. Circles

13.1 Circumference and Rotation

Key Terms and Concepts

A circle is a set of points in a plane that are equidistant from a given point, called the center.
A **radius** of a circle is a line segment that has endpoints at the center and on the circle.
A **diameter** of a circle passes through the center of the circle and has endpoints on the circle.

A radius (*plural, radii*) of a circle is half the length of its diameter. All diameters of the same circle are congruent, and all radii of the same circle are congruent.

Circumference represents the distance around the outside of a circle. The formulas for the circumference of a circle are:

$C = 2\pi r$ where r is the radius, or

$C = \pi d$ where d is the diameter

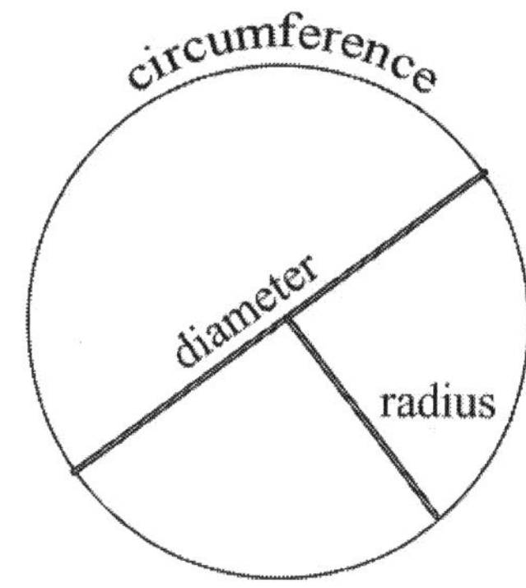

If you know the area or circumference of a circle, you can find its radius by solving for r:

$$r = \sqrt{\frac{A}{\pi}} \qquad r = \frac{C}{2\pi}$$

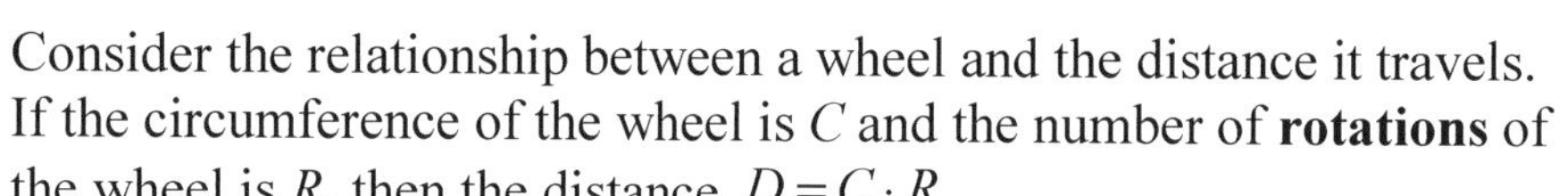

Consider the relationship between a wheel and the distance it travels. If the circumference of the wheel is C and the number of **rotations** of the wheel is R, then the distance $D = C \cdot R$.

Example: A robot with 2 inch diameter wheels must travel at least 10 feet. The wheel would need to make

$$R = \frac{D}{C} = \frac{120\,in}{2\pi\,in} \approx 19 \text{ rotations.}$$

Model Problem:

To use the machine below, you turn the crank, which turns the pulley wheel, which winds the rope and lifts the box. Through how many rotations, to the *nearest tenth*, must you turn the crank to lift the box 10 feet?

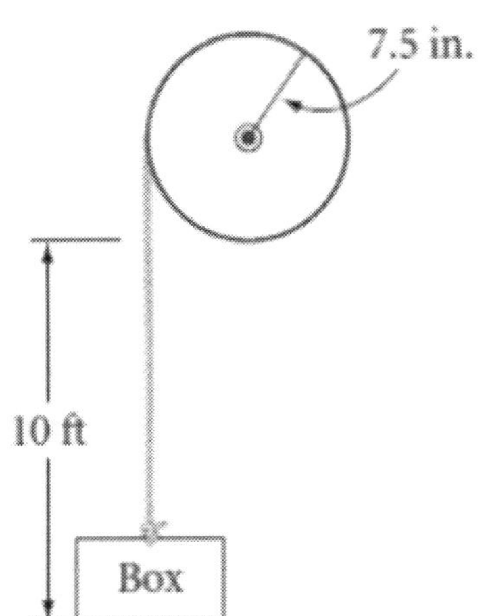

(not drawn to scale)

Solution:

$D = 10$ ft. $= 120$ in.

$C = 2\pi r = 15\pi$ in.

$R = \frac{D}{C} = \frac{120\,in}{15\pi\,in} \approx 2.5$ rotations

Explanation:

Use the formula $R = \frac{D}{C}$

Practice Problems

1. A manufacturer wants to frame the circumferences of 8 inch diameter clocks with decorative rope. How many clocks can be framed from 100 feet of rope?	2. The London Eye Ferris wheel has a diameter of 394 ft. If the Ferris wheel was to separate from its base and start rolling away, how many rotations, to the *nearest tenth*, would it take to travel the two miles to London Bridge? *(1 mile = 5,280 feet)*

REGENTS QUESTIONS

Multiple Choice

1. Every time the pedals go through a 360° rotation on a certain bicycle, the tires rotate three times. If the tires are 24 inches in diameter, what is the minimum number of complete rotations of the pedals needed for the bicycle to travel at least 1 mile? *(1 mile = 5,280 feet)*
   (1) 12 (3) 561
   (2) 281 (4) 5,280

2. (CC) A designer needs to create perfectly circular necklaces. The necklaces each need to have a radius of 10 cm. What is the largest number of necklaces that can be made from 1000 cm of wire?
   (1) 15 (3) 31
   (2) 16 (4) 32

Constructed Response

3. To measure the length of a hiking trail, a worker uses a device with a 2-foot-diameter wheel that counts the number of revolutions the wheel makes. If the device reads 1,100.5 revolutions at the end of the trail, how many miles long is the trail, to the *nearest tenth of a mile*? *(1 mile = 5,280 feet)*

4. A wheel has a radius of 5 feet. What is the minimum number of *complete* revolutions that the wheel must make to roll at least 1,000 feet?

13.2 Arcs and Chords

Key Terms and Concepts

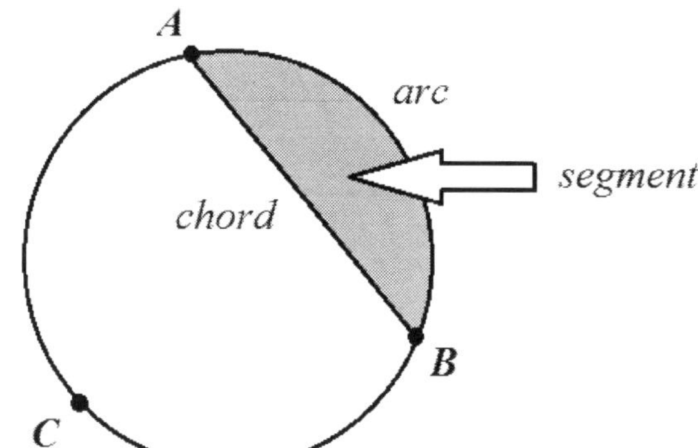

An **arc** is part of the circumference of a circle. A **chord** of a circle is a line segment whose endpoints both lie on the circle. A **segment** is the area of a circle bounded by a chord and the arc that shares the same endpoints.

If two arcs are congruent, their chords are congruent, and vice versa.

A **minor arc** is an arc of no more than half of a circle; a **major arc** is more than half of a circle. The symbol for the minor arc between points A and B on a circle is $\overset{\frown}{AB}$. To represent a major arc, a third point on the circle is needed; for example, in the diagram above, $\overset{\frown}{ACB}$ is a major arc.

An arc may be measured not only by its length, but also by its measure in degrees. $\mathrm{m}\overset{\frown}{AB}$ represents the measure of arc AB in degrees.

To determine the measure of an arc, multiply the fraction of the circle that the arc represents by 360°. For examples, a semicircle arc is $\frac{1}{2}360° = 180°$ and a quarter-circle arc is $\frac{1}{4}360° = 90°$.

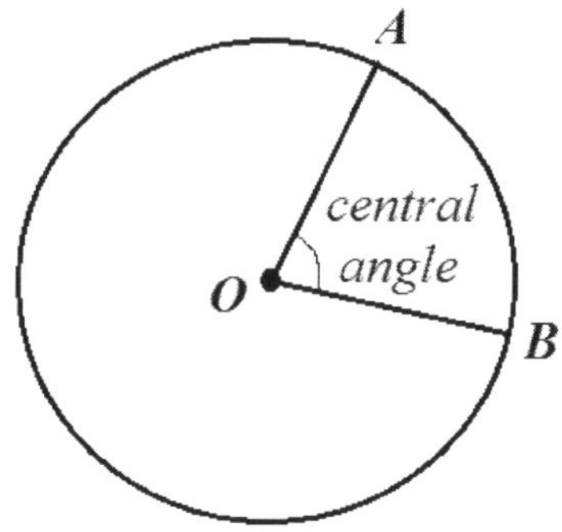

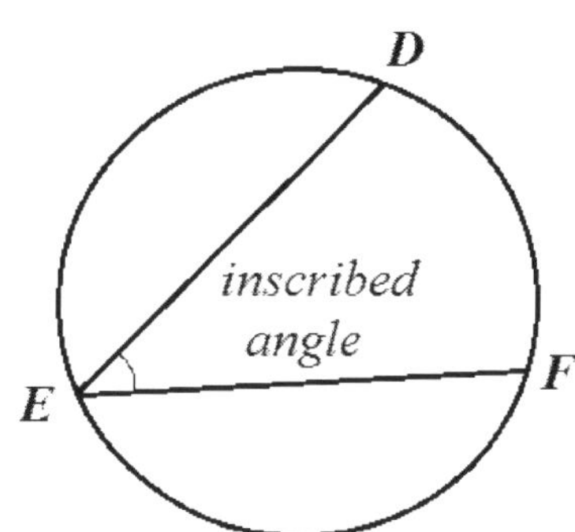

A **central angle** of a circle has its vertex at the center of the circle and sides that are radii.
A central angle is equal in measure to its intercepted arc.

Example: In the diagram above, $\mathrm{m}\angle AOB = \mathrm{m}\overset{\frown}{AB}$.

An **inscribed angle** of a circle has its vertex on the circle and sides that are chords.
An inscribed angle is equal to half the measure of its intercepted arc.

Example: In the diagram above, $\mathrm{m}\angle DEF = \frac{1}{2} \cdot \mathrm{m}\overset{\frown}{DF}$.

If two central angles are congruent, their chords are congruent, and vice versa.

Example: In circle E below, if $\angle AEB \cong \angle CED$, then $\overline{AB} \cong \overline{CD}$, and vice versa.

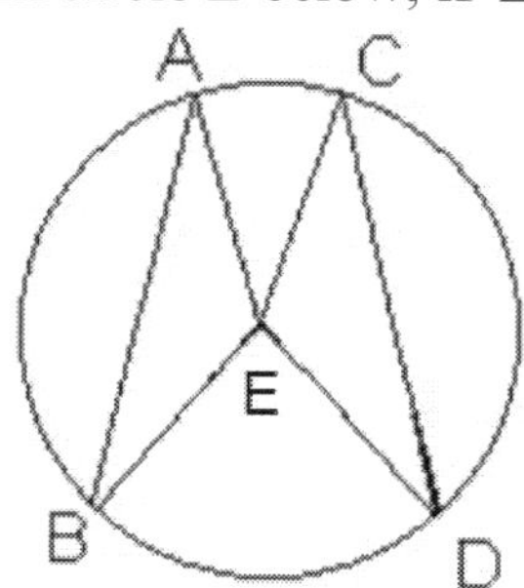

Since radii $\overline{AE} \cong \overline{BE} \cong \overline{CE} \cong \overline{DE}$, this is easy to prove:

If $\angle AEB \cong \angle CED$, then $\triangle AEB \cong \triangle CED$ by SAS, and $\overline{AB} \cong \overline{CD}$ by CPCTC.

Also, if $\overline{AB} \cong \overline{CD}$, then $\triangle AEB \cong \triangle CED$ by SSS, and $\angle AEB \cong \angle CED$ by CPCTC.

If two inscribed angles are congruent, their chords are congruent, and vice versa.

An angle inscribed in a semicircle is a right angle.

Example: If A, B and C are points on a circle where $\overline{AC}$ is a diameter of the circle, then $\angle ABC$ is a right angle. Since $\overset{\frown}{AC} = 180°$, it follows that inscribed $\angle ABC$ should measure 90° (half of 180°) and is therefore a right angle.

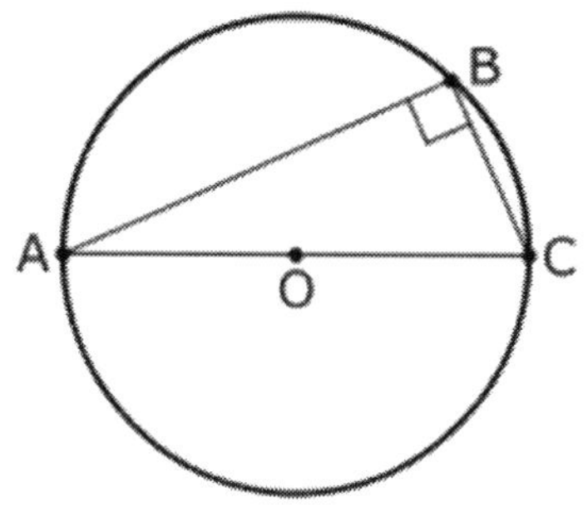

If two chords are parallel, they intercept congruent arcs between the chords.

Example: Given $\overline{AB} \parallel \overline{CD}$ below, then $\overset{\frown}{AC} \cong \overset{\frown}{BD}$. To prove this, start by drawing the transversal $\overline{AD}$. Two inscribed angles are formed, and since they are alternate interior angles, they are congruent. So, their intercepted arcs are congruent.

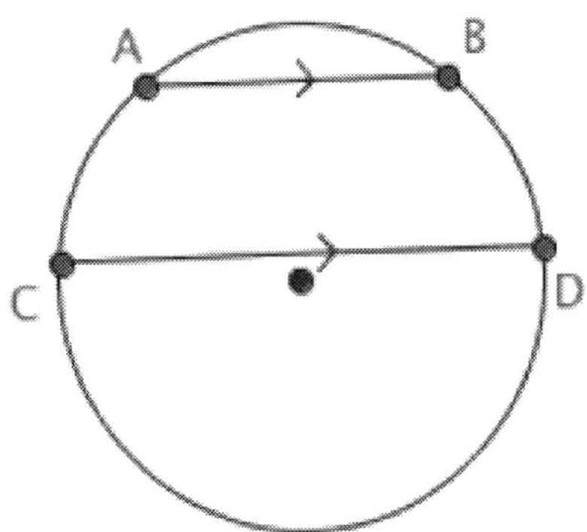

The perpendicular bisector of a chord bisects its arc and intersects the center of the circle.

Example: $\overleftrightarrow{WV}$ is the perpendicular bisector of chord $\overline{XY}$. Therefore, $\overset{\frown}{XW} \cong \overset{\frown}{YW}$ and $\overleftrightarrow{WV}$ passes through the center of the circle, D.

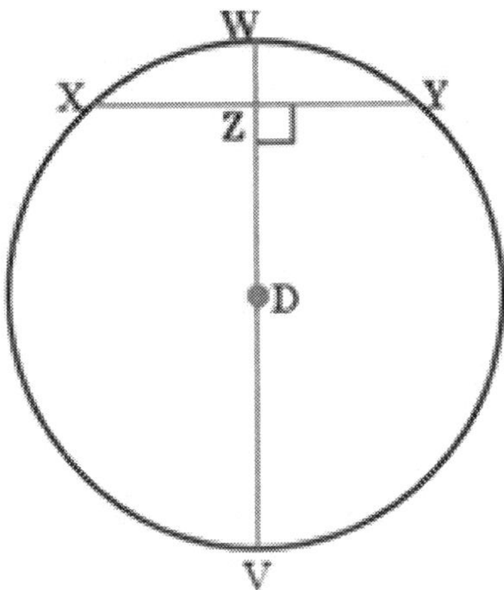

If a radius (or diameter) is perpendicular to a chord, it bisects the chord, and vice versa.

If a quadrilateral is inscribed in a circle, both pairs of opposite angles are supplementary. This can easily be shown. Given quadrilateral $ABCD$, let $x = \text{m}\angle A$. Since $\angle A$ is an inscribed angle, $\text{m}\overset{\frown}{BCD} = 2x$. Since $\overset{\frown}{BCD} + \overset{\frown}{BAD} = 360°$, it follows that $\text{m}\overset{\frown}{BAD} = 360 - 2x$. $\angle C$ is an inscribed angle, so $\text{m}\angle C = \frac{\text{m}\overset{\frown}{BAD}}{2} = \frac{360-2x}{2} = 180 - x$. So, $\angle A$ and $\angle C$ are supplementary.

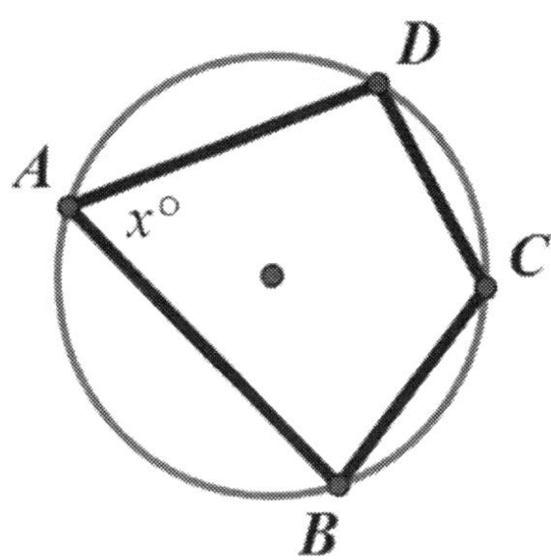

Two intersecting chords form a pair of vertical angles that are each equal in measure to the average of their intercepted arcs.

Example: In the circle below, $x = \dfrac{a+b}{2}$.

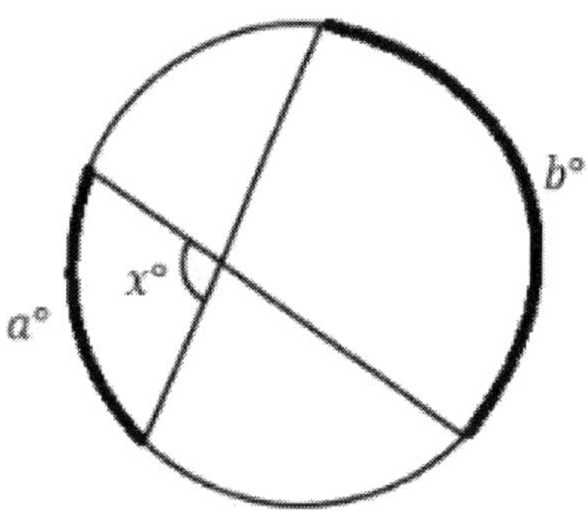

When two intersecting chords are drawn in a circle, and each endpoint of one chord is connected to a different endpoint of the other, two *similar triangles* are formed.

Example: In the diagram below of circle O, chords $\overline{AD}$ and $\overline{BC}$ intersect at E, and chords $\overline{AC}$ and $\overline{BD}$ are drawn. Since $\angle A$ and $\angle B$ are inscribed angles of the same arc, $\overset{\frown}{CD}$, they are congruent. Also, since $\angle C$ and $\angle D$ are inscribed angles of the same arc, $\overset{\frown}{AB}$, they are congruent. $\angle AEC$ is also congruent to $\angle BED$ since they are vertical angles. Therefore, $\triangle AEC \sim \triangle BED$.

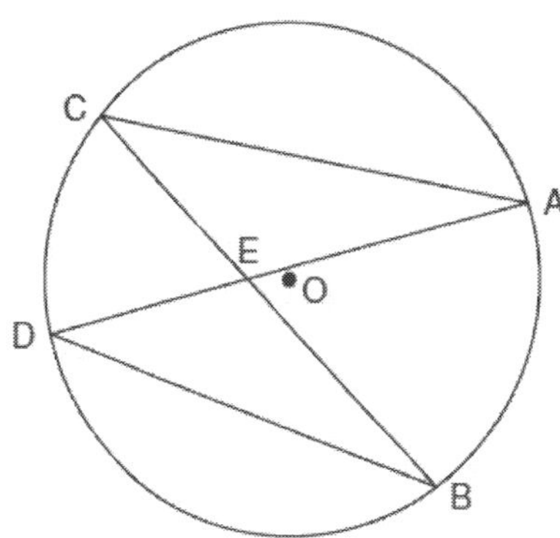

Because of this, we know $\dfrac{AE}{BE} = \dfrac{CE}{DE}$, or $(AE)(DE) = (BE)(CE)$, which leads to the next theorem.

When two chords intersect in a circle, the product of the segments of one chord equals the product of the segments of the other chord.

Example: In the circle to the right, $ab = cd$.

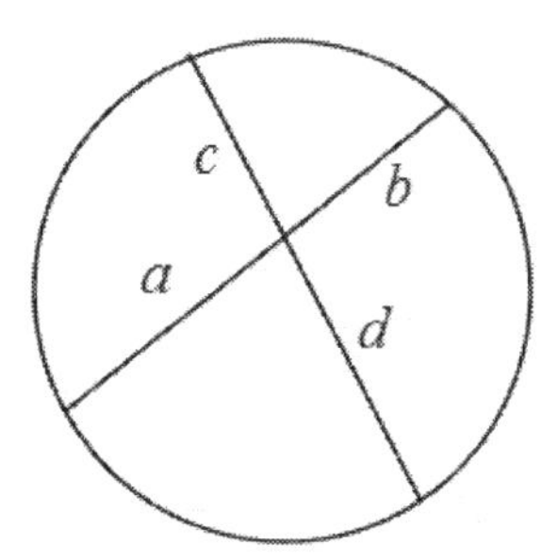

Model Problem:

Given: $\overline{CB}$ is a diameter of circle A and $m\angle CAD = 70°$.

Find: $m\overset{\frown}{CD}$, $m\angle CBD$, $m\angle BAD$, $m\overset{\frown}{BD}$, and $m\angle ADB$.

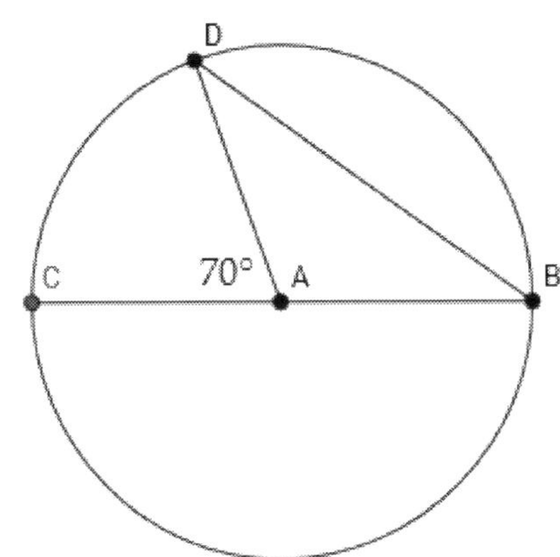

Solution:

(A) $m\overset{\frown}{CD} = m\angle CAD = 70°$

(B) $m\angle CBD = \frac{1}{2} m\overset{\frown}{CD} = 35°$

(C) $m\angle BAD = 180 - 70 = 110°$

(D) $m\overset{\frown}{BD} = m\angle BAD = 110°$

(E) $m\angle ADB = m\angle CBD = 35°$

Explanation of steps:

(A) A central angle is equal in measure to its intercepted arc.

(B) An inscribed angle is half the measure of its intercepted arc.

(C) A linear pair adds to 180°.

(D) A central angle is equal in measure to its intercepted arc.

(E) Base angles of an isosceles triangle are equal in measure. *[$\overline{AB}$ and $\overline{AD}$ are radii.]*

Practice Problems

1. Find the measure of the angle formed by the hands of the clock.

2. In circle M below, $m\angle AMC = 60°$. Find the measure of inscribed $\angle ABC$.

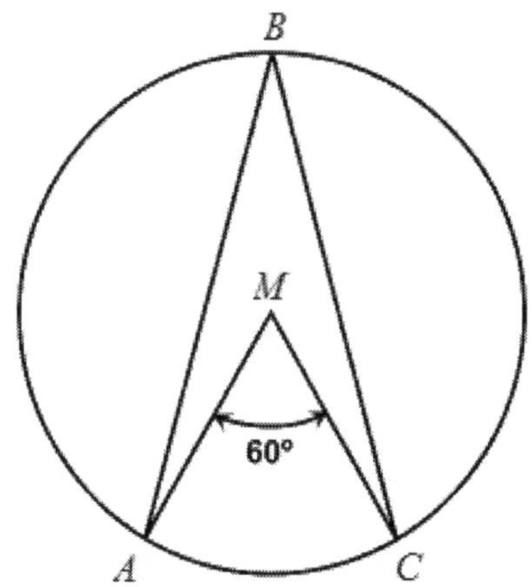

3. $\overline{BC}$ is a diameter of the circle below. If $m\angle B = 55°$, find $m\angle C$.

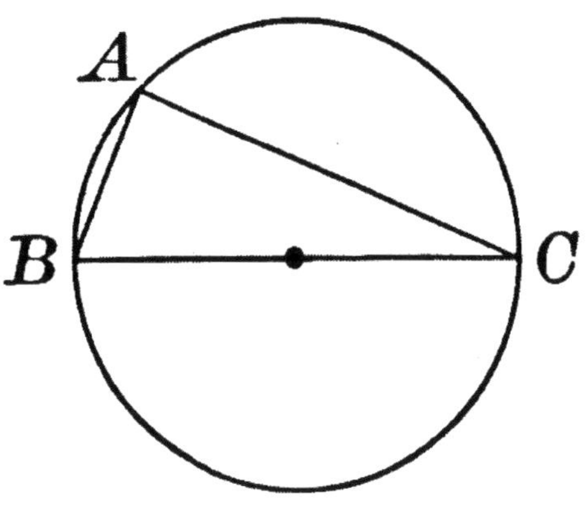

4. In the diagram, quadrilateral $ABCD$ is inscribed in circle O. If $m\widehat{AB} = 132$ and $m\widehat{BC} = 82$, find $m\angle ADC$.

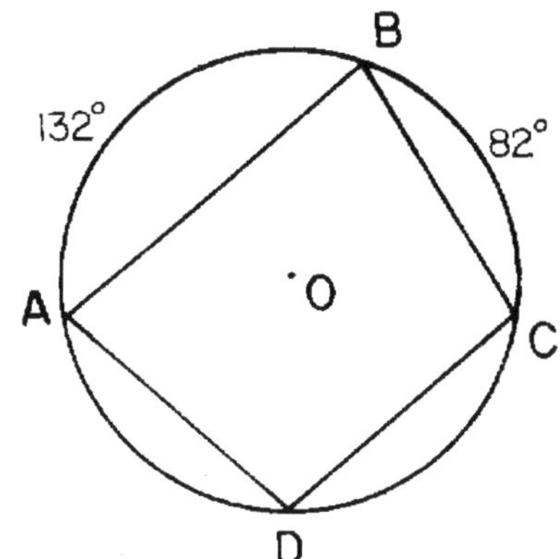

5. In the diagram of circle O, chords $\overline{DF}$, $\overline{DE}$, $\overline{FG}$, and $\overline{EG}$ are drawn such that $m\widehat{DF} : m\widehat{FE} : m\widehat{EG} : m\widehat{GD} = 5:2:1:7$. Identify one pair of inscribed angles that are congruent to each other and give their measure.

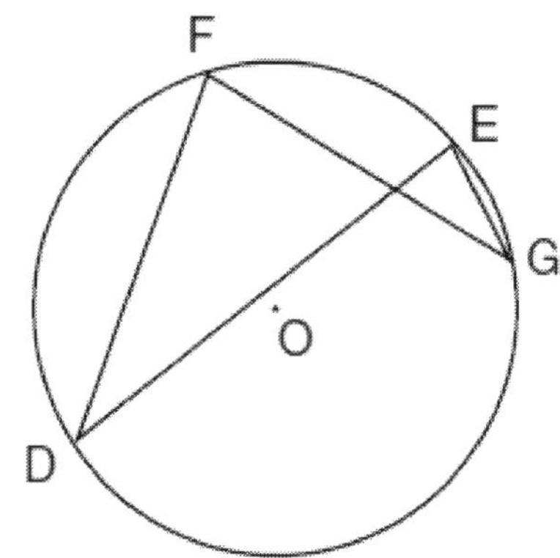

REGENTS QUESTIONS

Multiple Choice

1. In the diagram below, quadrilateral $JUMP$ is inscribed in a circle.

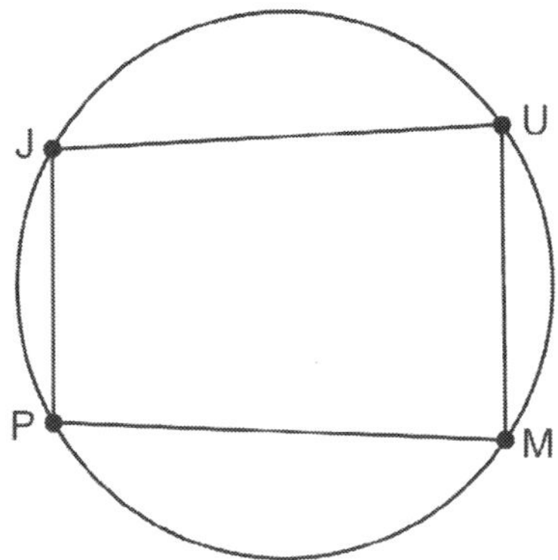

Opposite angles J and M must be

(1) right
(2) complementary
(3) congruent
(4) supplementary

2. In the diagram of circle O below, chord $\overline{CD}$ is parallel to diameter $\overline{AOB}$ and $m\overset{\frown}{CD} = 110$.

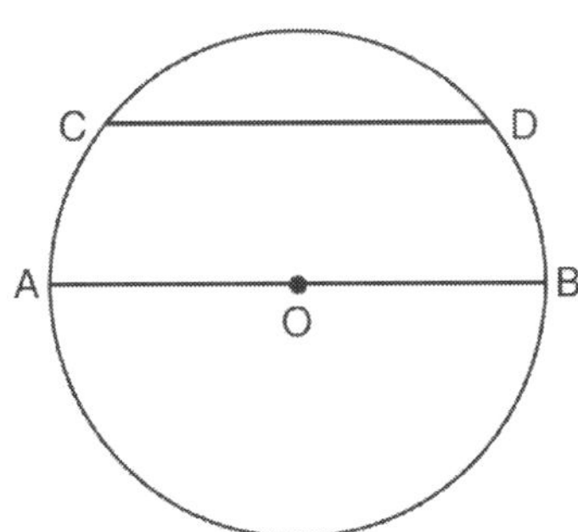

What is $m\overset{\frown}{DB}$?

(1) 35
(2) 55
(3) 70
(4) 110

3. In circle O, diameter $\overline{AB}$ intersects chord $\overline{CD}$ at E. If $CE = ED$, then $\angle CEA$ is which type of angle?

(1) straight
(2) obtuse
(3) acute
(4) right

4. In the diagram of the circle shown below, chords $\overline{AC}$ and $\overline{BD}$ intersect at Q, and chords $\overline{AE}$ and $\overline{BD}$ are parallel.

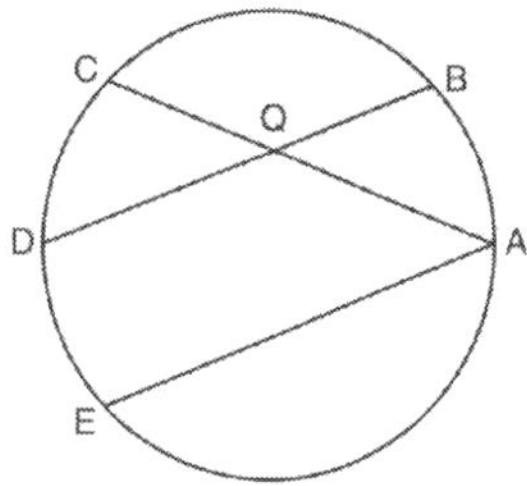

Which statement must always be true?

(1) $\overparen{AB} \cong \overparen{CD}$
(2) $\overparen{DE} \cong \overparen{CD}$
(3) $\overparen{AB} \cong \overparen{DE}$
(4) $\overparen{BD} \cong \overparen{AE}$

5. In the diagram below of circle O, diameter $\overline{AB}$ and chord $\overline{CD}$ intersect at E.

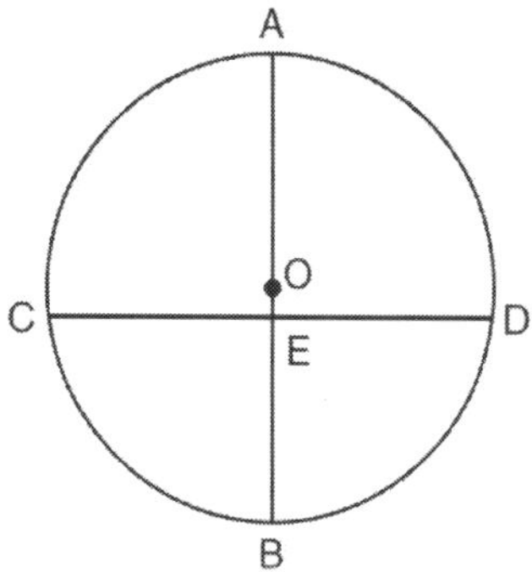

If $\overline{AB} \perp \overline{CD}$, which statement is always true?

(1) $\overparen{AC} \cong \overparen{BD}$
(2) $\overparen{BD} \cong \overparen{DA}$
(3) $\overparen{AD} \cong \overparen{BC}$
(4) $\overparen{CB} \cong \overparen{BD}$

6. In the diagram below of circle O, chord $\overline{AB}$ is parallel to chord $\overline{CD}$.

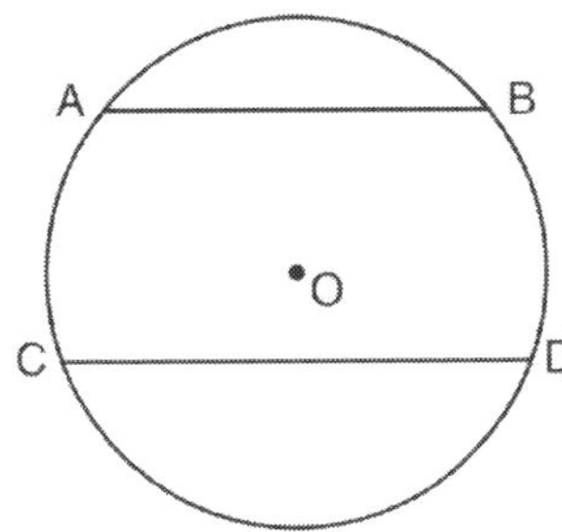

A correct justification for $m\overset{\frown}{AC} = m\overset{\frown}{BD}$ in circle O is
1) parallel chords intercept congruent arcs
2) congruent chords intercept congruent arcs
3) if two chords are parallel, then they are congruent
4) if two chords are equidistant from the center, then the arcs they intercept are congruent

7. In the diagram below of circle O, $m\angle ABC = 24$.

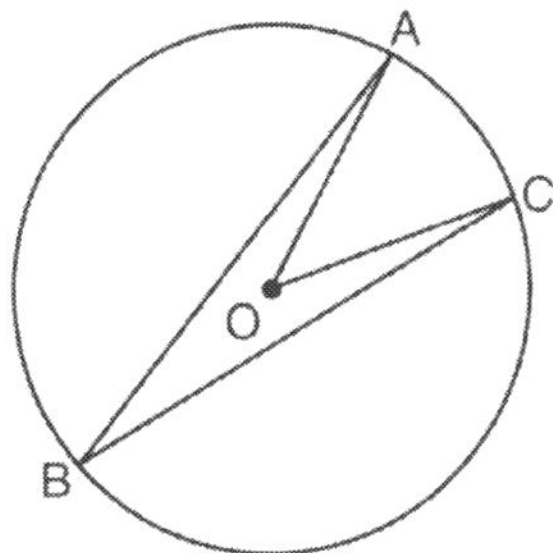

What is the $m\angle AOC$?

(1) 12
(2) 24
(3) 48
(4) 60

8. As shown in the diagram below, $\overline{AB}$ is a diameter of circle O, and chord $\overline{AC}$ is drawn.

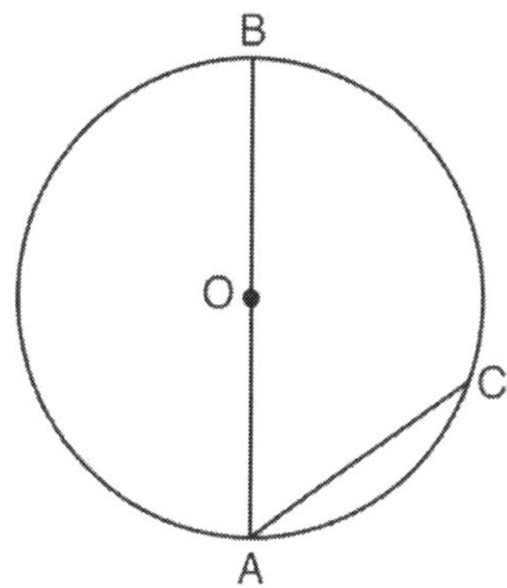

If $m\angle BAC = 70$, then $m\overset{\frown}{AC}$ is
(1) 40
(2) 70
(3) 110
(4) 140

9. In the diagram below, $\angle ABC$ is inscribed in circle O.

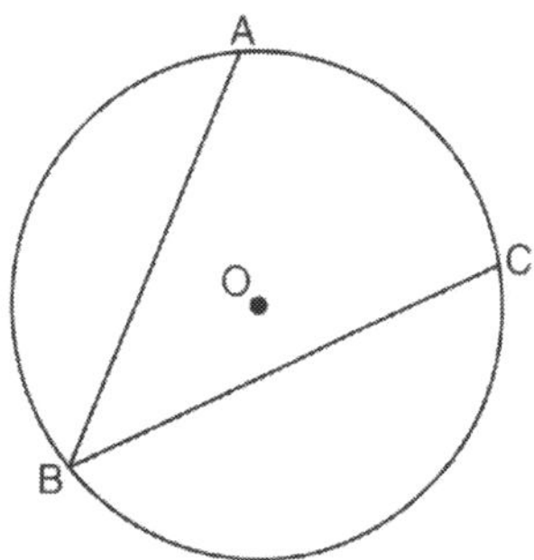

The ratio of the measure of $\angle ABC$ to the measure of $\overset{\frown}{AC}$ is
(1) 1 : 1
(2) 1 : 2
(3) 1 : 3
(4) 1 : 4

10. In circle O shown below, chord $\overline{AB}$ and diameter $\overline{CD}$ are parallel, and chords $\overline{AD}$ and $\overline{BC}$ intersect at point E.

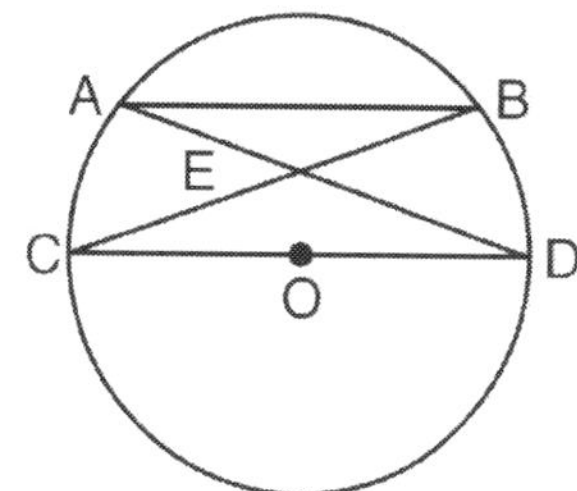

Which statement is *false*?
(1) $\overset{\frown}{AC} \cong \overset{\frown}{BD}$
(2) $BE = CE$
(3) $\triangle ABE \sim \triangle CDE$
(4) $\angle B \cong \angle C$

11. (CC) In the diagram of circle A shown below, chords $\overline{CD}$ and $\overline{EF}$ intersect at G, and chords $\overline{CE}$ and $\overline{FD}$ are drawn.

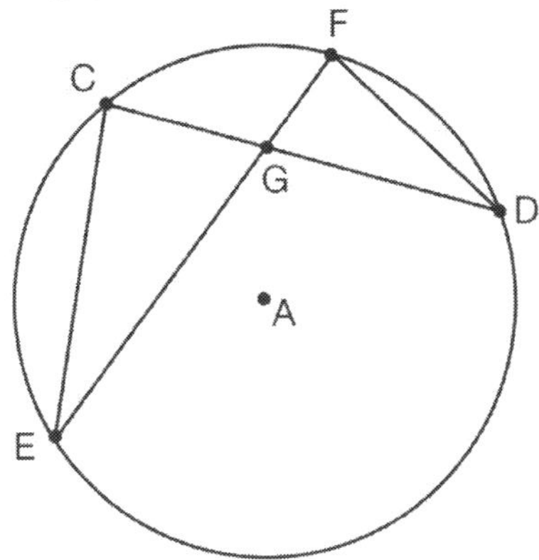

Which statement is *not* always true?

(1) $\overline{CG} \cong \overline{FG}$

(2) $\angle CEG \cong \angle FDG$

(3) $\frac{CE}{EG} = \frac{FD}{DG}$

(4) $\triangle CEG \sim \triangle FDG$

12. (CC) In the diagram below, quadrilateral $ABCD$ is inscribed in circle P.

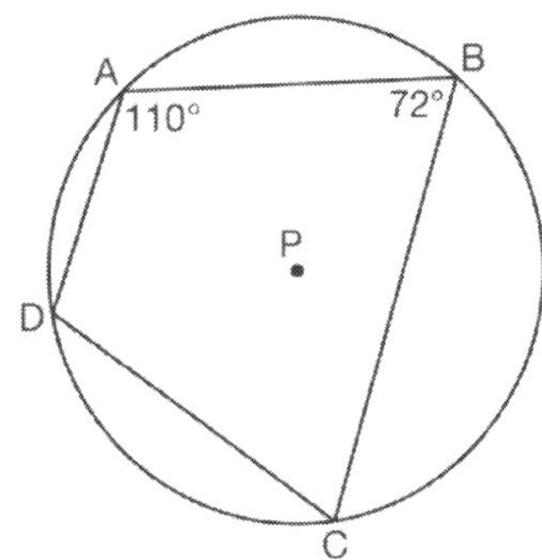

What is $m\angle ADC$?

(1) 70°

(2) 72°

(3) 108°

(4) 110°

13. (CC) In the diagram below of circle O, $\overline{OB}$ and $\overline{OC}$ are radii, and chords $\overline{AB}$, $\overline{BC}$, and $\overline{AC}$ are drawn.

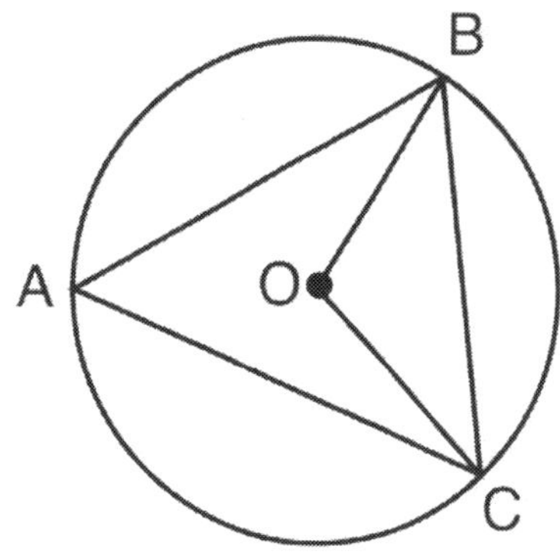

Which statement must always be true?

(1) $\angle BAC \cong \angle BOC$

(2) $m\angle BAC = \frac{1}{2} m\angle BOC$

(3) $\triangle BAC$ and $\triangle BOC$ are isosceles.

(4) The area of $\triangle BAC$ is twice the area of $\triangle BOC$.

14. (CC) In the diagram below, $\overline{BC}$ is the diameter of circle A.

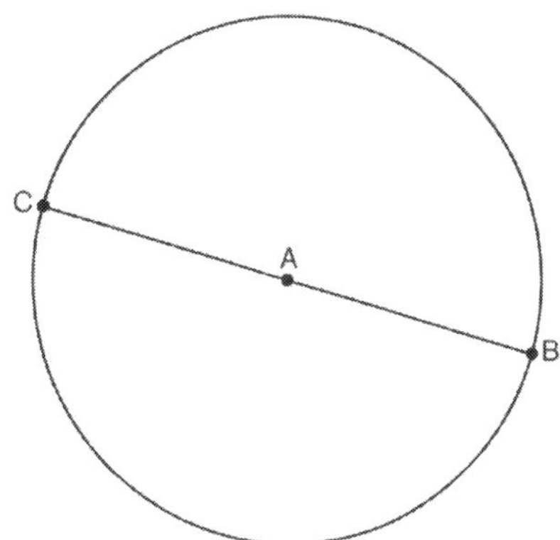

Point D, which is unique from points B and C, is plotted on circle A. Which statement must always be true?

(1) $\triangle BCD$ is a right triangle.

(2) $\triangle BCD$ is an isosceles triangle.

(3) $\triangle BAD$ and $\triangle CBD$ are similar triangles.

(4) $\triangle BAD$ and $\triangle CAD$ are congruent triangles.

Constructed Response

15. In the diagram below, trapezoid $ABCD$, with bases $\overline{AB}$ and $\overline{DC}$, is inscribed in circle O, with diameter $\overline{DC}$. If $m\widehat{AB} = 80$, find $m\widehat{BC}$.

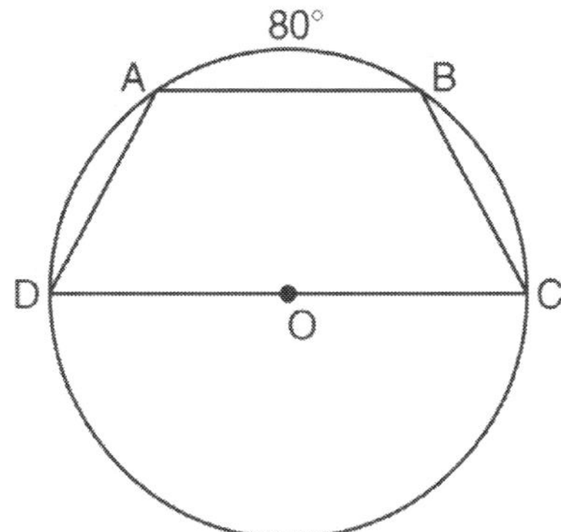

16. As shown in the diagram below, quadrilateral $DEFG$ is inscribed in a circle and $m\angle D = 86$.

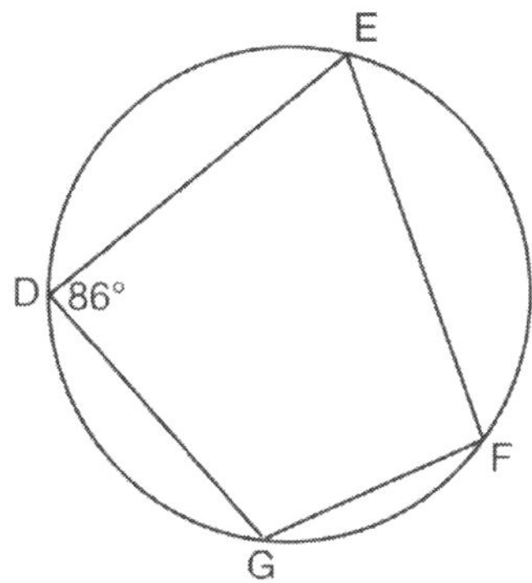

Determine and state $m\widehat{GFE}$. Determine and state $m\angle F$.

17. (CC) In the diagram below of circle O with diameter $\overline{BC}$ and radius $\overline{OA}$, chord $\overline{DC}$ is parallel to chord $\overline{BA}$.

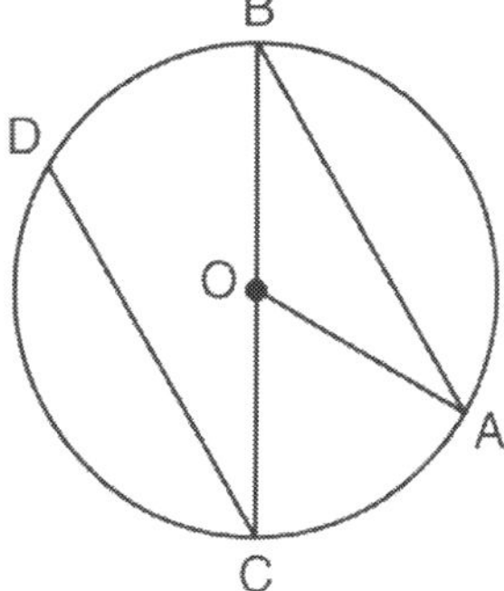

If $m\angle BCD = 30°$, determine and state $m\angle AOB$.

13.3 Tangents

Key Terms and Concepts

A **tangent** is a straight line that touches a circle at only one point, called the **point of tangency**.
Example: Line ℓ below is a tangent to circle O with P as the point of tangency.

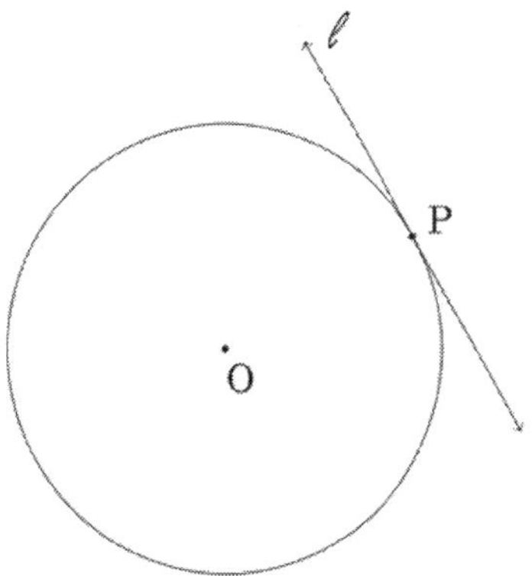

When a line is tangent to two circles, it is called a **common tangent**.
Examples:

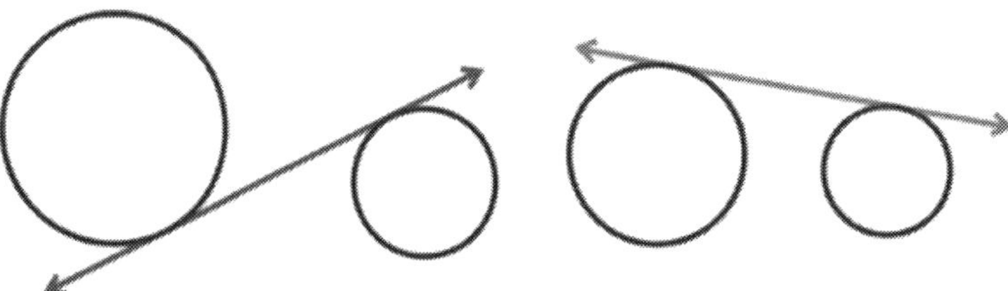

The tangent of a circle is perpendicular to a radius at the point of tangency.
Example: In the diagram below, tangent $\overleftrightarrow{AC}$ is perpendicular to radius $\overline{DB}$ at the point of tangency, B.

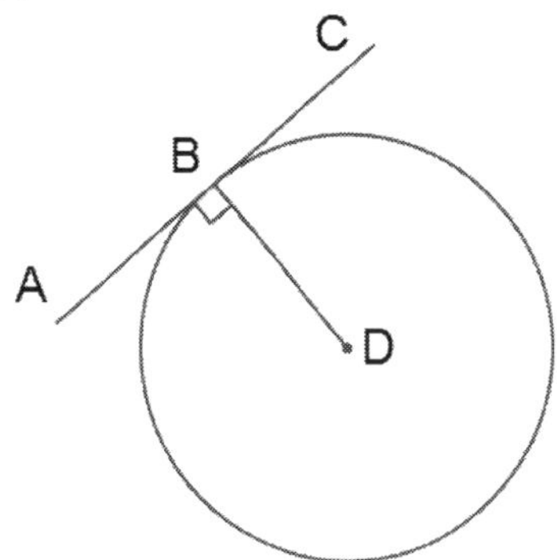

The angle formed by a tangent and a chord is half the measure of the intercepted arc.

Example: In the diagram below, tangent $\overleftrightarrow{AD}$ and chord $\overline{AB}$ intersect on the circle at A, forming $\angle BAD$. If $m\overset{\frown}{AB} = 124°$, then $m\angle BAD = \frac{1}{2}m\overset{\frown}{AB} = 62°$.

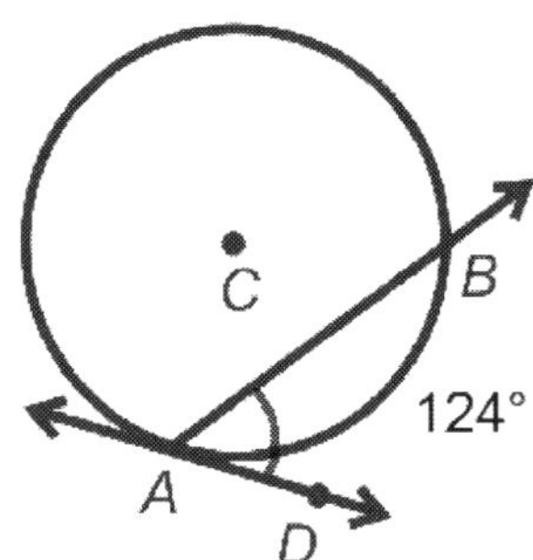

To prove this theorem, draw diameter $\overline{ACE}$ and chord $\overline{EB}$ to form $\triangle ABE$, as shown below.

1. $\angle DAE$ is a right angle (a tangent is perpendicular to a diameter)
2. $\angle ABE$ is a right angle (the inscribed angle of a semicircle is a right angle)
3. $\triangle ABE$ is a right triangle (definition of right triangle)
4. $m\angle BAD = 90° - m\angle BAE$ (adjacent angles that form a right angle are complementary)
5. $m\angle AEB = 90° - m\angle BAE$ (acute angles of a right triangle are complementary)
6. $m\angle BAD = m\angle AEB$ (substitution)
7. $m\angle AEB = \frac{1}{2}m\overset{\frown}{AB}$ (an inscribed angle is one-half the measure of its intercepting arc)
8. $m\angle BAD = \frac{1}{2}m\overset{\frown}{AB}$ (substitution)

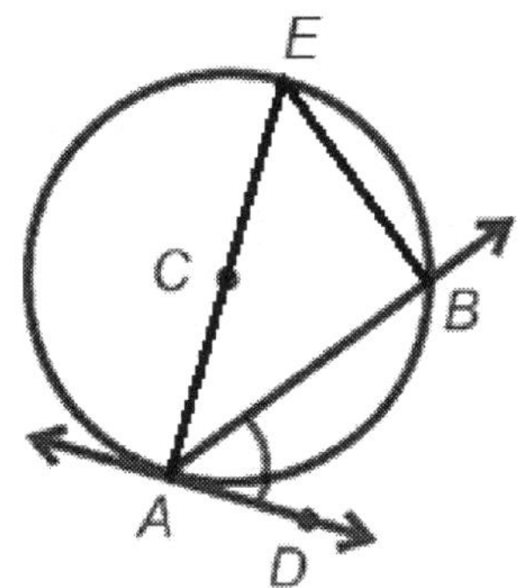

If two tangents intersect at a point outside a circle, then two segments between the points of tangency and the point of intersection are formed. **The two tangent segments drawn to a circle from the same external point are congruent.**

Example: In the diagram of circle O below, tangent segments $\overline{AP}$ and $\overline{BP}$ are congruent.

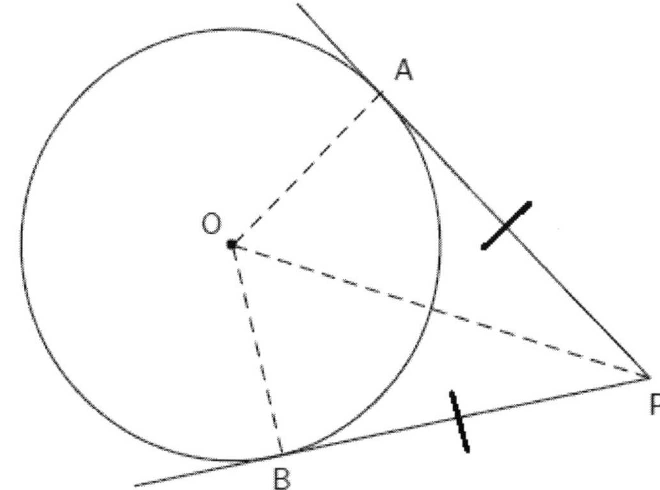

We can prove that these tangent segments are congruent by drawing segment $\overline{OP}$ and radii $\overline{AO}$ and $\overline{BO}$.

Two right triangles, OAP and OBP, are formed, with right angles at $\angle OAP$ and $\angle OBP$.

$\overline{OP} \cong \overline{OP}$ by the Reflexive Property, and since they are radii, $\overline{AO} \cong \overline{BO}$. Therefore, by HL, $\triangle OAP \cong \triangle OBP$.

So, $\overline{AP} \cong \overline{BP}$ by CPCTC.

A **circumscribed angle** is the angle formed when two tangents intersect at an external point. **A circumscribed angle is equal to 180° minus the measure of the minor arc it intercepts.**

Example: $\angle A$ is a circumscribed angle formed by tangents $\overline{AC}$ and $\overline{AB}$.

So, $m\angle A = 180 - m\overset{\frown}{BC}$. We can show this by first drawing radii $\overline{OB}$ and $\overline{OC}$.

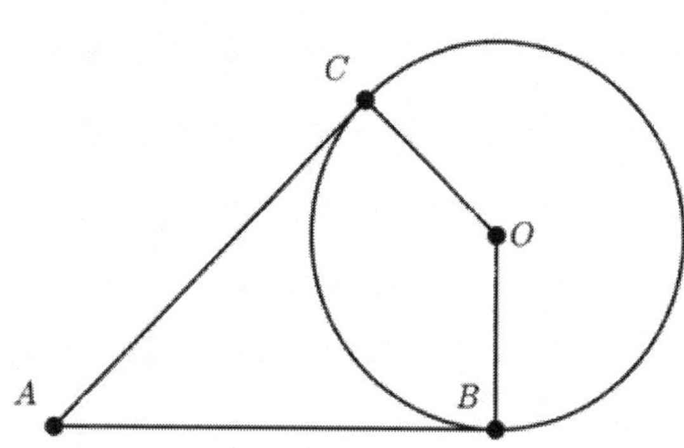

Since the measures of the angles of a quadrilateral add to 360°, $m\angle A = 360 - m\angle B - m\angle C - m\angle O$.

Since the tangents are perpendicular to the radii, $m\angle B = m\angle C = 90°$. So, by substitution, $m\angle A = 180 - m\angle O$.

$\angle O$ is a central angle, so $m\angle O = m\angle\overset{\frown}{BC}$. Therefore, $m\angle A = 180 - m\overset{\frown}{BC}$.

Model Problem:

$\overline{XZ}$ is a diameter of circle O. Chords $\overline{XY}$ and $\overline{YZ}$ and tangent $\overleftrightarrow{WZ}$ are drawn. $m\angle YXZ = 50°$. Find $m\angle YZW$.

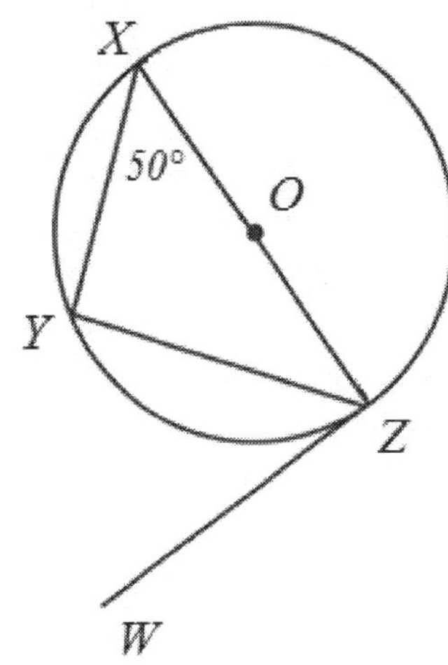

Solution:

(A) $m\angle XYZ = 90°$
(B) $m\angle XZY = 40°$
(C) $m\angle XZW = 90°$
(D) $m\angle YZW = 50°$

Explanation of steps:

(A) $\angle XYZ$ is an inscribed angle of a semicircle
[$m\angle XYZ = \frac{1}{2}180° = 90°$ *].*

(B) The sum of the measures of the angles of a triangle is 180°
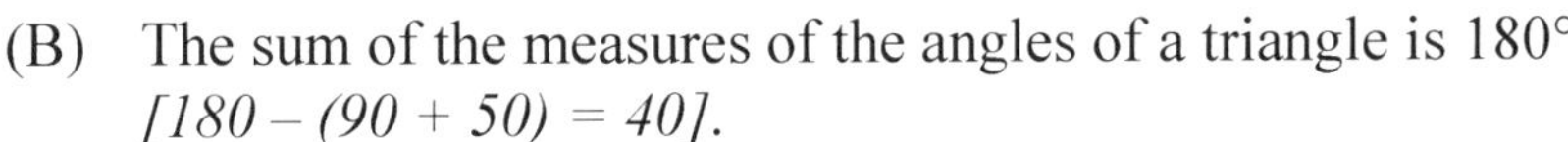
[180 – (90 + 50) = 40].

(C) The diameter of a circle is perpendicular to the tangent passing through its endpoint.

(D) Subtract *[* $m\angle XZW - m\angle XZY = 90° - 40° = 50°$ *].*

Practice Problems

1. Circle O is inscribed in $\triangle ABC$ with D, E, and F as points of tangency as shown. If $AE = 15$, $EC = 10$, and $BF = 14$, find the perimeter of $\triangle ABC$.

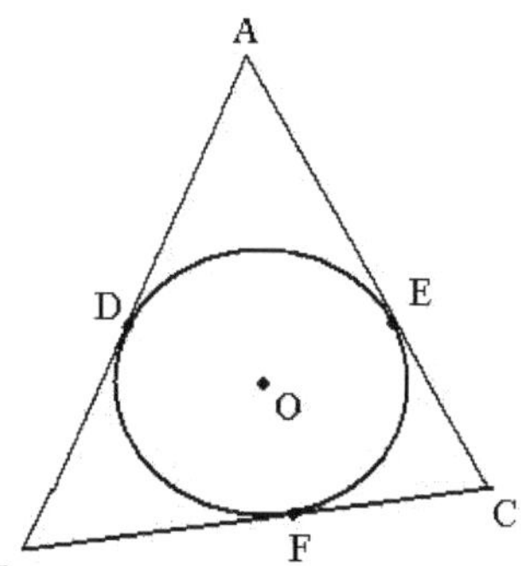

2. $\overline{CB}$ is tangent to circle A at B. Find the length of $\overline{CD}$.

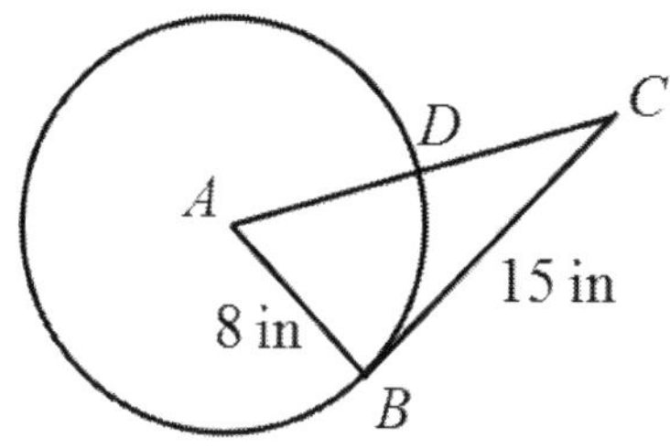

3. Rays QR and QS are tangents to circle P at R and S. If $m\angle Q = 54°$, find $m\widehat{RS}$.

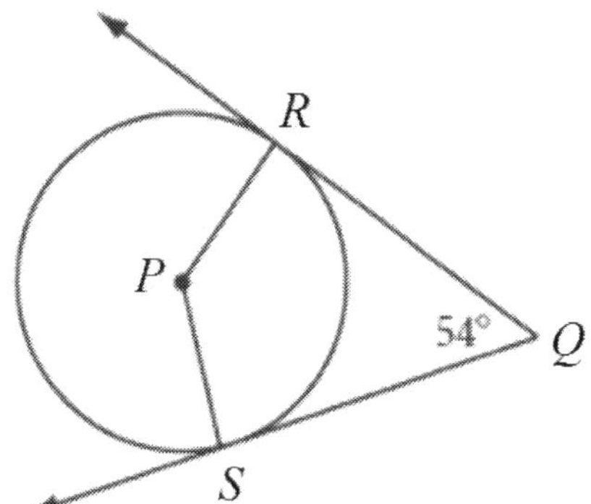

4. $\overleftrightarrow{ED}$ and $\overleftrightarrow{EC}$ are tangents to circle B at D and C. Find $m\angle DEC$.

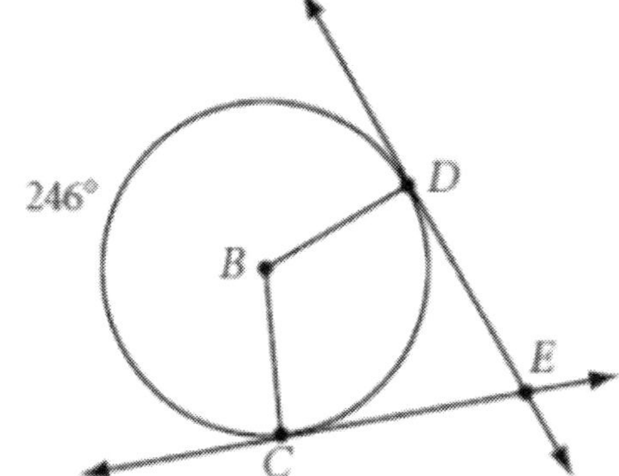

REGENTS QUESTIONS

Multiple Choice

1. In the diagram below, $\triangle ABC$ is circumscribed about circle O and the sides of $\triangle ABC$ are tangent to the circle at points D, E, and F.

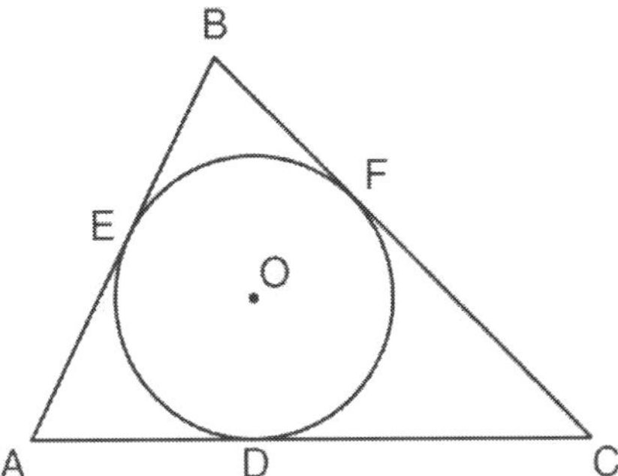

If $AB = 20$, $AE = 12$, and $CF = 15$, what is the length of $\overline{AC}$?
(1) 8 (3) 23
(2) 15 (4) 27

2. The angle formed by the radius of a circle and a tangent to that circle has a measure of
(1) 45° (3) 135°
(2) 90° (4) 180°

3. In the diagram below, $\overline{AC}$ and $\overline{BC}$ are tangent to circle O at A and B, respectively, from external point C.

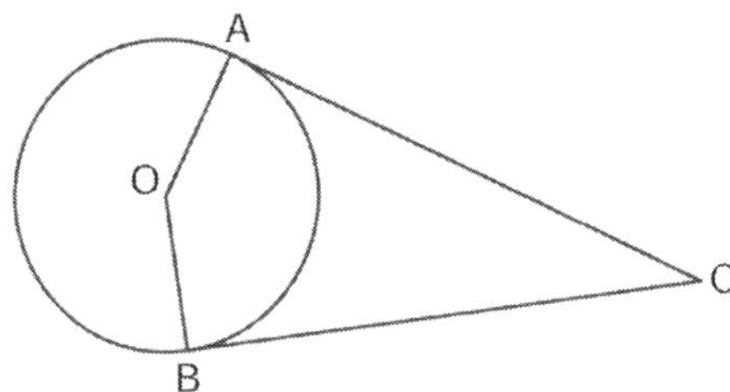

If $m\angle ACB = 38$, what is $m\angle AOB$?
(1) 71 (3) 142
(2) 104 (4) 161

4. How many common tangent lines can be drawn to the circles shown below?

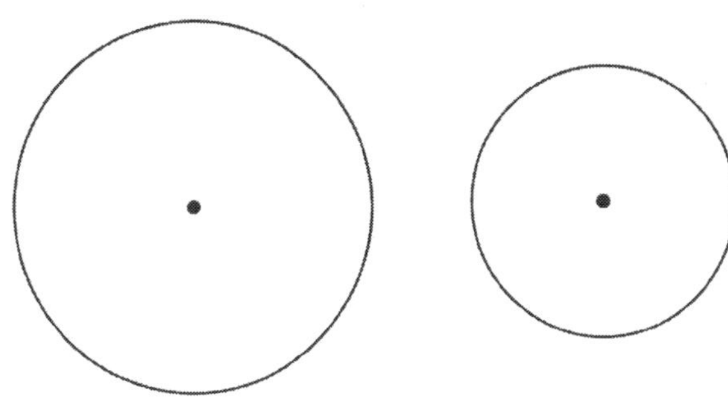

(1) 1
(2) 2
(3) 3
(4) 4

5. From external point A, two tangents to circle O are drawn. The points of tangency are B and C. Chord $\overline{BC}$ is drawn to form $\triangle ABC$. If $m\angle ABC = 66$, what is $m\angle A$?
(1) 33
(2) 48
(3) 57
(4) 66

6. (CC) In circle O shown below, diameter $\overline{AC}$ is perpendicular to $\overline{CD}$ at point C, and chords $\overline{AB}$, $\overline{BC}$, $\overline{AE}$, and $\overline{CE}$ are drawn.

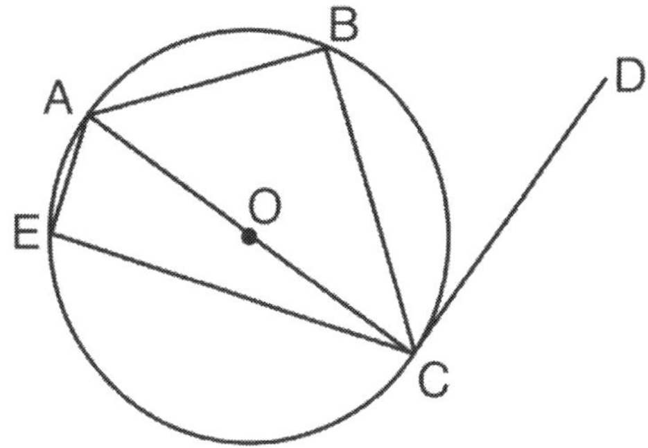

Which statement is *not* always true?
(1) $\angle ACB \cong \angle BCD$
(2) $\angle ABC \cong \angle ACD$
(3) $\angle BAC \cong \angle DCB$
(4) $\angle CBA \cong \angle AEC$

7. (CC) In the diagram shown below, $\overline{AC}$ is tangent to circle O at A and to circle P at C, $\overline{OP}$ intersects $\overline{AC}$ at B, $OA = 4$, $AB = 5$, and $PC = 10$.

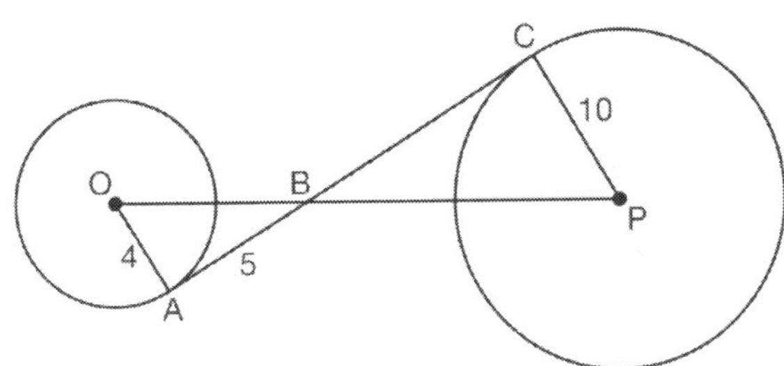

What is the length of $\overline{BC}$?

(1) 6.4
(2) 8
(3) 12.5
(4) 16

8. (CC) In the diagram below, $\overline{DC}$, $\overline{AC}$, $\overline{DOB}$, $\overline{CB}$, and $\overline{AB}$ are chords of circle O, $\overleftrightarrow{FDE}$ is tangent at point D, and radius $\overline{AO}$ is drawn. Sam decides to apply this theorem to the diagram: "An angle inscribed in a semi-circle is a right angle."

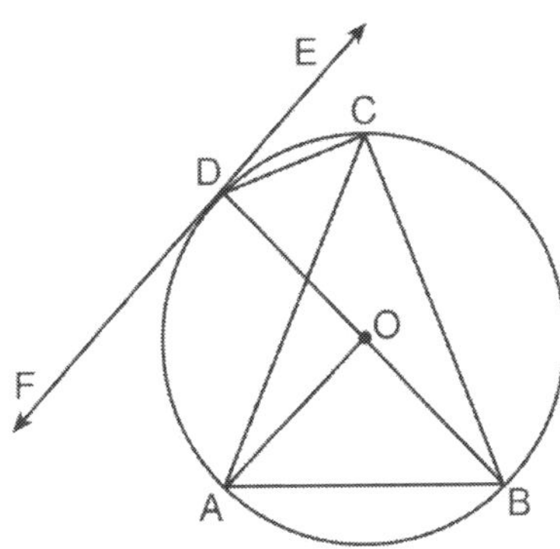

Which angle is Sam referring to?

(1) $\angle AOB$
(2) $\angle BAC$
(3) $\angle DCB$
(4) $\angle FDB$

Constructed Response

9. In the diagram below, circles A and B are tangent at point C and $\overline{AB}$ is drawn. Sketch all common tangent lines.

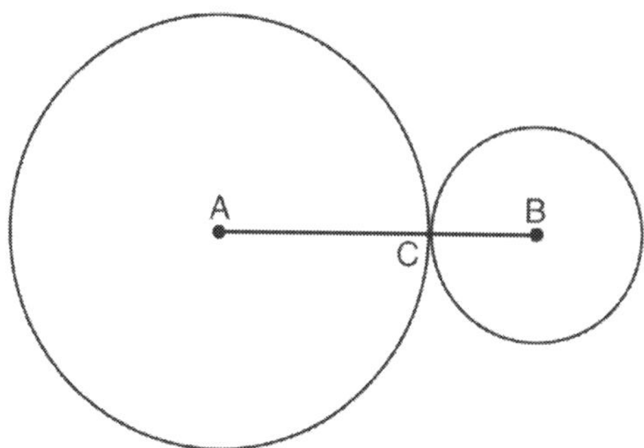

10. (CC) Lines AE and BD are tangent to circles O and P at A, E, B, and D, as shown in the diagram below. If $AC:CE = 5:3$, and $BD = 56$, determine and state the length of $\overline{CD}$.

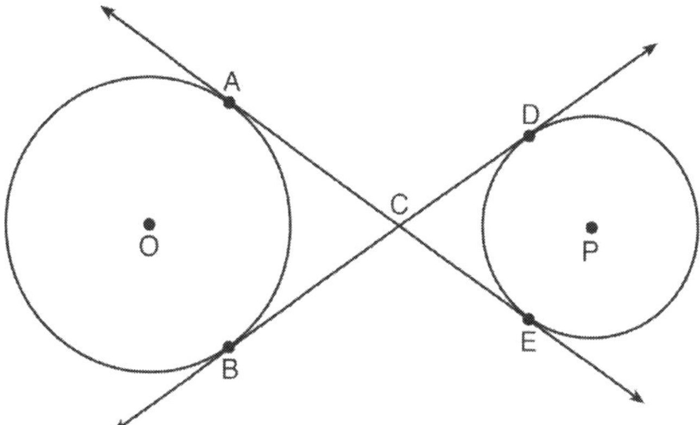

13.4 Secants

Key Terms and Concepts

A **secant** is a straight line that intersects a circle at two points. It is the extension of a chord.

Example: In the diagram below, line m is a secant intersecting the circle at points A and B, and line n is a tangent intersecting the circle at the point of tangency, C.

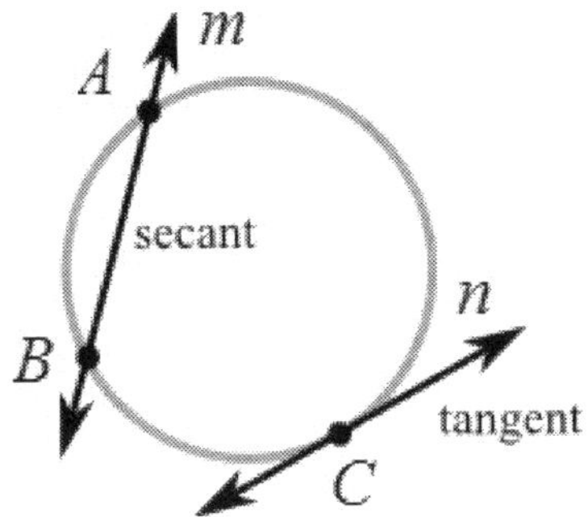

If two secants intersect at an external point, then the measure of the angle formed is half the difference of the intercepted arcs.

In other words, in the diagram to the right, $x = \dfrac{a-b}{2}$.

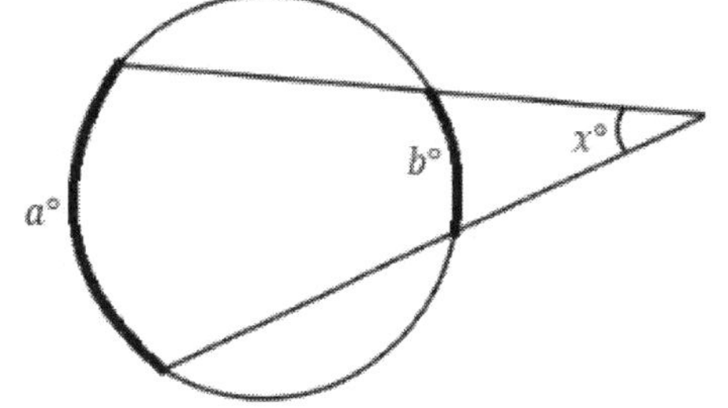

Below is an informal proof of this statement.

Given secants $\overline{JLN}$ and $\overline{KMN}$, draw $\overline{KL}$ to form $\triangle KLN$.

$m\overset{\frown}{JK} = a$, $m\overset{\frown}{LM} = b$, and $m\angle N = x$.

$m\angle JLK = m\angle K + m\angle N$ (exterior angle of a triangle)

$m\angle JLK = \dfrac{a}{2}$ and $m\angle K = \dfrac{b}{2}$ (inscribed angles)

$\dfrac{a}{2} = \dfrac{b}{2} + x$ (substitution)

$x = \dfrac{a-b}{2}$ (solve for x)

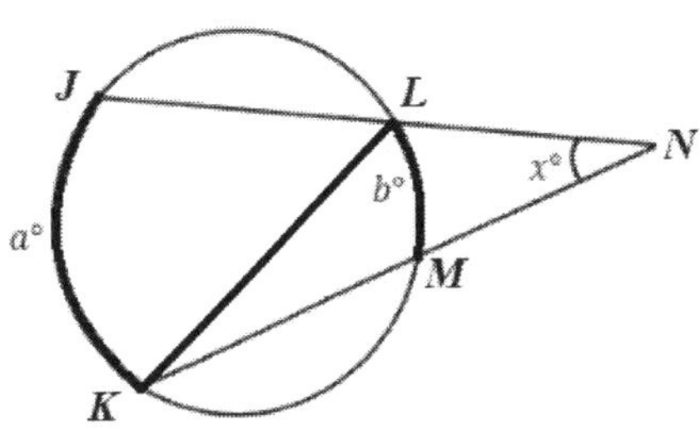

The same relationship exists when **two tangents**, or **a secant and a tangent** intersect at a point outside the circle. In each of the diagrams below, $x = \frac{a-b}{2}$.

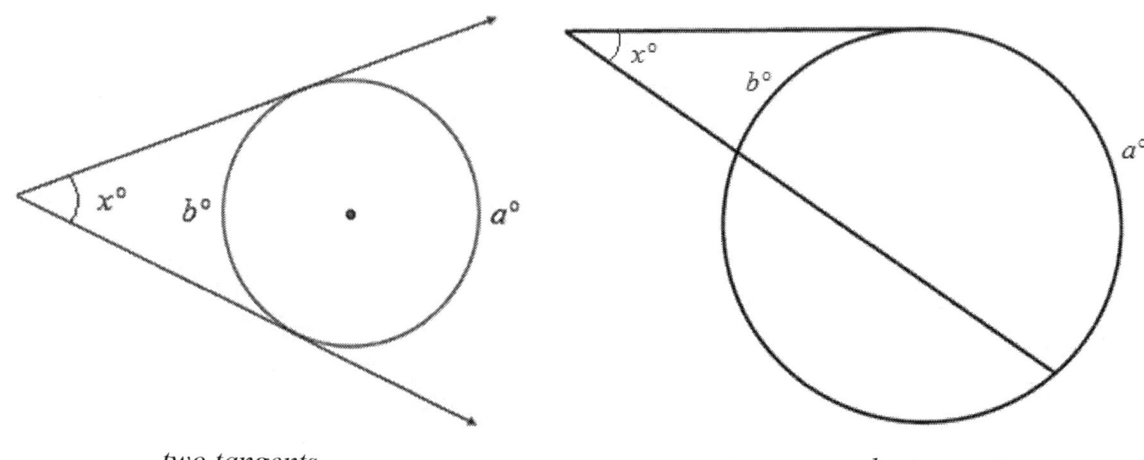

two tangents *a secant and a tangent*

If two secants are drawn to a circle from an external point, the product of the lengths of one secant segment and its external part equals the product of the lengths of the other secant segment and its external part.

Example: In the diagram below, $a(a+b) = c(c+d)$.

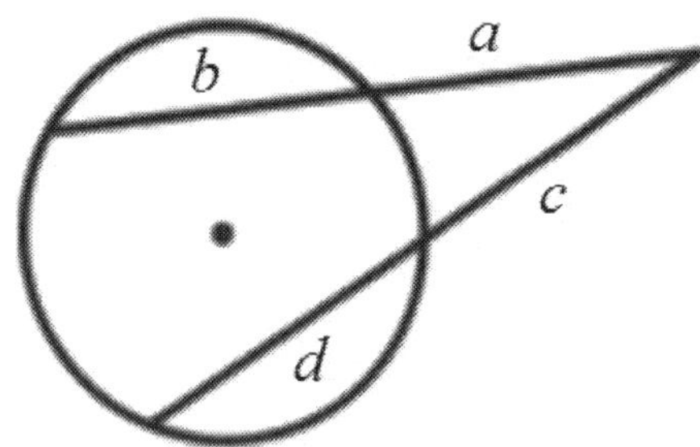

This can be proven by drawing chords connecting the points of intersection of the secants with the circle, as shown below.

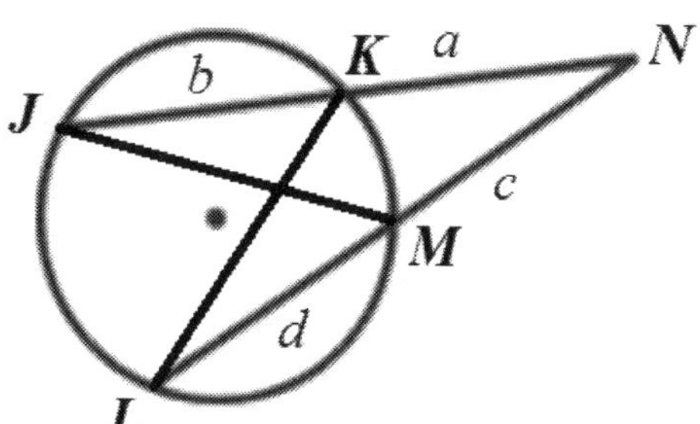

Since $\angle J \cong \angle L$ (inscribed angles of the same arc) and $\angle N \cong \angle N$ (reflexive), $\triangle NJM \sim \triangle NLK$ (AA Similarity).

So, $\frac{a}{c+d} = \frac{c}{a+b}$, which gives us $a(a+b) = c(c+d)$.

When a secant and tangent are drawn from an external point, the product of the lengths of the secant segment and its external part equals the square of the length of the tangent segment.

Example: In the diagram below, $a(a+b)=c^2$.

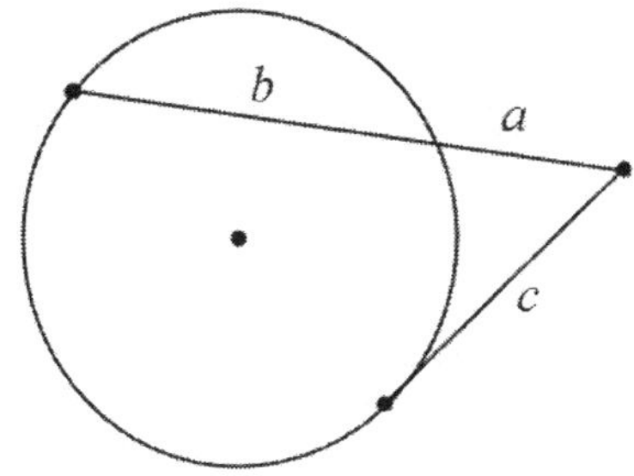

Model Problem:

In the accompanying diagram, $\overline{PAB}$ and $\overline{PCD}$ are secants drawn to circle O.
If $PA=8$, $PB=20$, and $PD=16$, find PC.

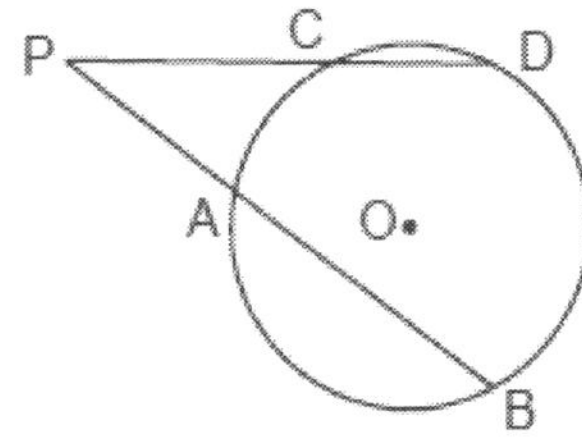

Solution:

$(PA)(PB)=(PC)(PD)$
$(8)(20)=(PC)(16)$
$PC=10$

Explanation of steps:

Use the formula $a(a+b)=c(c+d)$ where a and c are the external parts of the secants [PA and PC] and b and d are the internal parts of the secants [chords AB and CD].

Practice Problems

1. What is the name of each line or segment in relation to circle O?
(a) $\overleftrightarrow{HK}$ (b) $\overleftrightarrow{MN}$ (c) $\overline{BC}$ (d) $\overline{DF}$

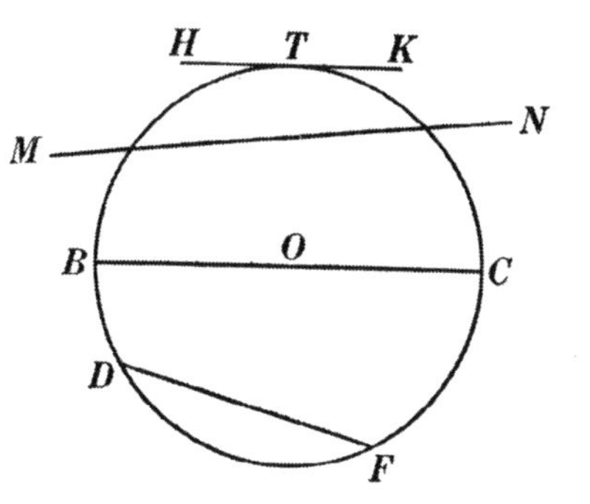

2. $\overline{PA}$ is tangent to circle O at A, secant $\overline{PBC}$ is drawn, $PB = 4$, and $BC = 12$. Find PA.

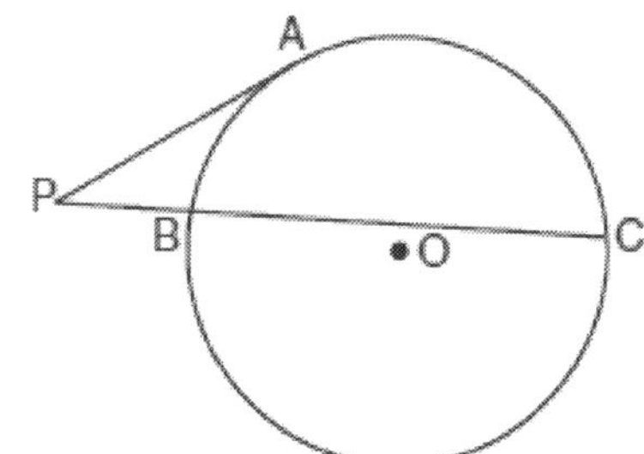

3. Find x.

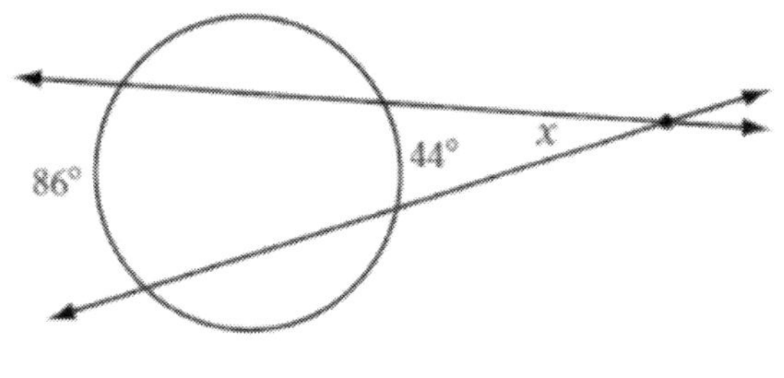

4. Find $m\widehat{DE}$.

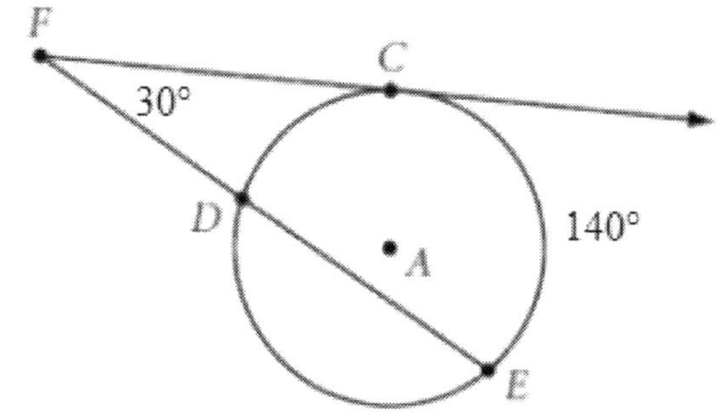

5. $\overline{CBA}$ and $\overline{CED}$ are secants. $AC = 12$, $BC = 3$, and $DC = 9$. Find EC.

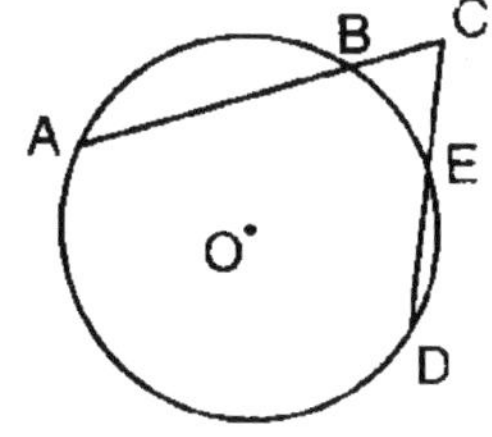

6. $\overline{PS}$ is a tangent to circle O at point S and $\overline{PQR}$ is a secant. If $PS = x$, $PQ = 3$, and $PR = x + 18$, find x.

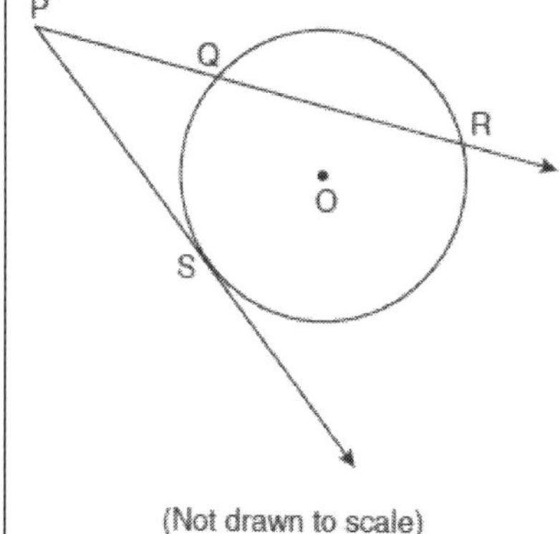

(Not drawn to scale)

REGENTS QUESTIONS

Multiple Choice

1. In the diagram below of circle O, $\overline{PAC}$ and $\overline{PBD}$ are secants.

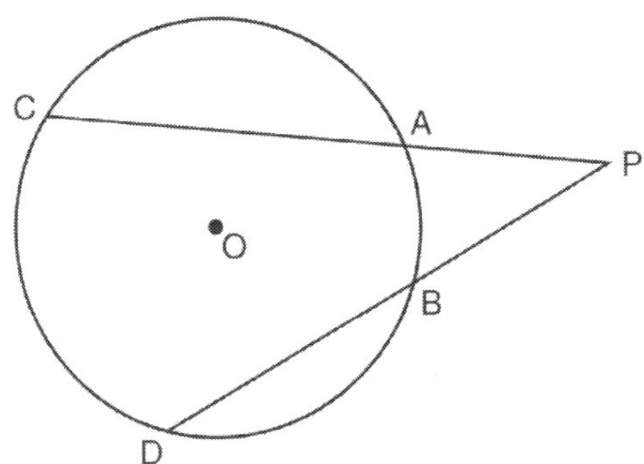

If $m\overset{\frown}{CD} = 70$ and $m\overset{\frown}{AB} = 20$, what is the degree measure of $\angle P$?
(1) 25
(2) 35
(3) 45
(4) 50

2. Secants $\overline{JKL}$ and $\overline{JMN}$ are drawn to circle O from an external point, J. If $JK = 8$, $LK = 4$, and $JM = 6$, what is the length of $\overline{JN}$?
(1) 16
(2) 12
(3) 10
(4) 8

3. In the diagram below of circle O with radius $\overline{OA}$, tangent $\overline{CA}$ and secant $\overline{COB}$ are drawn.

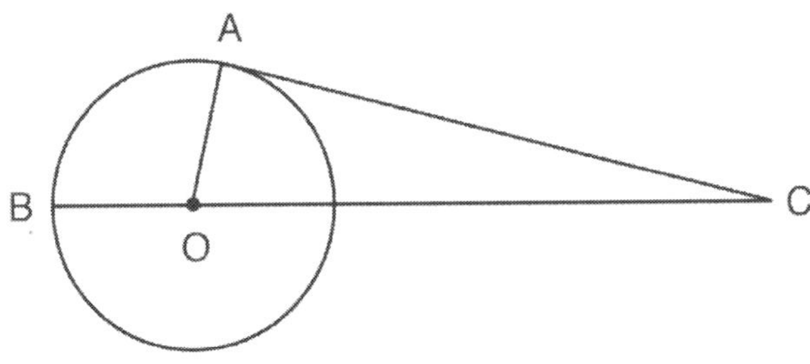

(Not drawn to scale)

If $AC = 20$ cm and $OA = 7$ cm, what is the length of $\overline{OC}$, to the *nearest centimeter*?
(1) 19
(2) 20
(3) 21
(4) 27

4. (CC) In circle O, secants $\overline{ADB}$ and $\overline{AEC}$ are drawn from external point A such that points D, B, E, and C are on circle O. If $AD = 8$, $AE = 6$, and EC is 12 more than BD, the length of $\overline{BD}$ is
(1) 6
(2) 22
(3) 36
(4) 48

Constructed Response

5. In the diagram below of circle O, chords $\overline{RT}$ and $\overline{QS}$ intersect at M. Secant $\overline{PTR}$ and tangent $\overline{PS}$ are drawn to circle O. The length of $\overline{RM}$ is two more than the length of $\overline{TM}$, $QM = 2$, $SM = 12$, and $PT = 8$.

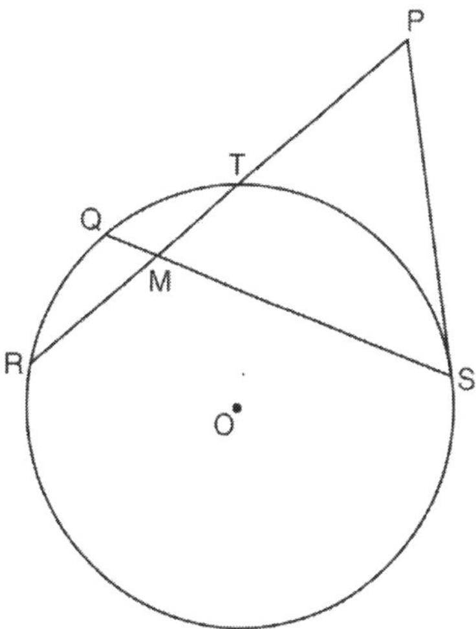

Find the length of $\overline{RT}$. Find the length of $\overline{PS}$.

6. Chords $\overline{AB}$ and $\overline{CD}$ intersect at E in circle O, as shown in the diagram below. Secant $\overline{FDA}$ and tangent $\overline{FB}$ are drawn to circle O from external point F and chord $\overline{AC}$ is drawn. The $m\overset{\frown}{DA} = 56$, $m\overset{\frown}{DB} = 112$, and the ratio of $m\overset{\frown}{AC} : m\overset{\frown}{CB} = 3:1$.

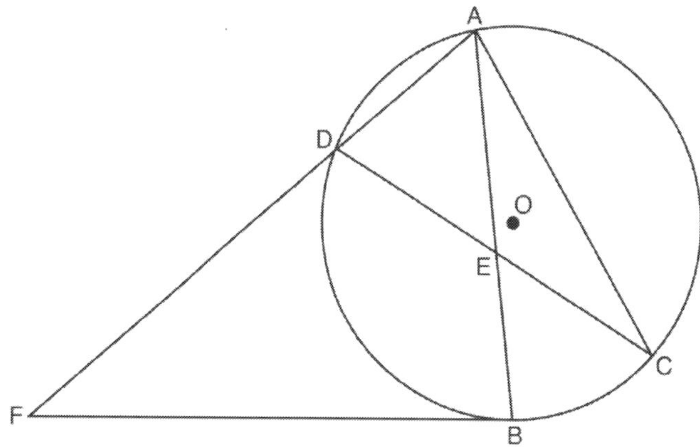

Determine $m\angle CEB$. Determine $m\angle F$. Determine $m\angle DAC$.

7. In the diagram below, secants $\overline{PQR}$ and $\overline{PST}$ are drawn to a circle from point P.

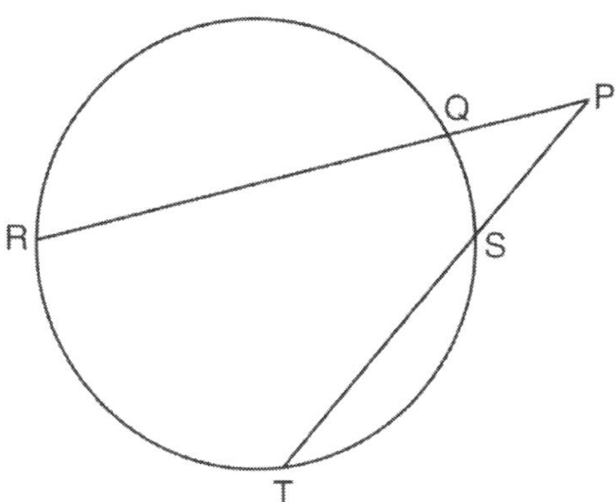

If $PR = 24$, $PQ = 6$, and $PS = 8$, determine and state the length of $\overline{PT}$.

8. (CC) In the diagram below, tangent $\overline{DA}$ and secant $\overline{DBC}$ are drawn to circle O from external point D, such that $\overparen{AC} \cong \overparen{BC}$.

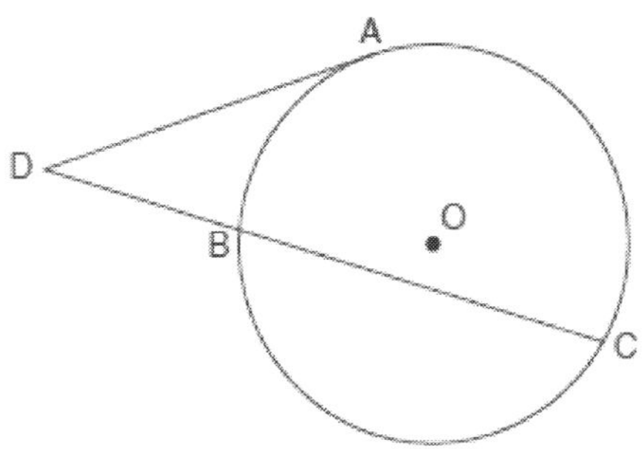

If $m\overparen{BC} = 152°$, determine and state $m\angle D$.

13.5 Circle Proofs

Key Terms and Concepts

For **proofs that involve circles**, any of the following theorems may be used:

- A central angle is equal in measure to its intercepted arc.
- If two arcs are congruent, their chords are congruent, and vice versa.
- If two arcs are congruent, their central angles are congruent, and vice versa.
- An inscribed angle is equal to half the measure of its intercepted arc.
- If two central angles are congruent, their chords are congruent, and vice versa.
- If two inscribed angles are congruent, their chords are congruent, and vice versa.
- An angle inscribed in a semicircle is a right angle.
- If two chords are parallel, they intercept congruent arcs between the chords.
- The perpendicular bisector of a chord bisects its arc and intersects the center of the circle.
- If a radius (or diameter) is perpendicular to a chord, it bisects the chord, and vice versa.
- If a quadrilateral is inscribed in a circle, both pairs of opposite angles are supplementary.
- Two intersecting chords form a pair of vertical angles that are each equal in measure to the average of their intercepted arcs.
- The tangent of a circle is perpendicular to a radius at the point of tangency.
- The angle formed by a tangent and a chord is half the measure of the intercepted arc.
- The two tangent segments drawn to a circle from the same external point are congruent.
- A circumscribed angle is equal to 180° minus the measure of the minor arc it intercepts.
- If two secants (or two tangents, or a secant and a tangent) intersect at an external point, then the measure of the angle formed is half the difference of the intercepted arcs.
- If two secants are drawn to a circle from an external point, the product of the lengths of one secant segment and its external part equals the product of the lengths of the other secant segment and its external part.
- When a secant and tangent are drawn from an external point, the product of the lengths of the secant segment and its external part equals the square of the length of the tangent segment.

Model Problem:

Given: Circle O, $\overline{DB}$ is tangent to the circle at B, $\overline{BC}$ and $\overline{BA}$ are chords, and C is the midpoint of $\overset{\frown}{AB}$.

Prove: $\angle ABC \cong \angle CBD$

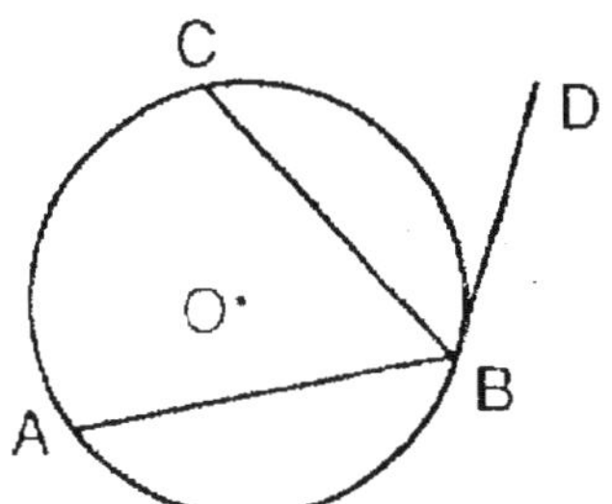

Solution:

Statements	*Reasons*
Circle O, $\overline{DB}$ is tangent to the circle at B, $\overline{BC}$ and $\overline{BA}$ are chords, and C is the midpoint of $\overset{\frown}{AB}$	Given
$m\overset{\frown}{AC} = m\overset{\frown}{BC}$	Definition of midpoint
$m\angle ABC = \frac{1}{2} m\overset{\frown}{AC}$	The measure of an inscribed angle is one-half the measure of its intercepted arc
$m\angle CBD = \frac{1}{2} m\overset{\frown}{BC}$	The measure of an angle formed by a tangent and a chord is one-half the measure of the intercepted arc
$\frac{1}{2} m\overset{\frown}{AC} = \frac{1}{2} m\overset{\frown}{BC}$	Multiplication property of equality
$m\angle ABC = m\angle CBD$	Substitution
$\angle ABC \cong \angle CBD$	Definition of congruence

Practice Problems

1. Given: Circle O, $\overset{\frown}{AB} \cong \overset{\frown}{AC}$
   Prove: $\triangle AOC \cong \triangle AOB$

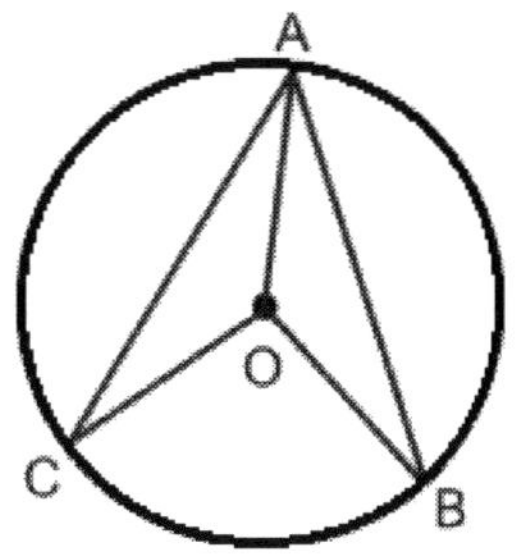

Statements	Reasons

2. Given: Circle Q, $\overline{PQR} \perp \overline{ST}$
   Prove: $\overline{PS} \cong \overline{PT}$

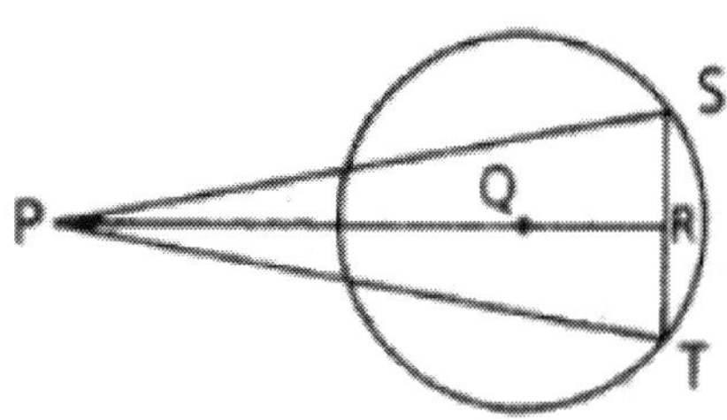

Statements	Reasons

REGENTS QUESTIONS

Constructed Response

1. In the accompanying diagram, $m\overset{\frown}{BR} = 70$, $m\overset{\frown}{YD} = 70$, and $\overline{BOD}$ is the diameter of circle O. Write an explanation or a proof that shows $\triangle RBD$ and $\triangle YBD$ are congruent.

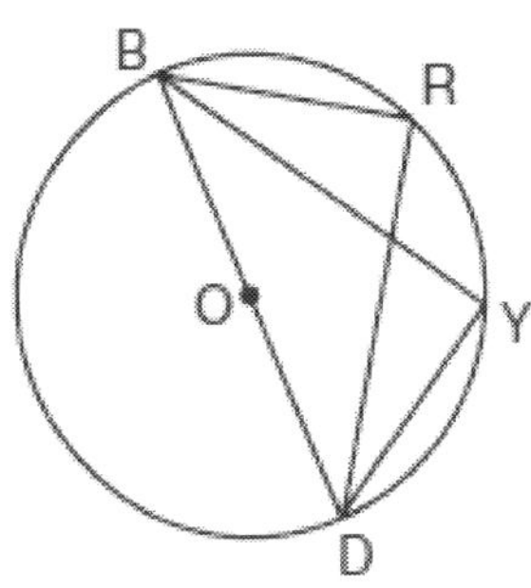

2. In the diagram below, quadrilateral $ABCD$ is inscribed in circle O, $\overline{AB} \parallel \overline{DC}$, and diagonals $\overline{AC}$ and $\overline{BD}$ are drawn. Prove that $\triangle ACD \cong \triangle BDC$.

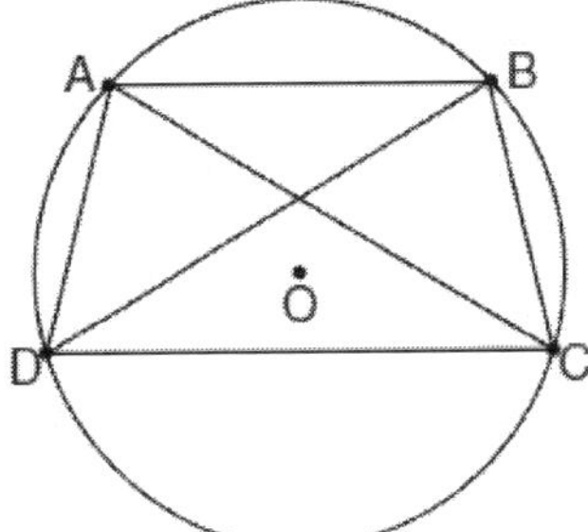

3. In the accompanying diagram of circle O, $\overline{AD}$ is a diameter with $\overline{AD}$ parallel to chord $\overline{BC}$, chords $\overline{AB}$ and $\overline{CD}$ are drawn, and chords $\overline{BD}$ and $\overline{AC}$ intersect at E.
Prove: $\overline{BE} \cong \overline{CE}$

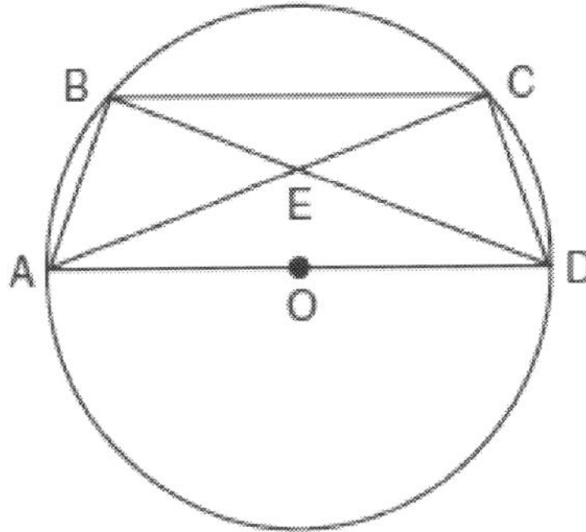

4. In the diagram below, $\overline{PA}$ and $\overline{PB}$ are tangent to circle O, $\overline{OA}$ and $\overline{OB}$ are radii, and $\overline{OP}$ intersects the circle at C. Prove: $\angle AOP \cong \angle BOP$.

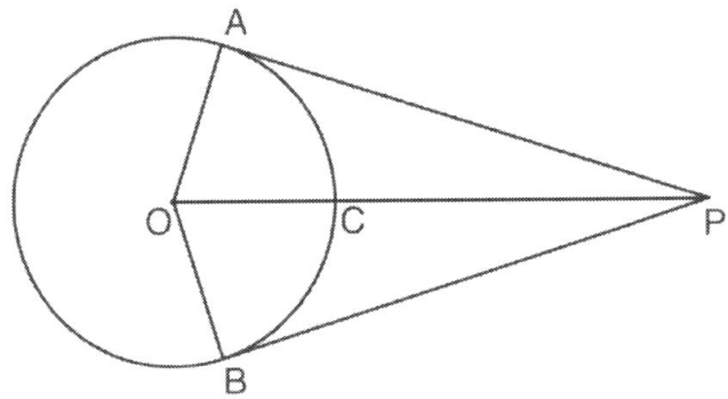

5. In the diagram of circle O below, diameter $\overline{RS}$, chord $\overline{AS}$, tangent $\overrightarrow{TS}$, and secant $\overline{TAR}$ are drawn.

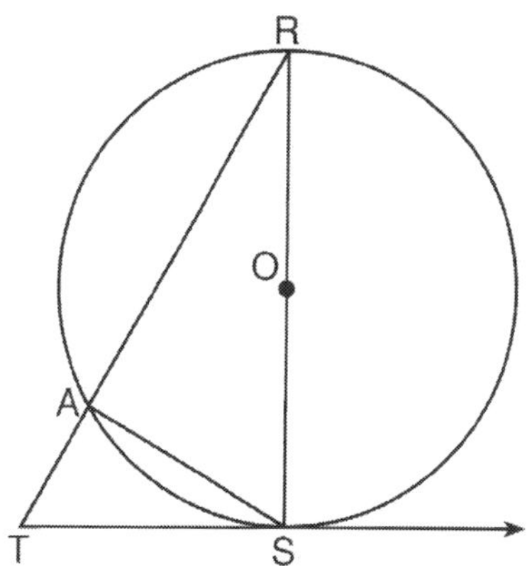

Complete the following proof to show $(RS)^2 = RA \cdot RT$.

Statements	Reasons
1. circle O, diameter $\overline{RS}$, chord $\overline{AS}$, tangent $\overrightarrow{TS}$, and secant $\overline{TAR}$	1. Given
2. $\overline{RS} \perp \overrightarrow{TS}$	2. ________________
3. $\angle RST$ is a right angle	3. $\perp$ lines form right angles
4. $\angle RAS$ is a right angle	4. ________________
5. $\angle RST \cong \angle RAS$	5. ________________
6. $\angle R \cong \angle R$	6. Reflexive property
7. $\triangle RST \sim \triangle RAS$	7. ________________
8. $\frac{RS}{RA} = \frac{RT}{RS}$	8. ________________
9. $(RS)^2 = RA \bullet RT$	9. ________________

6. (CC) In the diagram below, secant $\overline{ACD}$ and tangent $\overline{AB}$ are drawn from external point A to circle O.

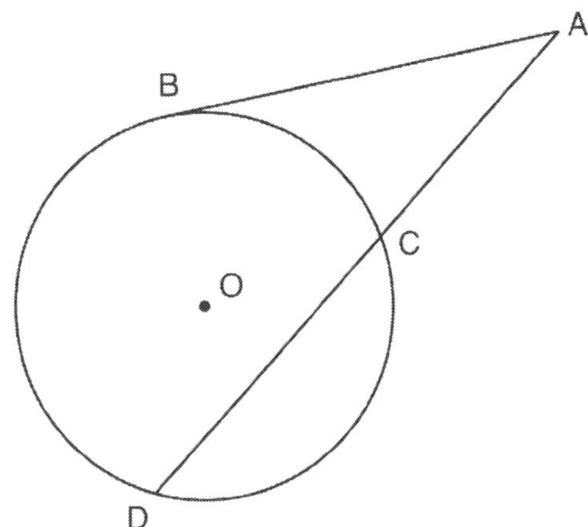

Prove the theorem:
If a secant and a tangent are drawn to a circle from an external point, the product of the lengths of the secant segment and its external segment equals the length of the tangent segment squared. ($AC \cdot AD = AB^2$)

7. (CC) Given: Circle O, chords $\overline{AB}$ and $\overline{CD}$ intersect at E

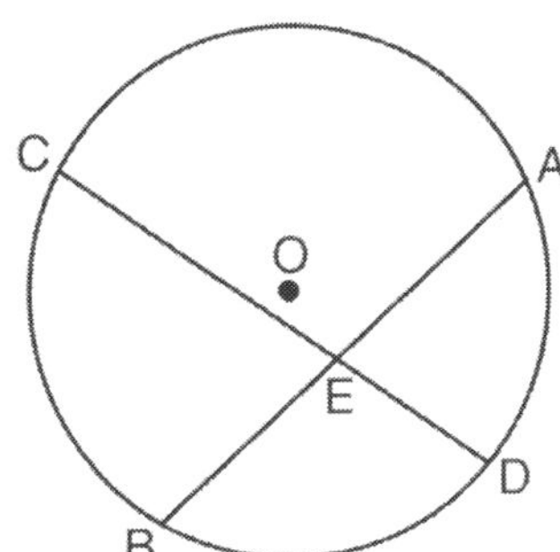

Theorem: If two chords intersect in a circle, the product of the lengths of the segments of one chord is equal to the product of the lengths of the segments of the other chord. Prove this theorem by proving $AE \cdot EB = CE \cdot ED$.

13.6 Arc Lengths and Sectors

Key Terms and Concepts

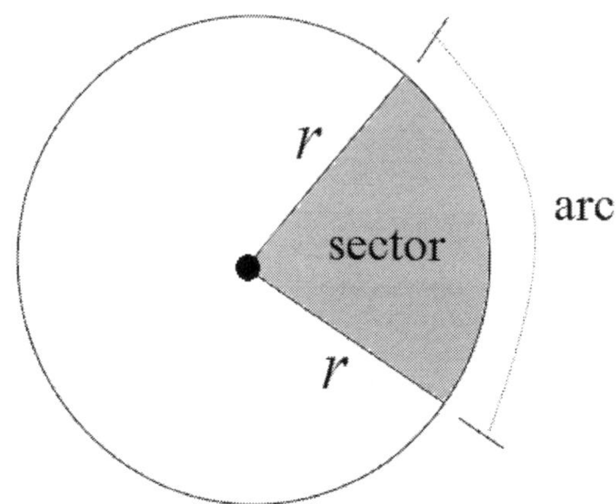

A **sector** is a figure enclosed by two radii of a circle and the arc between them.

To find the **length of an arc** or the **area of a sector**, you can set up a proportion:

Length of an arc
The central angle θ is to 360° as the length of the arc, L, is to the circumference of the full circle:

$$\frac{\theta}{360^\circ} = \frac{L}{2\pi r}$$

Area of a sector
The central angle θ is to 360° as the area of the sector, S, is to the area of the full circle:

$$\frac{\theta}{360^\circ} = \frac{S}{\pi r^2}$$

You can find the **area of a segment** but subtraction.

Example: We can find the area of the segment formed by chord $\overline{AB}$ below by subtracting the area of the sector minus the area of the triangle.

$$A = \frac{60}{360} \cdot 36\pi = 6\pi \qquad A = \tfrac{1}{2}(6)(6)\sin 60^\circ = 9\sqrt{3} \qquad A = 6\pi - 9\sqrt{3}$$

Model Problem:

In the diagram below, find the length of arc AB, to the *nearest tenth.*

Solution:

(A) $\frac{\theta}{360^\circ} = \frac{L}{2\pi r}$

(B) $\frac{45}{360} = \frac{L}{2\pi \cdot 12}$

(C) $360L = 1080\pi$

(D) $L = 3\pi \approx 9.4$

Explanation of steps:

(A) Write the proportion for the length of an arc.

(B) Substitute known values.

(C) Cross multiply.

(D) Solve.

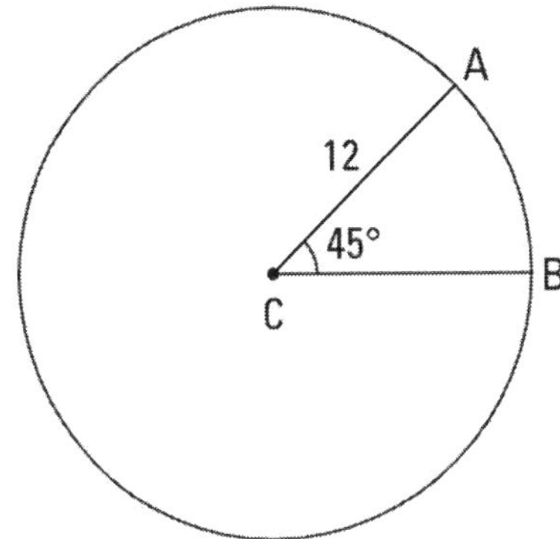

Practice Problems

1. Find the length of $\overset{\frown}{AB}$.

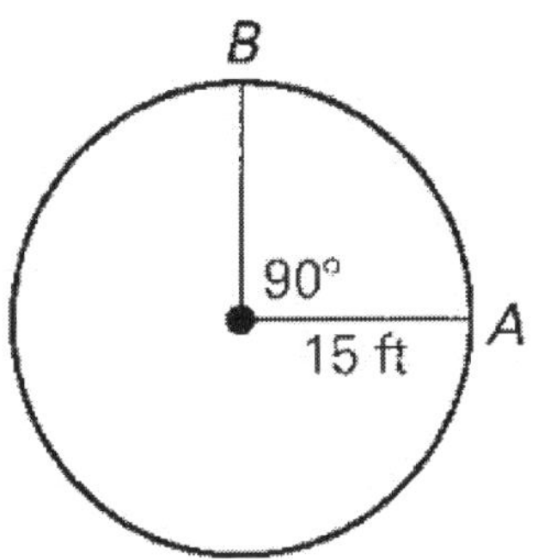

2. Find the length of $\overset{\frown}{AB}$.

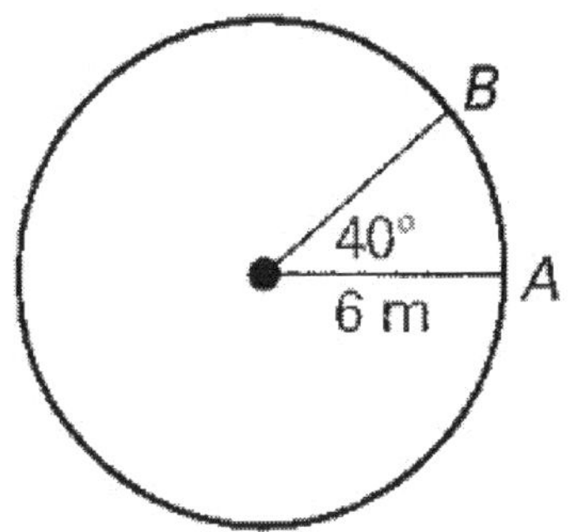

3. Diameter AC is 34 cm long. Find the length of $\overset{\frown}{BC}$ to the *nearest centimeter.*

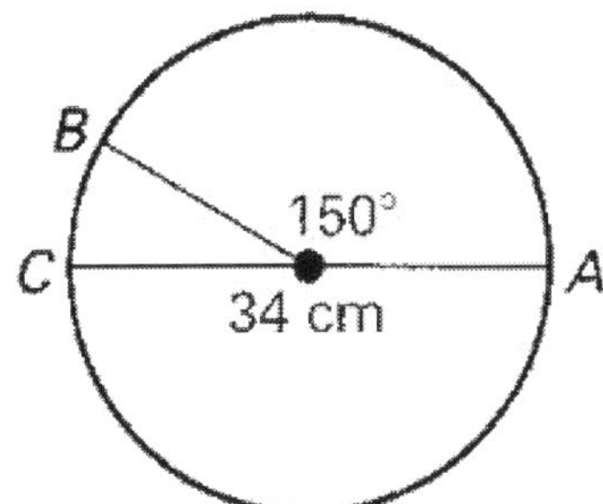

4. Find the measure of the central angle of an arc if its length is 14π and the radius is 18.

5. Find the area of the shaded sector.

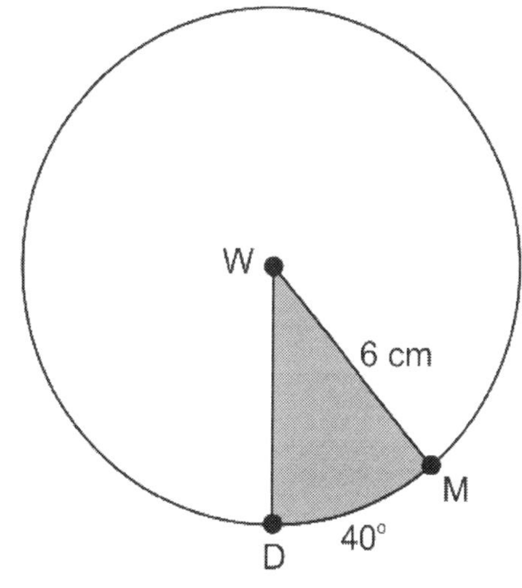

6. Find the area of the shaded sector.

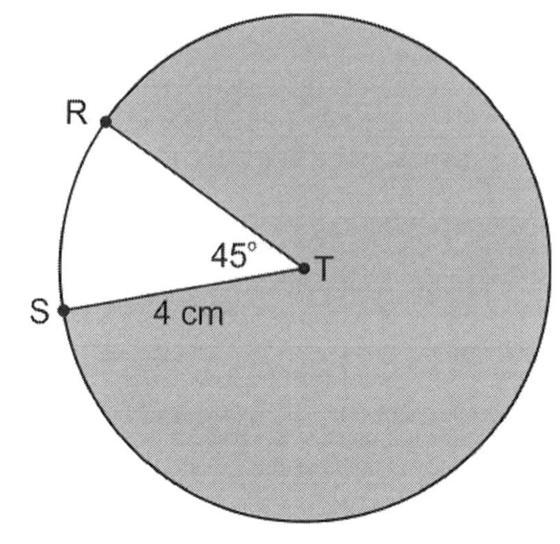

7. Find the area of the shaded segment.

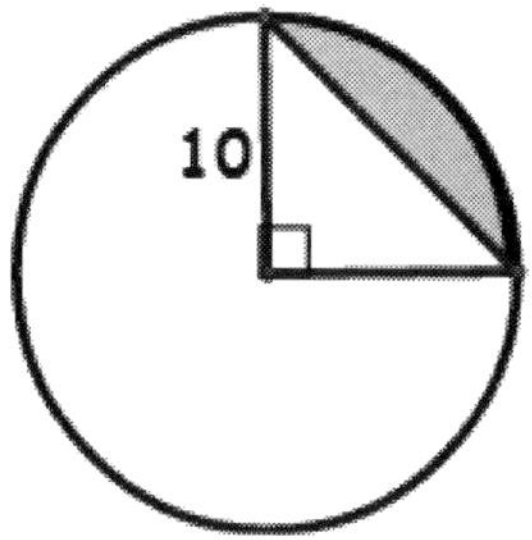

8. Find the area of the shaded segment.

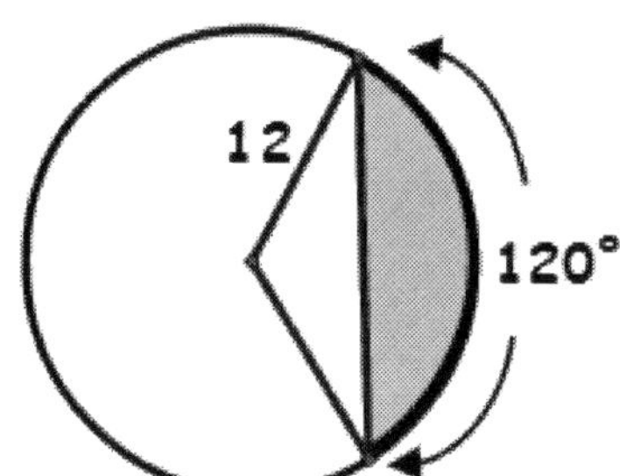

REGENTS QUESTIONS

Multiple Choice

1. Cerise waters her lawn with a sprinkler that sprays water in a circular pattern at a distance of 15 feet from the sprinkler. The sprinkler head rotates through an angle of 300°, as shown by the shaded area in the accompanying diagram.

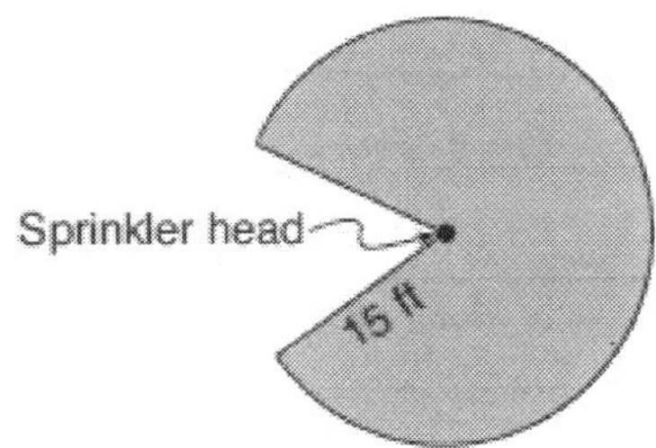

What is the area of the lawn, to the *nearest square foot*, that receives water from this sprinkler?

(1) 79
(2) 94
(3) 589
(4) 707

2. A circle is drawn to represent a pizza with a 12 inch diameter. The circle is cut into eight congruent pieces. What is the length of the outer edge of any one piece of this circle?

(1) $\frac{3\pi}{4}$
(2) π
(3) $\frac{3\pi}{2}$
(4) 3π

3. CC A circle with a radius of 5 was divided into 24 congruent sectors. The sectors were then rearranged, as shown in the diagram below.

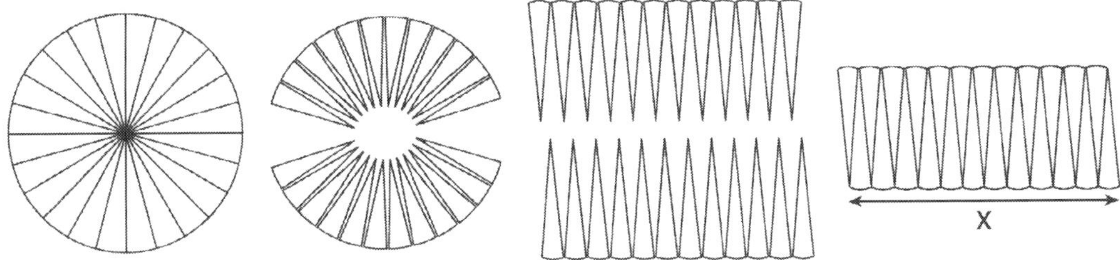

To the *nearest integer*, the value of x is

(1) 31
(2) 16
(3) 12
(4) 10

4. (CC) Triangle FGH is inscribed in circle O, the length of radius $\overline{OH}$ is 6, and $\overline{FH} \cong \overline{OG}$.

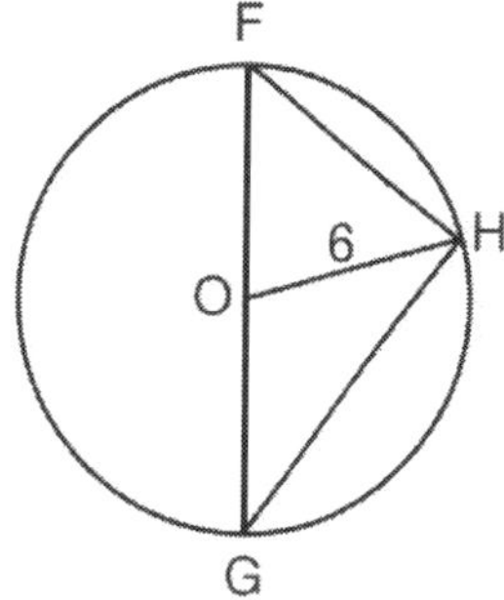

What is the area of the sector formed by angle FOH?

(1) 2π
(2) $\frac{3}{2}\pi$
(3) 6π
(4) 24π

5. (CC) In the diagram below of circle O, the area of the shaded sector LOM is $2\pi\ \text{cm}^2$.

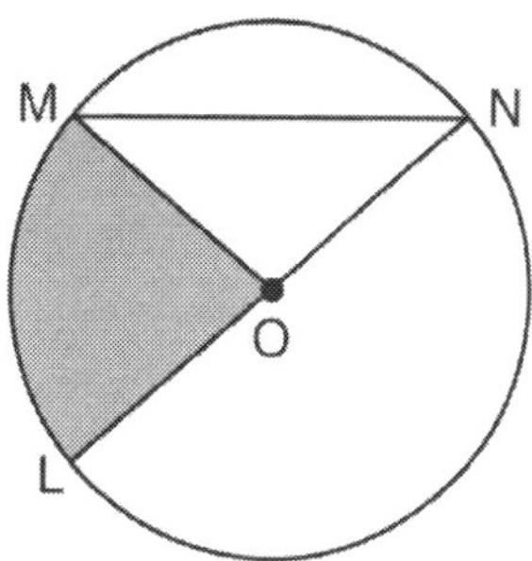

If the length of $\overline{NL}$ is 6 cm, what is $m\angle N$?

(1) 10°
(2) 20°
(3) 40°
(4) 80°

6. (CC) What is the area of a sector of a circle with a radius of 8 inches and formed by a central angle that measures 60°?

(1) $\frac{8\pi}{3}$
(2) $\frac{16\pi}{3}$
(3) $\frac{32\pi}{3}$
(4) $\frac{64\pi}{3}$

7. (CC) In circle O, diameter $\overline{AB}$, chord $\overline{BC}$, and radius $\overline{OC}$ are drawn, and the measure of arc BC is 108°.

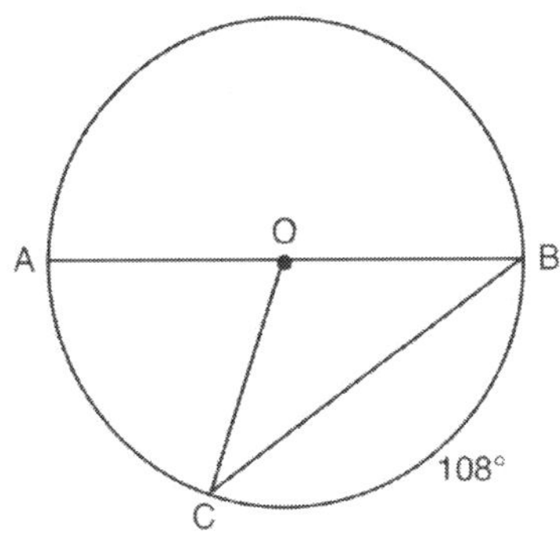

Some students wrote these formulas to find the area of sector COB:

Amy $\frac{3}{10} \cdot \pi \cdot (BC)^2$

Beth $\frac{108}{360} \cdot \pi \cdot (OC)^2$

Carl $\frac{3}{10} \cdot \pi \cdot \left(\frac{1}{2} AB\right)^2$

Dex $\frac{108}{360} \cdot \pi \cdot \frac{1}{2}(AB)^2$

Which students wrote correct formulas?

(1) Amy and Dex
(2) Beth and Carl
(3) Carl and Amy
(4) Dex and Beth

8. (CC) In the diagram below of circle O, $GO = 8$ and $m\angle GOJ = 60°$.

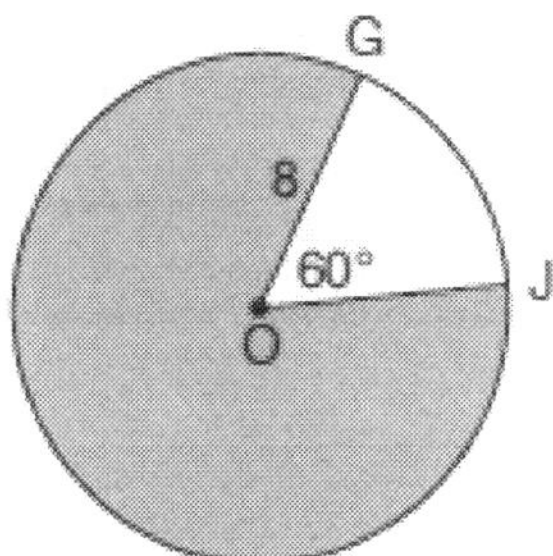

What is the area, in terms of π, of the shaded region?

(1) $\frac{4\pi}{3}$
(2) $\frac{20\pi}{3}$
(3) $\frac{32\pi}{3}$
(4) $\frac{160\pi}{3}$

Constructed Response

9. Cities *H* and *K* are located on the same line of longitude and the difference in the latitude of these cities is 9°, as shown in the accompanying diagram. If Earth's radius is 3,954 miles, how many miles north of city *K* is city *H* along arc *HK*? Round your answer to the *nearest tenth of a mile*.

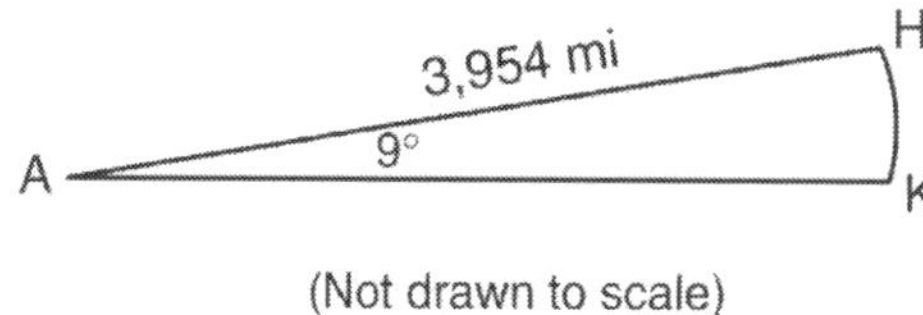

10. The accompanying diagram shows the path of a cart traveling on a circular track of radius 2.40 meters. The cart starts at point *A* and stops at point *B*, moving in a counterclockwise direction. What is the length of minor arc *AB*, over which the cart traveled, to the *nearest tenth of a meter*?

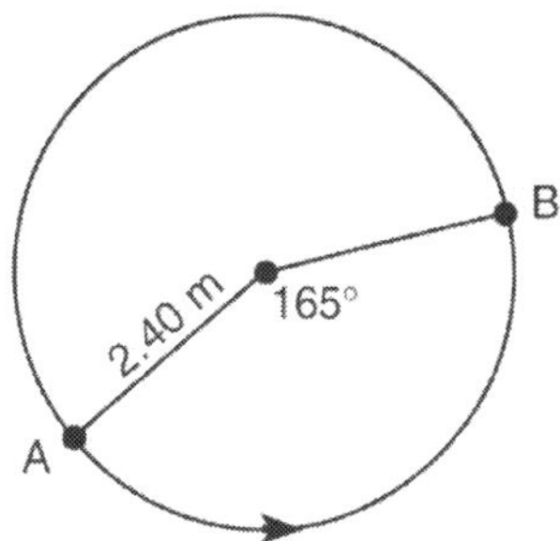

11. Kathy and Tami are at point *A* on a circular track that has a radius of 150 feet, as shown in the accompanying diagram. They run counterclockwise along the track from *A* to *S*, a distance of 247 feet. Find, to the *nearest degree*, the measure of minor arc *AS*.

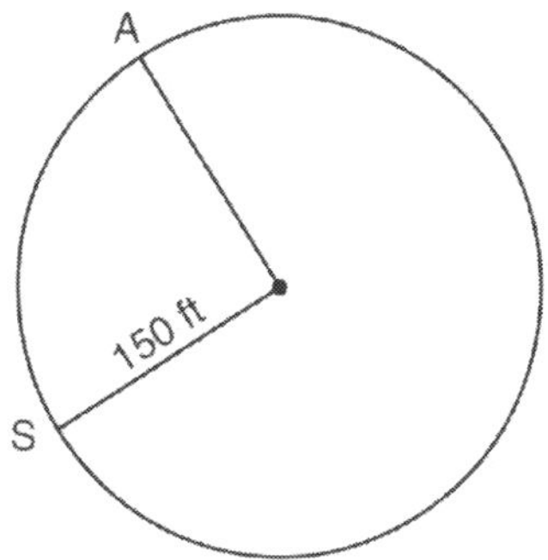

12. (CC) In the diagram below of circle O, diameter $\overline{AB}$ and radii $\overline{OC}$ and $\overline{OD}$ are drawn. The length of $\overline{AB}$ is 12 and the measure of $\angle COD$ is 20 degrees.

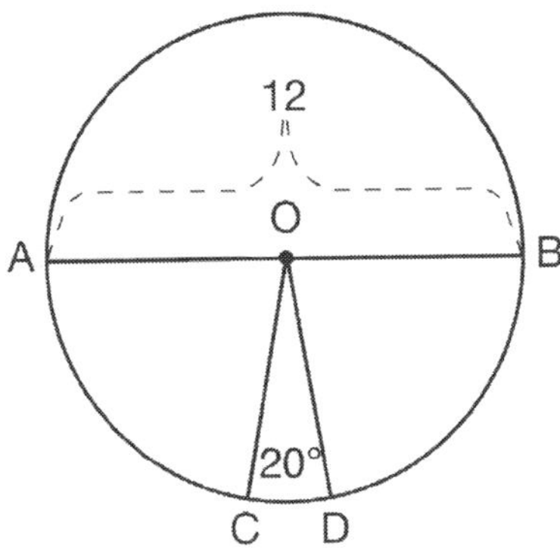

If $\overset{\frown}{AC} \cong \overset{\frown}{BD}$, find the area of sector BOD in terms of π.

13. (CC) In the diagram below of circle O, the area of the shaded sector AOC is 12π in^2 and the length of $\overline{OA}$ is 6 inches. Determine and state $m\angle AOC$.

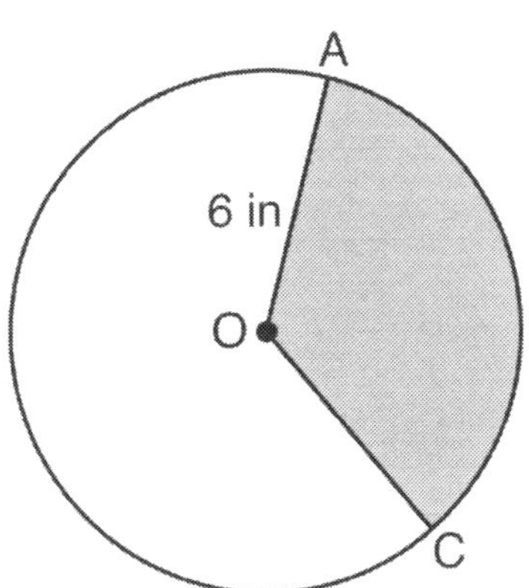

13.7 Radians (-)

In the Geometry Draft of Revised Standards for 2018-19, NYS has recommended ***removing*** *this topic.*

Key Terms and Concepts

Angles may be measured in degrees or in radians. A central angle of **one radian** intercepts an arc whose length is equal to the radius of the circle.

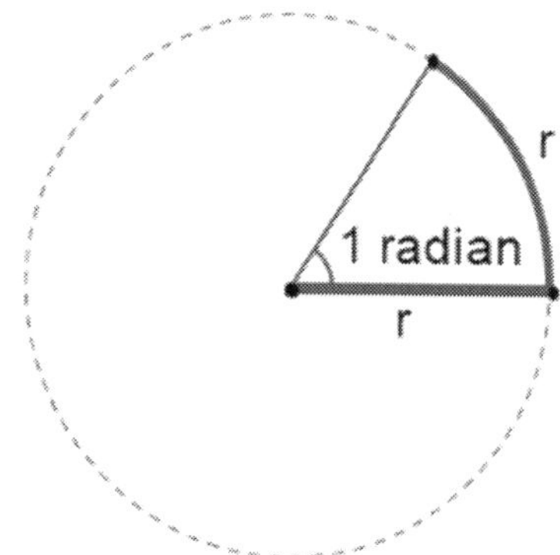

1 radian is approximately 57.2958°. The abbreviation for radians is rad.

The circumference of a circle is $2\pi r$, so there are 2π radians in a circle.
So, 2π radians = 360°, and π radians = 180°. Therefore:

$$1 \text{ radian} = \frac{180}{\pi} \text{ degrees} \qquad 1 \text{ degree} = \frac{\pi}{180} \text{ radians}$$

Note: The formulas for converting between degrees and radians are included on the Reference Sheet at the back of the Regents exam.

When a central angle θ is measured in radians, we can set up the proportion, $\frac{\theta}{2\pi} = \frac{L}{2\pi r}$.

Therefore, the length of the intercepted arc, $L = \theta r$.

Model Problem:

In a circle with a 3 inch radius, what is the measure of a central angle, in radians, that intercepts an arc of:

(a) 6 inches? (b) 3π inches?

Solution:

(a) $6 = 3 \cdot \theta$ $\theta = 2$ radians

(b) $3\pi = 3 \cdot \theta$ $\theta = \pi$ radians

Explanation of steps:

Substitute the known values *[the arc length L and the radius r]* into the formula $L = \theta r$ and solve.

Practice Problems

1. Convert to radians:

   a) 45°

   b) 270°

   c) 150°

2. Convert to degrees:

   a) $\frac{\pi}{6}$ rad

   b) $\frac{5\pi}{4}$ rad

   c) $\frac{3\pi}{5}$ rad

3. A circle has a radius of 12 inches. What is the length of the arc intercepting a central angle of $\frac{\pi}{4}$ rad?

4. In a circle with a 10 inch radius, an arc measures 8π inches in length. Find the measure of its central angle in radians.

5. Find the radius of a circle in which a central angle of 5 radians intercepts an arc of 65 feet.

6. In the circle with a radius of 9 cm, find the area of the shaded sector bounded by the central angle measuring 0.97 radians.

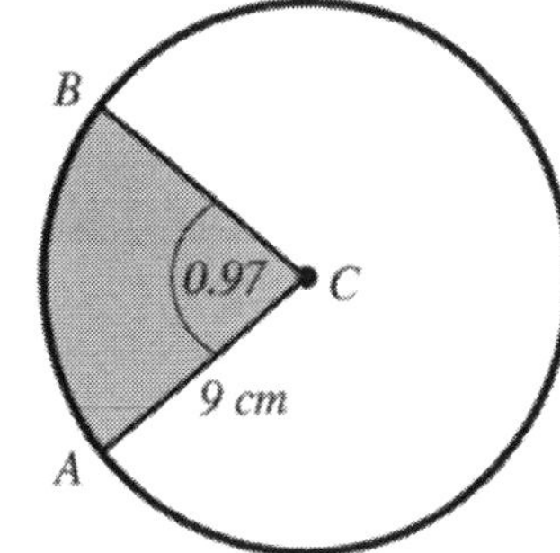

REGENTS QUESTIONS

Multiple Choice

1. A sprinkler system is set up to water the sector shown in the accompanying diagram, with angle ABC measuring 1 radian and radius $AB = 20$ feet.

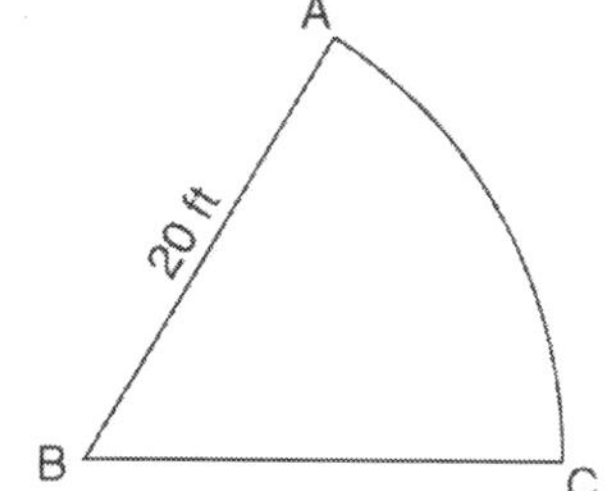

What is the length of arc AC, in feet?

(1) 63 (3) 20
(2) 31 (4) 10

2. A central angle of a circular garden measures 2.5 radians and intercepts an arc of 20 feet. What is the radius of the garden?

(1) 8 ft (3) 100 ft
(2) 50 ft (4) 125 ft

3. The number of degrees equal to $\frac{5}{9}\pi$ radians is

(1) 45 (3) 100
(2) 90 (4) 900

4. A circle has a radius of 4 inches. In inches, what is the length of the arc intercepted by a central angle of 2 radians?

(1) 2π (3) 8π
(2) 2 (4) 8

5. A central angle of $\frac{4\pi}{15}$ radians intercepts an arc whose degree measure is

(1) 48 (3) 96
(2) 72 (4) $\frac{4\pi}{15}$

6. What is the number of degrees in an angle whose radian measure is $\frac{11\pi}{12}$?

(1) 150 (3) 330
(2) 165 (4) 518

7. What is the radian measure of the smaller angle formed by the hands of a clock at 7 o'clock?

(1) $\frac{\pi}{2}$ (3) $\frac{5\pi}{6}$
(2) $\frac{2\pi}{3}$ (4) $\frac{7\pi}{6}$

8. What is the number of degrees in an angle whose measure is 2 radians?

1) $\frac{360}{\pi}$ (3) 360

2) $\frac{\pi}{360}$ (4) 90

9. What is the number of degrees in an angle whose radian measure is $\frac{8\pi}{5}$?

(1) 576 (3) 225
(2) 288 (4) 113

10. Approximately how many degrees does five radians equal?

(1) 286 (3) $\frac{\pi}{36}$

(2) 900 (4) 5π

11. A wheel has a radius of 18 inches. Which distance, to the *nearest inch*, does the wheel travel when it rotates through an angle of $\frac{2\pi}{5}$ radians?

(1) 45 (3) 13
(2) 23 (4) 11

12. In a circle with a diameter of 24 cm, a central angle of $\frac{4\pi}{3}$ radians intercepts an arc. The length of the arc, in centimeters, is

(1) 8π (3) 16π
(2) 9π (4) 32π

13. (CC) In the diagram below, the circle shown has radius 10. Angle B intercepts an arc with a length of 2π.

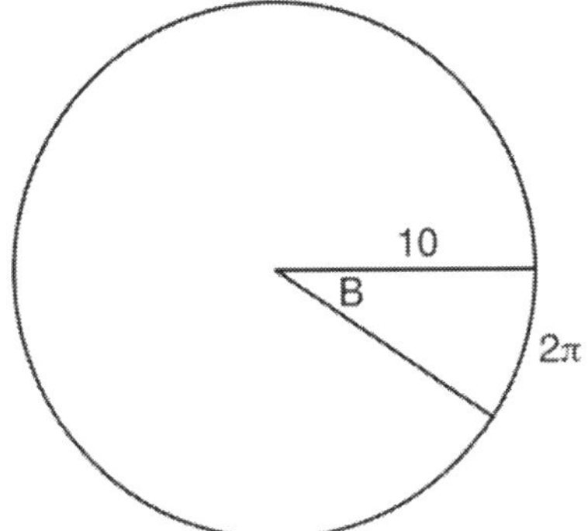

What is the measure of angle B, in radians?

(1) $10 + 2\pi$ (3) $\frac{\pi}{5}$

(2) 20π (4) $\frac{5}{\pi}$

Constructed Response

14. The tip of a pendulum describes an arc 18 centimeters long when the pendulum swings through an angle of $\frac{3}{4}$ of a radian. Find the length, in centimeters, of the pendulum.

15. Find, to the *nearest tenth of a degree*, the angle whose measure is 2.5 radians.

16. Find, to the *nearest tenth*, the radian measure of 216°.

17. A central angle whose measure is $\frac{2\pi}{3}$ radians intercepts an arc with a length of 4π feet. Find the radius of the circle, *in feet*.

18. (CC) In the diagram below, Circle 1 has radius 4, while Circle 2 has radius 6.5. Angle A intercepts an arc of length π, and angle B intercepts an arc of length $\frac{13\pi}{8}$.

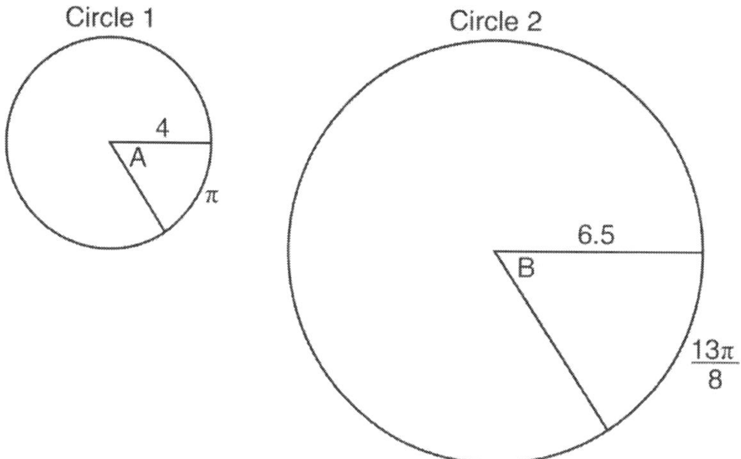

Dominic thinks that angles A and B have the same radian measure. State whether Dominic is correct or not. Explain why.

13.8 Circles in the Coordinate Plane

Key Terms and Concepts

The equation of a circle with its center at the origin and a radius of r is $x^2 + y^2 = r^2$. This equation is derived from the Pythagorean Theorem. Consider the circle below, with a random point on the circle at (x,y). As long as $x \neq 0$ and $y \neq 0$, then we can draw a right triangle whose hypotenuse is the radius draw to (x,y), a horizontal leg of length x and a vertical leg of length y. By the Pythagorean Theorem, $x^2 + y^2 = r^2$. This is true for any point (x,y) on the circle. *(In the special cases, if $x = 0$, then $y = r$, and if $y = 0$, then $x = r$, so the equation still holds.)*

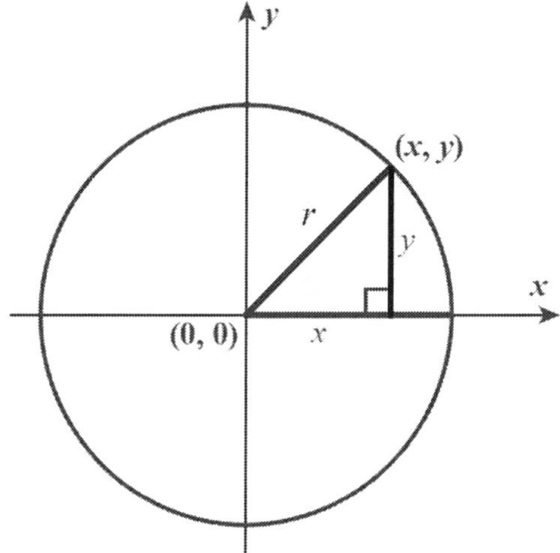

A circle need not have its center at the origin. The **general equation of a circle** with a center at (a,b) and a radius of r is $(x-a)^2 + (y-b)^2 = r^2$.

Example: A circle with the equation $(x-2)^2 + (y+3)^2 = 64$ has a center of $(2,-3)$ and a radius of 8.

We can derive the general equation of a circle with a center of (a,b) just as we did with a circle whose center is at the origin. The only difference is that we would apply the Pythagorean Theorem to the right triangle whose legs are $(x-a)$ and $(y-b)$ in length, as shown to the right.

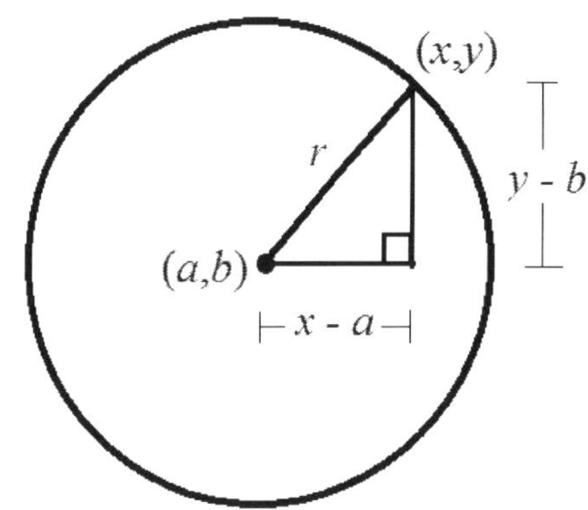

If the equation of a circle is not written in $(x-a)^2+(y-b)^2=r^2$ form, you may be able to transform the equation into this form by completing the square twice – once for the x terms and once for the y terms.

To convert an equation into $(x-a)^2+(y-b)^2=r^2$ form by completing the square twice:

1. Write the equation in the form, $x^2+hx+y^2+ky+c=0$, where h and k are coefficients.
2. Add the opposite of c to both sides.
3. Add $\left(\frac{h}{2}\right)^2$ to both sides. *(If $h=0$, skip steps 3 and 4.)*
4. Factor the trinomial involving x into a binomial squared.
5. Add $\left(\frac{k}{2}\right)^2$ to both sides. *(If $k=0$, skip steps 5 and 6.)*
6. Factor the trinomial involving y into a binomial squared.

Model Problem 1: *circle centered at the origin*

Does the point $\left(1,\sqrt{3}\right)$ lie on the circle centered at the origin and containing the point (0,2)?

Solution:

(A) $x^2+y^2=r^2$

$0^2+2^2=r^2$

$r=2$

(B) $1^2+\left(\sqrt{3}\right)^2=2^2$?

$1+3=4$ ✓

Explanation of steps:

(A) In the equation of a circle centered at the origin, substitute the coordinates of the known point in order to find the radius. *[Substitute 0 for x and 2 for y and solve for r.]*

(B) In the equation of a circle centered at the origin, substitute the coordinates of the point in question *[1 and $\sqrt{3}$]* and the radius *[2]* to determine whether the equation is true.

Model Problem 2: *rewriting an equation into the general form for a circle*

What is the center and radius of the circle whose equation is $x^2-2x+y^2+6y-6=0$?

Solution:

(A) $x^2-2x+y^2+6y-6=0$

(B) $x^2-2x+y^2+6y=6$

(C) $\left(\frac{h}{2}\right)^2=\left(\frac{-2}{2}\right)^2=1$

$(x^2-2x+1)+y^2+6y=6+1$

(D) $(x-1)^2+y^2+6y=7$

(E) $\left(\frac{k}{2}\right)^2=\left(\frac{6}{2}\right)^2=9$

$(x-1)^2+(y^2+6y+9)=7+9$

(F) $(x-1)^2+(y+3)^2=16$

(G) Center is $(1,-3)$ and radius is 4.

Explanation of steps:

(A) Write the equation in the form, $x^2+hx+y^2+ky+c=0$. *[The equation was already in this form.]*

(B) Add the opposite of c to both sides. *[Add 6 to both sides.]*

(C) Add $\left(\frac{h}{2}\right)^2$ to both sides. *[Write this value, 1, next to the x terms to create a trinomial.]*

(D) Factor the trinomial involving x into a binomial squared.

(E) Add $\left(\frac{k}{2}\right)^2$ to both sides. *[Write this value, 9, next to the y terms to create a trinomial.]*

(F) Factor the trinomial involving y into a binomial squared.

(G) Now that the equation is in the form $(x-a)^2+(y-b)^2=r^2$, state the center (a,b) and the radius r. *[$r^2=16$, so $r=\sqrt{16}=4$.]*

Practice Problems

1. A circle has the equation $x^2 + y^2 = 10$.
   (a) What are the coordinates of its center and the length of its radius?

   (b) Is the point (3,–1) on the circle?

2. A circle has the equation $(x-2)^2 + (y+3)^2 = 36$. What are the coordinates of its center and the length of its radius?

3. A circle has the equation $x^2 + (y-7)^2 = 32$. What are the coordinates of its center and the length of its radius in simplest radical form?

4. What is the equation of this circle, with the center and a point on the circle given?

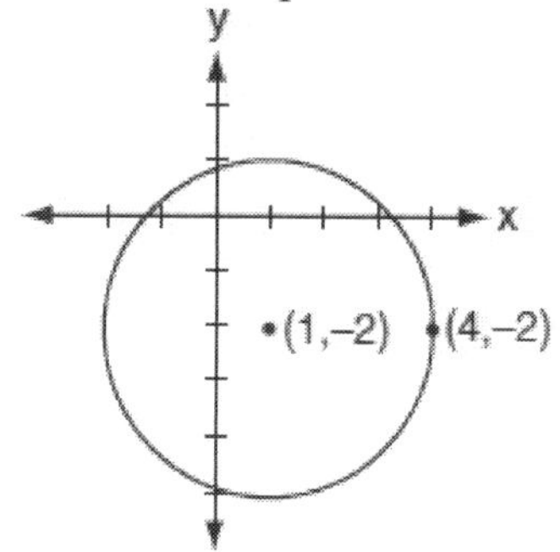

5. A circle has the equation $x^2 + y^2 + 6x - 4y = 12$. What are the coordinates of its center and the length of its radius in simplest radical form?

6. A circle has the equation $x^2 - 2x = -y^2 + 10y + 1$. What are the coordinates of its center and the length of its radius in simplest radical form?

REGENTS QUESTIONS

Multiple Choice

1. Which point is on the circle whose equation is $x^2+y^2=289$?
   (1) (–12,12)
   (2) (7,–10)
   (3) (–1,–16)
   (4) (8,–15)

2. Using a drawing program, a computer graphics designer constructs a circle on a coordinate plane on her computer screen. She determines that the equation of the circle's graph is $(x-3)^2+(y+2)^2=36$. She then dilates the circle with the transformation D_3. After this transformation, what is the center of the new circle?
   (1) (6,–5)
   (2) (–6,5)
   (3) (9,–6)
   (4) (–9,6)

3. The equation $x^2+y^2-2x+6y+3=0$ is equivalent to
   (1) $(x-1)^2+(y+3)^2=-3$
   (2) $(x-1)^2+(y+3)^2=7$
   (3) $(x+1)^2+(y+3)^2=7$
   (4) $(x+1)^2+(y+3)^2=10$

4. What are the coordinates of the center of a circle whose equation is $x^2+y^2-16x+6y+53=0$?
   (1) (–8,–3)
   (2) (–8,3)
   (3) (8,–3)
   (4) (8,3)

5. The coordinates of the endpoints of the diameter of a circle are (2,0) and (2,–8). What is the equation of the circle?
   (1) $(x-2)^2+(y+4)^2=16$
   (2) $(x+2)^2+(y-4)^2=16$
   (3) $(x-2)^2+(y+4)^2=8$
   (4) $(x+2)^2+(y-4)^2=8$

6. A circle whose center has coordinates (–3,4) passes through the origin. What is the equation of the circle?
   (1) $(x+3)^2+(y-4)^2=5$
   (2) $(x+3)^2+(y-4)^2=25$
   (3) $(x-3)^2+(y+4)^2=5$
   (4) $(x-3)^2+(y+4)^2=25$

7. The diagram below is a graph of circle O.

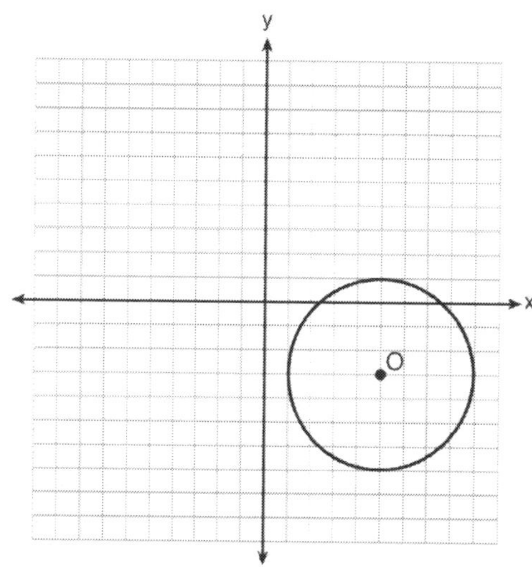

Which equation represents circle O?

(1) $(x-5)^2+(y+3)^2=4$

(2) $(x+5)^2+(y-3)^2=4$

(3) $(x-5)^2+(y+3)^2=16$

(4) $(x+5)^2+(y-3)^2=16$

8. Which equation represents the circle shown in the graph below?

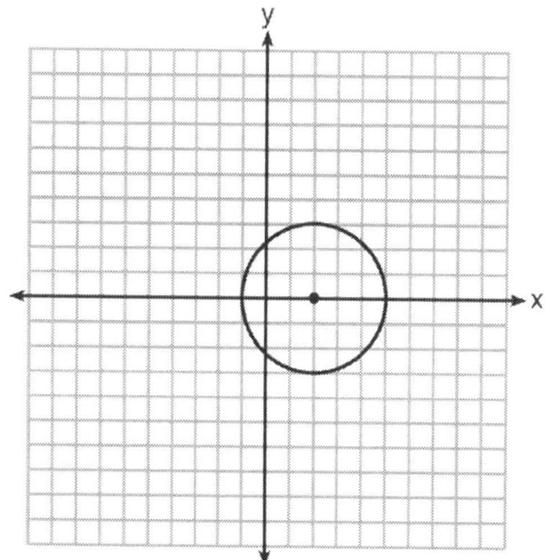

(1) $(x-2)^2+y^2=9$

(2) $(x+2)^2+y^2=9$

(3) $(x-2)^2+y^2=3$

(4) $(x+2)^2+y^2=3$

9. Students made four statements about a circle.
   - A: The coordinates of its center are (4,–3).
   - B: The coordinates of its center are (–4,3).
   - C: The length of its radius is $5\sqrt{2}$.
   - D: The length of its radius is 25.

   If the equation of the circle is $(x+4)^2+(y-3)^2=50$, which statements are correct?

   (1) A and C
   (2) A and D
   (3) B and C
   (4) B and D

10. Which equation represents a circle whose center is the origin and that passes through the point (–4,0)?

    (1) $x^2+y^2=8$
    (2) $x^2+y^2=16$
    (3) $(x+4)^2+y^2=8$
    (4) $(x+4)^2+y^2=16$

11. In a circle whose equation is $(x-1)^2+(y+3)^2=9$, the coordinates of the center and length of its radius are

    (1) (1,–3) and $r=81$
    (2) (–1,3) and $r=81$
    (3) (1,–3) and $r=3$
    (4) (–1,3) and $r=3$

12. Which equation represents the circle shown in the graph below?

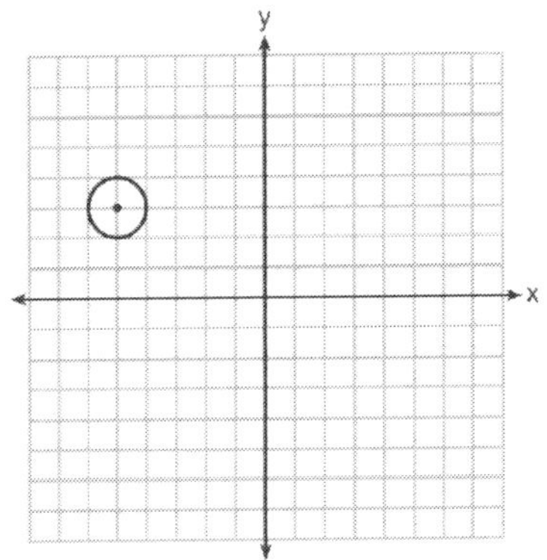

    (1) $(x-5)^2+(y+3)^2=1$
    (2) $(x+5)^2+(y-3)^2=1$
    (3) $(x-5)^2+(y+3)^2=2$
    (4) $(x+5)^2+(y-3)^2=2$

13. What are the center and radius of the circle whose equation is $x^2+y^2+4x=5$?

    (1) (2,0) and 1
    (2) (–2,0) and 1
    (3) (2,0) and 3
    (4) (–2,0) and 3

14. CC The equation of a circle is $x^2+y^2+6y=7$. What are the coordinates of the center and the length of the radius of the circle?

    (1) center (0,3) and radius 4
    (2) center (0,–3) and radius 4
    (3) center (0,3) and radius 16
    (4) center (0,–3) and radius 16

15. (CC) If $x^2+4x+y^2-6y-12=0$ is the equation of a circle, the length of the radius is

(1) 25
(2) 16
(3) 5
(4) 4

16. (CC) What are the coordinates of the center and length of the radius of the circle whose equation is $x^2+6x+y^2-4y=23$?

(1) (3,–2) and 36
(2) (3,–2) and 6
(3) (–3,2) and 36
(4) (–3,2) and 6

17. (CC) Kevin's work for deriving the equation of a circle is shown below.

$$x^2+4x=-(y^2-20)$$

STEP 1 $\quad x^2+4x=-y^2+20$

STEP 2 $\quad x^2+4x+4=-y^2+20-4$

STEP 3 $\quad (x+2)^2=-y^2+20-4$

STEP 4 $\quad (x+2)^2+y^2=16$

In which step did he make an error in his work?

(1) Step 1
(2) Step 2
(3) Step 3
(4) Step 4

18. (CC) The graph below shows $\overline{AB}$, which is a chord of circle O. The coordinates of the endpoints of $\overline{AB}$ are A(3,3) and B(3,–7). The distance from the midpoint of $\overline{AB}$ to the center of circle O is 2 units.

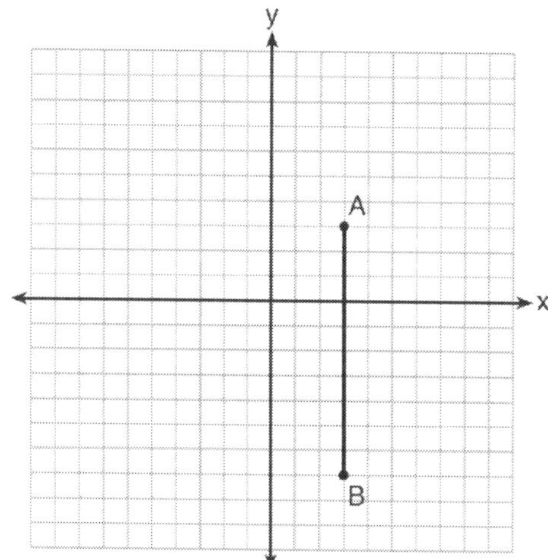

What could be a correct equation for circle O?

(1) $(x-1)^2+(y+2)^2=29$
(2) $(x+5)^2+(y-2)^2=29$
(3) $(x-1)^2+(y-2)^2=25$
(4) $(x-5)^2+(y+2)^2=25$

19. (CC) What are the coordinates of the center and the length of the radius of the circle represented by the equation $x^2 + y^2 - 4x + 8y + 11 = 0$?

(1) center (2,–4) and radius 3
(2) center (–2,4) and radius 3
(3) center (2,–4) and radius 9
(4) center (–2,4) and radius 9

20. (CC) The equation of a circle is $x^2 + y^2 - 6y + 1 = 0$. What are the coordinates of the center and the length of the radius of this circle?

(1) center (0,3) and radius = $2\sqrt{2}$
(2) center (0,–3) and radius = $2\sqrt{2}$
(3) center (0,6) and radius = $\sqrt{35}$
(4) center (0,–6) and radius = $\sqrt{35}$

21. (CC) A circle whose center is the origin passes through the point (–5,12). Which point also lies on this circle?

(1) (10,3)
(2) (–12,13)
(3) $\left(11, 2\sqrt{12}\right)$
(4) $\left(-8, 5\sqrt{21}\right)$

Constructed Response

22. The engineering office in the village of Whitesboro has a map of the village that is laid out on a rectangular coordinate system. A traffic circle located on the map is represented by the equation $(x+4)^2 + (y-2)^2 = 81$. The village planning commission asks that the transformation D_2 be applied to produce a new traffic circle, where the center of dilation is at the origin. Find the coordinates of the center of the new traffic circle.

Find the length of the radius of the new traffic circle.

23. (CC) A circle has a center at (1,–2) and radius of 4. Does the point (3.4,1.2) lie on the circle? Justify your answer.

Chapter 14. Solids

14.1 Volume

Key Terms and Concepts

Volume is a measure of capacity that gives the number of cubic units of space inside a three-dimensional solid.

A **polyhedron** is a solid figure that has polygons as surfaces, such as a prism or pyramid. Each polygon surface is called a **face**, each side of a face is called an **edge**, and each vertex of a face is also a **vertex** of the solid figure. Not all solid figures are polyhedrons; for example, cylinders, cones, and spheres are not, since they are not made up of polygonal surfaces.

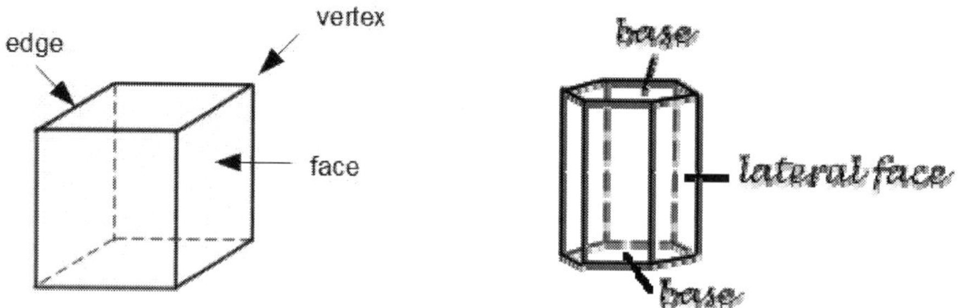

A **prism** is a solid with two congruent and parallel faces, called **bases**, that are any polygonal shape, with n additional faces, called **lateral faces**, joining the corresponding sides of the two bases, where n is the number of sides of the polygon in each base. A prism has a total of $n + 2$ faces. There are different types of prisms, each named after the shape of its bases, such as triangular prisms, rectangular prisms, pentagonal prisms, etc.
Examples: The following are a triangular prism, a rectangular prism, and a pentagonal prism.

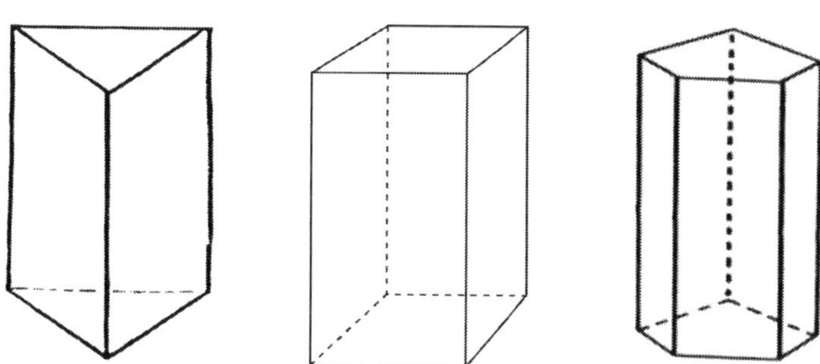

Although the bases of a prism may be any polygonal shape, the lateral faces are always parallelograms. When the lateral faces are rectangles, the solid is called a **right prism**. In right prisms, the lateral edges are perpendicular to the bases. When the lateral faces are not rectangles, it is called an **oblique prism**. The word "oblique" means slanted.
Example: Shown below are a right hexagonal prism and an oblique hexagonal prism.

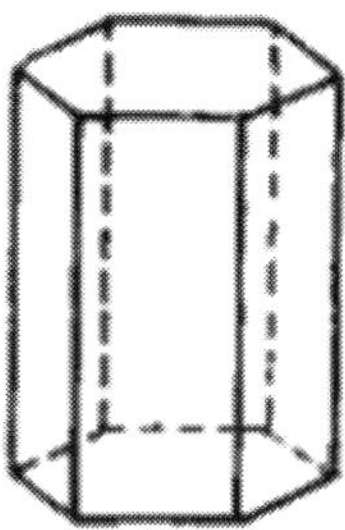

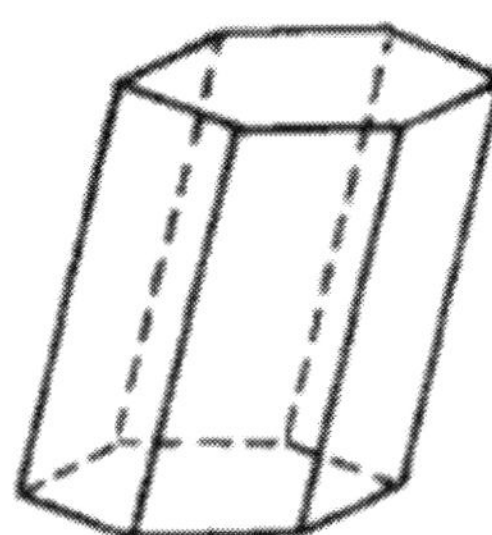

When the bases are regular polygons (ie, equilateral and equiangular), the prism is called a **regular prism**. A **cube** is a prism that has six square faces.

The method for finding the volume of any type of prism is the same: use the formula $V = Bh$, where B represents the area of a prism's base and h is the height. The height of a prism is the distance between the two planes on which the bases lie, often shown in a diagram as a line segment that is perpendicular to both planes.
Examples: (a) For a rectangular prism, the base is a rectangle with an area of length times width, so the volume of the prism is $V = Bh = (lw)h$.

(b) For a triangular prism, the base is a triangle with $A = \frac{1}{2}ba$ (where b and a are the lengths of the base and altitude of the triangle), so the volume of the prism is $V = Bh = (\frac{1}{2}ba)h$.

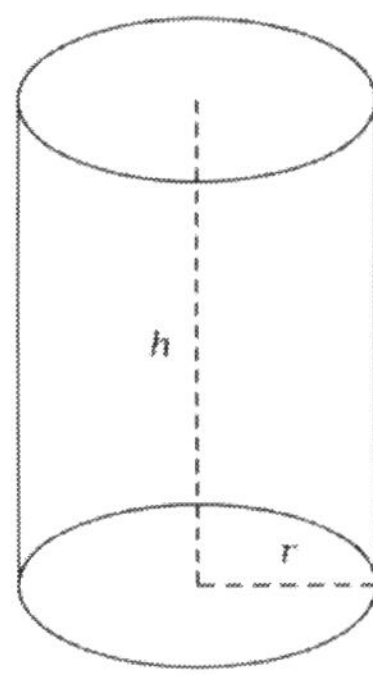

A **cylinder** is similar to a prism, except that instead of having polygon bases, its bases are circles. Also, instead of multiple lateral faces, it has only one curved lateral surface. Cylinders may also be right or oblique.

The volume of a cylinder is found using the same formula, $V = Bh$. However, since the base of a cylinder is always a circle with an area of πr^2, the formula is usually written as $V = \pi r^2 h$.

A **pyramid** is also similar to a prism, but instead of connecting two polygon bases, it connects one polygon base and a point, called an **apex**. As a result, instead of having lateral faces that are parallelograms, a pyramid's lateral faces are triangles.

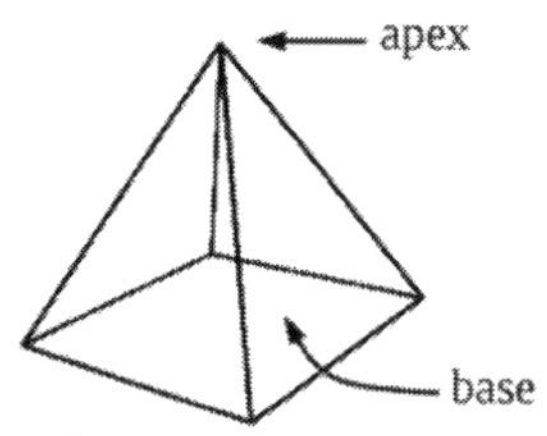

The base of a pyramid may be any polygon. The height of a pyramid is the distance from the apex to the plane on which the base lies.
Example: Shown below are a triangular, rectangular, and hexagonal pyramid.

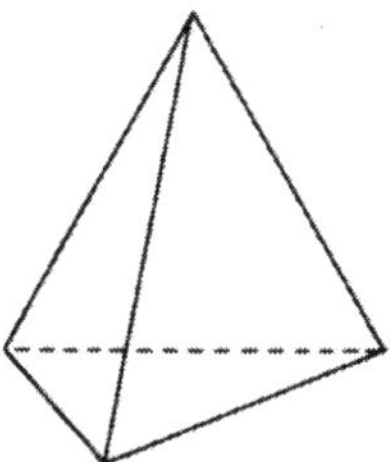
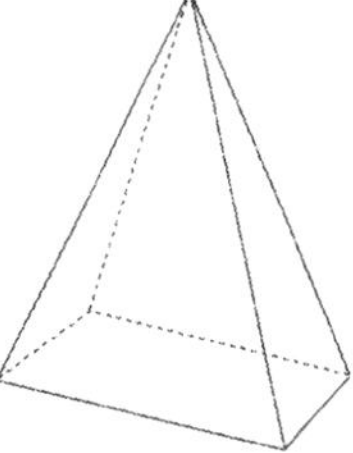
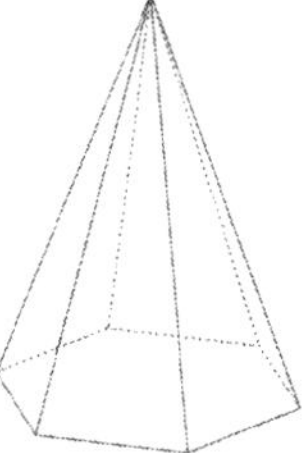

In a **right pyramid**, the lateral faces are congruent isosceles triangles. A **regular pyramid** is a right pyramid that has a regular polygon as its base.

If we compare a pyramid to a prism whose base is congruent to the pyramid's base and whose height is equal to the pyramid's height, we will find that the volume of the pyramid is one-third the volume of the prism. Therefore, we can use the formula $V=\frac{1}{3}Bh$ to calculate the volume of a pyramid, where B is the area of the polygonal base and h is the height.

In the same way that a pyramid is related to a prism, a cone is related to a cylinder. Whereas a cylinder has two congruent circular bases, a **cone** connects one circular base to a point, called the apex. The formula for the volume of a cone is the same as for a pyramid, $V=\frac{1}{3}Bh$. However, since the base of a cone is always a circle, it is usually written as $V=\frac{1}{3}\pi r^2 h$.

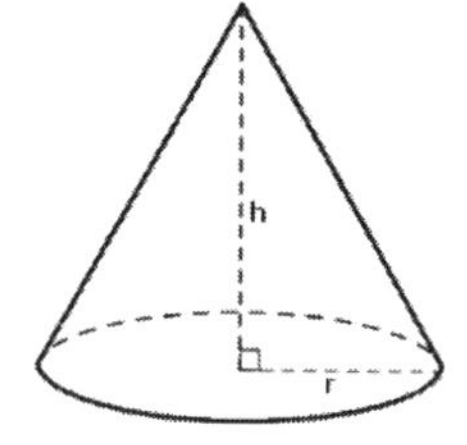

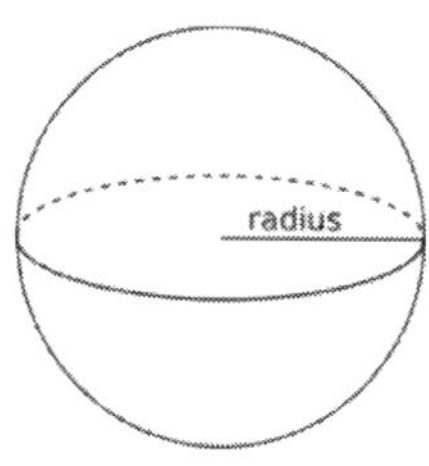

A **sphere** is a geometrical solid that is shaped like a ball. As you may recall, a circle is the set of all points *on a plane* that are equidistant from a point, where the point is the center of the circle and the distance is called the radius. Similarly, a sphere is the set of all points *in three-dimensional space* that are equidistant from a point, where the point is the center of the sphere and the distance is called the radius. The volume of a sphere is $V=\frac{4}{3}\pi r^3$.

To summarize, the formulas for the volumes of solids include:

Prism: $V=Bh$	Cylinder: $V=\pi r^2 h$	
Pyramid: $V=\frac{1}{3}Bh$	Cone: $V=\frac{1}{3}\pi r^2 h$	Sphere: $V=\frac{4}{3}\pi r^3$

Note: The formulas for the volume of a prism, cylinder, pyramid, cone, and sphere are included on the Reference Sheet at the back of the Regents exam.

Model Problem:

A water tank in the shape of a rectangular prism measures 6 feet by 5 feet by 4 feet and is completely filled with water. Drew needs to fill as many barrels as he can with water from the tank. Each barrel is shaped as a cylinder with a 1 foot radius and a height of 2 feet. What is the maximum number of whole barrels can he completely fill with water taken from the tank?

Solution:

(A) $V_{tank} = lwh = (6)(5)(4) = 120$ cubic feet

$V_{barrel} = \pi r^2 h = \pi(1)^2(2) = 2\pi \approx 6.28$ cubic feet

(B) Since $120 \div 6.28 \approx 19.11$, Drew can fill 19 whole barrels.

Explanation of steps:

(A) Calculate the volume of each solid by substituting for the variables in the appropriate formulas *[the tank is a rectangular prism and the barrel is a cylinder].*

(B) State the solution *[divide the volumes to determine how many barrels can be filled].*

Practice Problems

1. What is the volume, in cubic feet, of the rectangular prism shown below?

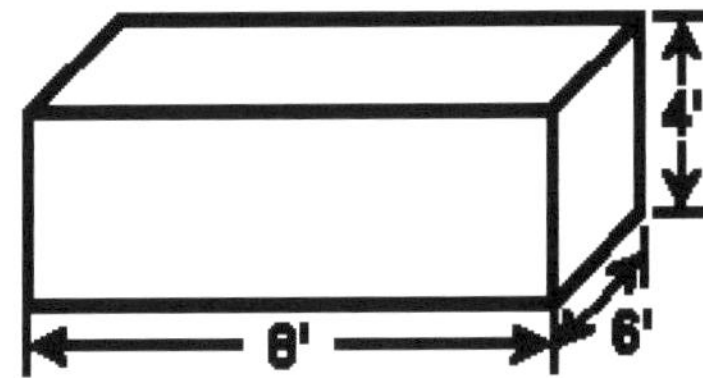

2. A sphere has a diameter of 18 meters. Find the volume of the sphere, in cubic meters, in terms of π.

3. What is the volume of this cylinder, to the *nearest hundredth of a cubic inch*?

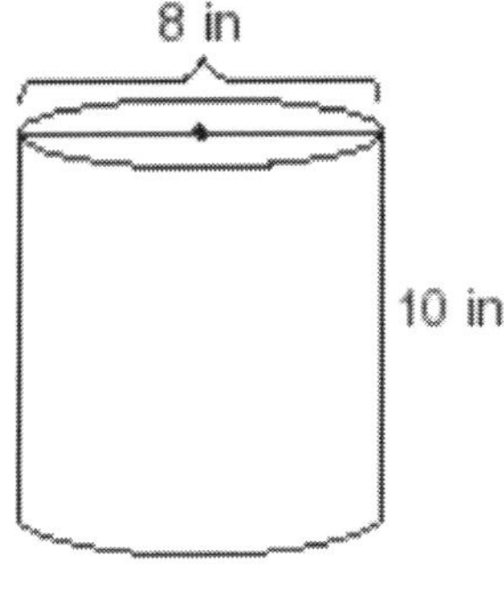

4. What is the volume of this rectangular pyramid, in cubic inches?

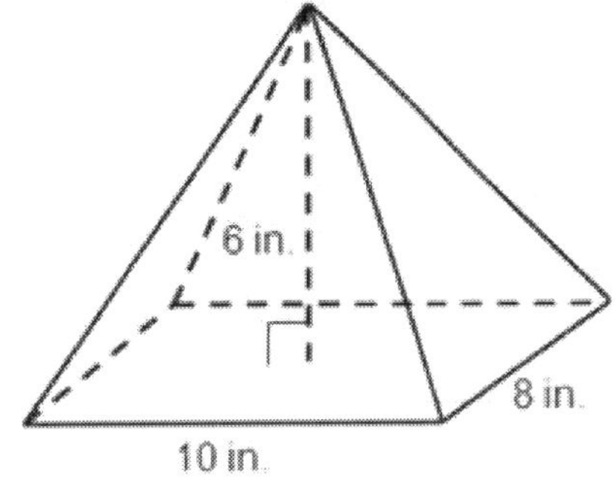

5. The volume of a cone is 8 cm^3. What is the volume of a cylinder with the same base as the cone and the same height as the cone?

6. A paper container in the shape of a right circular cone has a radius of 3 inches and a height of 8 inches. Determine and state the number of cubic inches in the volume of the cone, in terms of π.

7. A cardboard box has length $x-2$, width $x+1$, and height $2x$. Write the volume of the box as a polynomial in terms of x.

8. The solid below is made up of a cube and a square pyramid. Find its volume.

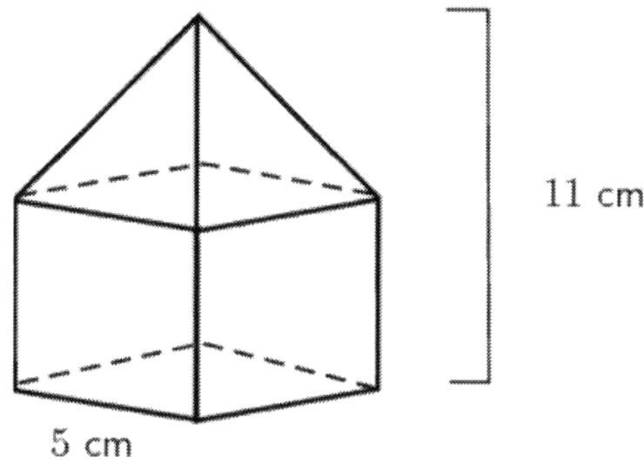

9. In the accompanying diagram, a rectangular container with the dimensions 10 inches by 15 inches by 20 inches is to be filled with water, using a cylindrical cup whose radius is 2 inches and whose height is 5 inches. What is the maximum number of full cups of water that can be placed into the container without the water overflowing the container?

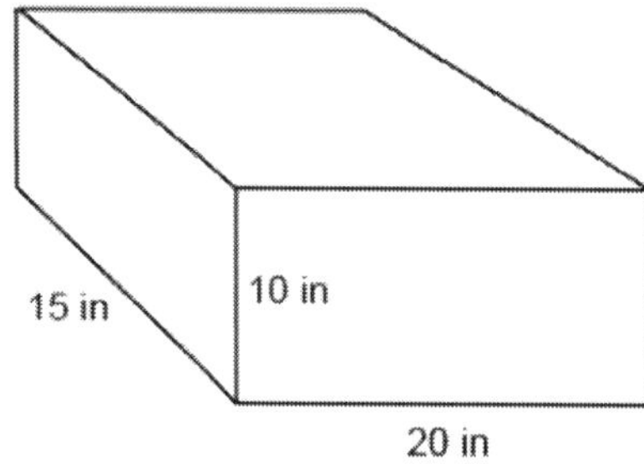

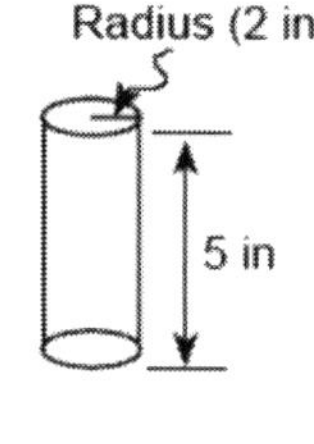

REGENTS QUESTIONS

Multiple Choice

1. A cylindrical container has a diameter of 12 inches and a height of 15 inches, as illustrated in the diagram below.

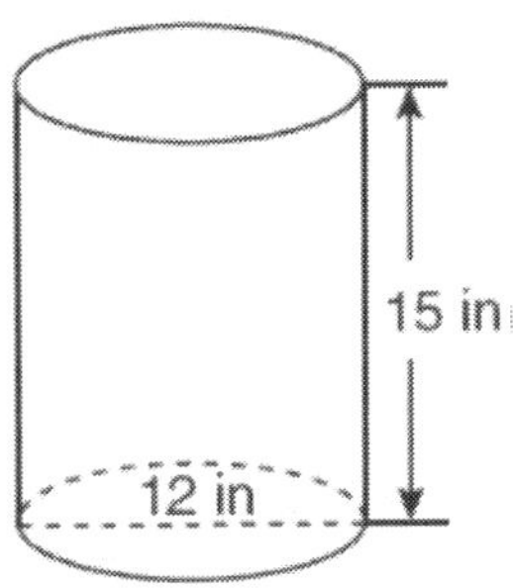

What is the volume of this container to the *nearest tenth* of a cubic inch?

(1) 6,785.8 (3) 2,160.0
(2) 4,241.2 (4) 1,696.5

2. The figure in the diagram below is a triangular prism.

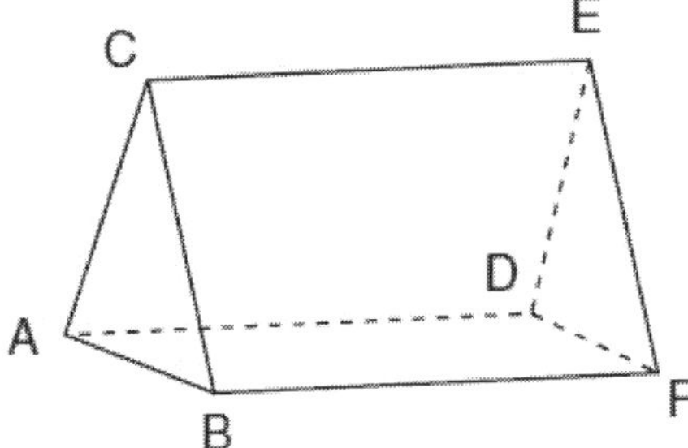

Which statement must be true?

(1) $\overline{DE} \cong \overline{AB}$ (3) $\overline{AD} \parallel \overline{CE}$
(2) $\overline{AD} \cong \overline{BC}$ (4) $\overline{DE} \parallel \overline{BC}$

3. Lenny made a cube in technology class. Each edge measured 1.5 cm. What is the volume of the cube in cubic centimeters?

(1) 2.25 (3) 9.0
(2) 3.375 (4) 13.5

4. The lateral faces of a regular pyramid are composed of

(1) squares (3) congruent right triangles
(2) rectangles (4) congruent isosceles triangles

5. A cylinder has a diameter of 10 inches and a height of 2.3 inches. What is the volume of this cylinder, to the *nearest tenth of a cubic inch*?

(1) 72.3 (3) 180.6
(2) 83.1 (4) 722.6

6. The volume of a cylindrical can is 32π cubic inches. If the height of the can is 2 inches, what is its radius, in inches?
   (1) 8 (3) 16
   (2) 2 (4) 4

7. How many cubes with 5-inch sides will completely fill a cube that is 10 inches on a side?
   (1) 50 (3) 8
   (2) 25 (4) 4

8. A regular pyramid has a height of 12 centimeters and a square base. If the volume of the pyramid is 256 cubic centimeters, how many centimeters are in the length of one side of its base?
   (1) 8 (3) 32
   (2) 16 (4) 64

9. CC The Great Pyramid of Giza was constructed as a regular pyramid with a square base. It was built with an approximate volume of 2,592,276 cubic meters and a height of 146.5 meters. What was the length of one side of its base, to the *nearest meter*?
   (1) 73 (3) 133
   (2) 77 (4) 230

10. CC A fish tank in the shape of a rectangular prism has dimensions of 14 inches, 16 inches, and 10 inches. The tank contains 1680 cubic inches of water. What percent of the fish tank is empty?
    (1) 10 (3) 50
    (2) 25 (4) 75

11. CC As shown in the diagram below, a regular pyramid has a square base whose side measures 6 inches.

6 in

If the altitude of the pyramid measures 12 inches, its volume, in cubic inches, is
    (1) 72 (3) 288
    (2) 144 (4) 432

12. CC The diameter of a basketball is approximately 9.5 inches and the diameter of a tennis ball is approximately 2.5 inches. The volume of the basketball is about how many times greater than the volume of the tennis ball?
    (1) 3591 (3) 55
    (2) 65 (4) 4

13. (CC) A company is creating an object from a wooden cube with an edge length of 8.5 cm. A right circular cone with a diameter of 8 cm and an altitude of 8 cm will be cut out of the cube. Which expression represents the volume of the remaining wood?

(1) $(8.5)^3 - \pi(8)^2(8)$
(2) $(8.5)^3 - \pi(4)^2(8)$
(3) $(8.5)^3 - \frac{1}{3}\pi(8)^2(8)$
(4) $(8.5)^3 - \frac{1}{3}\pi(4)^2(8)$

14. (CC) Tennis balls are sold in cylindrical cans with the balls stacked one on top of the other. A tennis ball has a diameter of 6.7 cm. To the *nearest cubic centimeter*, what is the minimum volume of the can that holds a stack of 4 tennis balls?

(1) 236
(2) 282
(3) 564
(4) 945

15. (CC) A solid metal prism has a rectangular base with sides of 4 inches and 6 inches, and a height of 4 inches. A hole in the shape of a cylinder, with a radius of 1 inch, is drilled through the entire length of the rectangular prism.

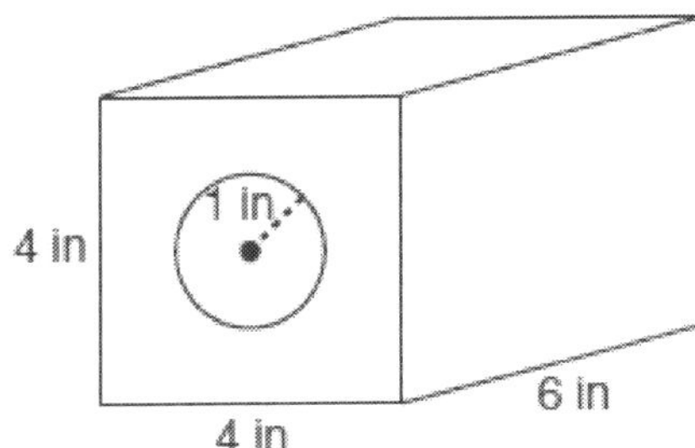

What is the approximate volume of the remaining solid, in cubic inches?

(1) 19
(2) 77
(3) 93
(4) 96

16. (CC) A water cup in the shape of a cone has a height of 4 inches and a maximum diameter of 3 inches. What is the volume of the water in the cup, to the *nearest tenth of a cubic inch*, when the cup is filled to half its height?

(1) 1.2
(2) 3.5
(3) 4.7
(4) 14.1

Constructed Response

17. A soup can is in the shape of a cylinder. The can has a volume of $342\,\text{cm}^3$ and a diameter of 6 cm. Express the height of the can in terms of π. Determine the maximum number of soup cans that can be stacked on their base between two shelves if the distance between the shelves is exactly 36 cm. Explain your answer.

18. The diagram below represents Joe's two fish tanks.

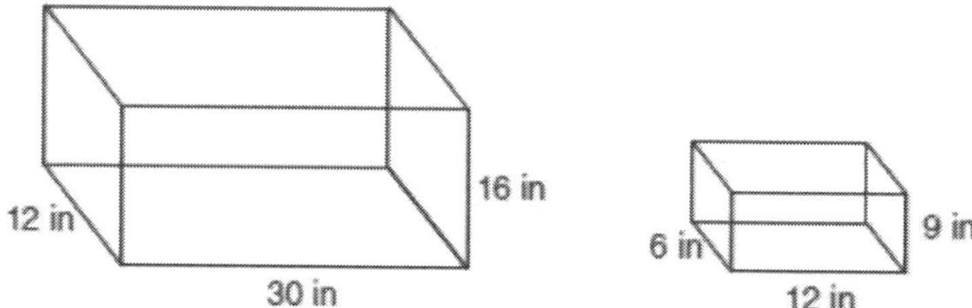

Joe's larger tank is completely filled with water. He takes water from it to completely fill the small tank. Determine how many cubic inches of water will remain in the larger tank.

19. A regular pyramid with a square base is shown in the diagram below.

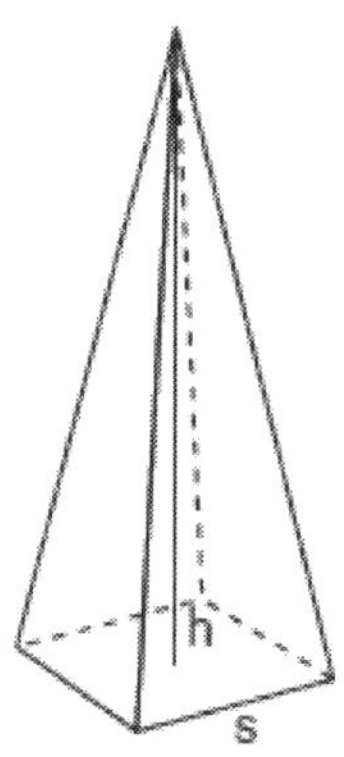

A side, s, of the base of the pyramid is 12 meters, and the height, h, is 42 meters. What is the volume of the pyramid in cubic meters?

20. The base of a pyramid is a rectangle with a width of 6 cm and a length of 8 cm. Find, in centimeters, the height of the pyramid if the volume is 288 cm^3.

21. Mike buys his ice cream packed in a rectangular prism-shaped carton, while Carol buys hers in a cylindrical-shaped carton. The dimensions of the prism are 5 inches by 3.5 inches by 7 inches. The cylinder has a diameter of 5 inches and a height of 7 inches. Which container holds more ice cream? Justify your answer.

Determine, to the *nearest tenth of a cubic inch*, how much more ice cream the larger container holds.

22. Oatmeal is packaged in a cylindrical container, as shown in the diagram below.

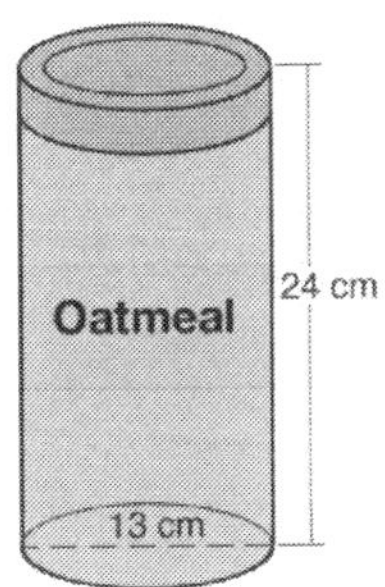

The diameter of the container is 13 centimeters and its height is 24 centimeters. Determine, in terms of π, the volume of the cylinder, in cubic centimeters.

23. (CC) A barrel of fuel oil is a right circular cylinder where the inside measurements of the barrel are a diameter of 22.5 inches and a height of 33.5 inches. There are 231 cubic inches in a liquid gallon. Determine and state, to the *nearest tenth*, the gallons of fuel that are in a barrel of fuel oil.

24. (CC) A water glass can be modeled by a truncated right cone (a cone which is cut parallel to its base) as shown below.

The diameter of the top of the glass is 3 inches, the diameter at the bottom of the glass is 2 inches, and the height of the glass is 5 inches.

The base with a diameter of 2 inches must be parallel to the base with a diameter of 3 inches in order to find the height of the cone. Explain why.

Determine and state, in inches, the height of the larger cone.

Determine and state, to the *nearest tenth of a cubic inch*, the volume of the water glass.

25. (CC) A candle maker uses a mold to make candles like the one shown below.

The height of the candle is 13 cm and the circumference of the candle at its widest measure is 31.416 cm. Use modeling to approximate how much wax, to the *nearest cubic centimeter*, is needed to make this candle. Justify your answer.

14.2 Density

Key Terms and Concepts

The **weight** (or **mass**) of a solid, or of the contents of a three-dimensional container, depends on both the volume and density of the solid or its contents. **Density** is a measure of weight per cubic unit. For example, the density of water is 1 g/cm^3, or 1,000 kg/m^3.

The formula for the weight, W, of a solid, or its contents, is:
$W = VD$, where V is volume and D is density

Therefore, to calculate the weight, in kilograms, of 5 cubic meters of water:

$$W = 5\,\text{m}^3 \cdot \frac{1{,}000\,\text{kg}}{1\,\text{m}^3} = 5{,}000\,\text{kg}$$

Sometimes, you may need to convert among units of measure, using methods learned in algebra.
Example: To find the weight *in pounds* of 5 cubic meters of water, include the conversion fraction in the product.

$$W = 5\,\cancel{\text{m}^3} \cdot \frac{1{,}000\,\cancel{\text{kg}}}{1\,\cancel{\text{m}^3}} \cdot \frac{2.2\,\text{lbs}}{1\,\cancel{\text{kg}}} = 11{,}000\ \text{lbs}$$

If you need to calculate density, you can rearrange the formula to $D = \frac{W}{V}$.

Be careful in your conversions when working with cubic measures. When a conversion equation compares linear measures, you'll need to cube the values to arrive at an appropriate equation to compare cubic measures.
Example: Even though 1 m = 100 cm, it is *not* true that 1 m^3 = 100 cm^3.
In fact, 1 m^3 = 100^3 cm^3 = 1,000,000 cm^3.

Note: In physical sciences, weight and mass have different meanings in that weight also depends on gravity. For questions on the Regents exam, however, you may assume Earth's gravity and treat the terms weight and mass equally.

Model Problem:

The Longball baseball factory needs to limit the content weight of a case of baseballs to a maximum of 10 pounds. A baseball is a sphere with a diameter of 2.9 inches and a density of 0.4 ounces per cubic inch. Determine and state the maximum number of baseballs that a case may contain.

Solution:

(A) $V = \frac{4}{3}\pi r^3 = \frac{4}{3}\pi(1.45)^3 \approx 12.77 \text{ in}^3$

(B) $W = VD = 12.77 \cancel{\text{in}^3} \cdot \frac{0.4 \cancel{\text{oz}}}{1 \cancel{\text{in}^3}} \cdot \frac{1 \text{ lb}}{16 \cancel{\text{oz}}} \approx 0.32 \text{ lb}$

(C) $\frac{10}{0.32} = 31.25$, so a maximum of 31 baseballs may be packed in a case.

Explanation of steps:

(A) Find the volume of the object. *[For one baseball, use the formula for a sphere, substituting the radius 1.45, which is half the diameter.]*

(B) Determine the weight of the object as the product of the volume and the density. *[We also need to include a conversion fraction here, to convert from ounces to pounds, since the maximum load for a case of baseballs is given in terms of pounds.]*

(C) State the solution. *[Divide 10 lbs. by the weight of each baseball.]*

Practice Problems

1. A rectangular prism, measuring 5 cm by 10 cm by 2 cm, is made of lead, which has a density of 11.34 grams per cubic cm. Find the weight of the item, in grams.

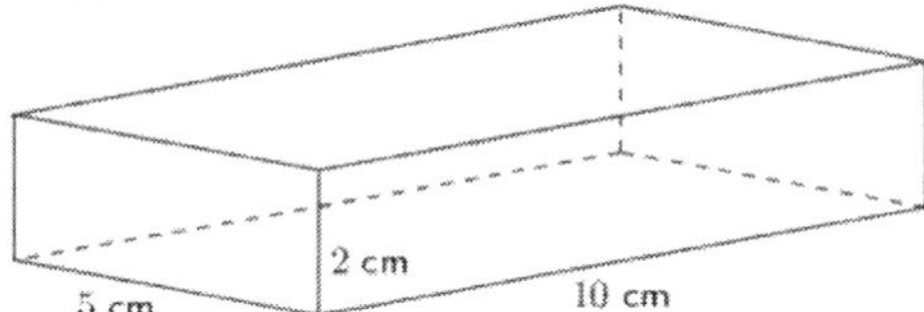

2. A 10 cm^3 sample of copper weighs 89.6 g. What is the density of copper?

3. A sample of iron in the shape of a rectangular prism has dimensions of 2 cm by 3 cm by 2 cm. If the sample weighs 94.44 g, what is the density of iron?

4. Gasoline has a density of 0.7 g/mL. Find the volume, in mL, of 9.8 g of gasoline.

5. A triangular pyramid, with a right triangle as its base, is shown below. The pyramid is made of a plastic which has a density of 2 g/cm^3. Find the weight, in pounds, of the pyramid, to the *nearest tenth of a pound.*

   1 inch = 2.54 cm
   1 kg = 2.2 lbs

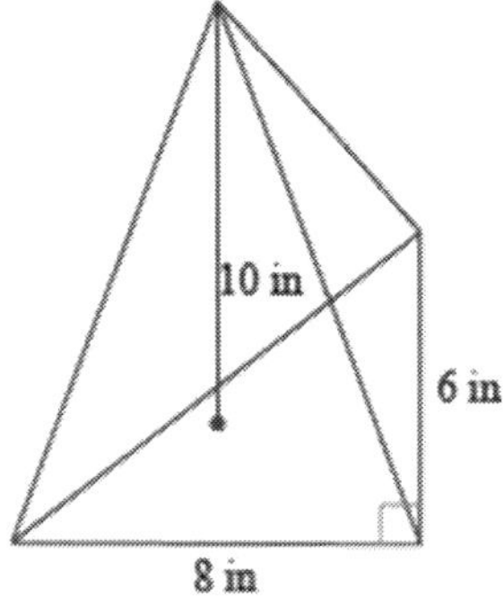

REGENTS QUESTIONS

Multiple Choice

1. (CC) A shipping container is in the shape of a right rectangular prism with a length of 12 feet, a width of 8.5 feet, and a height of 4 feet. The container is completely filled with contents that weigh, on average, 0.25 pound per cubic foot. What is the weight, in pounds, of the contents in the container?

   (1) 1,632 (3) 102
   (2) 408 (4) 92

2. (CC) A hemispherical tank is filled with water and has a diameter of 10 feet. If water weighs 62.4 pounds per cubic foot, what is the total weight of the water in a full tank, to the *nearest pound*?

   (1) 16,336 (3) 130,690
   (2) 32,673 (4) 261,381

3. (CC) Molly wishes to make a lawn ornament in the form of a solid sphere. The clay being used to make the sphere weighs .075 pound per cubic inch. If the sphere's radius is 4 inches, what is the weight of the sphere, to the *nearest pound*?

   (1) 34 (3) 15
   (2) 20 (4) 4

4. (CC) Seawater contains approximately 1.2 ounces of salt per liter on average. How many gallons of seawater, to the *nearest tenth of a gallon*, would contain 1 pound of salt?

   (1) 3.3 (3) 4.7
   (2) 3.5 (4) 13.3

5. (CC) A hemispherical water tank has an inside diameter of 10 feet. If water has a density of 62.4 pounds per cubic foot, what is the weight of the water in a full tank, to the *nearest pound*?

   (1) 16,336 (3) 130,690
   (2) 32,673 (4) 261,381

6. (CC) The density of the American white oak tree is 752 kilograms per cubic meter. If the trunk of an American white oak tree has a circumference of 4.5 meters and the height of the trunk is 8 meters, what is the approximate number of kilograms of the trunk?

   (1) 13 (3) 13,536
   (2) 9694 (4) 30,456

Constructed Response

7. (CC) A contractor needs to purchase 500 bricks. The dimensions of each brick are 5.1 cm by 10.2 cm by 20.3 cm, and the density of each brick is 1920 kg/m^3. The maximum capacity of the contractor's trailer is 900 kg. Can the trailer hold the weight of 500 bricks? Justify your answer.

8. (CC) Trees that are cut down and stripped of their branches for timber are approximately cylindrical. A timber company specializes in a certain type of tree that has a typical diameter of 50 cm and a typical height of about 10 meters. The density of the wood is 380 kilograms per cubic meter, and the wood can be sold by mass at a rate of $4.75 per kilogram. Determine and state the minimum number of whole trees that must be sold to raise at least $50,000.

9. (CC) The water tower in the picture below is modeled by the two-dimensional figure beside it. The water tower is composed of a hemisphere, a cylinder, and a cone. Let *C* be the center of the hemisphere and let *D* be the center of the base of the cone.

Source: http://en.wikipedia.org

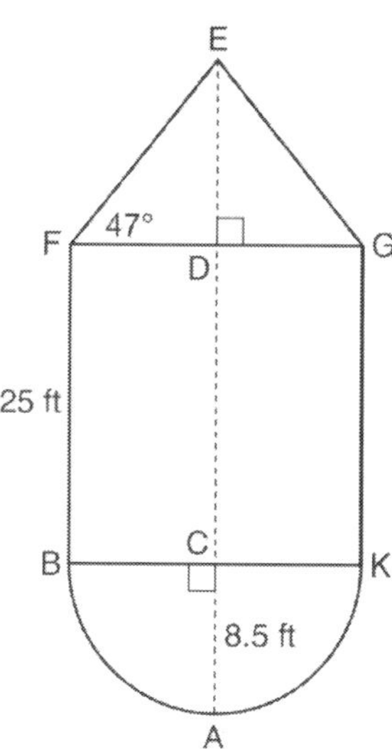

If AC = 8.5 feet, BF = 25 feet, and $m\angle EFD = 47°$, determine and state, to the *nearest cubic foot*, the volume of the water tower.

The water tower was constructed to hold a maximum of 400,000 pounds of water. If water weighs 62.4 pounds per cubic foot, can the water tower be filled to 85% of its volume and *not* exceed the weight limit? Justify your answer.

10. (CC) A wooden cube has an edge length of 6 centimeters and a mass of 137.8 grams. Determine the density of the cube, to the *nearest thousandth*. State which type of wood the cube is made of, using the density table below.

Type of Wood	Density (g/cm^3)
Pine	0.373
Hemlock	0.431
Elm	0.554
Birch	0.601
Ash	0.638
Maple	0.676
Oak	0.711

11. (CC) Walter wants to make 100 candles in the shape of a cone for his new candle business. The mold shown below will be used to make the candles. Each mold will have a height of 8 inches and a diameter of 3 inches. To the *nearest cubic inch*, what will be the total volume of 100 candles?

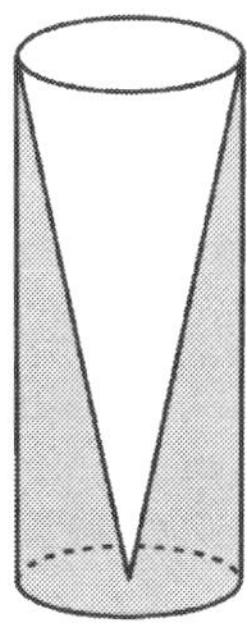

Walter goes to a hobby store to buy the wax for his candles. The wax costs \$0.10 per ounce. If the weight of the wax is 0.52 ounce per cubic inch, how much will it cost Walter to buy the wax for 100 candles?

If Walter spent a total of \$37.83 for the molds and charges \$1.95 for each candle, what is Walter's profit after selling 100 candles?

12. (CC) During an experiment, the same type of bacteria is grown in two petri dishes. Petri dish *A* has a diameter of 51 mm and has approximately 40,000 bacteria after 1 hour. Petri dish *B* has a diameter of 75 mm and has approximately 72,000 bacteria after 1 hour.

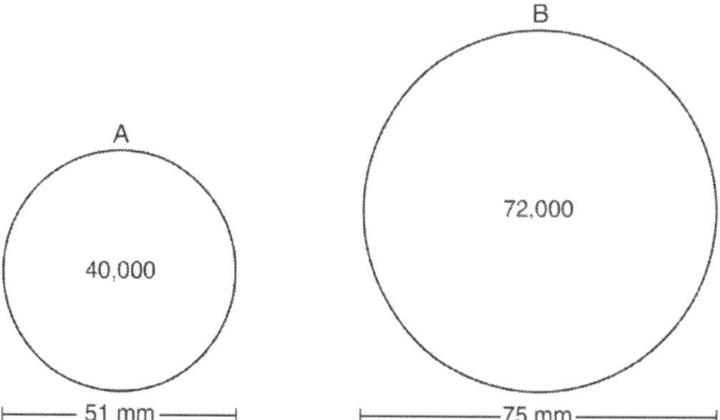

Determine and state which petri dish has the greater population density of bacteria at the end of the first hour.

13. (CC) A snow cone consists of a paper cone completely filled with shaved ice and topped with a hemisphere of shaved ice, as shown in the diagram below. The inside diameter of both the cone and the hemisphere is 8.3 centimeters. The height of the cone is 10.2 centimeters.

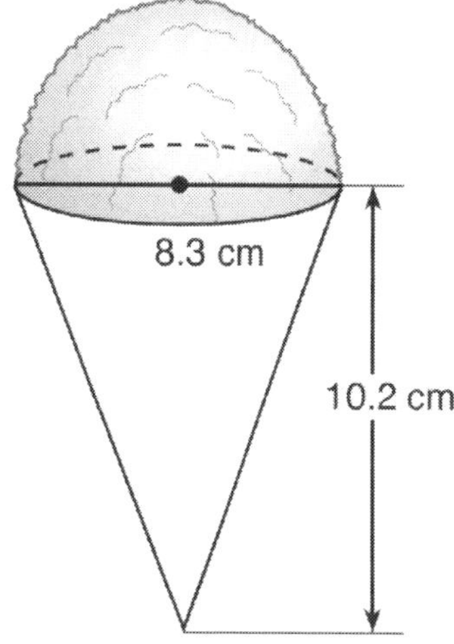

The desired density of the shaved ice is 0.697 g/cm^3, and the cost, per kilogram, of ice is \$3.83. Determine and state the cost of the ice needed to make 50 snow cones.

14. (CC) New streetlights will be installed along a section of the highway. The posts for the streetlights will be 7.5 m tall and made of aluminum. The city can choose to buy the posts shaped like cylinders or the posts shaped like rectangular prisms. The cylindrical posts have a hollow core, with aluminum 2.5 cm thick, and an outer diameter of 53.4 cm. The rectangular-prism posts have a hollow core, with aluminum 2.5 cm thick, and a square base that measures 40 cm on each side. The density of aluminum is 2.7 g/cm3, and the cost of aluminum is $0.38 per kilogram.

If all posts must be the same shape, which post design will cost the town less?

How much money will be saved per streetlight post with the less expensive design?

14.3 Lateral Area and Surface Area

Key Terms and Concepts

The **lateral area** of a solid is the sum of the areas of its lateral faces (or lateral surface), excluding the bases, in square units.

The **surface area** of a solid is the sum of the areas of all of its surfaces, including the bases.

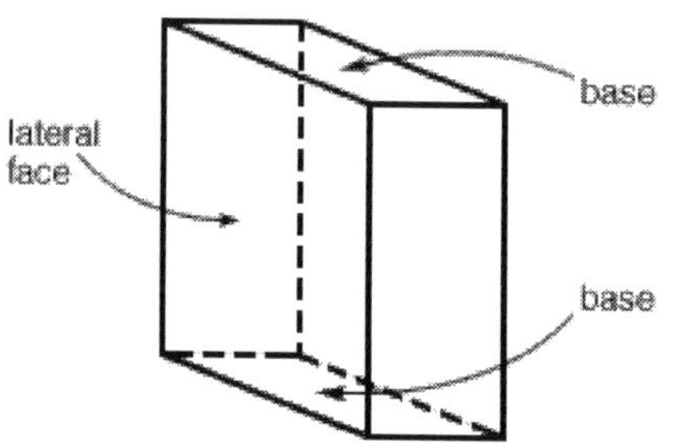

For example, a rectangular prism has 4 lateral faces. Its lateral area can be calculated as $LA = 2lh + 2wh$, where h is the height of the prism and l and w are the length and width of its base. To find the surface area, we include the areas of the two bases, so $SA = 2lw + 2lh + 2wh$.

Cylinders:
For a right cylinder, the lateral surface is really a curved rectangle. We can imagine cutting the lateral surface vertically and laying it flat on a surface, as shown below. The length of the resulting rectangle would equal the circumference of the cylinder's base, and the width would equal the cylinder's height. Therefore, the lateral area of a cylinder is $LA = \pi dh$, or $LA = 2\pi rh$.

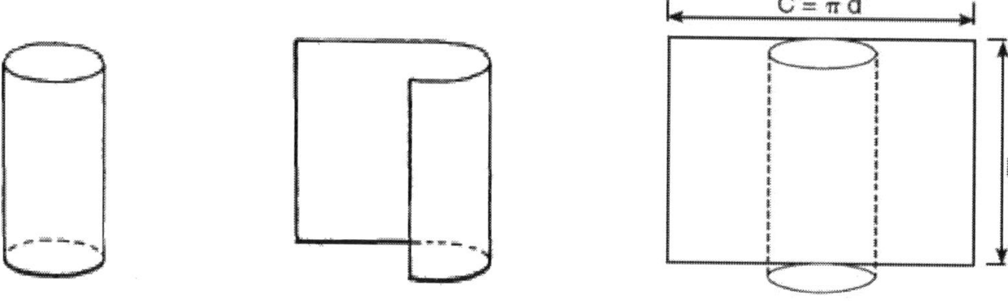

To find the cylinder's surface area, just add the areas of the two circular bases to the lateral area: $SA = 2\pi r^2 + 2\pi rh$.

Pyramids:

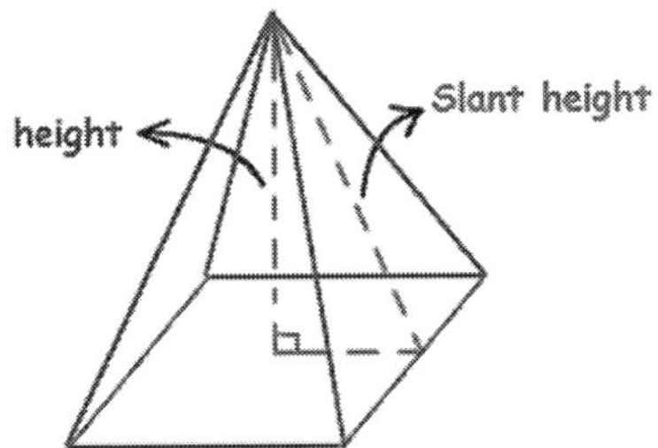

In a pyramid, the lateral faces are triangles. The lateral area is the sum of the areas of these triangles. Each triangle has a base and height, but to distinguish the height of the triangle from the height of the pyramid itself, we call it the **slant height**.

In a *regular pyramid*, the triangular faces are congruent, which makes it easier to calculate the *lateral area*. For a regular pyramid with a base of *n* sides, with each side *b* units in length, the $LA = n\left(\frac{1}{2}b\ell\right)$, where ℓ is the slant height.

Example: The base of a regular pyramid is a square with sides of 6 inches each. The slant height of the pyramid is 5 inches. Therefore, $LA = n\left(\frac{1}{2}b\ell\right) = 4\left(\frac{1}{2}\cdot 6\cdot 5\right) = 60 \text{ in}^2$.

We can find the *surface area of a regular pyramid* by adding the area of the base, *B*, to the lateral area, *LA*.

Example: For the pyramid above, $SA = B + LA = 6^2 + 4\left(\frac{1}{2}\cdot 6\cdot 5\right) = 96 \text{ in}^2$.

Cones:

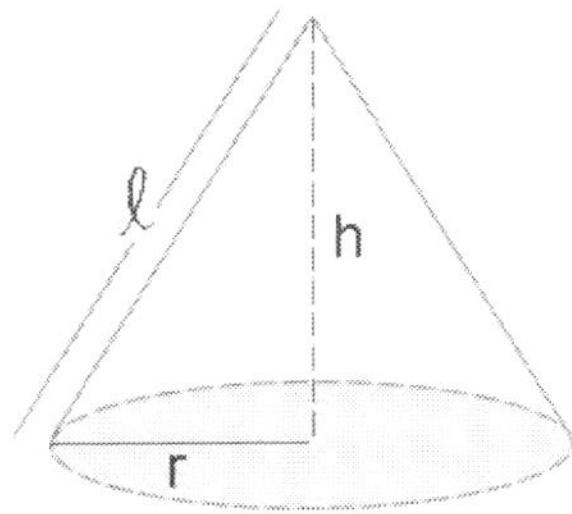

The formula for the *lateral area of a right cone* is $LA = \pi r\ell$, where *r* is the radius of the base and ℓ is the slant height.

The formula for the *surface area of a right cone* is $SA = B + LA = \pi r^2 + \pi r\ell$.

Where does the formula for the lateral area of a cone come from? If you cut the lateral surface of a right cone along its slant height, as shown below, and lay it flat on a surface, you'll get the sector of a circle. The radius of this sector is the slant height of the cone, ℓ, and the arc length is the circumference of the base of the cone, $2\pi r$. The sector is a fraction of a full circle. The area of the sector is the lateral area, LA, of the cone.

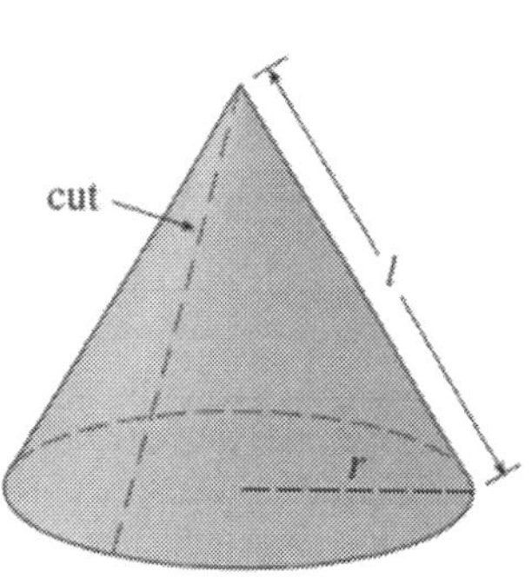

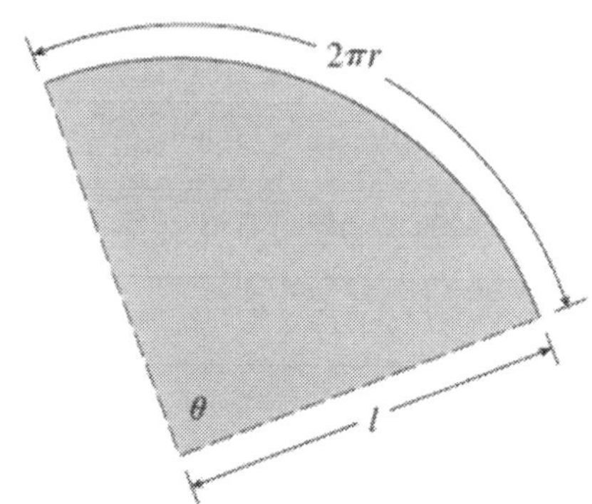

We can set up a proportion:

$$\frac{\text{area of sector}}{\text{area of circle}} = \frac{\text{measure of arc}}{\text{circumference of circle}}, \text{ or } \frac{LA}{\pi\ell^2} = \frac{2\pi r}{2\pi\ell}.$$

This gives us $LA = \pi\ell^2 \cdot \dfrac{2\pi r}{2\pi\ell} = \pi r\ell$

Spheres:

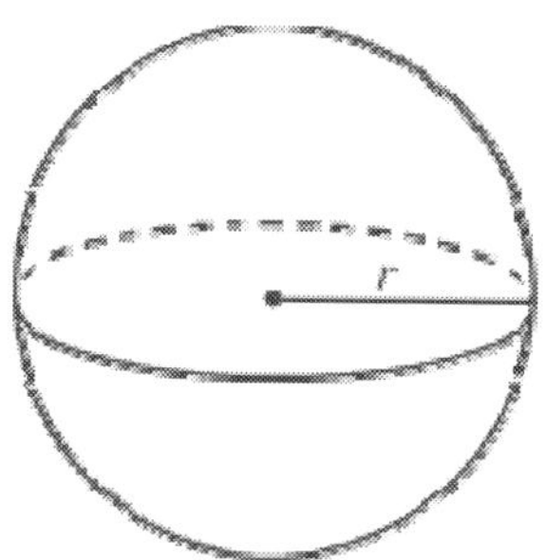

The formula for the *surface area of a sphere* is $SA = 4\pi r^2$.

Model Problem:

How many square centimeters of vinyl would it take to completely cover the surface of the cylinder shown below, to the *nearest square centimeter*?

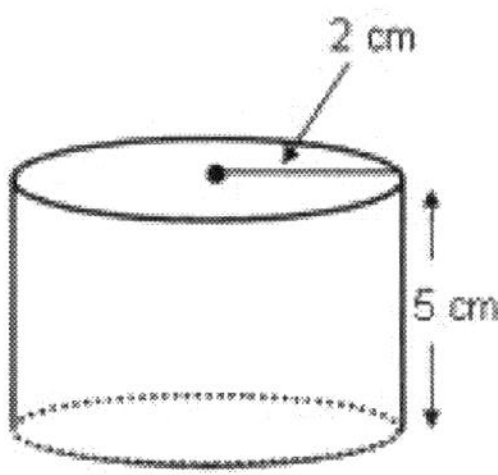

Solution:

(A) $SA = 2\pi r^2 + 2\pi rh$

(B) $= 2\pi(2)^2 + 2\pi(2)(5)$

(C) $= 8\pi + 20\pi = 28\pi \approx 88\, sq\, cm$

Explanation of steps:

(A) Write the appropriate formula for the surface area of the given solid *[cylinder]*.

(B) Substitute the given values for the variables *[r = 2 and h = 5]*.

(C) Simplify, and round the result.

Practice Problems

1. What is the surface area, in square feet, of the rectangular prism shown below?

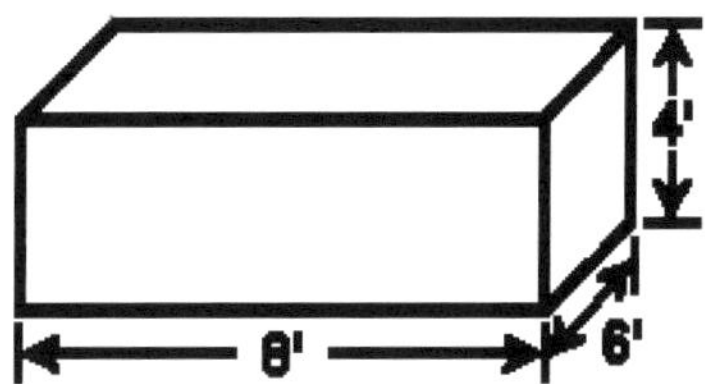

2. What are the lateral area and surface area, in terms of π, of the cylinder shown below?

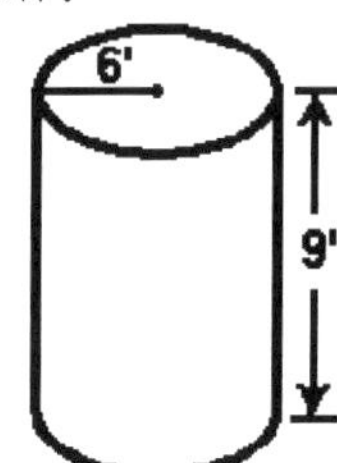

3. A cylinder has a height of 11 feet and a radius of 5 feet. What is the surface area, in square feet, of the cylinder, to the *nearest tenth*?

4. The volume of a cube is 64 cubic inches. What is its total surface area, in square inches?

5. Three different cylinders, with radii of 2, 4, and 6, have equal volumes of 144π each. Make a table showing the height and surface area of the three cylinders. Which cylinder has the least surface area?

REGENTS QUESTIONS

Multiple Choice

1. Mrs. Ayer is painting the outside of her son's toy box, including the top and bottom. The toy box measures 3 feet long, 1.5 feet wide, and 2 feet high. What is the total surface area she will paint?

   (1) 9.0 ft^2 (3) 22.5 ft^2
   (2) 13.5 ft^2 (4) 27.0 ft^2

2. How many square inches of wrapping paper are needed to entirely cover a box that is 2 inches by 3 inches by 4 inches?

   (1) 18 (3) 26
   (2) 24 (4) 52

3. The rectangular prism shown below has a length of 3.0 cm, a width of 2.2 cm, and a height of 7.5 cm.

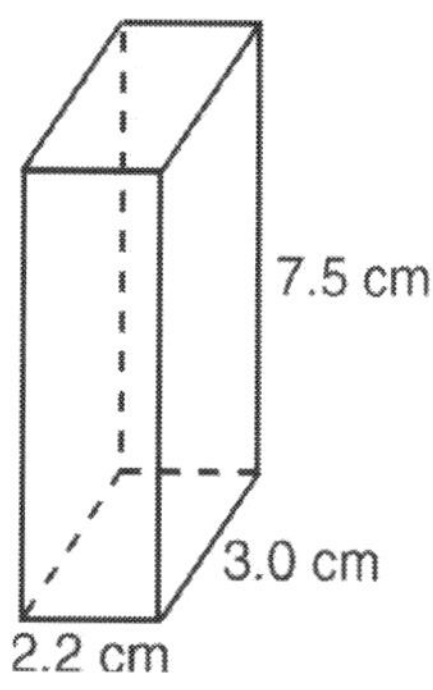

   What is the surface area, in square centimeters?

   (1) 45.6 (3) 78.0
   (2) 49.5 (4) 91.2

4. If the volume of a cube is 8 cubic centimeters, what is its surface area, in square centimeters?

   (1) 32 (3) 12
   (2) 24 (4) 4

5. (CC) A gallon of paint will cover approximately 450 square feet. An artist wants to paint all the outside surfaces of a cube measuring 12 feet on each edge. What is the *least* number of gallons of paint he must buy to paint the cube?

   (1) 1 (3) 3
   (2) 2 (4) 4

Constructed Response

6. Find the volume, in cubic centimeters, and the surface area, in square centimeters, of the rectangular prism shown below.

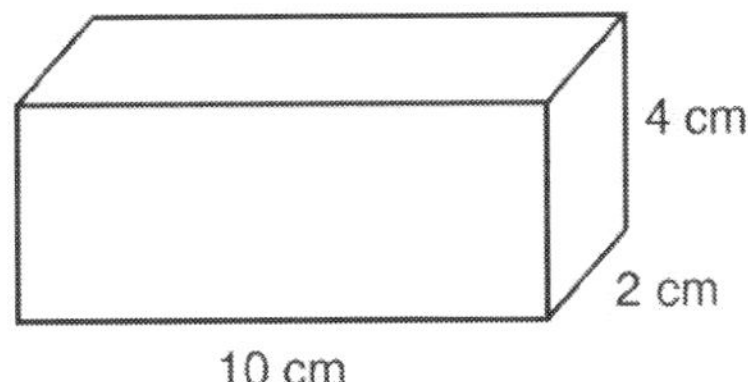

7. A plastic storage box in the shape of a rectangular prism has a length of $x+3$, a width of $x-4$, and a height of 5. Represent the surface area of the box as a trinomial in terms of x.

8. The length and width of the base of a rectangular prism are 5.5 cm and 3 cm. The height of the prism is 6.75 cm. Find the *exact* value of the surface area of the prism, in square centimeters.

14.4 Rotations of Two-Dimensional Objects

Key Terms and Concepts

If you spin a dime on a tabletop, the image of the coin rotating looks like a sphere. This is a demonstration of how a circle, when continuously rotated around its diameter, creates a sphere.

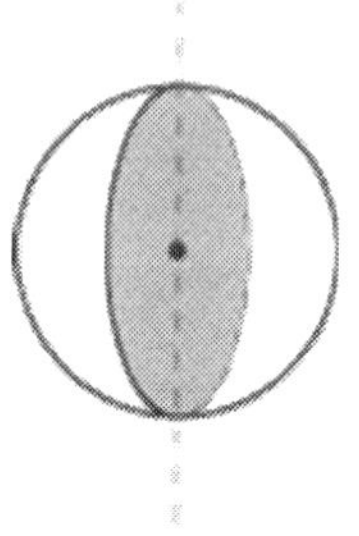

When we **rotate a two-dimensional object** around a rotation axis, we can generate a three-dimensional object. The **rotation axis** is the line around which the object rotates, just as the Earth rotates around its own axis every day. In the coin example above, the rotation axis is the diameter of the circle.

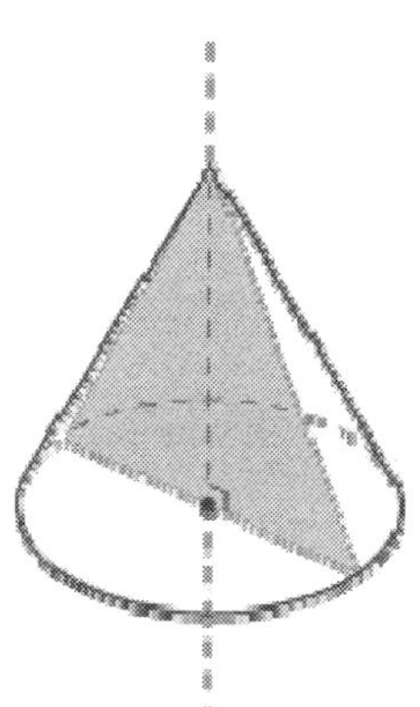

Rotating different shapes will create different figures.

Rotating a *rectangle* around its central axis would create a *cylinder*. The central axis of a rectangle is the line which intersects the midpoints of opposite sides. The diameter and height of the resulting cylinder would equal the lengths of the sides of the rectangle.

If we rotate a rectangle using a side as the rotation axis, the result is also a cylinder. The height of the cylinder would still equal the length of the side used as the axis, but now the length of the other side of the rectangle would equal the radius of the cylinder.

Rotating an *isosceles triangle*, using the altitude of the triangle that is perpendicular to the base as the rotation axis, the result is a *cone*. The altitude of the triangle would equal the height of the cone, and the length of the triangle's base would equal the diameter of the cone's base.

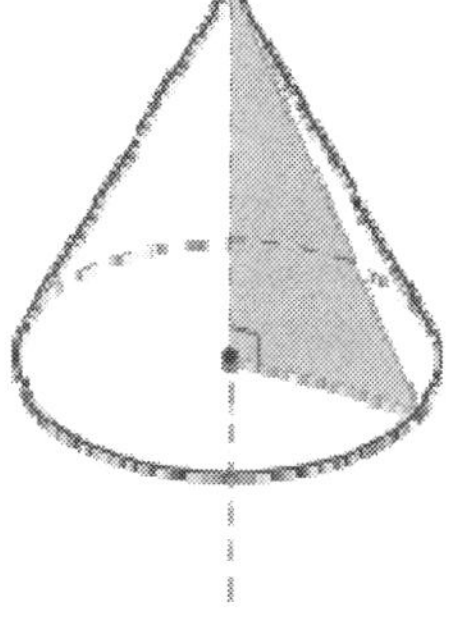

Rotating a *right triangle* around one of its legs would also create a *cone*. The height of the cone would equal the length of the leg which lies on the axis, and the radius of the base of the cone, would equal the length of the other leg.

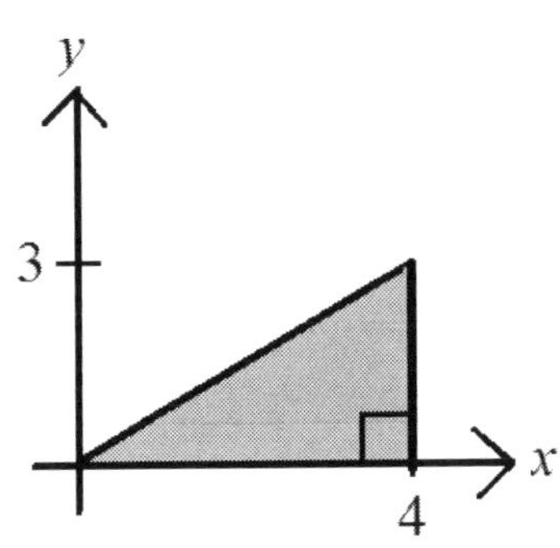

Model Problem:

Describe the object formed when the right triangle to the left is rotated about the y-axis. What is the volume of the object?

Solution:

(A) The object is a cylinder with a cone cutout, as shown below.

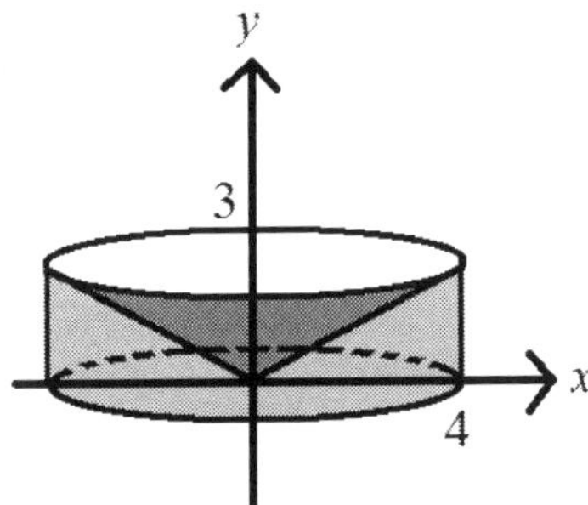

(B)
$$V_{object} = V_{cylinder} - V_{cone}$$
$$= \pi r^2 h - \tfrac{1}{3}\pi r^2 h = \tfrac{2}{3}\pi r^2 h$$
$$= \tfrac{2}{3}\pi (4)^2 (3) = 32\pi$$

Explanation of steps:

(A) Visualize the rotation of the figure. *[You can imagine a rectangle (below) rotating around the axis, forming a cylinder, and the darker triangle rotating around the axis, forming a cone cutout.]*

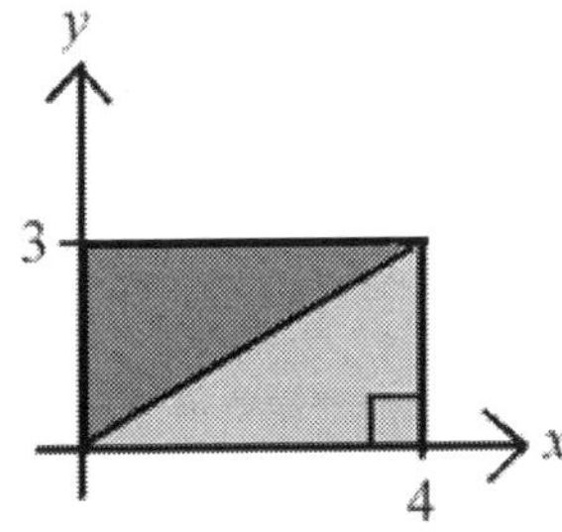

(B) To find the volume of a solid with a cutout, subtract the volume of the cutout from the solid. *[For both the cylinder and cone, the radius is 4 and the height is 3.]*

Practice Problems

1. What type of object is formed when the circle is rotated about the line?	2. What type of object is formed when the square is rotated about the line?
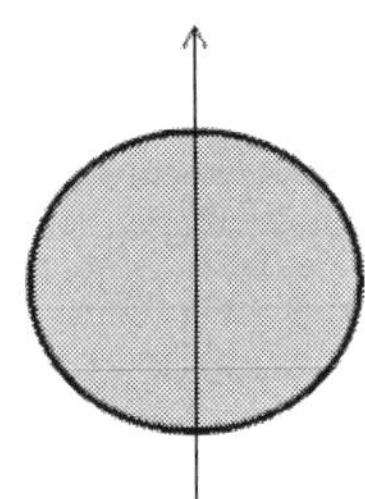	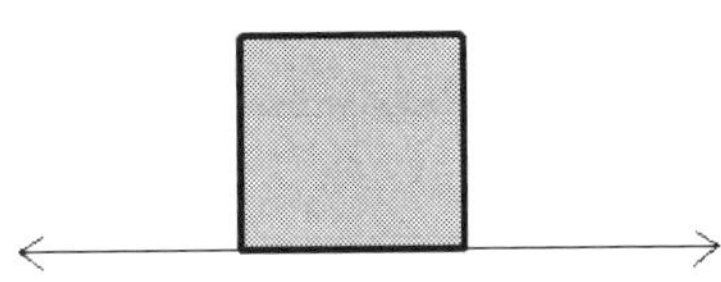 If each side of the square measures 5 inches, find the volume of the object formed by the rotation.

3. Describe the object formed when the figure below is rotated about the line. What is the volume of the object?

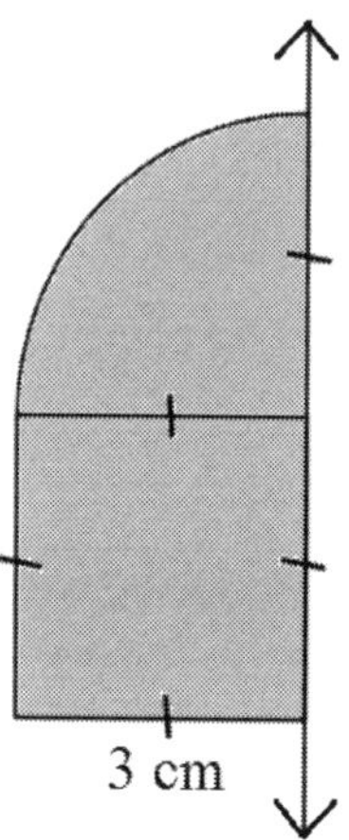

4. The area between two concentric semicircles is shaded below. Describe the object formed when the shaded figure is rotated about their extended diameter. What is the volume of the object?

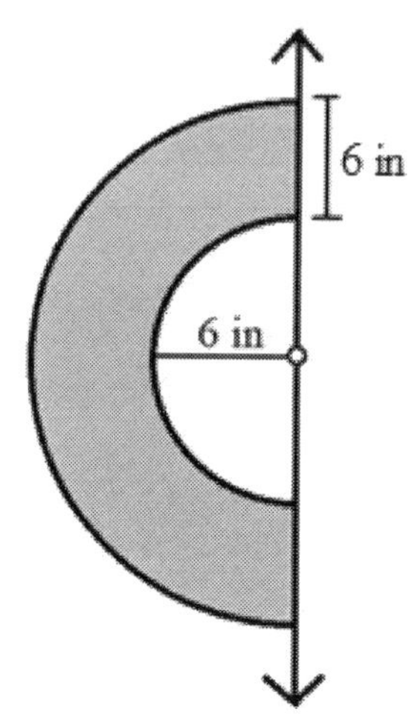

5. The arrow figure below consists of an isosceles triangle with legs 13 units long and a 5 by 12 unit rectangle. The height of the figure is 24 units.

Describe the object formed by rotating the figure around its axis of symmetry. What is the volume of the object?

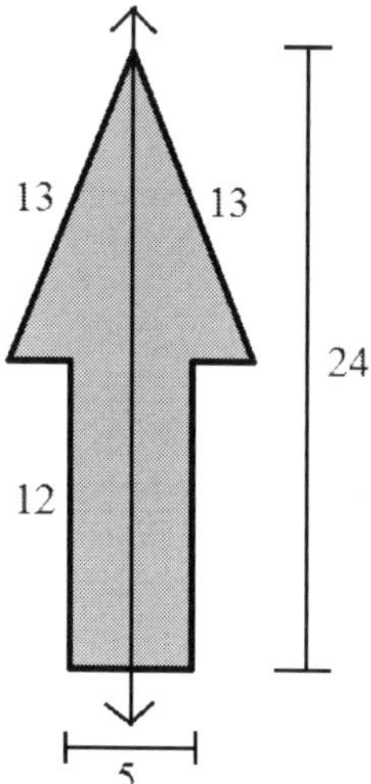

REGENTS QUESTIONS

Multiple Choice

1. Triangle ABC represents a metal flag on pole AD, as shown in the accompanying diagram. On a windy day the triangle spins around the pole so fast that it looks like a three-dimensional shape.

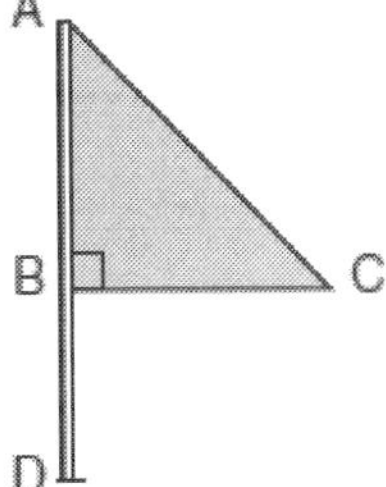

Which shape would the spinning flag create?

(1) sphere
(2) pyramid
(3) right circular cylinder
(4) cone

2. (CC) Which object is formed when right triangle RST shown below is rotated around leg $\overline{RS}$?

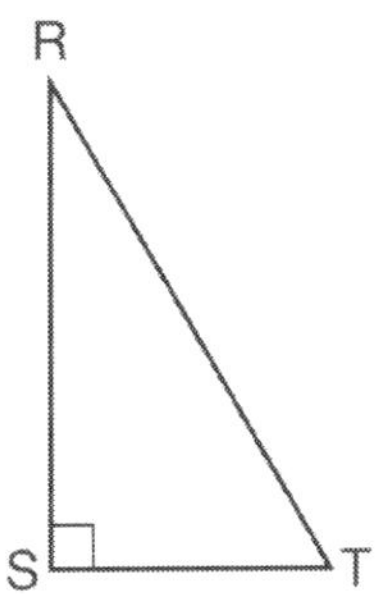

(1) a pyramid with a square base
(2) an isosceles triangle
(3) a right triangle
(4) a cone

3. (CC) If the rectangle below is continuously rotated about side w, which solid figure is formed?

(1) pyramid
(2) rectangular prism
(3) cone
(4) cylinder

4. (CC) A student has a rectangular postcard that he folds in half lengthwise. Next, he rotates it continuously about the folded edge. Which three-dimensional object below is generated by this rotation?

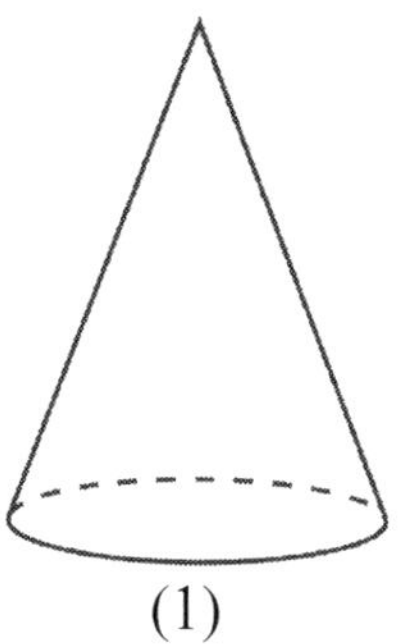

(1)

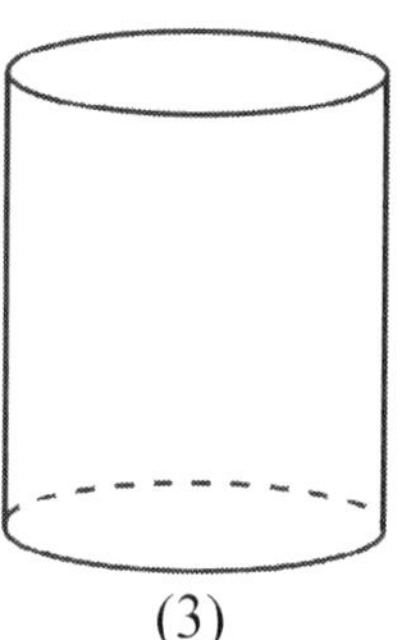

(3)

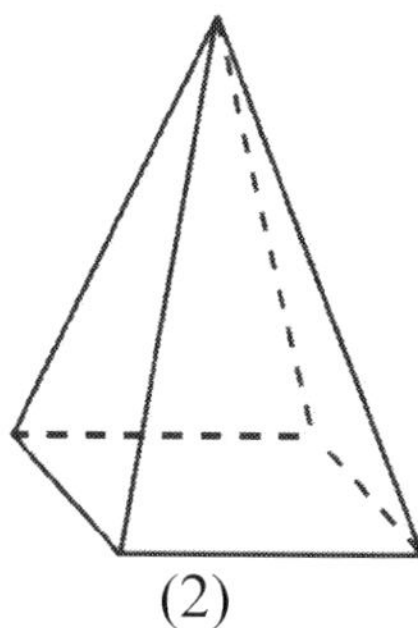

(2)

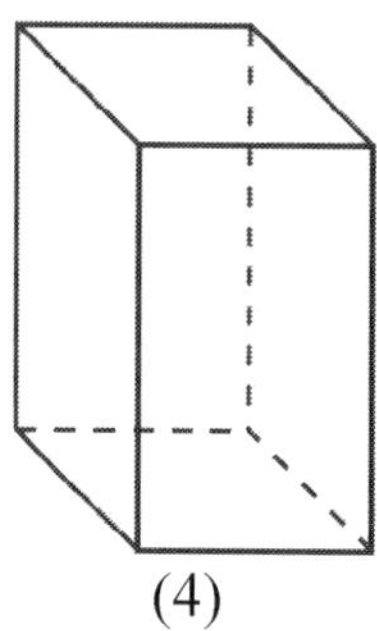

(4)

5. (CC) If an equilateral triangle is continuously rotated around one of its medians, which 3-dimensional object is generated?
   (1) cone
   (2) pyramid
   (3) prism
   (4) sphere

14.5 Cross Sections

Key Terms and Concepts

When we intersect a plane with a three-dimensional object, the resulting intersection, a two-dimensional figure, is called the **cross section** or **plane section**. The plane needs to cut straight through the object, but at any angle.

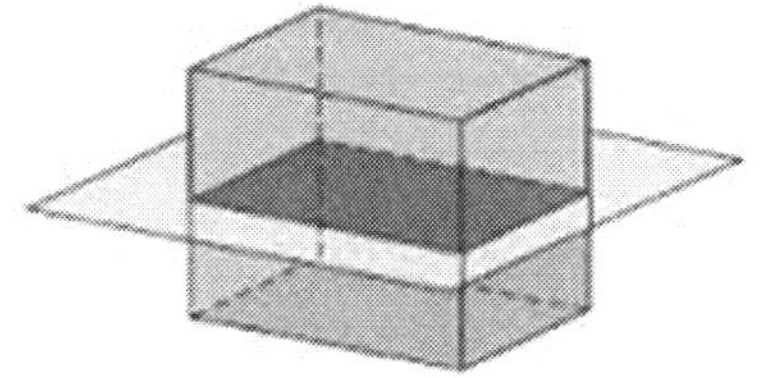

In a rectangular prism, a plane that is parallel to the bases will create a cross section that is a rectangle; in fact, this rectangle would be congruent to the bases. If a plane that is perpendicular to the base intersects a rectangular prism, it would also create a rectangular cross section.

However, other cross sections of a rectangular prism, with planes cutting diagonally, can create different shapes. For examples, the graphics below show that cross sections may be triangles, trapezoids, or even hexagons.

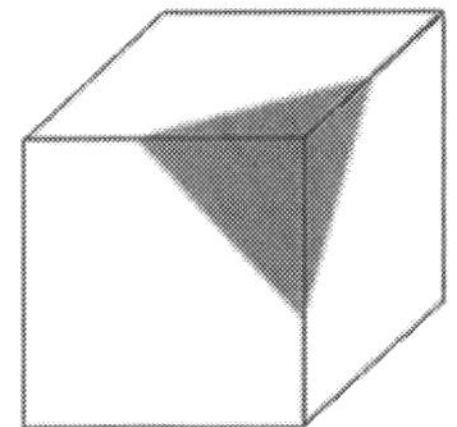
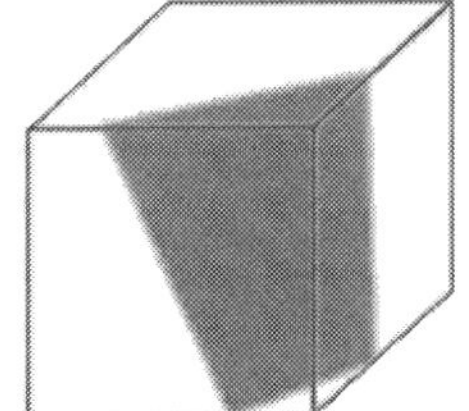
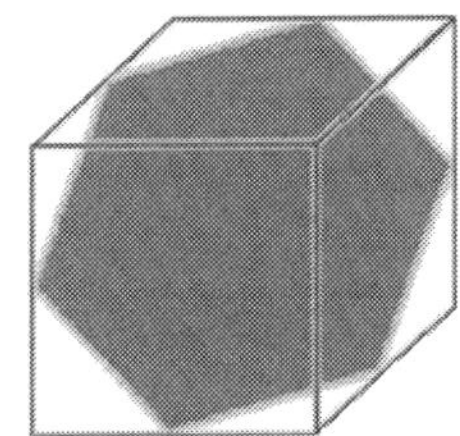

A cross section of a triangular prism can be a triangle or rectangle, as shown below, or other shapes, such as a trapezoid. But there are shapes that obviously *cannot* be the cross section of a prism; for example, since there are no curved surfaces, there are no cross sections that are circles.

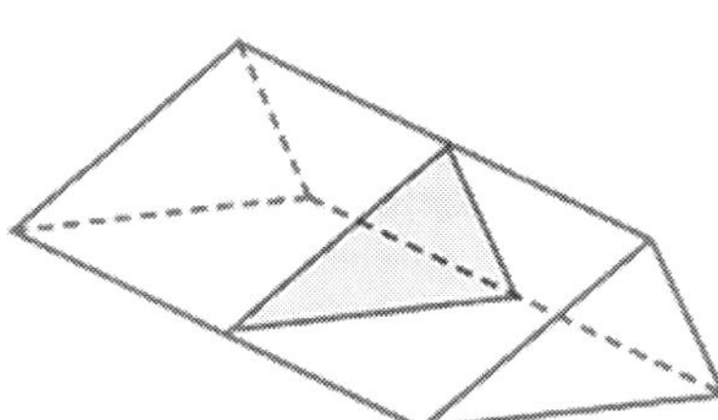
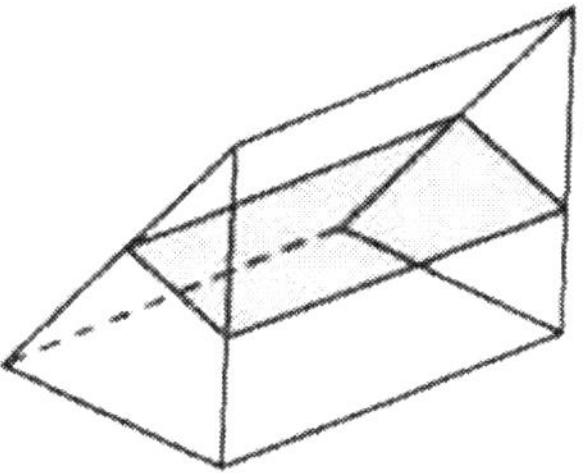

On the other hand, the cross section of a sphere will always be a circle. We also get a circular cross section when we cut a cylinder or cone with a plane that is parallel to its base.

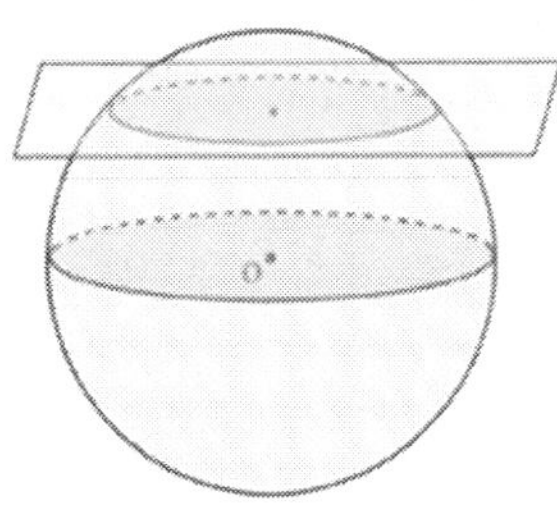

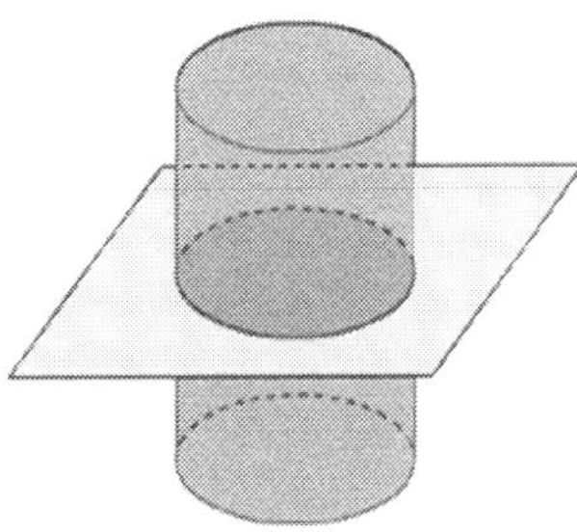
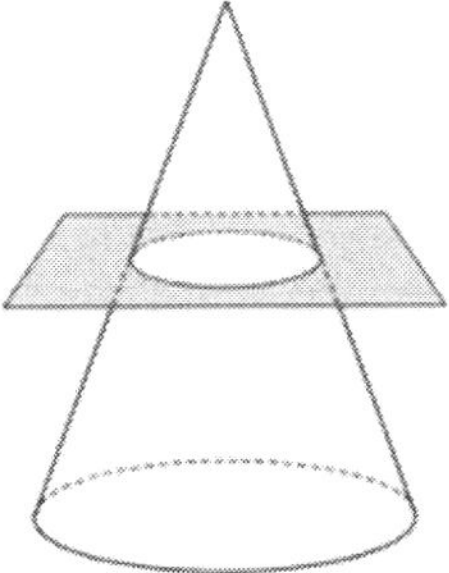

When a plane cuts perpendicular to the base of a right cylinder, the cross section is a rectangle. A cross section of a cone through its apex is a triangle. Both the cylinder and cone can have cross sections that are ellipses.

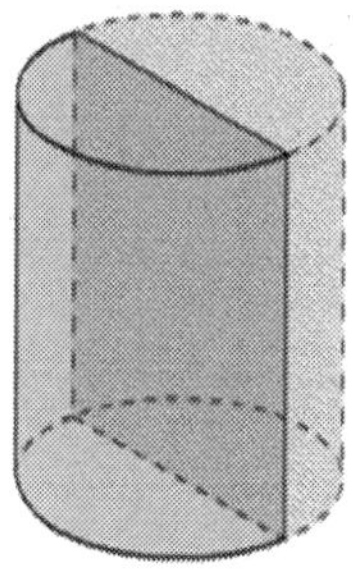
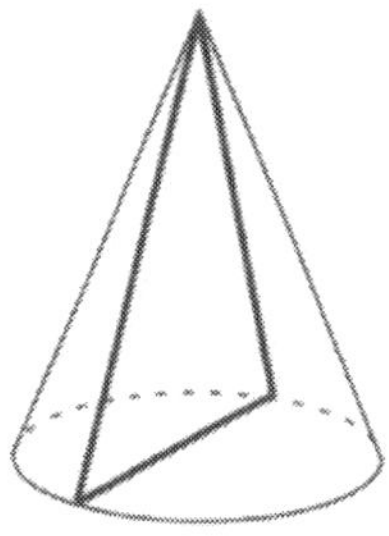
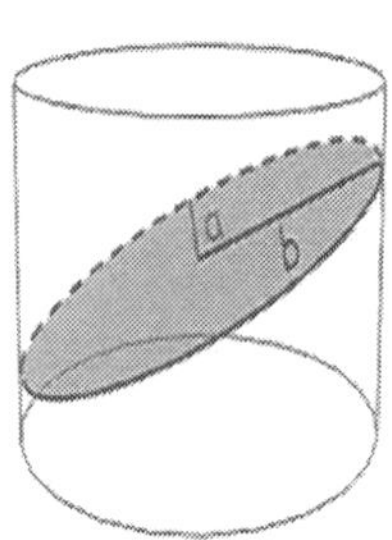

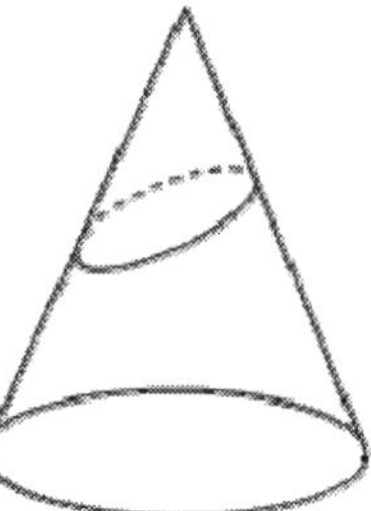

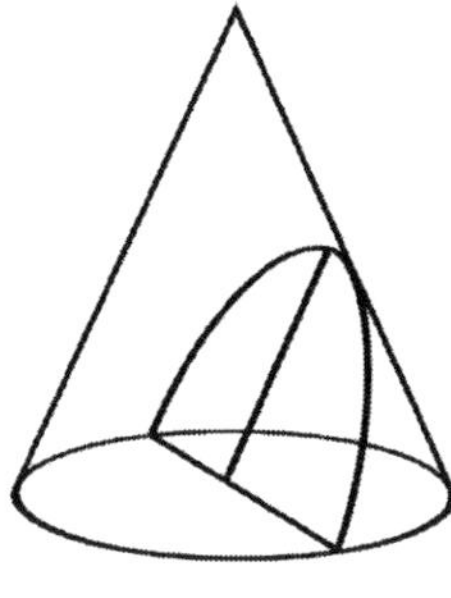

The cone can even have a parabola as a cross section!

Model Problem:

A cube with sides of length $\sqrt{8}$ is cut by a plane that passes through three of the cube's vertices as shown. What type of triangle is the cross section? What is the area of the triangle?

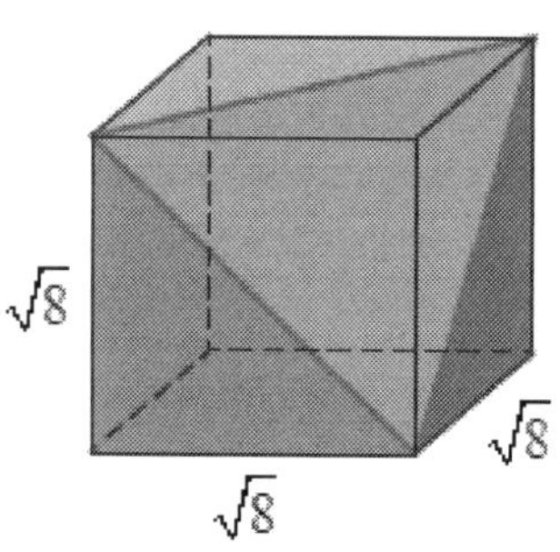

Solution:

(A) $c^2 = \left(\sqrt{8}\right)^2 + \left(\sqrt{8}\right)^2 = 16$

$c = 4$

Cross section is an equilateral triangle with sides of length 4.

(B) Draw a height, h, of the triangle, which bisects a base.

$h^2 + 2^2 = 4^2$

$h = \sqrt{12} = 2\sqrt{3}$

$Area = \frac{1}{2}bh = \frac{1}{2}(4)(2\sqrt{3}) = 4\sqrt{3}$

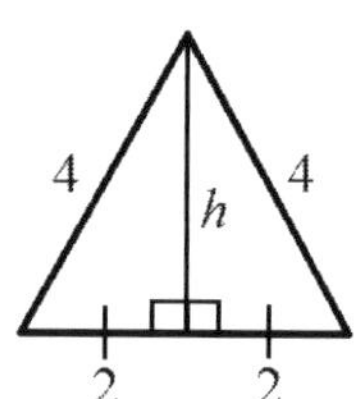

Explanation of steps:

(A) The sides are hypotenuses of congruent isosceles right triangles *[with legs of* $\sqrt{8}$ *]*, so the sides are congruent and the triangle is equilateral. We can find the length of a side using the Pythagorean Theorem.

(B) The height of an equilateral triangle bisects the triangle into two congruent right triangles. We can find the height using the Pythagorean Theorem, and then calculate the area. *[An alternative method of finding the area of the triangle is to use the SAS sine formula,* $Area = \frac{1}{2}ab\sin C = \frac{1}{2}(4)(4)\sin 60° = 4\sqrt{3}$. *]*

Practice Problems

1. The cross section of the cone shown below is a
   (1) triangle (3) semicircle
   (2) square (4) circle

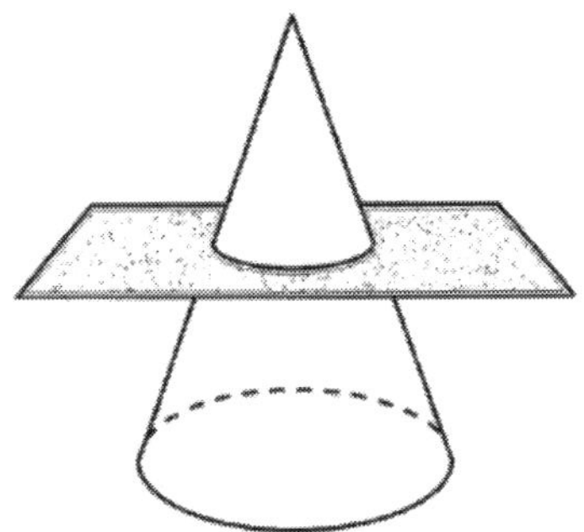

2. The cross section of the square pyramid shown below is a
   (1) triangle (3) rectangle
   (2) square (4) trapezoid

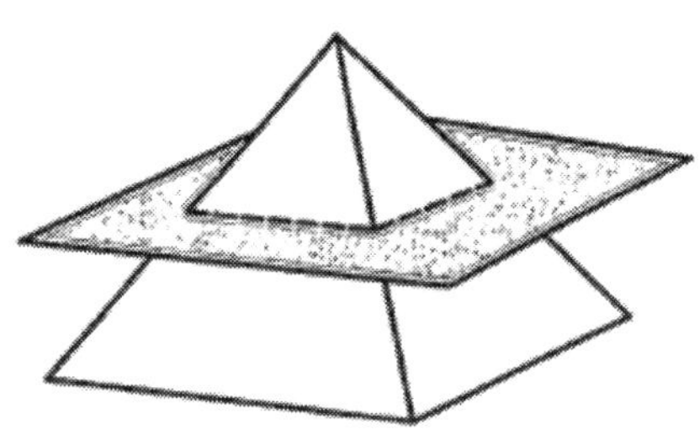

3. The cross section of the triangular pyramid shown below is a
   (1) triangle (3) rectangle
   (2) square (4) trapezoid

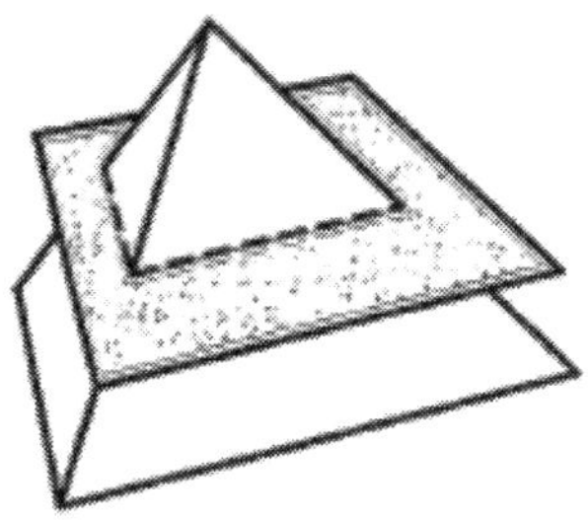

4. The cross section of the square pyramid shown below is a
   (1) triangle (3) rectangle
   (2) square (4) trapezoid

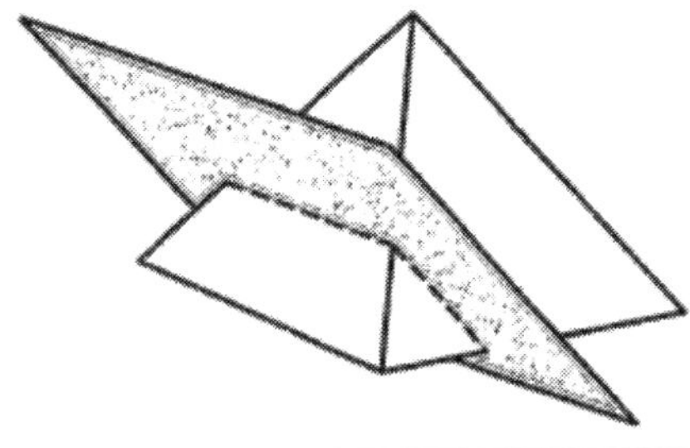

5. The cross section of the rectangular prism shown below is a
   (1) triangle (3) rectangle
   (2) square (4) hexagon

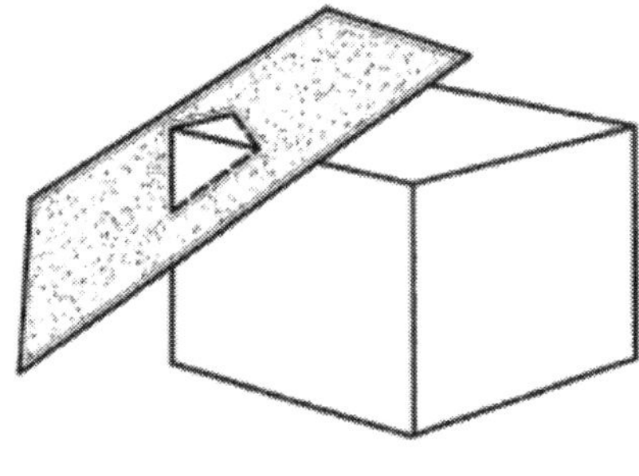

6. What is the shape of the cross section when the right pentagonal prism below is cut by a plane that is
   (a) parallel to the pentagonal bases?
   (b) perpendicular to the bases?

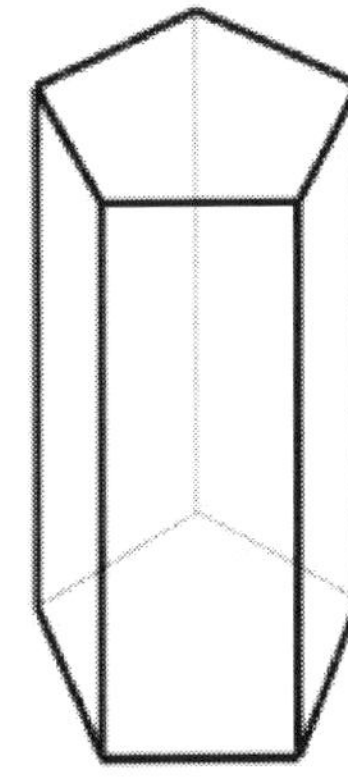

REGENTS QUESTIONS

Multiple Choice

1. (CC) Which figure can have the same cross section as a sphere?

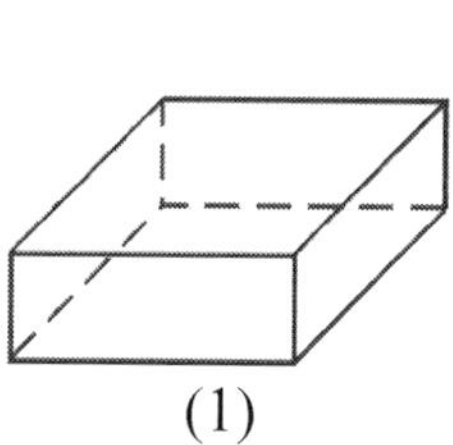
(1)

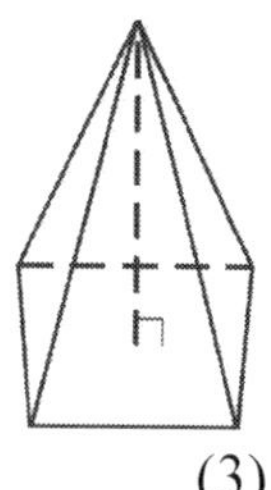
(3)

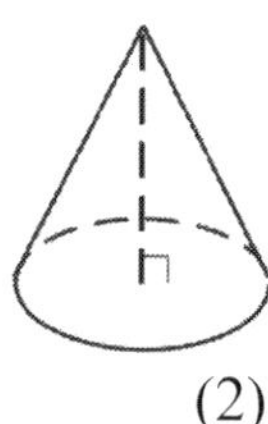
(2)

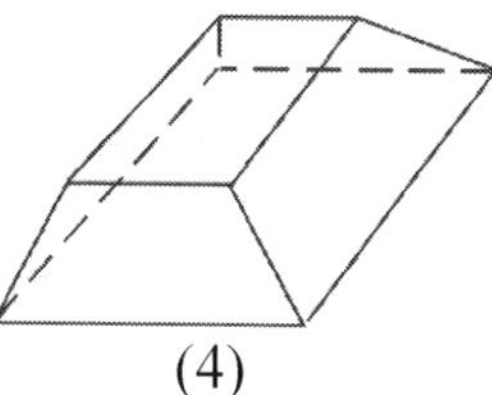
(4)

2. (CC) William is drawing pictures of cross sections of the right circular cone below.

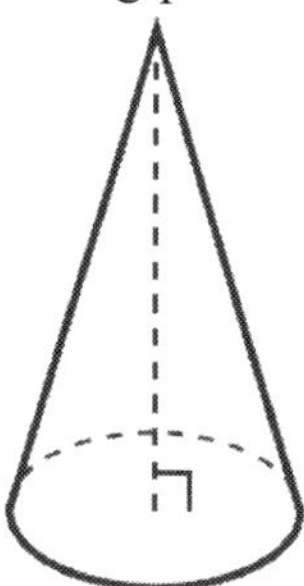

Which drawing can *not* be a cross section of a cone?

(1)

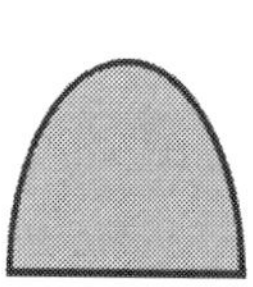
(3)

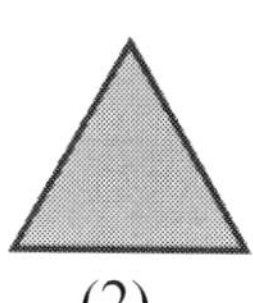
(2)

(4)

3. (CC) The cross section of a regular pyramid contains the altitude of the pyramid. The shape of this cross section is a

(1) circle (3) triangle
(2) square (4) rectangle

4. (CC) A plane intersects a hexagonal prism. The plane is perpendicular to the base of the prism. Which two-dimensional figure is the cross section of the plane intersecting the prism?

(1) triangle (3) hexagon
(2) trapezoid (4) rectangle

14.6 Cavalieri's Principle (-)

*In the Geometry Draft of Revised Standards for 2018-19, NYS has recommended **removing** this topic.*

Key Terms and Concepts

Cavalieri's Principle states that if two solids have the same height, and all corresponding pairs of cross sections have the same area, then the two solids have the same volume. When we look at the set of parallel planes that cut the two solids at every height level, cross sections are corresponding if they are cut by the same plane.

Example: The diagram of the right triangular prism and oblique rectangular prism below shows a shaded pair of corresponding cross sections that are cut by the same plane which is parallel to their bases. The two solids have the same height (8 cm), and this and every pair of corresponding cross sections have the same area (24 cm^2). So, by Cavalieri's Principal, both solids have the same volume.

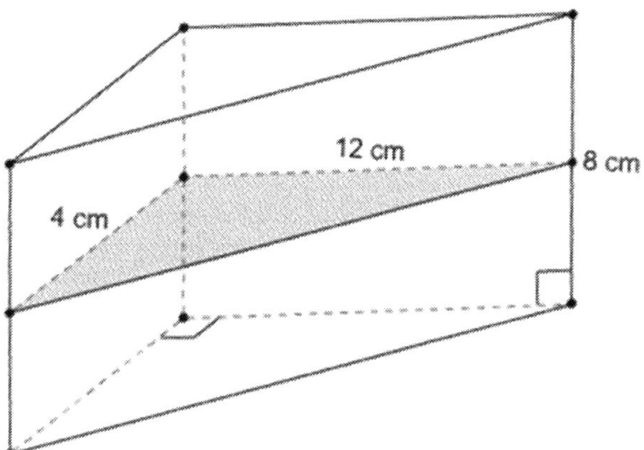

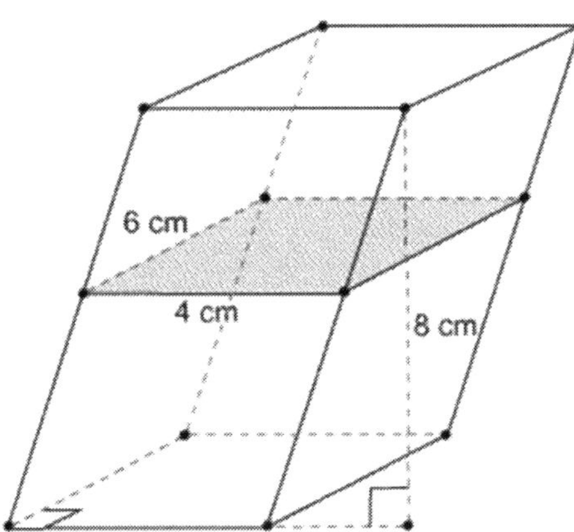

Model Problem:

Explain why the volume of a right prism is equal to the volume of an oblique prism with the same base and height.

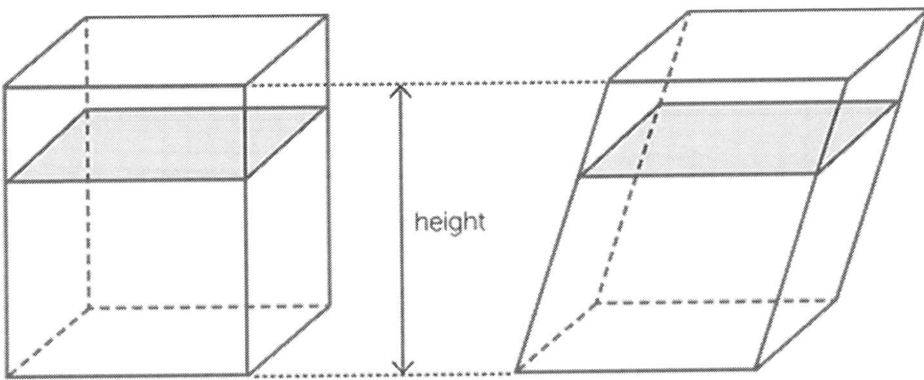

Solution:
Since each cross section parallel to the bases of the right and oblique prisms are the same size and shape (and the same area), then by Cavalieri's Principle, they have the same volume.

Practice Problems

1. Use Cavalieri's Principle to find the volume of the oblique prism. 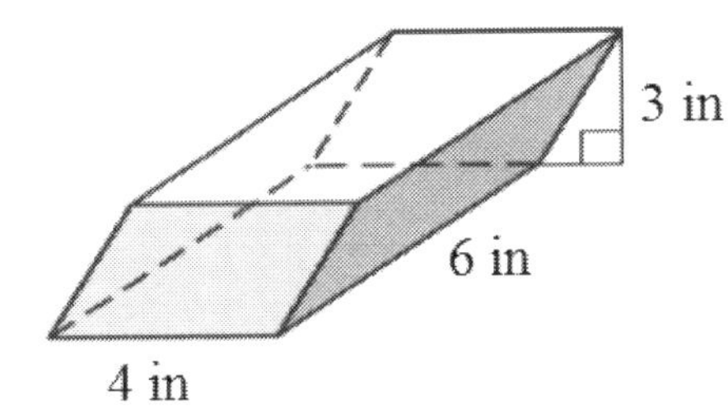	2. Use Cavalieri's Principle to find the volume of the oblique cylinder. 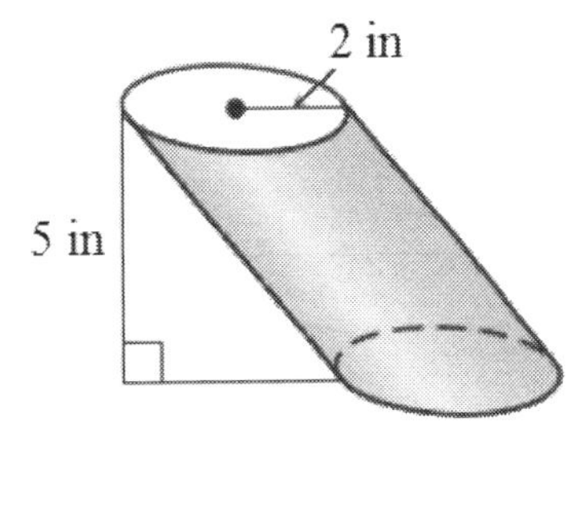

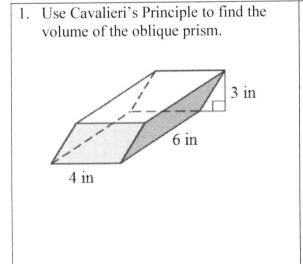

REGENTS QUESTIONS

Constructed Response

1. (CC) Two stacks of 23 quarters each are shown below. One stack forms a cylinder but the other stack does not form a cylinder.

Use Cavalieri's principle to explain why the volumes of these two stacks of quarters are equal.

Chapter 15. Constructions

15.1 Copy Segments, Angles, and Triangles

Key Terms and Concepts

The ancients Greeks worked extensively with geometric figures. Much of what they learned was documented by Euclid, around 300 B. C., in a series of books called the *Elements*. **Constructions**, or diagrams which were made using only a compass, pencil, and straightedge, were an essential ingredient in his books.

The **compass** is used to draw arcs and circles or mark off congruent segments. A stationary point, usually a pointed or rubber tip of the compass, is placed on the paper and represents the center of a circle. *The distance between the tip and the pencil represents the radius of the circle.* Once this distance is determined, the compass can be tightened so that it remains rigid. This allows the user to draw any number of arcs of the same circle, all equally distant from the stationary point. When solving a construction problem on an exam, it is important to *draw an arc in each step that the compass is used* and to leave those arcs on the drawing.

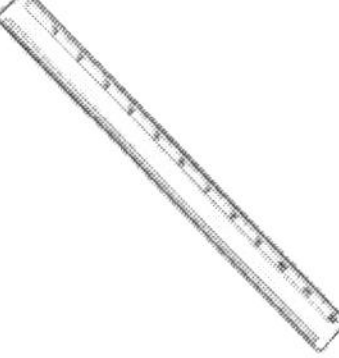

To draw a straight line, the **straightedge** is used. Although most students will use a ruler as a straightedge, it is important to accept that none of the measurement markings on the ruler may be used in a construction; the straightedge may only be used to draw straight lines.

The simplest construction is to **copy a line segment**. Given a segment such as $\overline{AB}$ below, you can construct a congruent segment $\overline{CD}$ by following these steps:

1) draw endpoint C, and use the straightedge to draw a ray from C
2) stretch the compass to match the distance from A to B, and tighten the compass
3) place the tip on C and draw an arc that intersects the ray, and label it D

Since the compass was used to set the lengths equal, $\overline{AB} \cong \overline{CD}$.

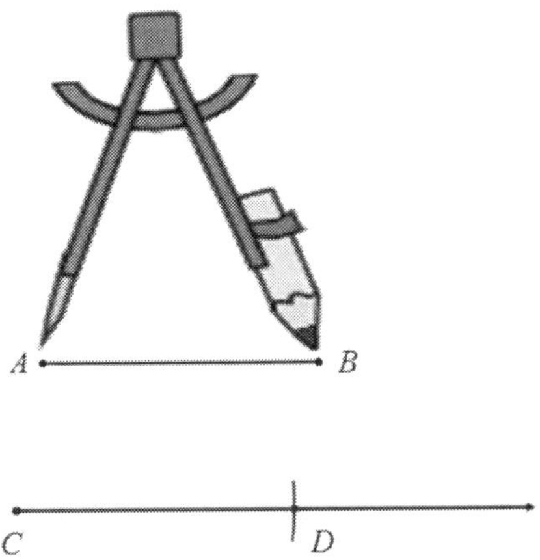

Another basic construction is to **copy an angle**. Given $\angle ABC$, follow these steps to construct congruent $\angle DEF$.

1) Use the straightedge to draw ray $\overrightarrow{EF}$.

2) With the compass set to remain rigid, place the tip on B and draw an arc that intersects both rays of $\angle ABC$. Then place the tip on E and draw an arc that is at least as long, using the same radius, as shown.

3) Set the compass to the distance between the two points where the arc intersects $\angle ABC$, by placing the tip where it intersects $\overrightarrow{BC}$ and drawing an arc where it intersects $\overrightarrow{BA}$. Then place the tip where the arc intersects $\overrightarrow{EF}$ and draw an arc that intersects the larger arc already drawn, as shown.

4) Use the straightedge to draw $\overrightarrow{ED}$ through the point of intersection of the two arcs. $\angle ABC \cong \angle DEF$.

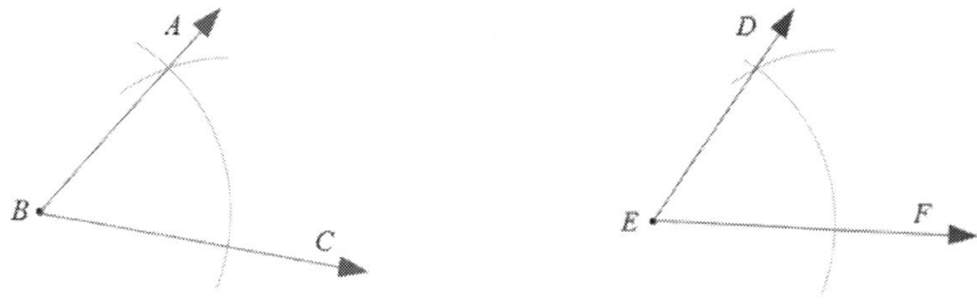

To **copy a triangle**, simply copy the sides, and the new triangle will be congruent to the given triangle by SSS.

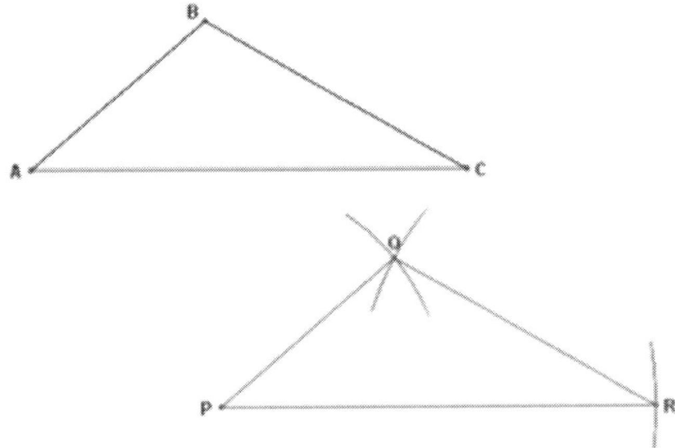

To construct $\triangle PQR$ that is congruent to given $\triangle ABC$ by SSS, as shown above:

1) Set the compass to the length of $\overline{AC}$, then place the tip on P and draw an arc at that distance. Draw a segment from P to that arc and label the point of intersection R.
2) Reset the compass to the length of $\overline{CB}$, place the tip on R and draw an arc in the region where vertex Q will be.
3) Reset the compass to the length of $\overline{AB}$, place the tip on P and draw an arc to intersect the arc just drawn, and label the point of intersection Q.
4) Use the straightedge to draw $\overline{RQ}$ and $\overline{PQ}$. $\triangle ABC \cong \triangle PQR$.

Alternatively, the triangle could have been constructed using one of the other congruence rules. For example, we could use SAS by copying side $\overline{AC}$ as $\overline{PR}$, then copy $\angle BAC$ as $\angle QPR$, and then copy side $\overline{AB}$ as $\overline{PQ}$.

Model Problem:

Construct a triangle with sides of lengths a, b, and c, as shown below. Be sure the longest side of your triangle lies on $\overline{PQ}$ and that point P is one of the triangle's vertices.

a

b

c

Solution:

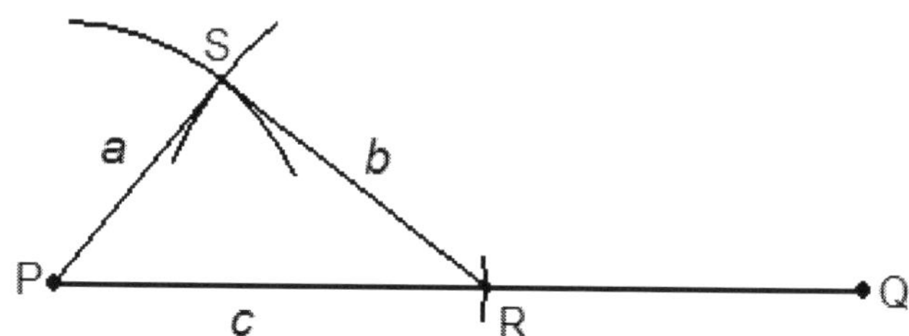

Explanation of steps:

(A) Copy segment c onto $\overline{PQ}$ with P as an endpoint. Label the other endpoint R.
(B) Set the compass to the length of a and draw an arc from P.
(C) Set the compass to the length of b and draw an arc from R. Label the intersection of the arcs point S.
(D) Use the straightedge to draw $\overline{PS}$ and $\overline{RS}$.

Practice Problems

1. Construct $\overline{EF}$ such that $EF = AB + CD$.

2. Construct a copy of right angle ABC.

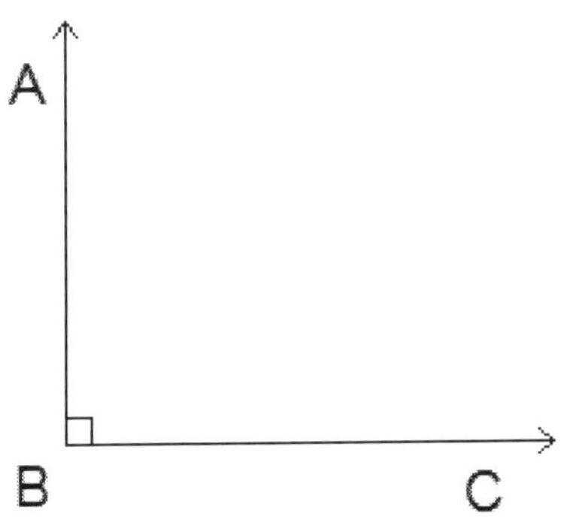

3. Based on the construction marks, which congruence rule was used to construct this triangle as congruent to a given triangle, not shown?

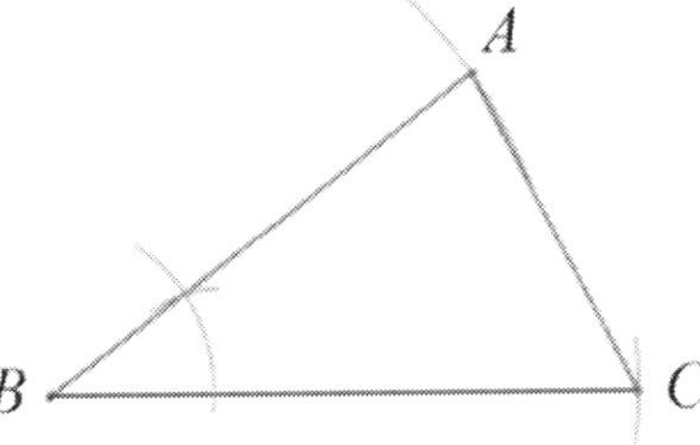

4. Based on the construction marks, which congruence rule was used to construct this triangle as congruent to a given triangle, not shown?

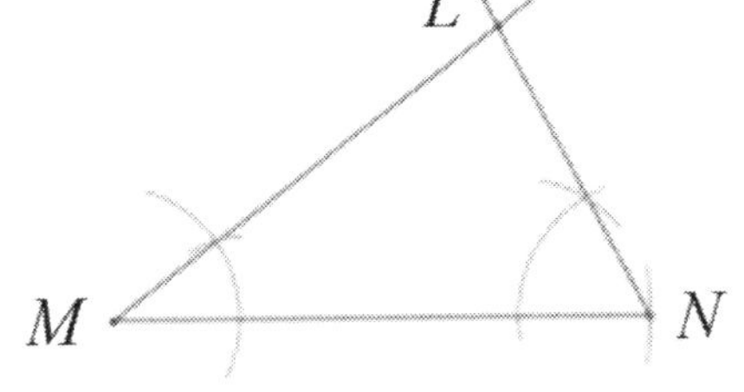

REGENTS QUESTIONS

Constructed Response

1. In the diagram below, $\triangle ABC$ is equilateral. Using a compass and straightedge, construct a new equilateral triangle congruent to $\triangle ABC$ in the space below.
[Leave all construction marks.]

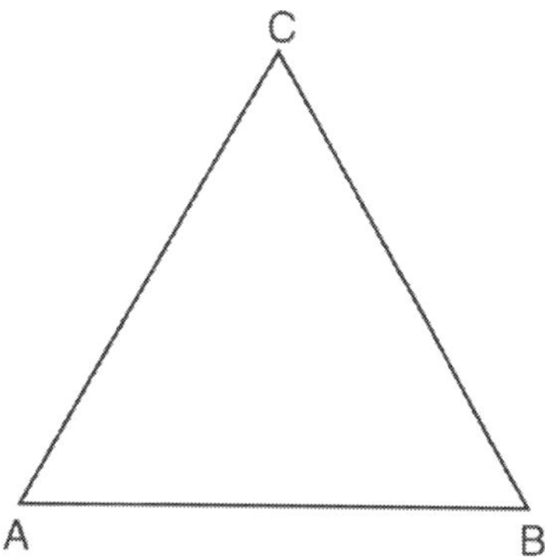

2. (CC) Triangle XYZ is shown below. Using a compass and straightedge, on the line below, construct and label $\triangle ABC$, such that $\triangle ABC \cong \triangle XYZ$. [Leave all construction marks.]

Based on your construction, state the theorem that justifies why $\triangle ABC$ is congruent to $\triangle XYZ$.

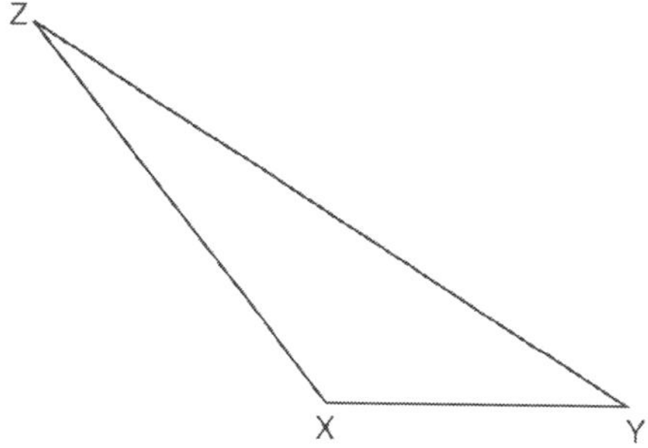

3. (CC) Using a compass and straightedge, construct and label $\triangle A'B'C'$, the image of $\triangle ABC$ after a dilation with a scale factor of 2 and centered at *B*. [Leave all construction marks.]

Describe the relationship between the lengths of $\overline{AC}$ and $\overline{A'C'}$.

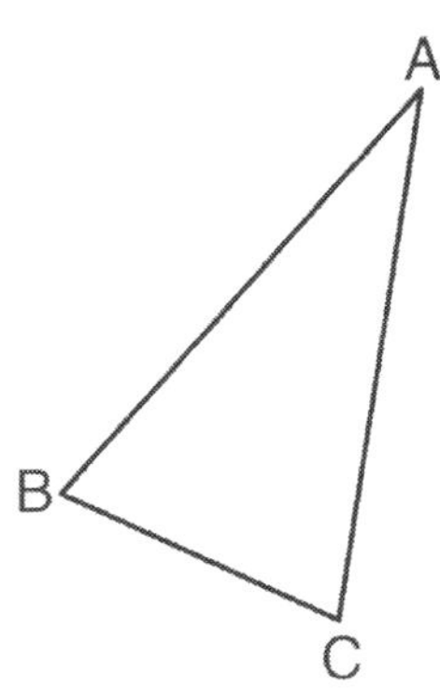

15.2 Construct an Equilateral Triangle

Key Terms and Concepts

To **construct an equilateral triangle** with a given side, simply set the compass to the length of that side and use this length to determine the location of the third vertex.

For example, given side $\overline{AB}$ shown to the right, equilateral triangle ABC can be constructed by following these steps:

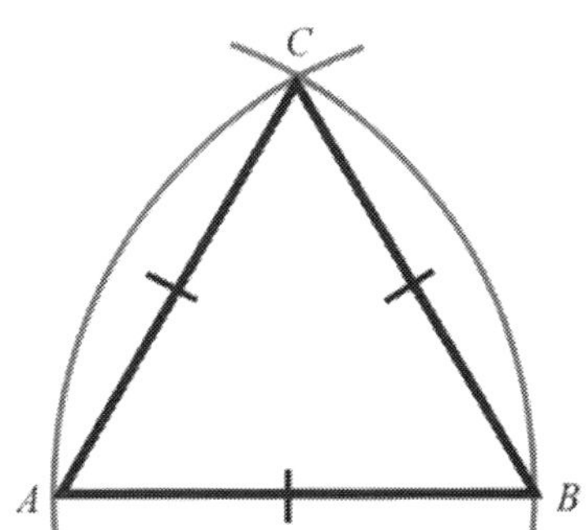

1) Place the tip of the compass on A, set the compass to the length of $\overline{AB}$, and draw a large arc. (Angle A will be $60°$, so the arc will need to be at least $\frac{1}{6}$ of a full circle.)
2) Place the tip of the compass on B, and with the compass still set to the length of $\overline{AB}$, draw another arc to intersect the first arc. Label the point of intersection as vertex C.
3) With the straightedge, draw $\overline{AC}$ and $\overline{BC}$.

Note that the construction drawing does not need to show the full arcs, but must at least show the essential parts of the arc to demonstrate how the point of intersection was determined. Example: The construction may look more like the diagram below.

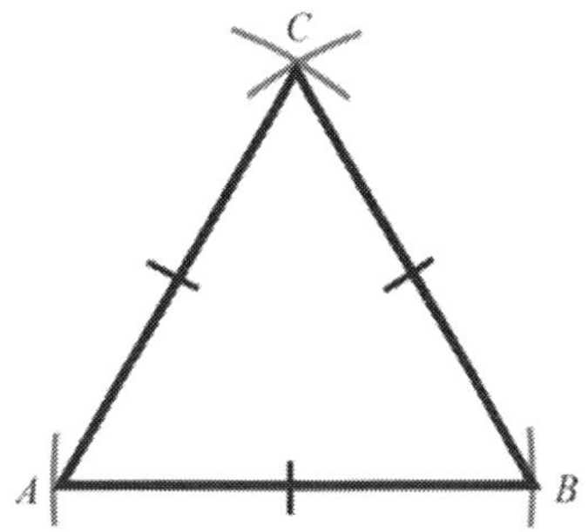

Model Problem:

Construct an equilateral triangle with $\overline{RS}$ as one side.

Solution:

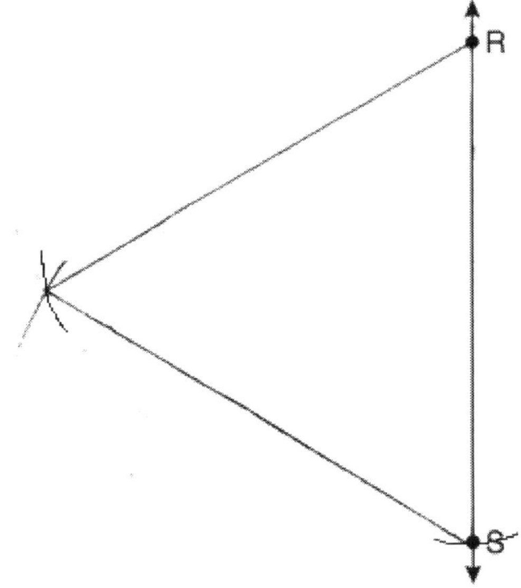

Explanation of steps:

(A) Set the compass to the length of $\overline{RS}$.
(B) Draw an arc from R and an arc from S with radii equal to RS.
(C) Using the intersection of the arcs as the third vertex, draw the sides.

Practice Problems

1. Construct a 60° angle with its vertex at point P.

P

2. Construct a 120° angle with its vertex at point P.
   Hint: Construct two adjacent 60° angles.

P

REGENTS QUESTIONS

Multiple Choice

1. Which diagram shows the construction of an equilateral triangle?

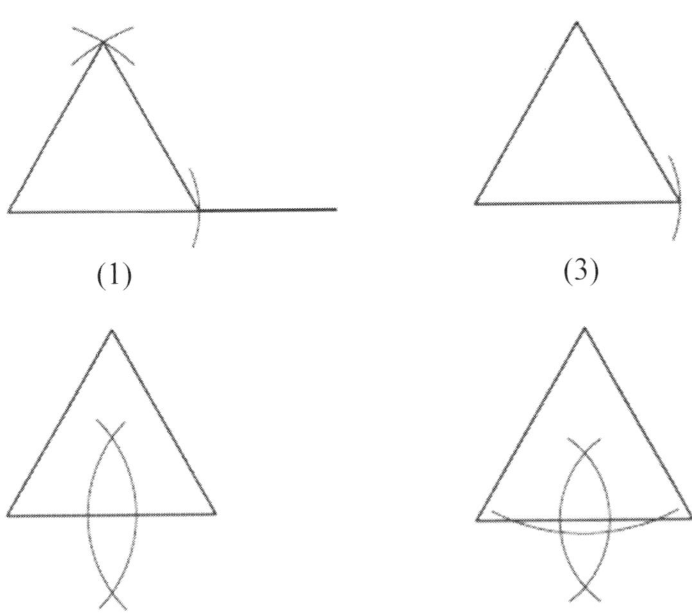

2. Which diagram represents a correct construction of equilateral $\triangle ABC$, given side $\overline{AB}$?

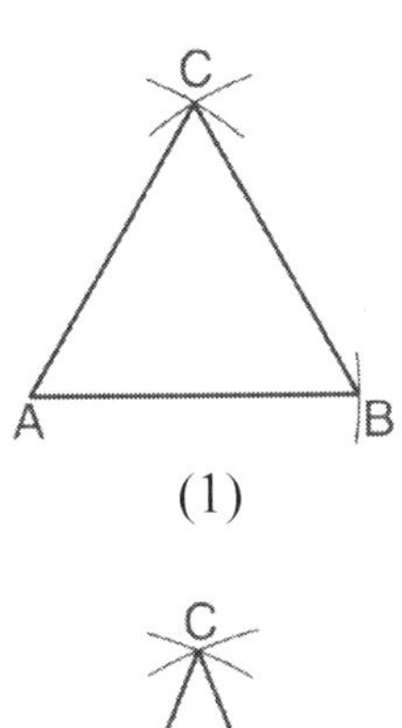

(1)

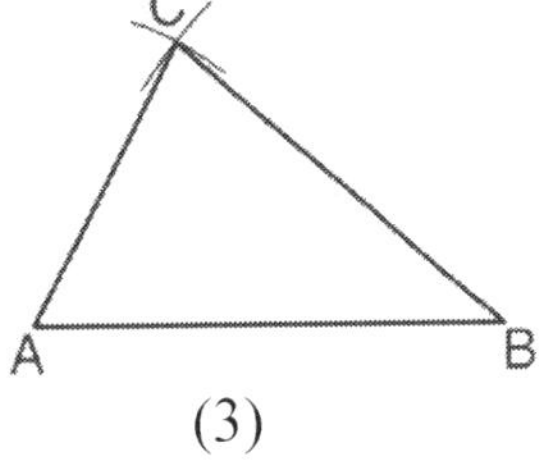

(3)

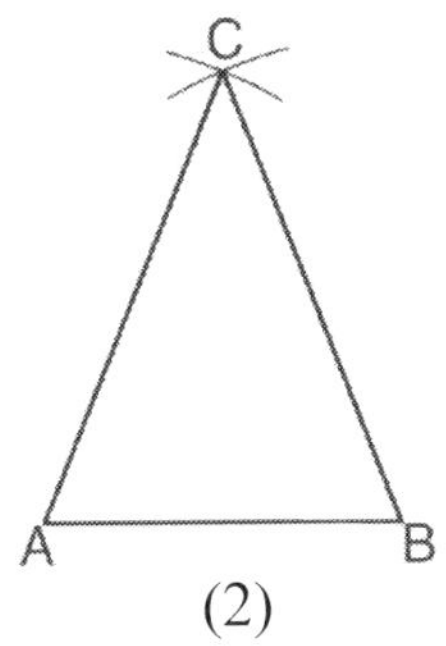

(2)

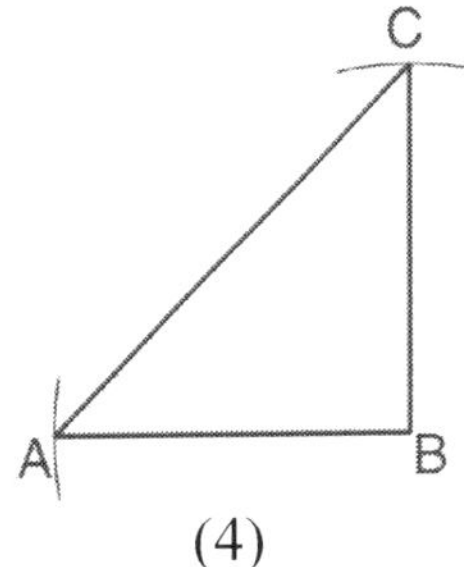

(4)

3. The diagram below shows the construction of an equilateral triangle.

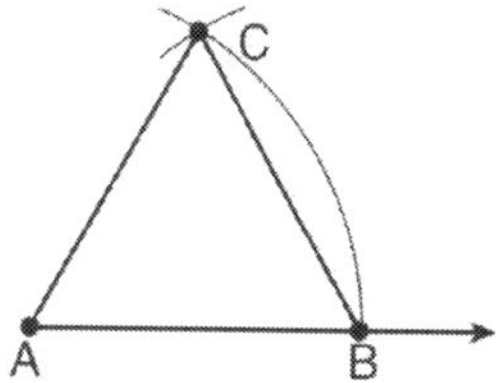

Which statement justifies this construction?

(1) $\angle A + \angle B + \angle C = 180$
(2) $m\angle A = m\angle B = m\angle C$
(3) $AB = AC = BC$
(4) $AB + BC > AC$

Constructed Response

4. On the line segment below, use a compass and straightedge to construct equilateral triangle *ABC*. [Leave all construction marks.]

5. On the ray drawn below, using a compass and straightedge, construct an equilateral triangle with a vertex at *R*. The length of a side of the triangle must be equal to a length of the diagonal of rectangle *ABCD*.

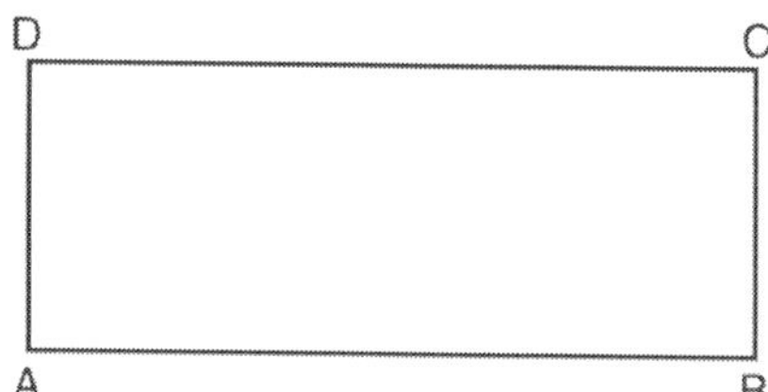

R

15.3 Construct an Angle Bisector

Key Terms and Concepts

Given $\angle ABC$, we can use the following steps to **construct an angle bisector**.

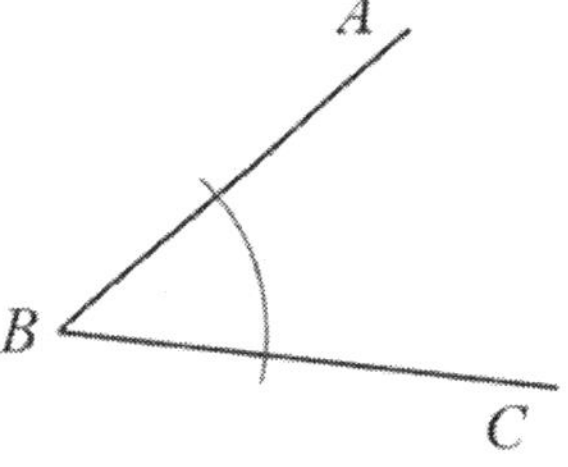

1) Place the compass tip on B and draw an arc that intersects both $\overrightarrow{BA}$ and $\overrightarrow{BC}$.

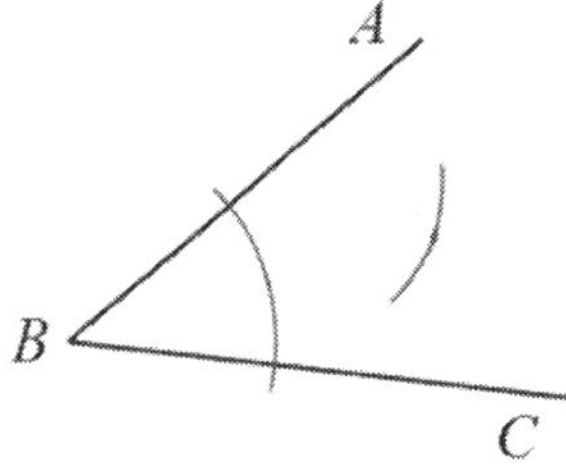

2) Place the compass tip on the point where the drawn arc intersects $\overrightarrow{BA}$, and draw an arc within the center of the angle.

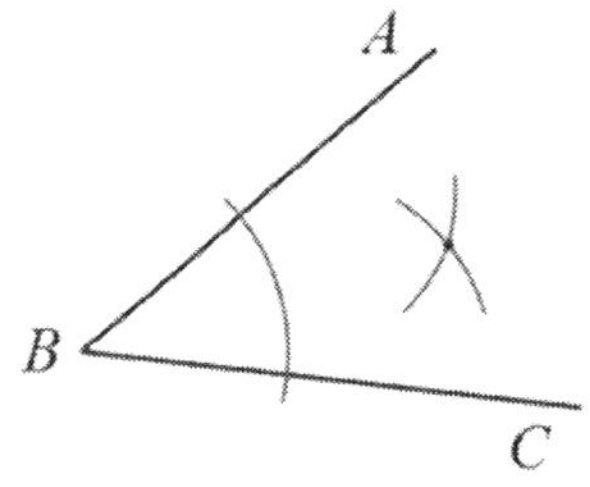

3) Using the same compass setting, place the tip on the point where the arc intersects $\overrightarrow{BC}$, and draw an arc that intersects the arc drawn in step 2.

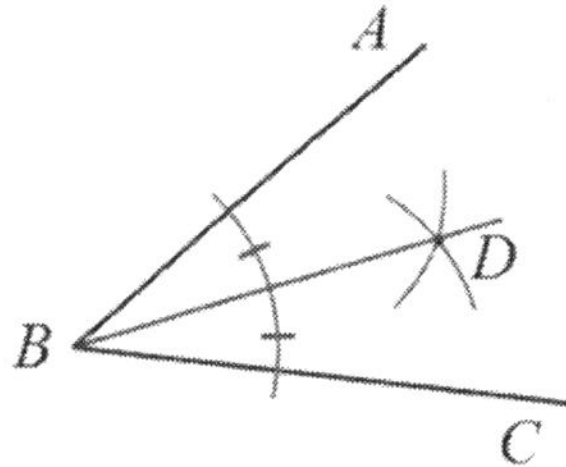

4) Use the straightedge to draw a ray from point B through the intersection of the two arcs and label it $\overrightarrow{BD}$.
$\angle ABD \cong \angle CBD$

Model Problem:

Construct the bisector of the angle shown below.

Solution:

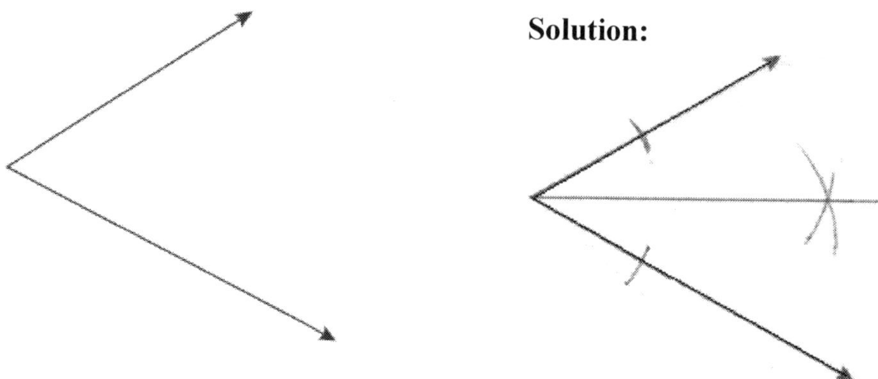

Explanation of steps:

(A) From the vertex and using the same radius, draw arcs onto both rays.
(B) From the intersection of one arc with a ray, draw an arc midway between the rays.
(C) From the intersection of the other arc and ray, and using the same radius, draw another arc midway between the rays.
(D) Use a straightedge to draw a line from the vertex through the point of intersection of the two arcs.

Practice Problems

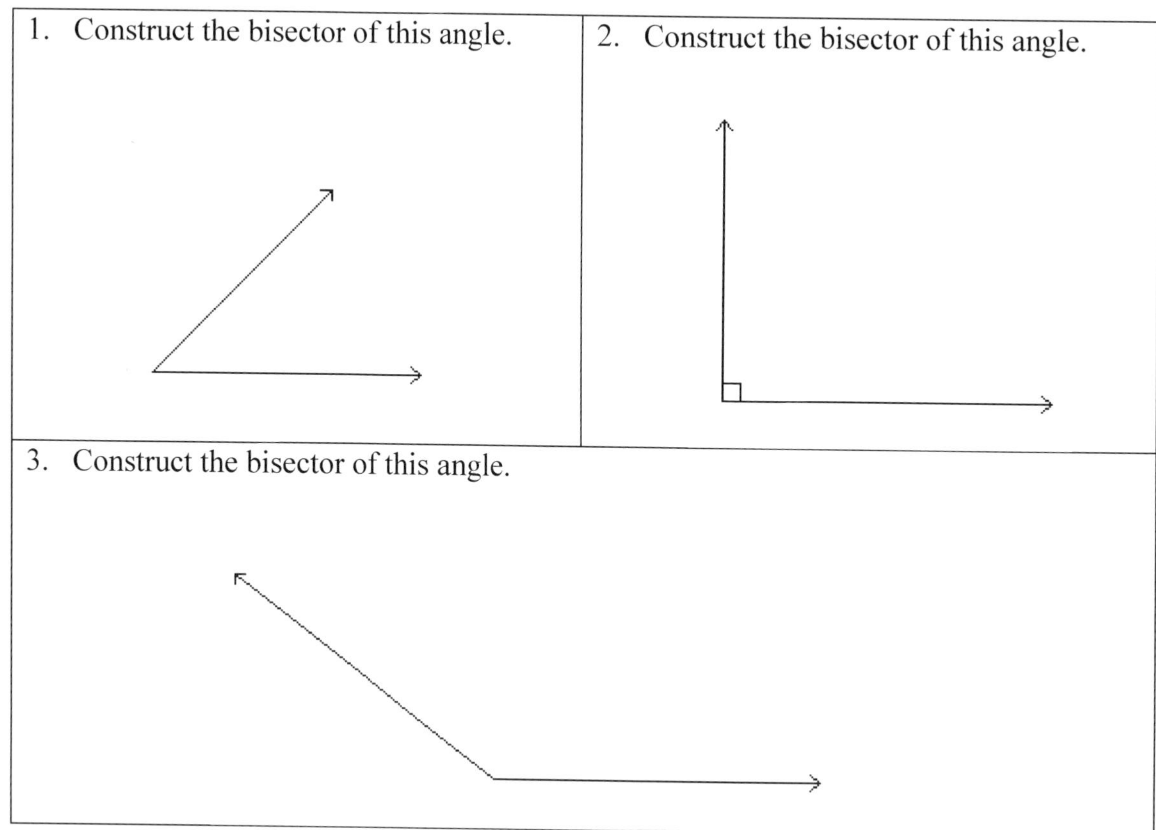

1. Construct the bisector of this angle.
2. Construct the bisector of this angle.
3. Construct the bisector of this angle.

REGENTS QUESTIONS

Multiple Choice

1. Which illustration shows the correct construction of an angle bisector?

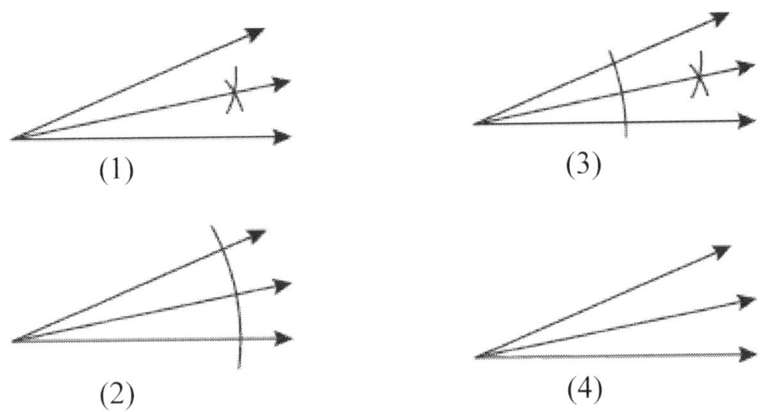

2. As shown in the diagram below of $\triangle ABC$, a compass is used to find points D and E, equidistant from point A. Next, the compass is used to find point F, equidistant from points D and E. Finally, a straightedge is used to draw $\overrightarrow{AF}$. Then, point G, the intersection of $\overrightarrow{AF}$ and side $\overline{BC}$ of $\triangle ABC$, is labeled.

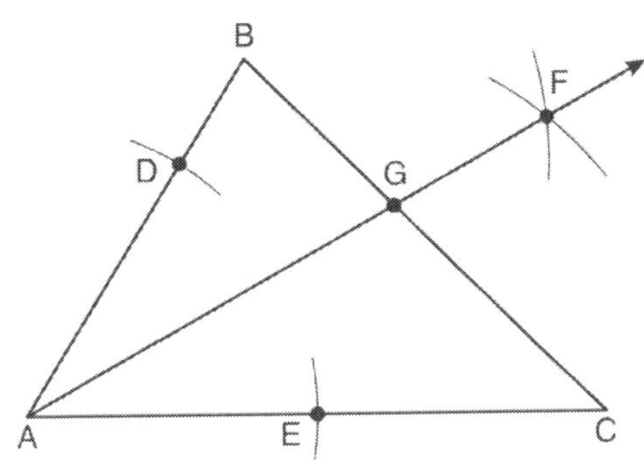

Which statement must be true?

(1) $\overrightarrow{AF}$ bisects side $\overline{BC}$
(2) $\overrightarrow{AF}$ bisects $\angle BAC$
(3) $\overrightarrow{AF} \perp \overline{BC}$
(4) $\triangle ABG \sim \triangle ACG$

3. A student used a compass and a straightedge to construct $\overline{CE}$ in $\triangle ABC$ as shown below.

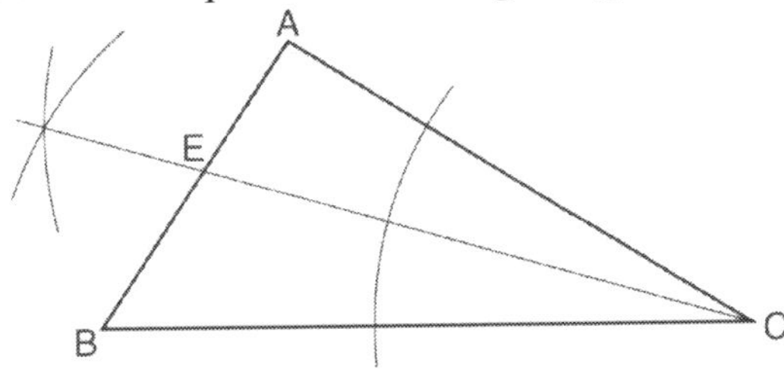

Which statement must always be true for this construction?

(1) $\angle CEA \cong \angle CEB$
(2) $\angle ACE \cong \angle BCE$
(3) $\overline{AE} \cong \overline{BE}$
(4) $\overline{EC} \cong \overline{AC}$

Constructed Response

4. On the diagram below, use a compass and straightedge to construct the bisector of $\angle ABC$. [Leave all construction marks.]

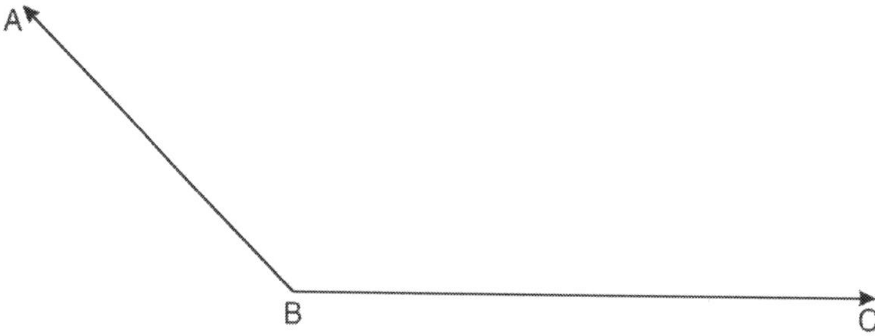

5. On the diagram below, use a compass and straightedge to construct the bisector of $\angle XYZ$. [Leave all construction marks.]

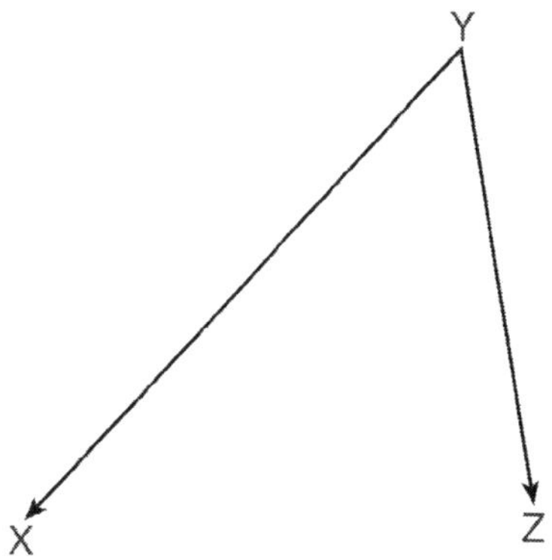

6. Using a compass and straightedge, construct the bisector of $\angle CBA$. [Leave all construction marks.]

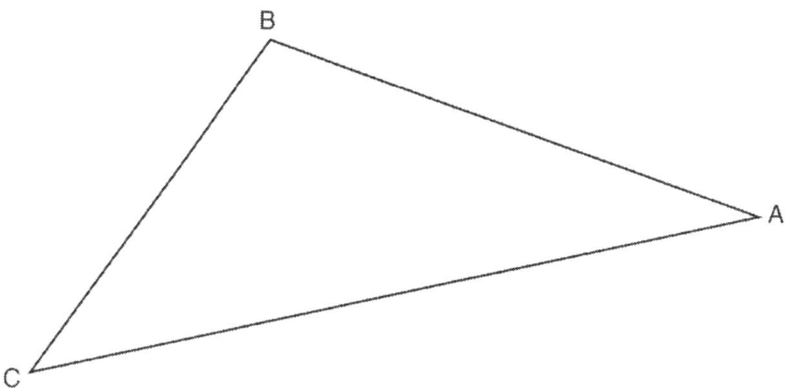

7. Using a compass and straightedge, construct the bisector of $\angle MJH$.
[Leave all construction marks.]

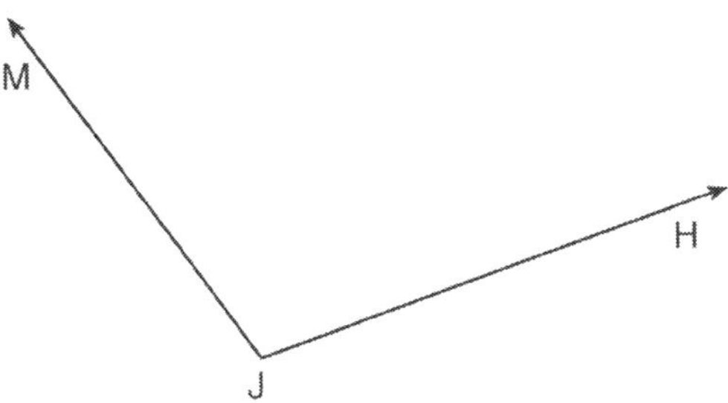

8. Using a compass and straightedge, construct an equilateral triangle with $\overline{AB}$ as a side. Using this triangle, construct a 30° angle with its vertex at A. [Leave all construction marks.]

9. Using a compass and a straightedge, construct the bisector of $\angle CDE$.
[Leave all construction marks.]

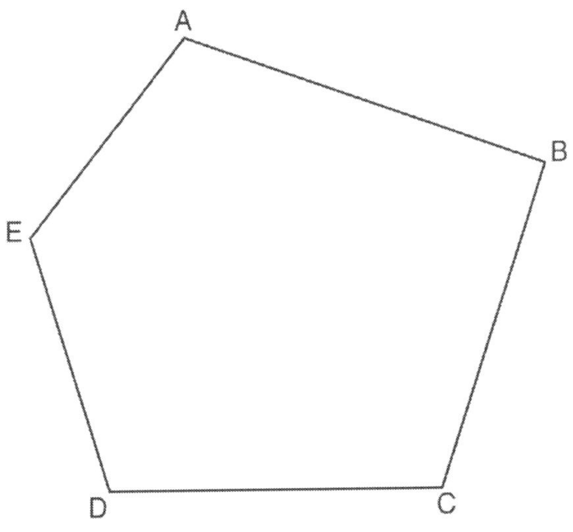

15.4 Construct a Perpendicular Bisector

Key Terms and Concepts

Given line segment $\overline{AB}$, we can **construct a perpendicular bisector** using the following steps:

1) Set the compass width to more than half the length of the segment, then place the compass tip on point A, and draw a large arc that extends above and below the segment, as shown.

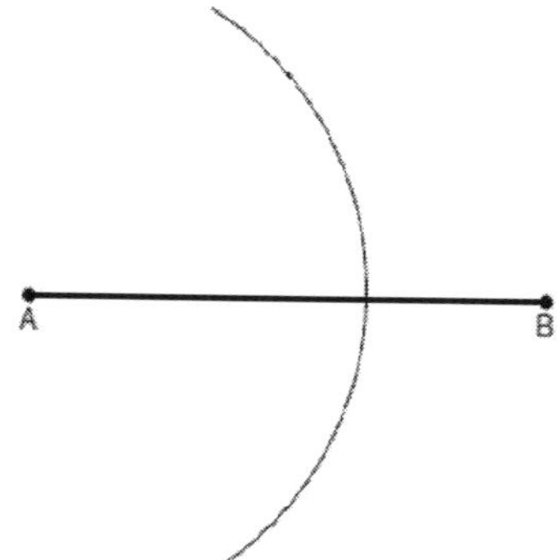

2) Using the same compass setting, place the tip on point B and draw an arc that intersects the first arc both above and below the segment.

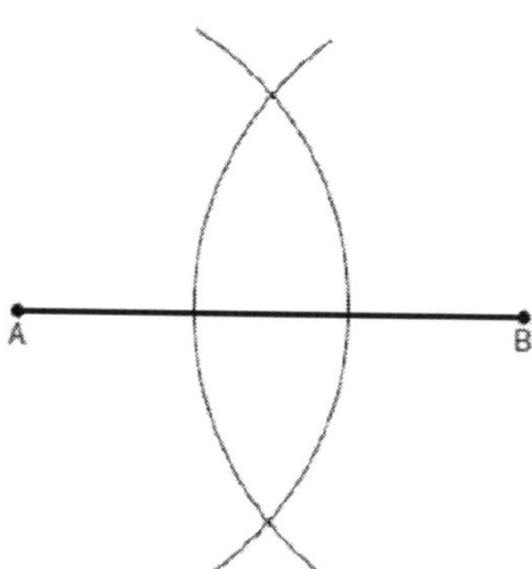

3) Use the straightedge to draw the line through both intersections. This line will intersect $\overline{AB}$ at its midpoint, C, and will be perpendicular to $\overline{AB}$. (It is also true that C is the midpoint between the two points where the arcs intersect.)

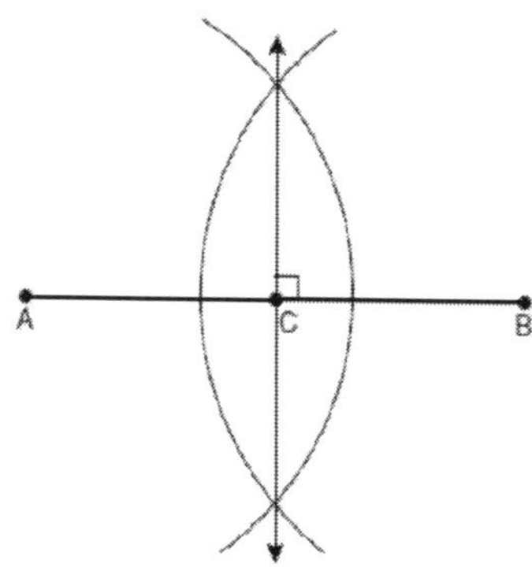

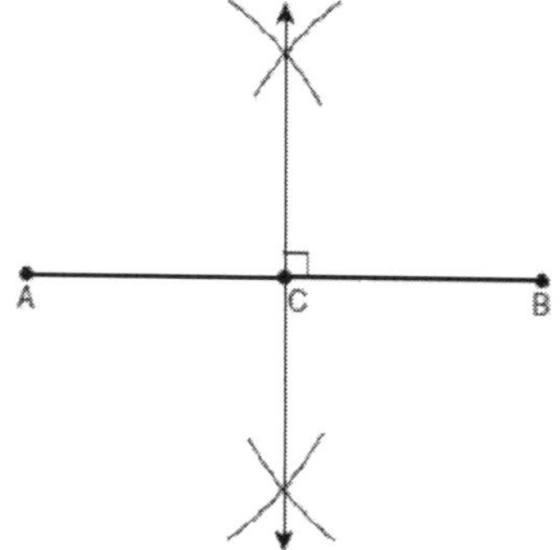

If you prefer, you may draw just those parts of the arcs where the intersections occur, above and below the middle of the segment. The construction would look as shown to the left.

Model Problem:
Construct a 45° angle.

Solution:

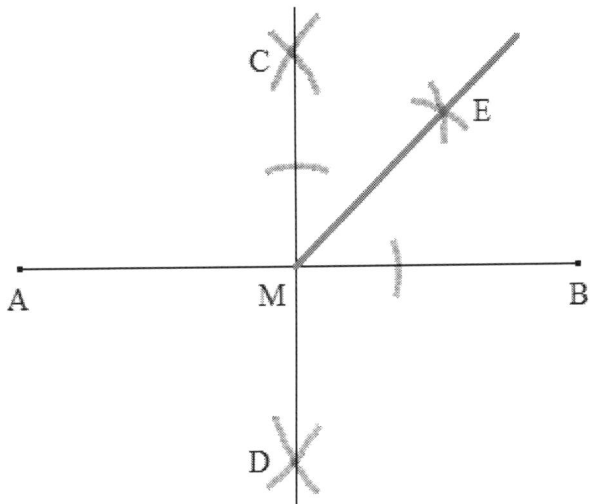

Explanation of steps:

(A) Draw $\overline{AB}$. Construct $\overline{CD}$, the perpendicular bisector of $\overline{AB}$. Label point M, the midpoint of $\overline{AB}$. $m\angle CMB = 90°$.

(B) Construct $\overrightarrow{ME}$, the bisector of $\angle CMB$. $m\angle EMB = \frac{1}{2} m\angle CMB = 45°$.

Practice Problems

1. Construct the perpendicular bisector of $\overline{AB}$.

A B

2. Use construction methods to divide $\overline{XY}$ into four congruent parts.

X Y

REGENTS QUESTIONS

Multiple Choice

1. Which diagram shows the construction of the perpendicular bisector of $\overline{AB}$?

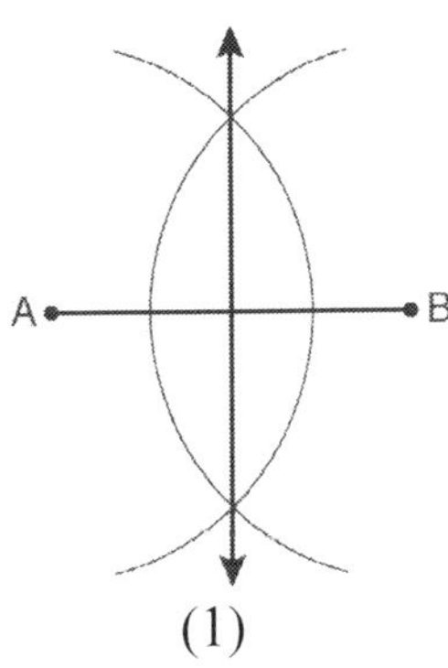

(1)

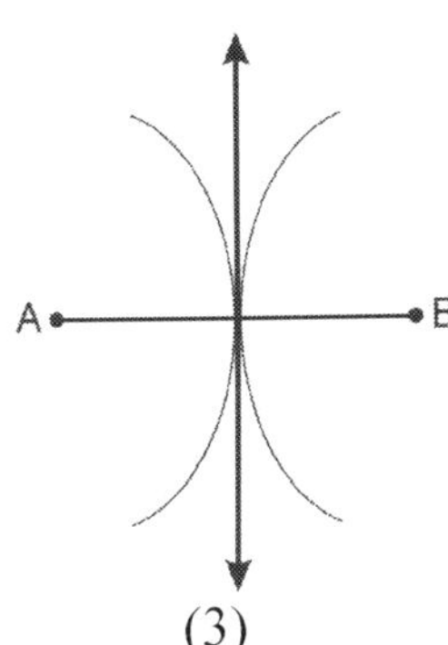

(3)

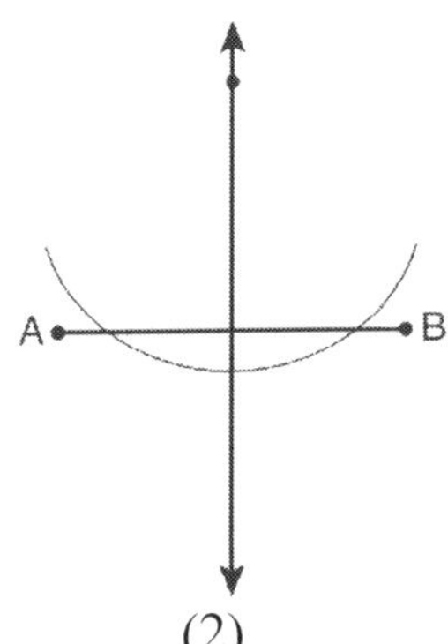

(2)

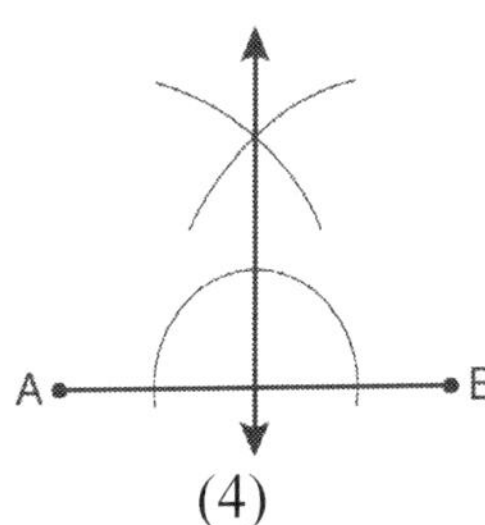

(4)

2. Based on the construction below, which conclusion is *not* always true?

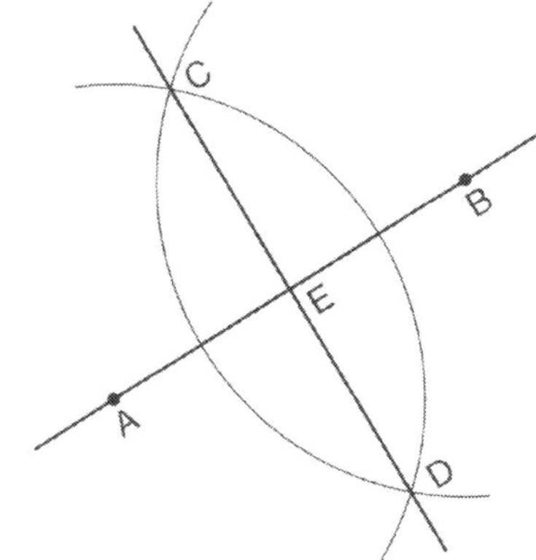

(1) $\overline{AB} \perp \overline{CD}$
(2) $AB = CD$
(3) $AE = EB$
(4) $CE = DE$

3. Which diagram shows the construction of a 45° angle?

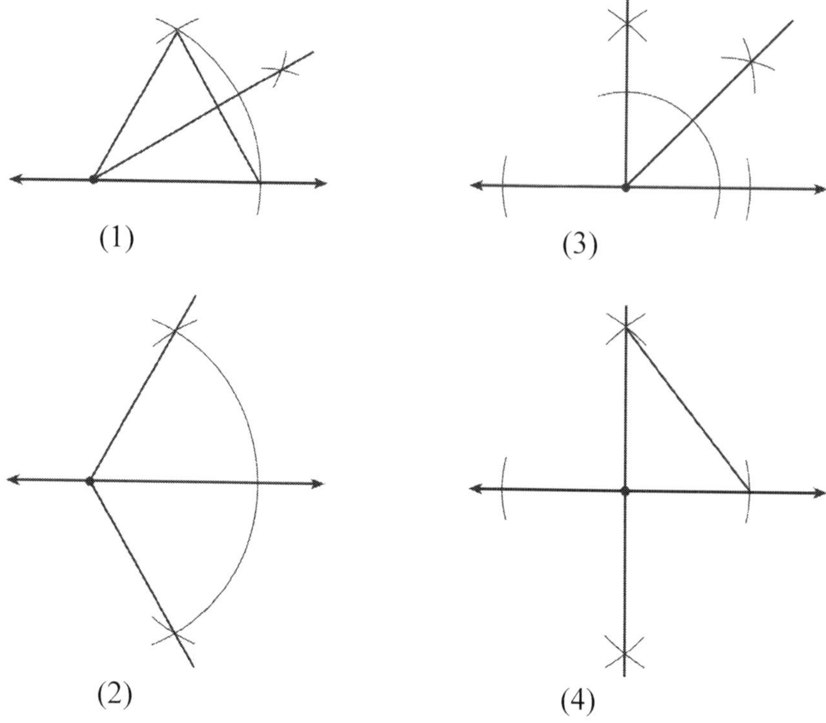

Constructed Response

4. On the diagram of $\triangle ABC$ shown below, use a compass and straightedge to construct the perpendicular bisector of $\overline{AC}$. [Leave all construction marks.]

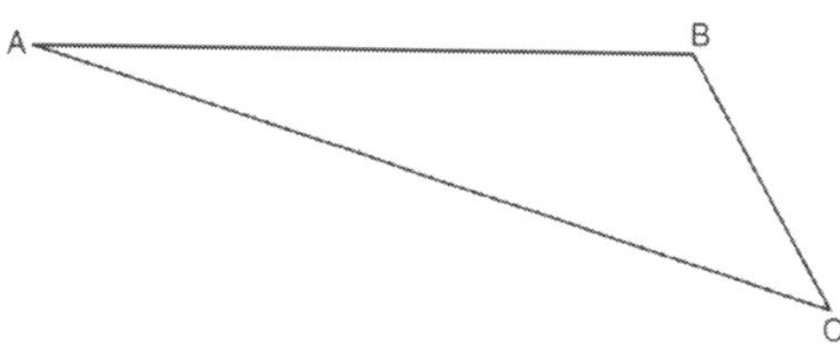

5. Using a compass and straightedge, construct the perpendicular bisector of $\overline{AB}$. [Leave all construction marks.]

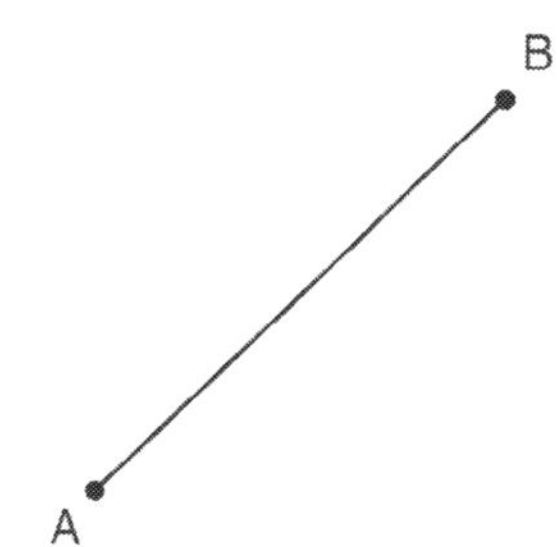

6. Use a compass and straightedge to divide line segment AB below into four congruent parts. [Leave all construction marks.]

7. Using a compass and straightedge, construct the perpendicular bisector of side $\overline{AR}$ in $\triangle ART$ shown below. [Leave all construction marks.]

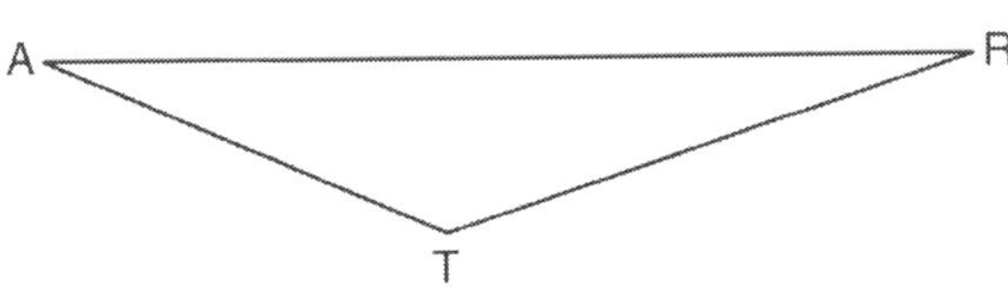

8. Using a compass and straightedge, locate the midpoint of $\overline{AB}$ by construction. [Leave all construction marks.]

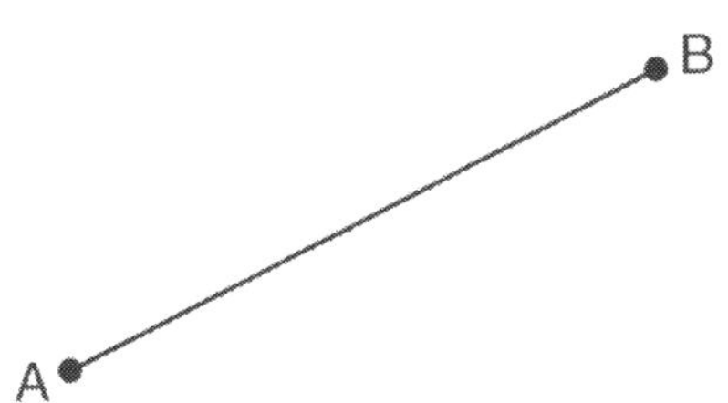

9. CC In the diagram of $\triangle ABC$ shown below, use a compass and straightedge to construct the median to $\overline{AB}$. [Leave all construction marks.]

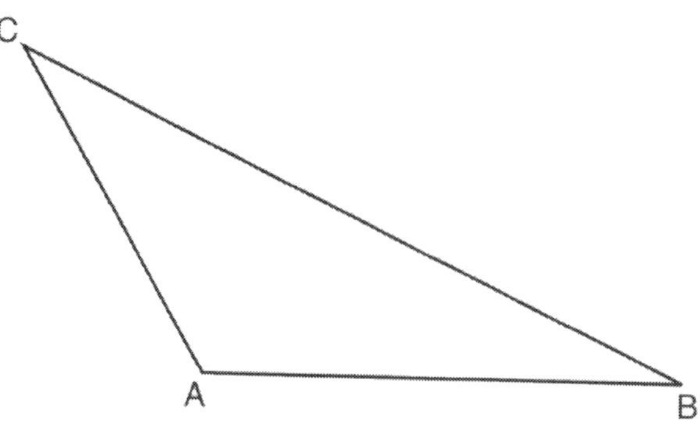

10. CC Using a compass and straightedge, construct the line of reflection over which triangle RST reflects onto triangle $R'S'T'$. [Leave all construction marks.]

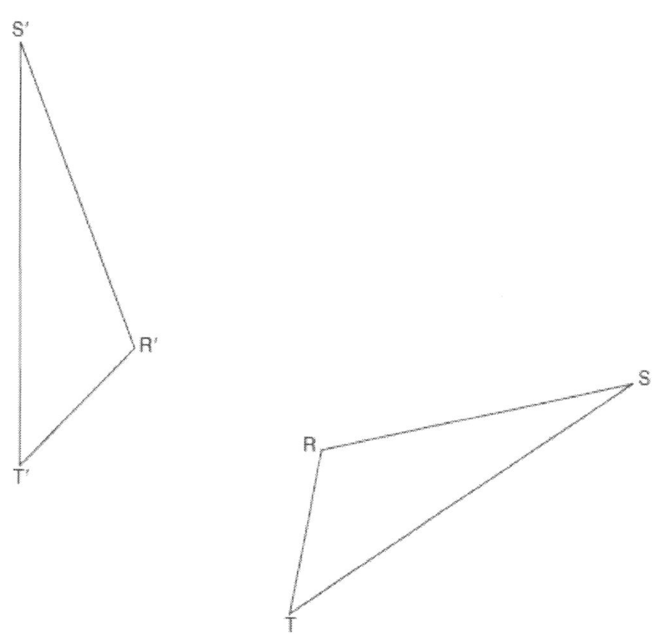

15.5 Construct Lines Through a Point

Key Terms and Concepts

We can apply our construction skills to a number of new construction tasks.

Constructing a line through a point that is parallel to a given line applies a skill we've already learned – copying an angle. When two lines are cut by a transversal so that corresponding angles are congruent, the lines must be parallel. So, we can follow these steps:

1) Given $\overleftrightarrow{AB}$ and point P, draw a line through P that intersects $\overleftrightarrow{AB}$ at M. This will be the transversal.

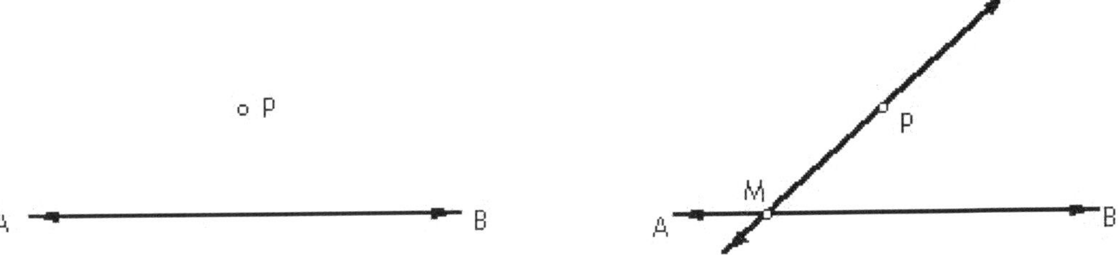

2) Copy $\angle PMB$ to create a congruent angle at vertex P, with a new line drawn through P. These two corresponding angles are congruent, so the two lines are parallel.

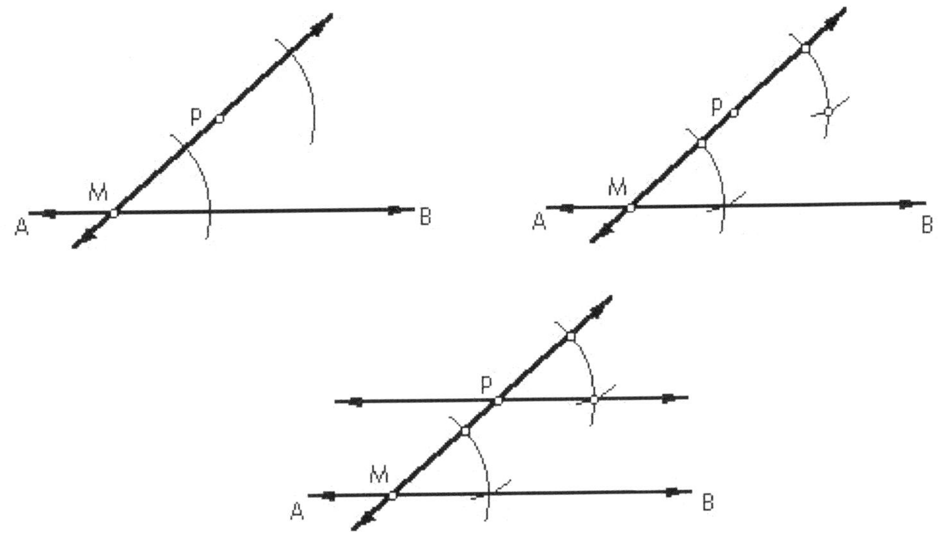

If you prefer, the angle could be copied to the other side of the transversal to create a pair of alternate interior angles that are congruent, as shown in the construction to the right.

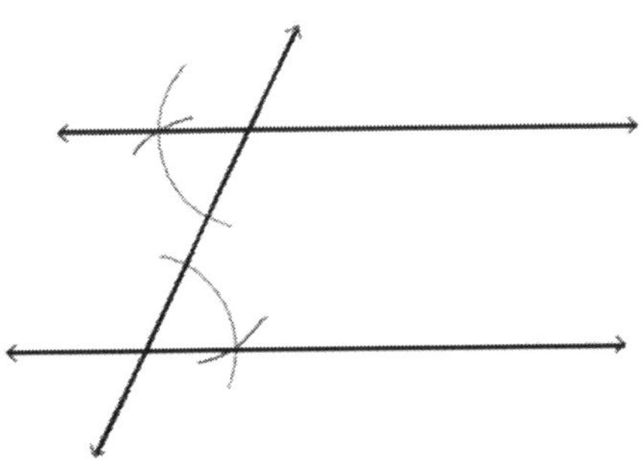

To construct a perpendicular line though a point on a given line, we can use a method similar to constructing a perpendicular bisector. Suppose P is a point on $\overleftrightarrow{AB}$.

1) Placing the compass tip at P, draw arcs at equal distances from P on $\overleftrightarrow{AB}$ that intersect the line at points C and D. P is now the midpoint of $\overline{CD}$.

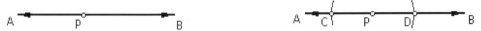

2) Now construct the perpendicular bisector of $\overline{CD}$. We can actually simplify this part of the construction by drawing only the intersecting arcs above the line. Since we know the line must pass through P, we don't need to draw the arcs below the line.

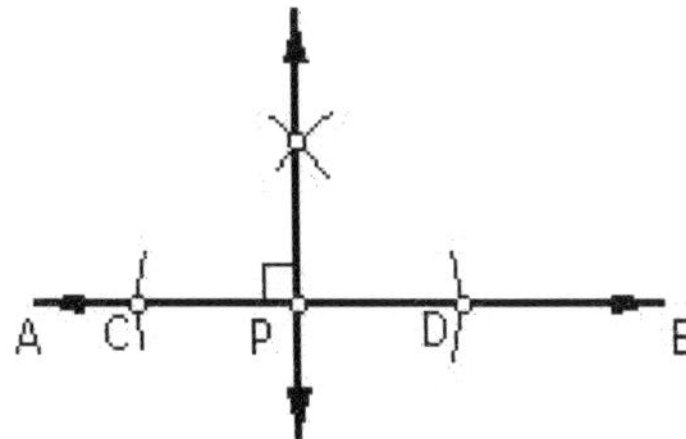

If we need to **construct a perpendicular line through a point that is not on the given line**, the steps are essentially the same, as shown below. Although P is not the midpoint of $\overline{CD}$, it is still equidistant to points C and D. Be sure to draw the intersecting arcs on the other side of $\overleftrightarrow{AB}$, opposite from P. This will allow you to more accurately draw the straight line.

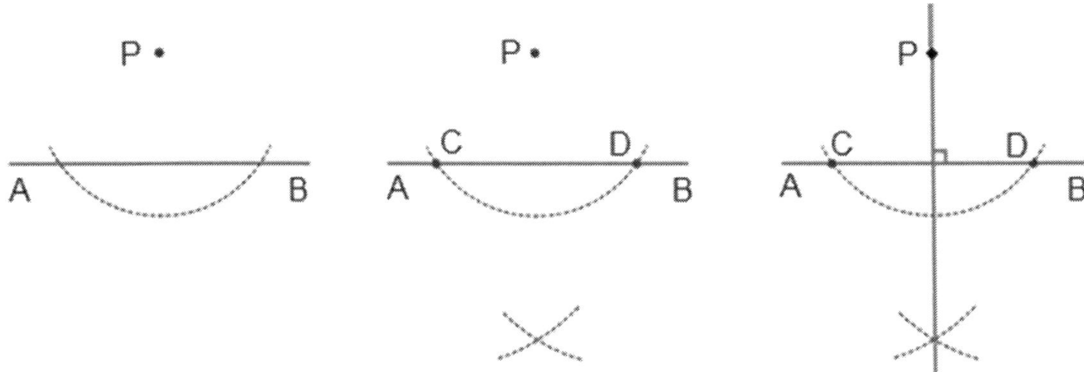

Model Problem:

Given segments of lengths a and b below, construct a trapezoid $WXYZ$ in which $WZ = a$, $XY = b$, and $\overline{WZ} \parallel \overline{XY}$.

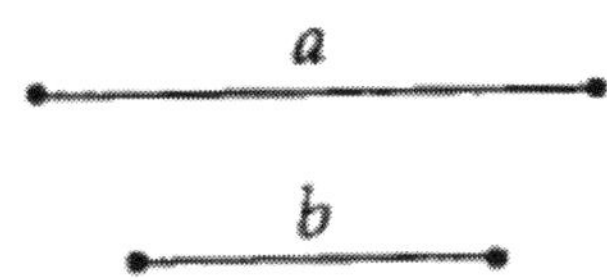

Solution:

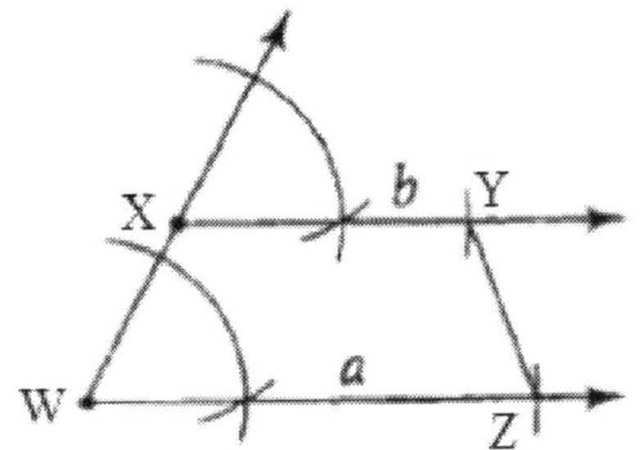

Explanation of steps:

(A) Construct $\overline{WZ}$ with length a.

(B) Draw point X not on $\overleftrightarrow{WZ}$. Then draw $\overrightarrow{WX}$.

(C) Construct a ray through X parallel to $\overleftrightarrow{WZ}$ by copying congruent corresponding angles.

(D) Construct $\overline{XY}$ with length b, then draw $\overline{YZ}$.

Practice Problems

1. Construct the line through point P that is perpendicular to the given line.

P

2. Construct the line through point P that is parallel to the given line.

P

3. Match each construction with one of the diagrams below.

_____ (a) angle bisector
_____ (b) perpendicular through a point on a line
_____ (c) perpendicular from a point to a line
_____ (d) perpendicular bisector
_____ (e) equilateral triangle
_____ (f) altitude of a triangle

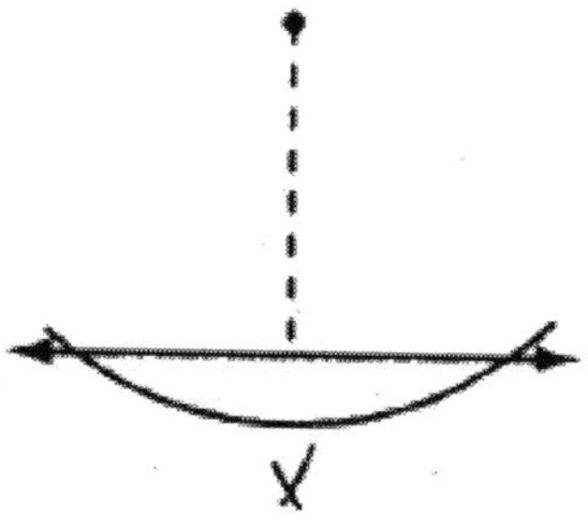

(1)

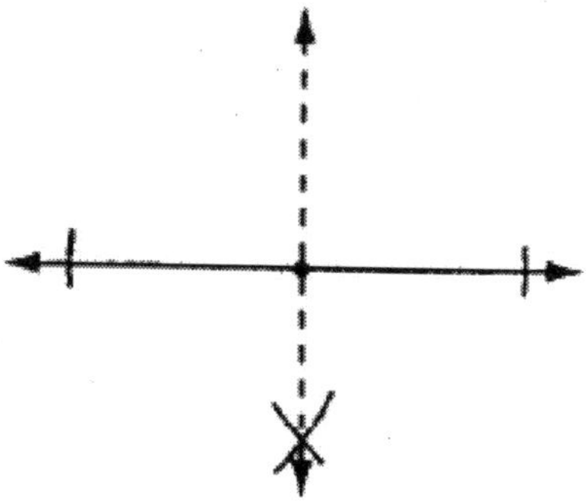

(2)

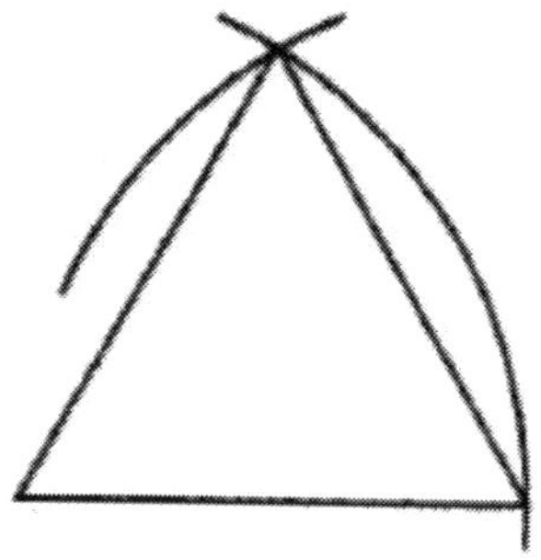

(3)

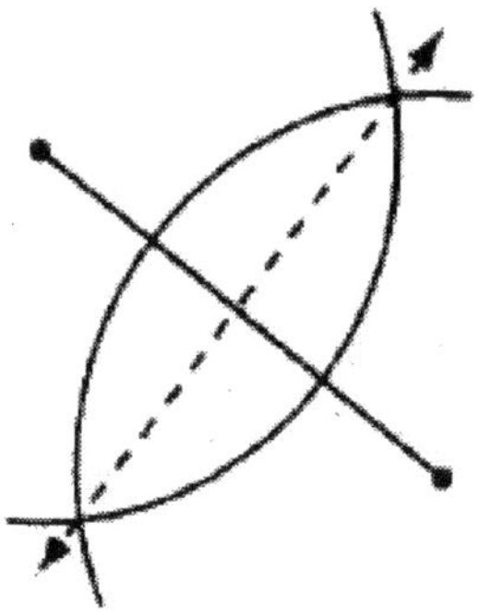

(4)

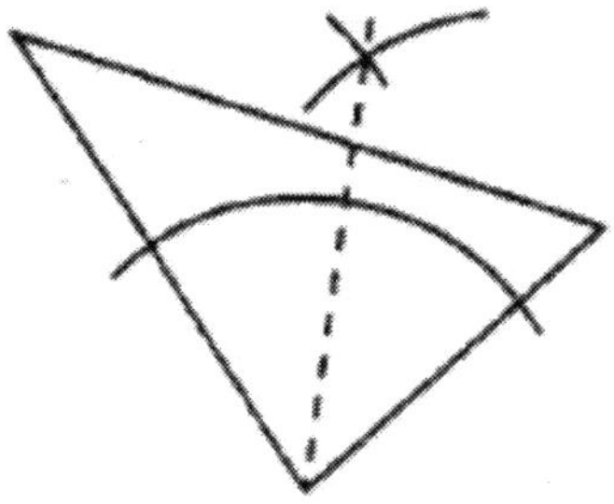

(5)

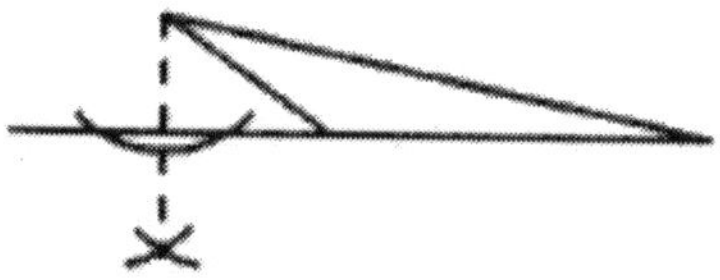

(6)

REGENTS QUESTIONS

Multiple Choice

1. The diagram below shows the construction of $\overleftrightarrow{AB}$ through point P parallel to $\overleftrightarrow{CD}$.

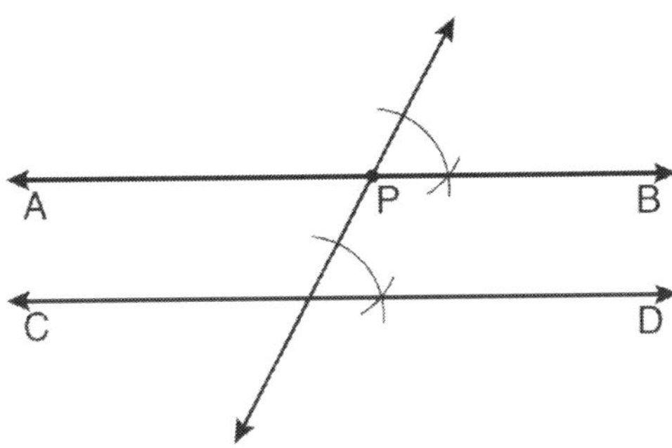

Which theorem justifies this method of construction?
(1) If two lines in a plane are perpendicular to a transversal at different points, then the lines are parallel.
(2) If two lines in a plane are cut by a transversal to form congruent corresponding angles, then the lines are parallel.
(3) If two lines in a plane are cut by a transversal to form congruent alternate interior angles, then the lines are parallel.
(4) If two lines in a plane are cut by a transversal to form congruent alternate exterior angles, then the lines are parallel.

2. Which diagram illustrates a correct construction of an altitude of $\triangle ABC$?

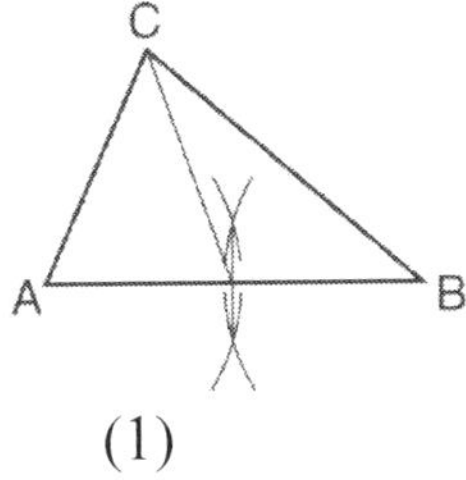

(1)

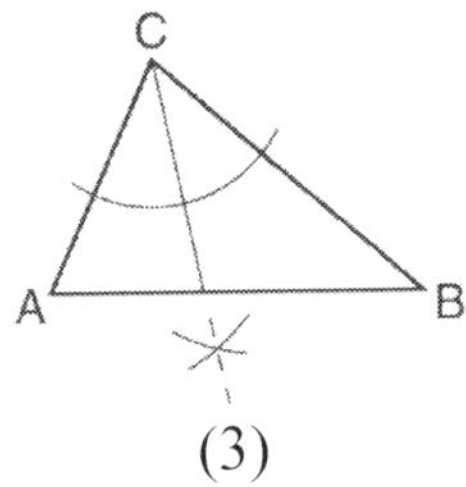

(3)

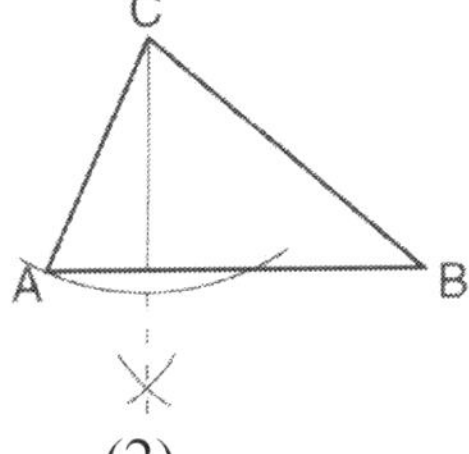

(2)

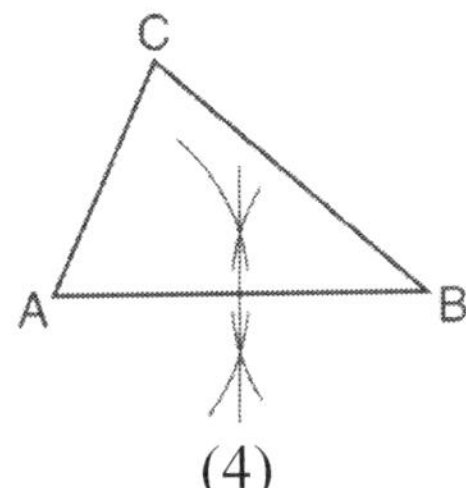

(4)

3. Which construction of parallel lines is justified by the theorem "If two lines are cut by a transversal to form congruent alternate interior angles, then the lines are parallel"?

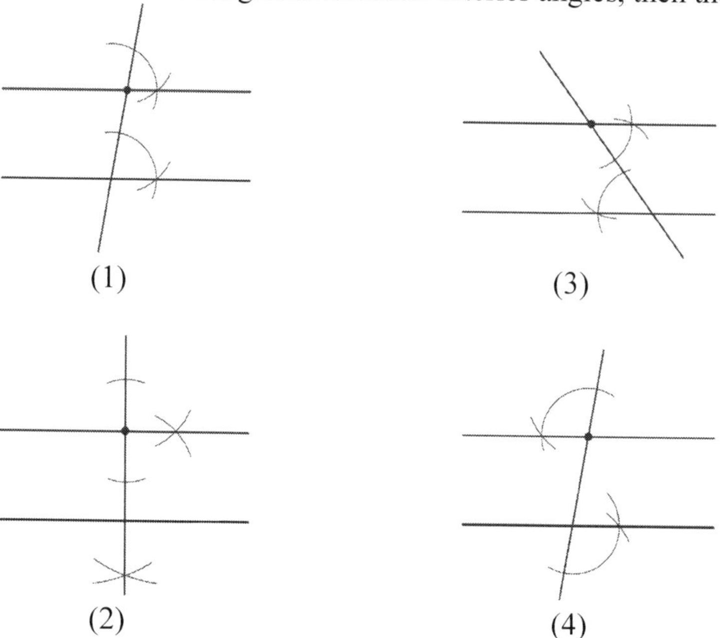

Constructed Response

4. Using a compass and straightedge, construct a line perpendicular to $\overline{AB}$ through point P. [Leave all construction marks.]

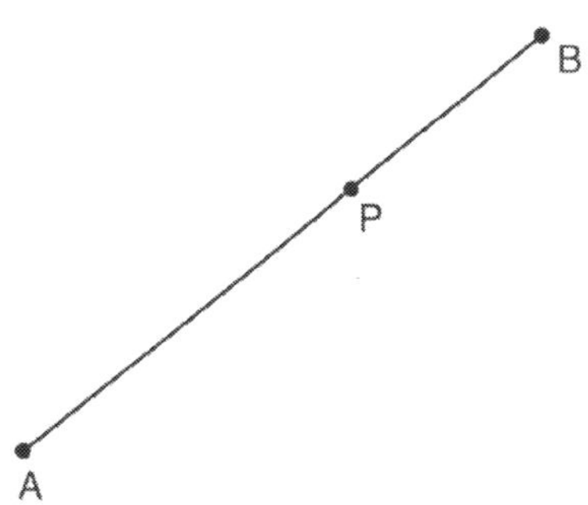

5. Using a compass and straightedge, construct a line perpendicular to line ℓ through point P. [Leave all construction marks.]

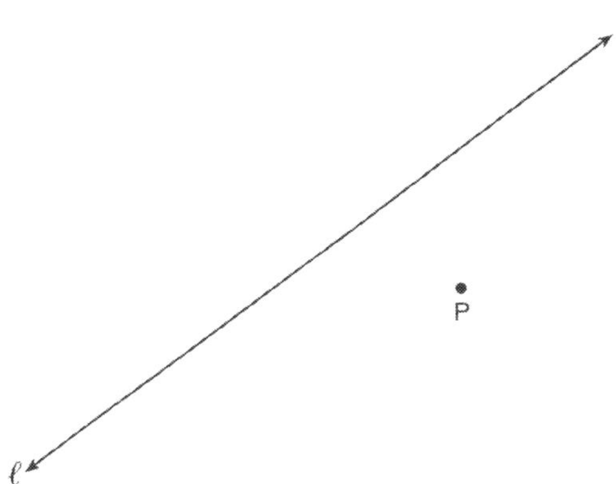

6. (CC) Using a compass and straightedge, construct an altitude of triangle ABC below. [Leave all construction marks.]

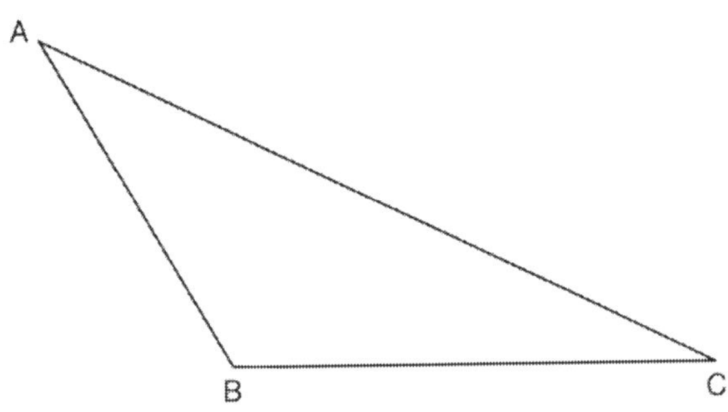

7. (CC) In the diagram below, radius $\overline{OA}$ is drawn in circle O. Using a compass and a straightedge, construct a line tangent to circle O at point A. [Leave all construction marks.]

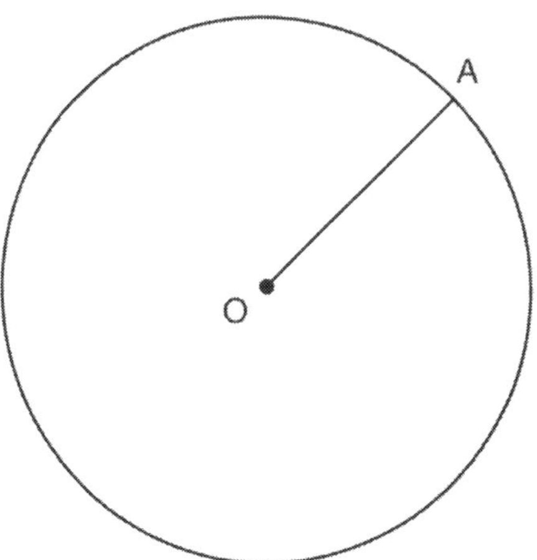

15.6 Construct Tangents to a Circle (+)

*In the Geometry Draft of Revised Standards for 2018-19, NYS has recommended **adding** this topic.*

Key Terms and Concepts

From a point outside a circle, we can **construct tangents** to the circle.

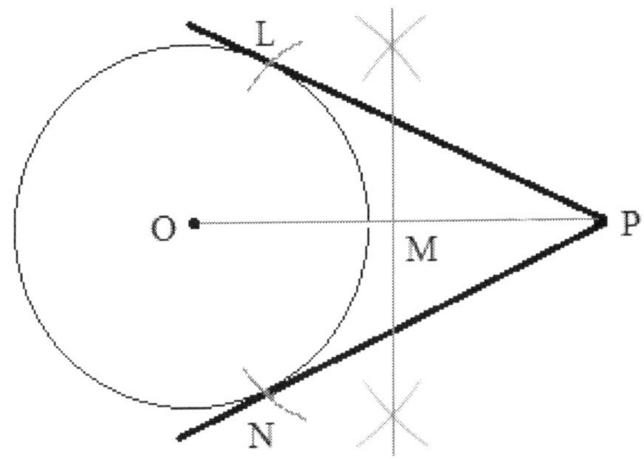

Given circle O and external point P, draw $\overline{OP}$. Then construct the perpendicular bisector of $\overline{OP}$ and label the midpoint M.

Place the tip of the compass on M, set the compass to the length of $\overline{OM}$, and construct two arcs that intercept the circle. These mark the points of tangency, L and N. Construct tangents $\overrightarrow{PL}$ and $\overrightarrow{PN}$.

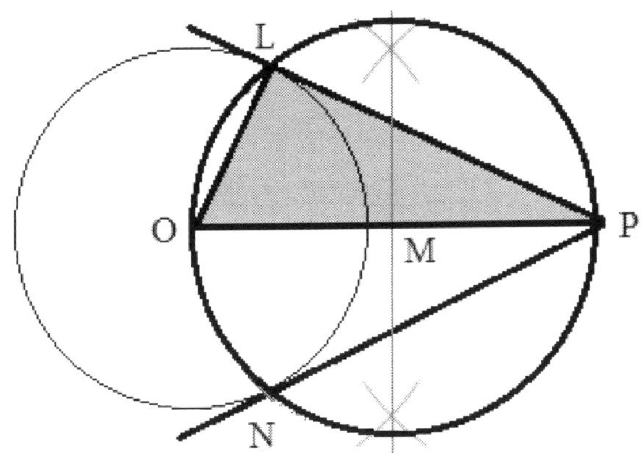

The reason we know $\overrightarrow{PL}$ is a tangent is that L is a point on the circle with the center at M and a diameter of $\overline{OP}$. Therefore, as an inscribed angle of a semicircle, $m\angle OLP = 90°$.

Since $\overline{OL}$ is a radius of circle O and $\overline{OL} \perp \overrightarrow{PL}$, then $\overrightarrow{PL}$ must be a tangent.

Similarly, we can show $\overrightarrow{PN}$ is a tangent.

Model Problem:

Sharon is looking for a different way to construct a tangent $\overrightarrow{PQ}$ to circle O from an exterior point P and has written the following directions. Will her directions accomplish the task?

1. Draw $\overleftrightarrow{OP}$.
2. Construct radius $\overline{OQ}$ such that $\overline{OQ} \perp \overleftrightarrow{OP}$.
3. Draw $\overleftrightarrow{PQ}$.

Solution:

No, the directions will not construct a tangent line. The tangent is always perpendicular to the radius at the point of tangency. If both the tangent and $\overleftrightarrow{OP}$ are perpendicular to the radius $\overline{OQ}$, then the tangent would be parallel to $\overleftrightarrow{OP}$. If the tangent is parallel to $\overleftrightarrow{OP}$, then they cannot intersect at P.

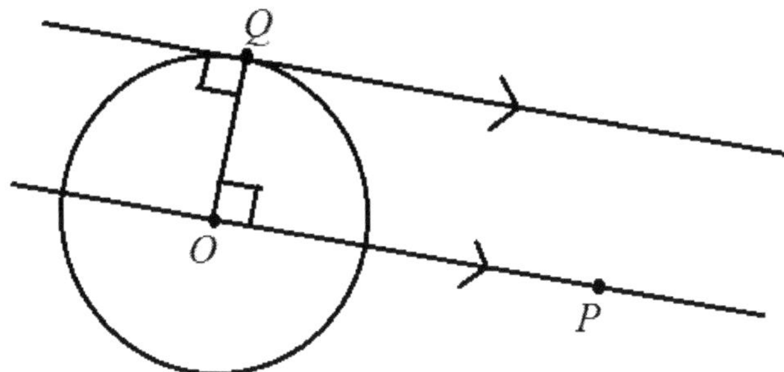

Practice Problems

1. Construct the two tangent lines to circle A through point B.

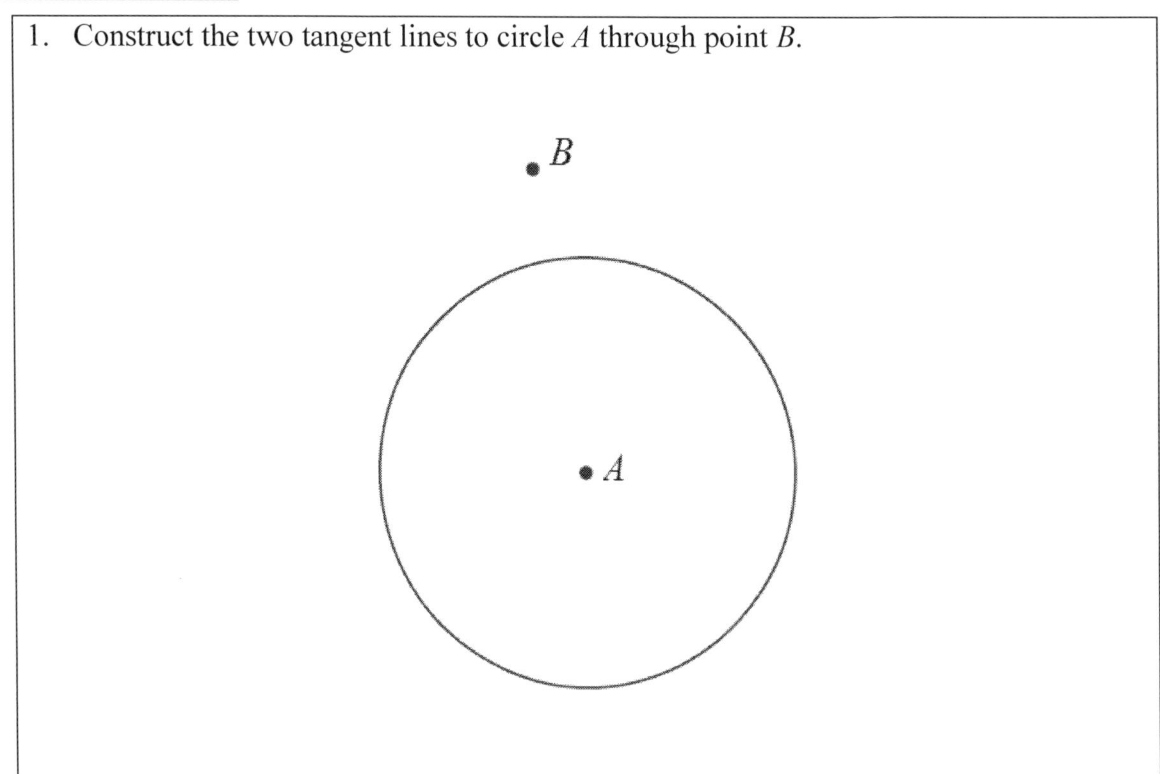

15.7 Construct Inscribed Regular Polygons

Key Terms and Concepts

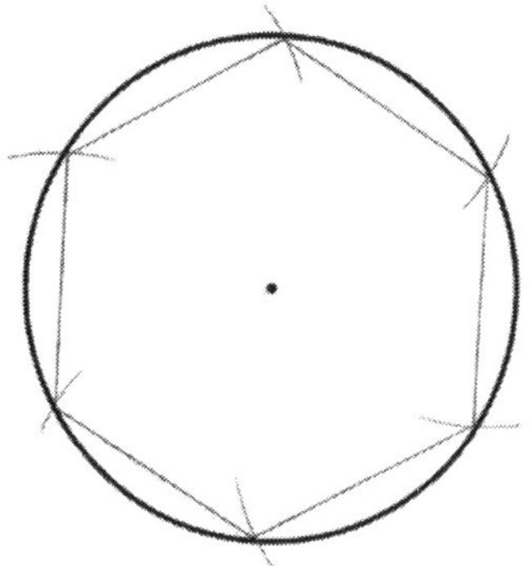

We can use a compass and straightedge to inscribe a regular polygon inside a circle.

For example, to **inscribe a regular hexagon**, we can set the compass width to the length of the radius. Then we select any starting point on the circle and mark off a chord of that length. We then move the compass to the other endpoint of the chord and draw another chord. We repeat this process around the entire circle, resulting in six congruent chords, as shown.

How do we know we'll end up at the starting point, and after exactly six chords are drawn? Think of these chords as the sides of triangles, with the center of the circle as the third vertex of each triangle, as shown to the right. Since each chord is congruent to the radius, these are all equilateral triangles, with angle measures of $60°$ at each vertex. We therefore have sectors with central angles of $60°$ each, and so their arcs are each exactly one-sixth of the full circle.

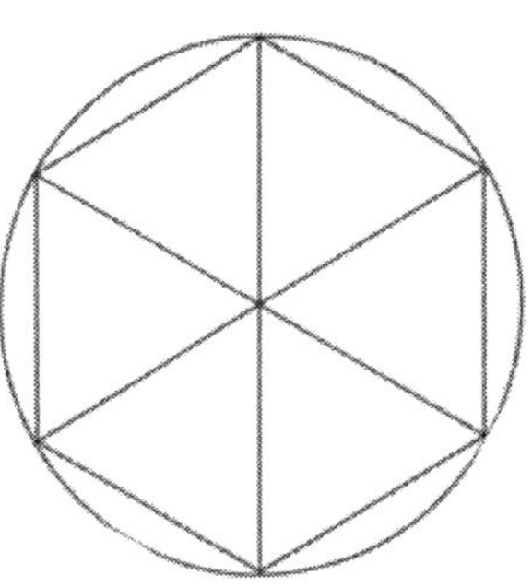

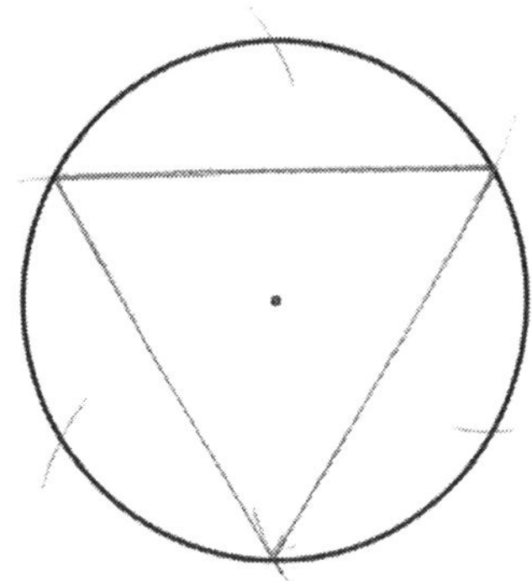

We can **inscribe an equilateral triangle** in a circle in almost the same way, except that we draw the chords connecting every second arc drawn by the compass. This results in three chords of equal length, rather than six, and therefore gives us an equilateral triangle.

To construct an **inscribed square**, we'll use a different approach. Dividing a circle into four equal arcs can be accomplished, but not as easily as dividing into six parts. Fortunately, there is an easier way that is based on the fact that an inscribed angle of a diameter measures $90°$.

First, draw any diameter through the center of the circle. Then, construct another diameter that is perpendicular to it, as shown to the right. Connect the endpoints of these diameters as four chords, and you're done!

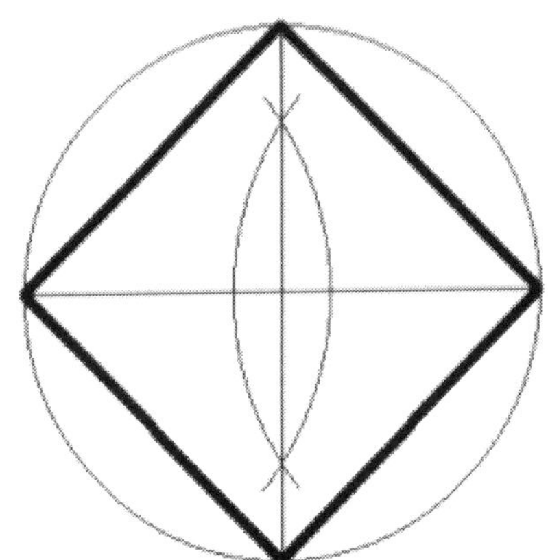

Model Problem:

A dodecagon is a twelve-sided polygon. Construct a regular dodecagon inscribed within a circle.

Solution:

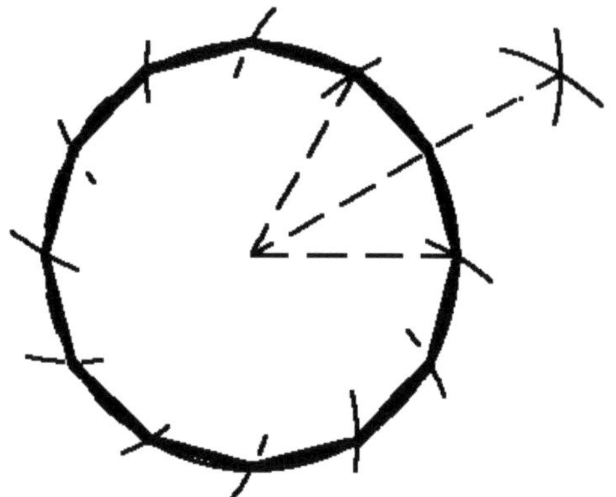

Explanation of steps:

A dodecagon has twice as many sides as a hexagon.

(A) Set the compass to the radius of the circle and draw six equidistant marks of this length around the circle.

(B) Draw the radii to two consecutive marks and bisect the angle between them, which will also bisect the arc between them.

(C) Reset the compass to the radius of the circle and place the tip at the point of intersection of the angle bisector and the circle.

(D) Draw six additional equidistant marks around the circle.

(E) The twelve marks represent the vertices of the dodecagon. Draw the chords between them.

Practice Problems

1. Construct an equilateral triangle inscribed within the circle below.

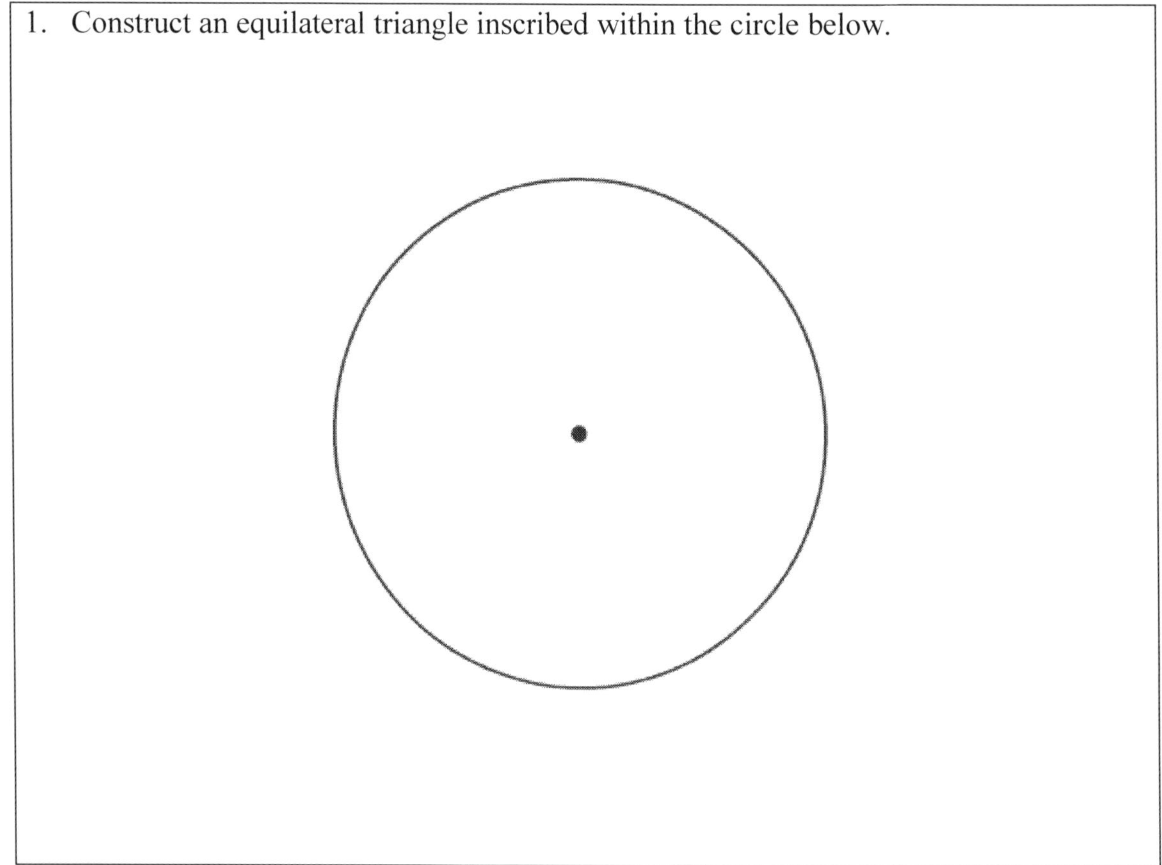

2. Construct a square inscribed within the circle below.

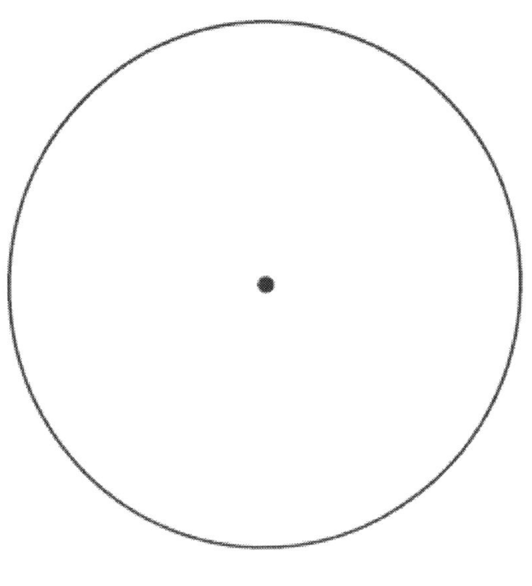

3. Inscribe a regular octagon inside the circle below.
Hint: an octagon has twice as many sides as a square.

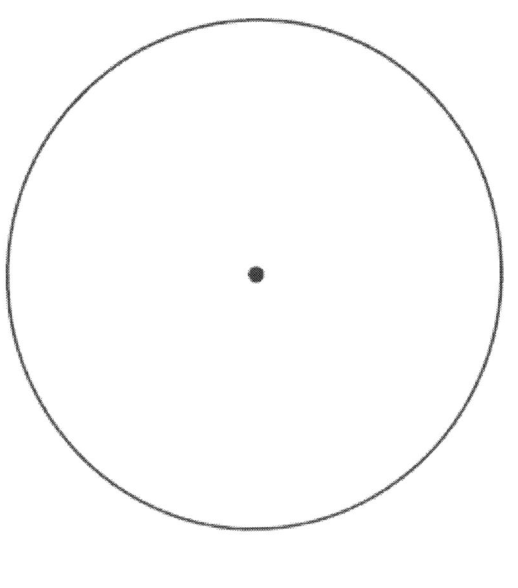

REGENTS QUESTIONS

Constructed Response

1. (CC) Using a straightedge and compass, construct a square inscribed in circle O below. [Leave all construction marks.]

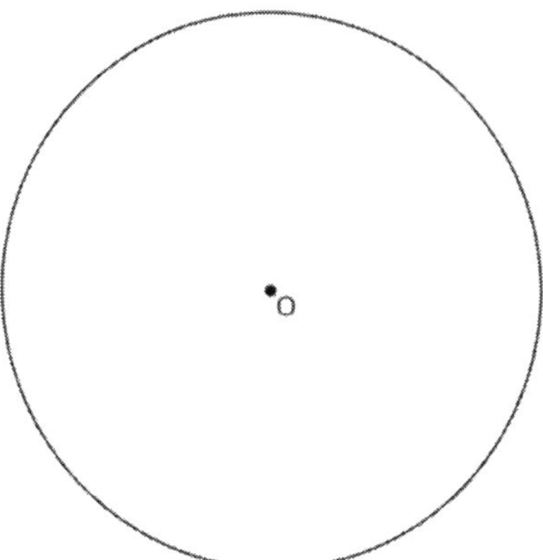

Determine the measure of the arc intercepted by two adjacent sides of the constructed square. Explain your reasoning.

2. (CC) Use a compass and straightedge to construct an inscribed square in circle T shown below. [Leave all construction marks.]

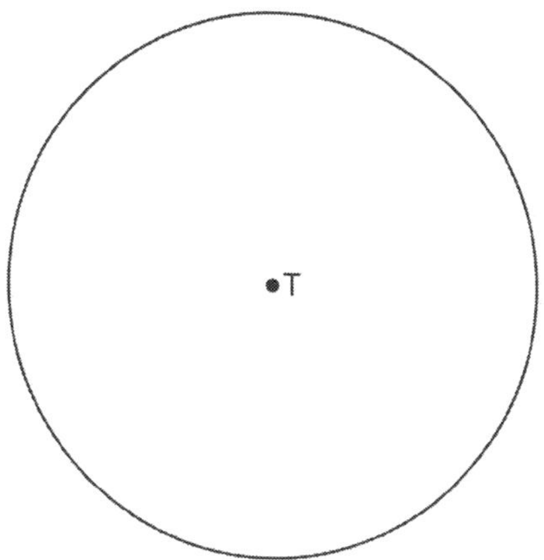

3. (CC) Construct an equilateral triangle inscribed in circle T shown below. [Leave all construction marks.]

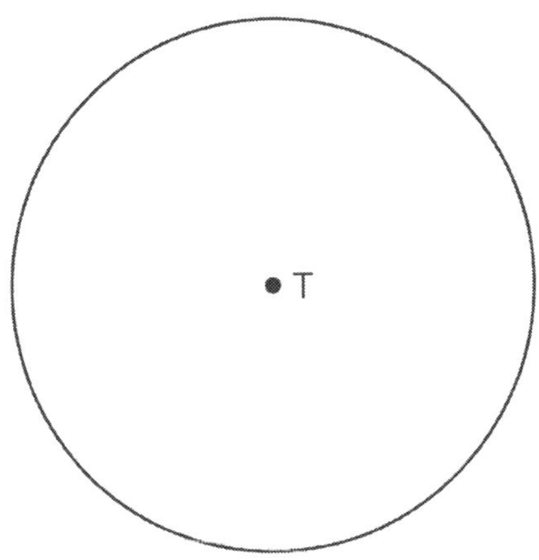

4. (CC) Using a compass and straightedge, construct a regular hexagon inscribed in circle O below. Label it $ABCDEF$. [Leave all construction marks.]

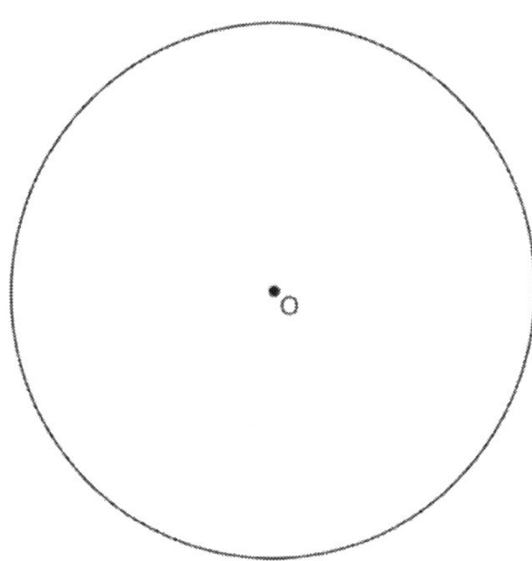

If chords $\overline{FB}$ and $\overline{FC}$ are drawn, which type of triangle, according to its angles, would $\triangle FBC$ be? Explain your answer.

15.8 Construct Points of Concurrency

Key Terms and Concepts

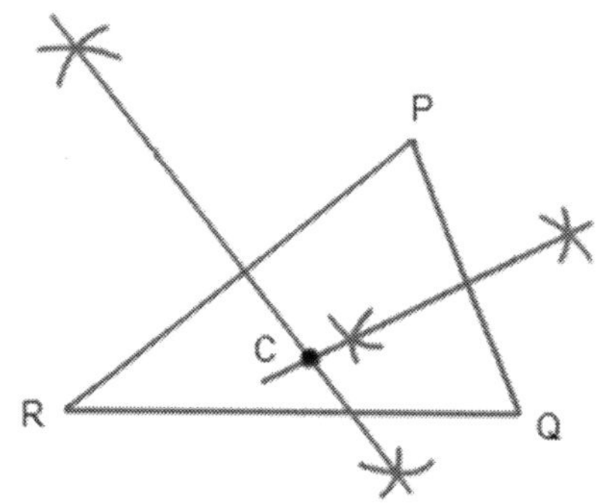

To construct the circumcenter of a triangle:,
The *circumcenter* is the point of intersection of the *perpendicular bisectors* of the three sides of the triangle. To find the circumcenter, we really need to construct perpendicular bisectors of only two sides, as shown, and label the point of intersection C. If we drew the perpendicular bisector of the third side, it would also pass through the circumcenter, C.

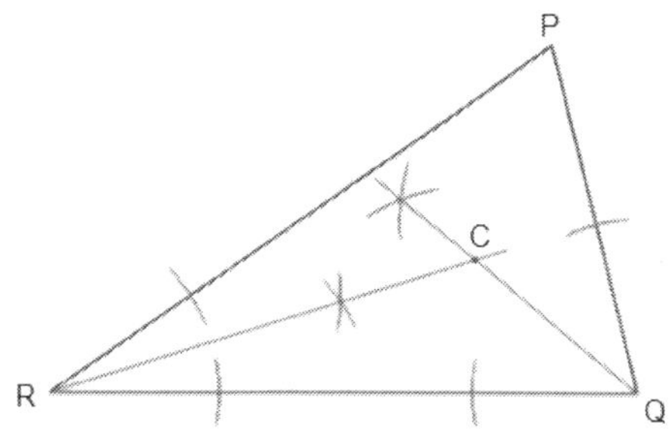

To construct the incenter of a triangle:
The *incenter* is the point of intersection of the *angle bisectors* of the three angles of the triangle. Again, we only really need to construct two angle bisectors, as shown, and label the point of intersection C. If we drew the bisector of the third angle, it would also pass through the incenter, C.

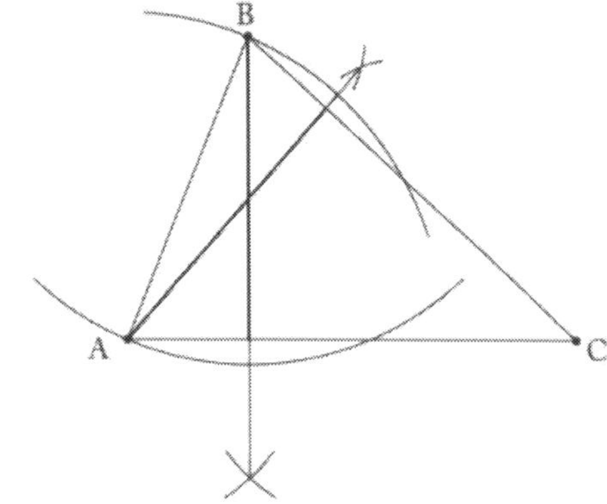

To construct the orthocenter of a triangle:
The *orthocenter* is the point of intersection of the three *altitudes* of the triangle. Here as well, we can construct just two of the altitudes and know that the third altitude would also intersect at the orthocenter.

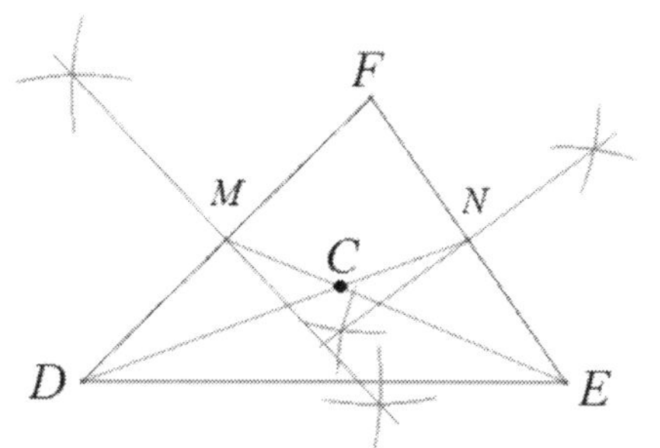

To construct the centroid of a triangle:
The *centroid* is the point of intersection of the three *medians* of the triangle. To draw a median, we must first construct a perpendicular bisector in order to find the midpoint of a side. For example, to the left, the bisector of $\overline{DF}$ is drawn to find midpoint M, and the bisector of $\overline{EF}$ is drawn to find midpoint N. We then connect both midpoints to their opposite vertices, and label the point of intersection C. If we drew the third median, it would also pass through the centroid, C.

Practice Problems

1. Construct the circumcenter of triangle ABC.

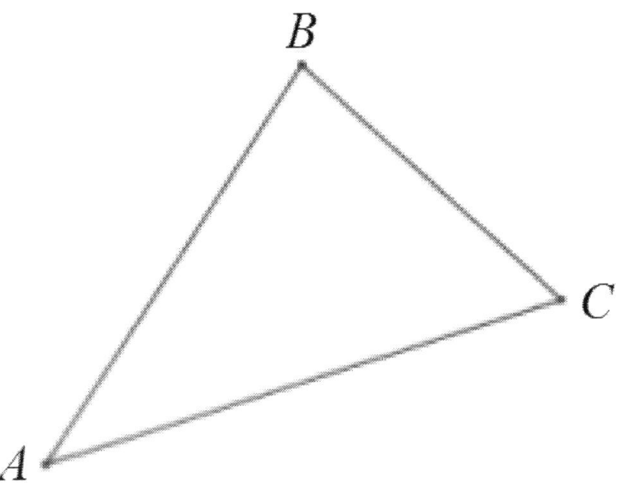

2. Construct the orthocenter of triangle DEF.

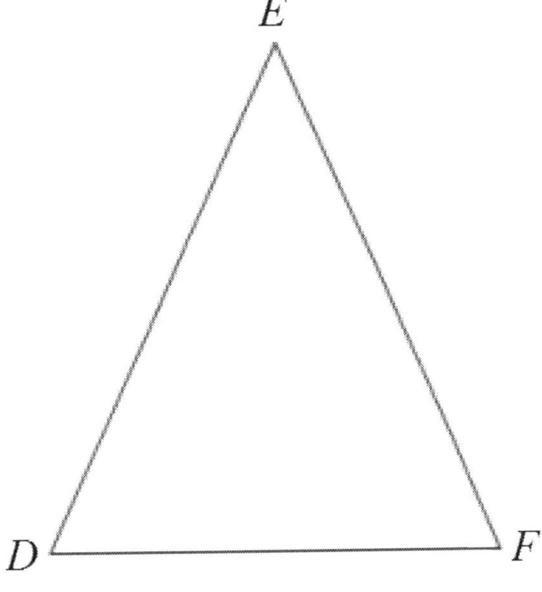

15.9 Construct Circles of Triangles

Key Terms and Concepts

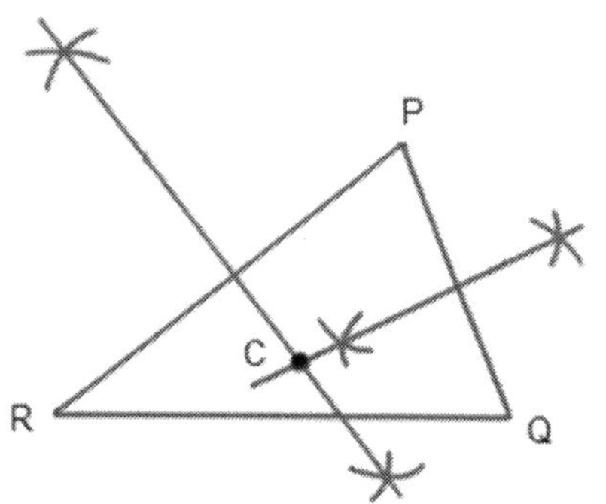

To circumscribe a circle around a triangle, we first need to find the *circumcenter* of the triangle, as this will be the center of the circumscribed circle. The circumcenter is the point of intersection of the *perpendicular bisectors* of the three sides of the triangle.

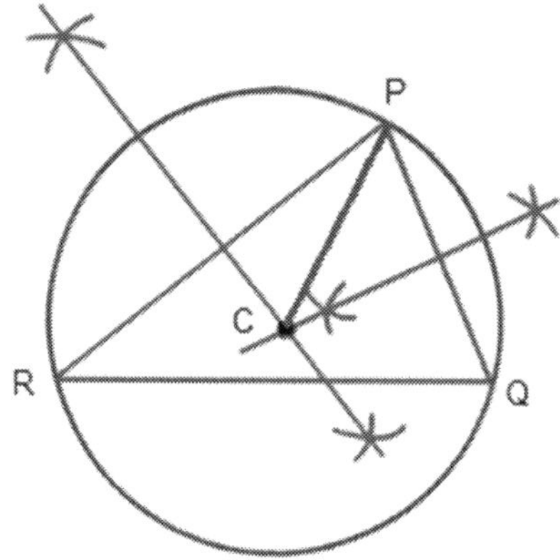

Once the circumcenter C is found, the radius of the circle will be the distance from C to any vertex of the triangle. So, set the compass width to that radius and draw the circle with C as its center.

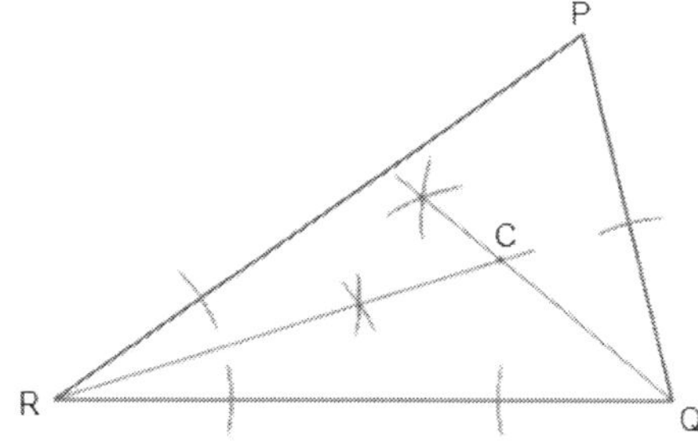

To inscribe a circle inside a triangle, the steps are similar. We first need to find the *incenter* of the triangle, as this will be the center of the inscribed circle. The incenter is the point of intersection of the *angle bisectors* of the three angles of the triangle.

Point C will be the center of the circle. So, the next step is to find the radius of the circle. All three sides of the triangle will be tangents to the inscribed circle. So, we just need to pick a side, then construct a perpendicular line to that side through point C. The distance from C to the intersection of the perpendicular lines is the radius of the circle. Once we have the radius, set the compass to that width, and draw the circle with C as its center.

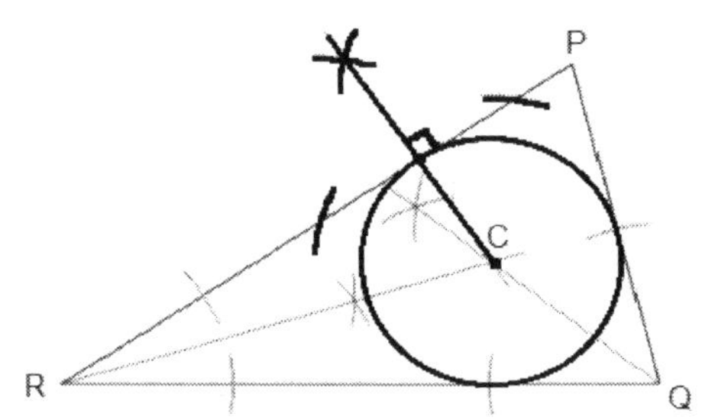

Practice Problems

1. Match each construction with one of the diagrams below.

   _____ (a) orthocenter
   _____ (b) incenter
   _____ (c) centroid
   _____ (d) circumcenter
   _____ (e) inscribed circle

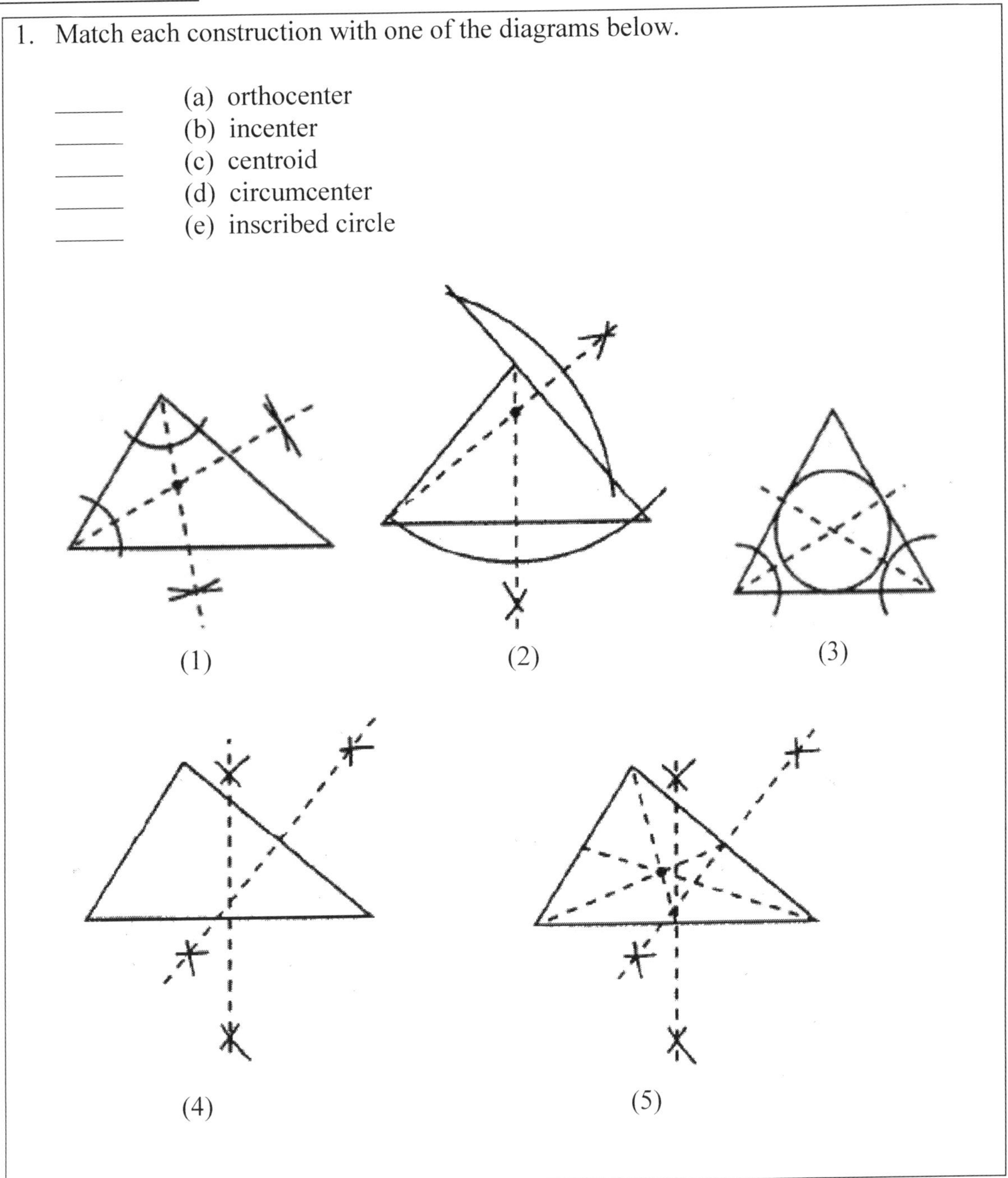

REGENTS QUESTIONS

Multiple Choice

1. The diagram below shows the construction of the center of the circle circumscribed about $\triangle ABC$.

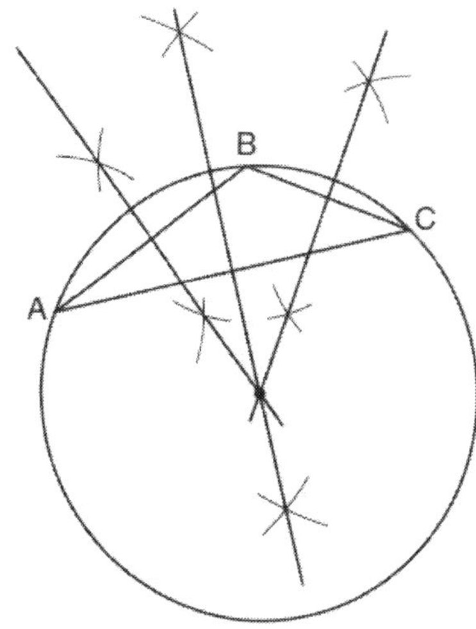

This construction represents how to find the intersection of
(1) the angle bisectors of $\triangle ABC$
(2) the medians to the sides of $\triangle ABC$
(3) the altitudes to the sides of $\triangle ABC$
(4) the perpendicular bisectors of the sides of $\triangle ABC$

2. Which geometric principle is used in the construction shown below?

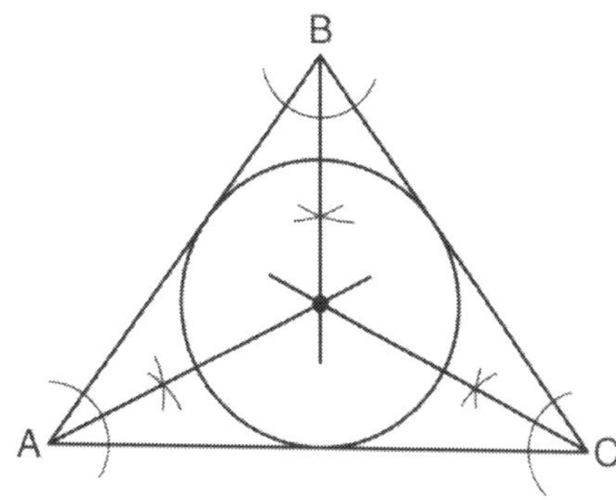

(1) The intersection of the angle bisectors of a triangle is the center of the inscribed circle.
(2) The intersection of the angle bisectors of a triangle is the center of the circumscribed circle.
(3) The intersection of the perpendicular bisectors of the sides of a triangle is the center of the inscribed circle.
(4) The intersection of the perpendicular bisectors of the sides of a triangle is the center of the circumscribed circle.

Appendix I. Standards

Geometry Common Core Standards (New York)

Code	Current Standard	Revised Standard Recommendation
G.CO.1	Know precise definitions of angle, circle, perpendicular line, parallel line, and line segment, based on the undefined notions of point, line, distance along a line, and distance around a circular arc.	Know precise definitions of angle, circle, perpendicular lines, parallel lines, and line segment, based on the undefined notions of point, line, distance along a line, and distance around a circular arc as these exist within a plane.
G.CO.2	Represent transformations in the plane using, e.g., transparencies and geometry software; describe transformations as functions that take points in the plane as inputs and give other points as outputs. Compare transformations that preserve distance and angle to those that do not (e.g., translation versus horizontal stretch).	Represent transformations as geometric functions that take points in the plane as inputs and give other points as outputs. Compare transformations that preserve distance and angle measure to those that do not (e.g., translation versus horizontal stretch). Note: Use a variety of strategies which include transparencies and software programs.
G.CO.3	Given a rectangle, parallelogram, trapezoid, or regular polygon, describe the rotations and reflections that carry it onto itself.	Given a regular or irregular polygon, describe the rotations and reflections that carry it onto itself.
G.CO.4	Develop definitions of rotations, reflections, and translations in terms of angles, circles, perpendicular lines, parallel lines, and line segments.	Develop definitions of rotations, reflections, and translations in terms of points, angles, circles, perpendicular lines, parallel lines, and line segments.
G.CO.5	Given a geometric figure and a rotation, reflection, or translation, draw the transformed figure using, e.g., graph paper, tracing paper, or geometry software. Specify a sequence of transformations that will carry a given figure onto another.	Given a geometric figure and a rotation, reflection, or translation, draw the transformed figure. Specify a sequence of transformations that will carry a given figure onto another. Note: Drawing tools, which could include graph paper, tracing paper and geometry software.

G.CO.6	Use geometric descriptions of rigid motions to transform figures and to predict the effect of a given rigid motion on a given figure; given two figures, use the definition of congruence in terms of rigid motions to decide if they are congruent.	Use geometric descriptions of rigid motions to transform figures and to predict the effect of a given rigid motion on a given figure. Given two figures, use the definition of congruence in terms of rigid motions to decide if they are congruent. Note: With rotations, the center of the transformation must be specified.
G.CO.7	Use the definition of congruence in terms of rigid motions to show that two triangles are congruent if and only if corresponding pairs of sides and corresponding pairs of angles are congruent.	*NO CHANGE*
G.CO.8	Explain how the criteria for triangle congruence (ASA, SAS, and SSS) follow from the definition of congruence in terms of rigid motions.	*NO CHANGE*
G.CO.9	Prove theorems about lines and angles. Theorems include: vertical angles are congruent; when a transversal crosses parallel lines, alternate interior angles are congruent and corresponding angles are congruent; points on a perpendicular bisector of a line segment are exactly those equidistant from the segment's endpoints.	Prove and apply theorems about lines and angles. Note: Include multi-step proofs and algebraic problems built upon these concepts. a. Prove and apply theorems about relationships, specifically: i. Vertical angles. ii. Angles created by a transversal intersecting parallel lines. iii. Points on a perpendicular bisector of a line segment are exactly those equidistant from the segment's endpoints.

G.CO.10	Prove theorems about triangles. Theorems include: measures of interior angles of a triangle sum to 180°; base angles of isosceles triangles are congruent; the segment joining midpoints of two sides of a triangle is parallel to the third side and half the length; the medians of a triangle meet at a point.	Prove and apply theorems about triangles. Note: Include multi-step proofs and algebraic problems built upon these concepts. a. Prove and apply theorems about angle relationships, specifically: i. Interior angles sum to 180 degrees. ii. Exterior angles sum to 360 degrees. iii. The measure of an exterior angle of a triangle is equal to the sum of the measures of its two non-adjacent interior angles of the triangle. b. Prove and apply theorems about isosceles triangles. c. Prove and apply theorems about the mid-segment of a triangle (parallel to the third side and half the length).
G.CO.11	Prove theorems about parallelograms. Theorems include: opposite sides are congruent, opposite angles are congruent, the diagonals of a parallelogram bisect each other, and conversely, rectangles are parallelograms with congruent diagonals.	Prove and apply theorems about parallelograms. Note: Include multi-step proofs and algebraic problems built upon these concepts. Note: Based on the inclusive definition of a trapezoid, a parallelogram is a trapezoid. a. Prove and apply theorems about properties which include opposite sides are congruent, opposite angles are congruent and that the diagonals bisect each other. b. Prove and apply theorems about special parallelograms and the properties that distinguish them.

G.CO.12	Make formal geometric constructions with a variety of tools and methods (compass and straightedge, string, reflective devices, paper folding, dynamic geometric software, etc.). Copying a segment; copying an angle; bisecting a segment; bisecting an angle; constructing perpendicular lines, including the perpendicular bisector of a line segment; and constructing a line parallel to a given line through a point not on the line.	Make formal geometric constructions while developing fluency with the use of construction tools. Note: Use a variety of tools and methods for construction, which include compass and straightedge, string, reflective devices, paper folding, dynamic geometric software, etc. a. Copy segments and angles. b. Bisect segments and angles. c. Construct perpendicular lines including through a point on or off a given line. d. Construct a line parallel to a given line through a point not on the line. e. Construct an isosceles triangle with given lengths. f. Construct points of concurrency of a triangle (centroid, circumcenter, and incenter).
G.CO.13	Construct an equilateral triangle, a square, and a regular hexagon inscribed in a circle.	*NO CHANGE*

G.SRT.1	Verify experimentally the properties of dilations given by a center and a scale factor: a. A dilation takes a line not passing through the center of the dilation to a parallel line, and leaves a line passing through the center unchanged. b. The dilation of a line segment is longer or shorter in the ratio given by the scale factor.	*NO CHANGE*
G.SRT.2	Given two figures, use the definition of similarity in terms of similarity transformations to decide if they are similar; explain using similarity transformations the meaning of similarity for triangles as the equality of all corresponding pairs of angles and the proportionality of all corresponding pairs of sides.	Given two figures, use the definition of similarity in terms of similarity transformations to decide if they are similar. Explain using similarity transformations the meaning of similarity for triangles as the equality of all corresponding pairs of angles and the proportionality of all corresponding pairs of sides. Note: With dilations or rotations, the center of the transformation must be specified.
G.SRT.3	Use the properties of similarity transformations to establish the AA criterion for two triangles to be similar.	*NO CHANGE*
G.SRT.4	Prove theorems about triangles. Theorems include: a line parallel to one side of a triangle divides the other two proportionally, and conversely; the Pythagorean Theorem proved using triangle similarity.	Prove and apply theorems about triangles. Note: Include multi-step proofs and algebraic problems built upon these concepts. a. Prove that a line parallel to one side of a triangle divides the other two proportionally, and conversely. b. Prove that the length of the altitude drawn from the vertex of the right angle of a right triangle to its hypotenuse is the geometric mean between the lengths of the two segments of the hypotenuse. c. Prove the Pythagorean Theorem using triangle similarity.

G.SRT.5	Use congruence and similarity criteria for triangles to solve problems and to prove relationships in geometric figures.	Use congruence and similarity criteria for triangles with fluency to: a. solve problems algebraically and geometrically. b. prove relationships in geometric figures. Note: ASA, SAS, SSS, AAS, and Hypotenuse-Leg (HL) theorems are valid criteria for triangle congruence. AA, SAS, and SSS are valid criteria for triangle similarity.
G.SRT.6	Understand that by similarity, side ratios in right triangles are properties of the angles in the triangle, leading to definitions of trigonometric ratios for acute angles.	Understand that by similarity, side ratios in right triangles are properties of the angles in the triangle, leading to definitions of sine, cosine and tangent ratios for acute angles.
G.SRT.7	Explain and use the relationship between the sine and cosine of complementary angles.	*NO CHANGE*
G.SRT.8	Use trigonometric ratios and the Pythagorean Theorem to solve right triangles in applied problems.	Use sine, cosine and tangent as well as the Pythagorean Theorem to solve right triangles in applied problems.
G.SRT.9	(+) Derive the formula $A = \frac{1}{2}ab\sin C$ for the area of a triangle by drawing an auxiliary line from a vertex perpendicular to the opposite side. *(+) standards are optional*	Explore the derivation of the formula $A = \frac{1}{2}ab\sin C$ for the area of a triangle by drawing an auxiliary line from a vertex perpendicular to the opposite side. Apply the formula $A = \frac{1}{2}ab\sin C$ to find the area of any triangle. *Eliminate (+)*
G.SRT.10	(+) Prove the Law of Sines and Cosines and use them to solve problems. *(+) standards are optional*	Explore the proofs and apply the Laws of Sines* and Cosines to solve problems. *The ambiguous case for Law of Sines (given one angle and two sides, find the other angle) is NOT addressed in this course. *Eliminate (+)*
G.SRT.11	(+) Understand and apply the Law of Sines and Cosines to find unknown measurements in right and non-right triangles (e.g., surveying problems, resultant forces). *(+) standards are optional*	Understand and apply the Law of Sines and the Law of Cosines to find unknown measurements in any triangle. At this level, force diagrams should not be included. *Eliminate (+)*

F.TF.1-2	*Covered in Algebra 2*	
F.TF.3	(+) Use special triangles to determine geometrically the values of sine, cosine, tangent for $\frac{\pi}{3}$, $\frac{\pi}{4}$ and $\frac{\pi}{6}$ and use the unit circle to express the values of sine, cosine, and tangent for $\pi - x$, $\pi + x$ and $2\pi - x$ in terms of their values for x, where x is any real number. *(+) standards are optional*	Use special triangles to determine geometrically the values of sine, cosine and tangent for 30, 45 and 60 degrees. Use the special triangles with the unit circle to find the values for sine, cosine and tangent of 30, 45, 60, 120, 135 and 150 degrees. Note: Side lengths could be given in radical form. *Eliminate (+)*
F.TF.4-8	*Covered in Algebra 2 and Precalculus*	

G.C.1	Prove that all circles are similar.	*NO CHANGE*
G.C.2	Identify and describe relationships among inscribed angles, radii, and chords. Include the relationship between central, inscribed, and circumscribed angles; inscribed angles on a diameter are right angles; the radius of a circle is perpendicular to the tangent where the radius intersects the circle.	Identify, describe and apply geometric properties of circles. a. Identify, describe and apply relationships among angles and intercepted arcs, specifically: i. central ii. inscribed iii. circumscribed iv. angles and arcs formed by any combination of intersecting tangents, secants or chords. b. Identify, describe and apply relationships among segments, specifically: i. radii ii. chords iii. tangents iv. secants
G.C.3	Construct the inscribed and circumscribed circles of a triangle, and prove properties of angles for a quadrilateral inscribed in a circle.	Prove properties of angles for a quadrilateral inscribed in a circle.
G.C.4	(+) Construct a tangent line from a point outside a given circle to the circle. *(+) standards are optional*	Construct a tangent line from a point outside a given circle to the circle. *Eliminate (+)*
G.C.5	Derive using similarity the fact that the length of the arc intercepted by an angle is proportional to the radius, and define the radian measure of the angle as the constant of proportionality; derive the formula for the area of a sector.	Using proportionality, find one of the following given two others: the central angle, arc length, radius or area of sector.

G.GPE.1	Derive the equation of a circle of given center and radius using the Pythagorean Theorem; complete the square to find the center and radius of a circle given by an equation.	Derive the equation of a circle of given center and radius using the Pythagorean Theorem. Complete the square to find the center and radius of a circle given by an equation.
G.GPE.2-3	*Covered in Algebra 2 and Precaclulus*	
G.GPE.4	Use coordinates to prove simple geometric theorems algebraically. For example, prove or disprove that a figure defined by four given points in the coordinate plane is a rectangle; prove or disprove that the point $(1, \sqrt{3})$ lies on the circle centered at the origin and containing the point (0, 2).	On the coordinate plane algebraically prove and apply with fluency geometric theorems and properties. a. Given points and/or characteristics, prove or disprove a polygon is a specified quadrilateral or triangle based on its properties. b. Given a point that lies on a circle centered at the origin, prove or disprove that a specified point lies on the same circle. Note: coordinates of points could be given in radical form.
G.GPE.5	Prove the slope criteria for parallel and perpendicular lines and use them to solve geometric problems (e.g., find the equation of a line parallel or perpendicular to a given line that passes through a given point).	On the coordinate plane: i. explore the proof for the relationship between slopes of parallel and perpendicular lines; ii. fluently determine if lines are parallel, perpendicular, or neither, based on their slopes; and iii. fluently apply properties of parallel and perpendicular lines to solve geometric problems.
G.GPE.6	Find the point on a directed line segment between two given points that partitions the segment in a given ratio.	*NO CHANGE*
G.GPE.7	Use coordinates to compute perimeters of polygons and areas of triangles and rectangles, e.g., using the distance formula.	Use coordinates with fluency to compute perimeters of polygons and areas of triangles and rectangles. Note: Values may be given or computed in radical form.

G.GMD.1	Give an informal argument for the formulas for the circumference of a circle, area of a circle, volume of a cylinder, pyramid, and cone. Use dissection arguments, Cavalieri's principle, and informal limit arguments.	Explore informal arguments for the formulas for the circumference of a circle, area of a circle, volume of a cylinder, pyramid, and cone.
G.GMD.2	*Covered in Calculus*	
G.GMD.3	Use volume formulas for cylinders, pyramids, cones, and spheres to solve problems.	*NO CHANGE*
G.GMD.4	Identify the shapes of two-dimensional cross-sections of three-dimensional objects, and identify three-dimensional objects generated by rotations of two-dimensional objects.	Identify the shapes of plane-sections of three-dimensional objects, and identify three-dimensional objects generated by rotations of two-dimensional objects. Note: Plane sections are not limited to being parallel or perpendicular to the base.
G.MG.1	Use geometric shapes, their measures, and their properties to describe objects (e.g., modeling a tree trunk or a human torso as a cylinder).	Use geometric shapes, their measures, and their properties to describe objects.
G.MG.2	Apply concepts of density based on area and volume in modeling situations (e.g., persons per square mile, BTUs per cubic foot).	Apply concepts of density based on area and volume in modeling situations using geometric figures.
G.MG.3	Apply geometric methods to solve design problems (e.g., designing an object or structure to satisfy physical constraints or minimize cost; working with typographic grid systems based on ratios).	Apply geometric methods to solve design problems. Note: Applications could include designing an object or structure to satisfy physical constraints or minimize cost, or to investigate applications of classical geometric problems like the Golden Ratio.

Appendix II. Pacing Calendar

Chapter	(CC) Regents Questions *	Days
1. PREREQUISITE TOPICS REVIEW	0	0-5
2. PERIMETER AND AREA	1	2
3. LINES, ANGLES AND PROOFS	4	4
4. TRIANGLES	28	20
5. RIGHT TRIANGLES AND TRIGONOMETRY	33	24
6. OBLIQUE TRIANGLES AND TRIGONOMETRY	0	0-5
7. QUADRILATERALS	16	10
8. COORDINATE GEOMETRY	15	10
9. POLYGONS IN THE COORDINATE PLANE	10	8
10. RIGID MOTIONS	10	8
11. DILATIONS	15	10
12. TRANSFORMATIONAL PROOFS	31	22
13. CIRCLES	33	24
14. SOLIDS	35	25
15. CONSTRUCTIONS	10	8
	Totals: 241	**175 – 185 days**

* number of questions appearing on Geometry Common Core Regents exams, through Jan. '17.

Appendix III. Index

Q

R

S

T

V

W

Made in the USA
Columbia, SC
09 January 2018